Antioxidant
Food Supplements
in Human Health

Antioxidant Food Supplements in Human Health

Edited by

Lester Packer

Department of Molecular and Cell Biology
University of California
Berkeley, California

Midori Hiramatsu

Division of Medical Science
Institute for Life Support Technology
Yamagata Technopolis Foundation
Yamagata, Japan

Toshikazu Yoshikawa

First Department of Medicine
Kyoto Prefectural University of Medicine
Kyoto, Japan

Academic Press
San Diego   London   Boston   New York   Sydney   Tokyo   Toronto

Academic Press
A Division of Harcourt Brace & Company
525 B Street, Suite 1900, San Diego, California 92101-4495, USA
http://www.apnet.com

Academic Press
24-28 Oval Road, London NW1 7DX
http://ww.hbuk.co.uk/ap/

Library of Congress Cataloging-in-Publication Data
Antioxidant food supplements in human health / edited by Lester
    Packer, Midori Hiramatsu, and Toshikazu Yoshikawa.
        p.   cm.
    Includes bibliographical references and index.
    ISBN 0-12-543590-8 (alk. paper)
    1. Antioxidants—Health aspects.   I. Packer, Lester.
II. Hiramatsu, Midori.   III. Yoshikawa, Toshikazu.
RB170.A573   1999
613.2'8—dc21   98-48062
                                                          CIP

PRINTED IN THE UNITED STATES OF AMERICA
99 00 01 02 03 04 MM 9 8 7 6 5 4 3 2 1

Contents

I Health Effects of Antioxidant Nutrients

1 Vitamin E and Lung Cancer Prevention

Paul Knekt

2 Natural Antioxidants in the Protection against Cigarette Smoke Injury

Garry G. Duthie

II Nutrients

A. Vitamins C and E

3 Emerging Role of Nutrition in Chronic Disease Prevention: A Look at the Data, with an Emphasis on Vitamin C

Gladys Block

4 Biological Activities of Tocotrienols and Tocopherols

Maret G. Traber, Elena A. Serbinova, and Lester Packer

B. *Vitamin E and Selenium*

C. α-*Lipoic Acid*

12 Coenzyme-Q Redox Cycle as an Endogenous Antioxidant

Takeo Kishi, Takayuki Takahashi, and Tadashi Okamoto

E. *Carotenoids*

13 Carotenoids: Occurrence, Biochemical Activities, and Bioavailability

Wilhelm Stahl and Helmut Sies

14 Dietary Carotenoids and Their Metabolites as Potentially Useful Chemoprotective Agents against Cancer

*Frederick Khachik, John S. Bertram, Mou-Tuan Huang,
Jed W. Fahey, and Paul Talalay*

15   Cancer Prevention by Natural Carotenoids

H. Nishino

F. Flavonoids

16   Screening of Phenolics and Flavonoids for Antioxidant Activity

Catherine Rice-Evans

III. Natural Source Antioxidants

A. Pine Bark

20 From Ancient Pine Bark Uses to Pycnogenol

G. Drehsen

21 Procyanidins from *Pinus maritima* Bark: Antioxidant Activity, Effects on the Immune System, and Modulation of Nitrogen Monoxide Metabolism

F. Virgili, H. Kobuchi, Y. Noda, E. Cossins, and L. Packer

D. *Uyaku and Carica papaya*

Contributors

Numbers in parentheses indicate the pages on which the authors' contributions begin.

Fernando Antunes (143), Department of Molecular Pharmacology and Toxicology, School of Pharmacy, University of Southern California, Los Angeles, California 90033

Kazunori Anzai (471), National Institute of Radiological Sciences, Chiba, Japan

Annika Assmann (87), Institut für Physiologische Chemie I und Biologisch-Medizinisches Forschungszentrum, Heinrich-Heine-Universität Düsseldorf, D-40001 Düsseldorf, Germany

Angelo Azzi (73), Institut für Biochemie und Molekularbiologie, Universität Bern, 3012 Bern, Switzerland

Gary R. Beecher (269), Food Composition Laboratory, Beltsville Human Nutrition Research Center, Agricultural Research Service, USDA, Beltsville, Maryland 20705

John S. Bertram (203), Cancer Research Center of Hawaii, University of Hawaii at Manoa, Honolulu, Hawaii 96813

Gladys Block (45), Department of Epidemiology and Public Health Nutrition, University of California, Berkeley, California 94720

Yoshio Boku (461), First Department of Medicine, Kyoto Prefectural University of Medicine, Kyoto 602-8566, Japan

Alberto Boveris (143), Department of Biophysics, School of Pharmacy and Biochemistry, University of Buenos Aires, Buenos Aires, Argentina

Karlis Briviba (87), Institut für Physiologische Chemie I und Biologisch-Medizinisches Forschungszentrum, Heinrich-Heine-Universität Düsseldorf, D-40001 Düsseldorf, Germany

Enrique Cadenas (143), Department of Molecular Pharmacology and Toxicology, School of Pharmacy, University of Southern California, Los Angeles, California 90033

E. Cossins (323), Department of Molecular and Cell Biology, University of California, Berkeley, California 94720

G. Drehsen (311), Basel, Switzerland

Marie-Thérèse Droy-Lefaix (343, 359), Department of Pharmacology, ISPEN Institute, 75781 Paris cedex 16, France

Garry G. Duthie (35), Rowett Research Institute, Scotland, United Kindgom

Jed W. Fahey (203), Brassica Chemoprotection Laboratory, Department of Pharmacology and Molecular Science, Johns Hopkins University School of Medicine, Baltimore, Maryland 21205

Edwin N. Frankel (385), Department of Food Science and Technology, University of California, Davis, California 95616

Elisabeth Franzini (371), INSERM U479, CHU Xavier Bichat, Paris, France

Takaaki Fujii (461), First Department of Medicine, Kyoto Prefectural University of Medicine, Kyoto 602-8566, Japan

Alexia Gozin (371), INSERM U479, CHU Xavier Bichat, Paris, France

Yukihiko Hara (429), Food Research Institute, Mitsui Norin Co., Ltd., Fujieda City 426-0133, Japan

Nobuya Haramaki (481), Membrane Bioenergetics Group, Department of Molecular and Cell Biology, University of California, Berkeley, California 94720-3200

Midori Hiramatsu (411), Division of Medical Science, Institute for Life Support Technology, Yamagata Technopolis Foundation, Yamagata 990-2473, Japan

Mou-Tuan Huang (203) Department of Chemical Biology, Laboratory for Cancer Research, College of Pharmacy, Rutgers, The State University of New Jersey, Piscataway, New Jersey 08854

Frederick Khachik (203), Department of Chemistry and Biochemistry, Joint Institute for Food Safety and Applied Nutrition (JIFSAN), University of Maryland, College Park, Maryland 20742

Takeo Kishi (165), Department of Biochemistry, Faculty of Pharmaceutical Sciences, Kobe Gakuin University, Kobe 651-21, Japan

Lars-Oliver Klotz (87), Institut für Physiologische Chemie I und Biologische-Medizinisches Forschungszentrum, Heinrich-Heine-Universität Düsseldorf, D-40001 Düsseldorf, Germany

Paul Knekt (3), National Public Health Institute, 00300 Helsinki, Finland

Hirotsugu Kobuchi (323, 359, 481), Department of Molecular and Cell Biology, University of California, Berkeley, California 94720; Environmental Energy Technologies Division, University of California, Berkeley, California 94720

Makiko Komatsu (411), Institute for Life Support Technology, Yamagata Technopolis Foundation, Yamagata 990-2473, Japan

Motoharu Kondo (461), First Department of Medicine, Kyoto Prefectural University of Medicine, Kyoto 602-8566, Japan

Tetsuya Konishi (135), Department of Radiochemistry-Biophysics, Niigata College of Pharmacy, Niigata 950-21, Japan

Onno Korver (445), Unilever Nutrition Centre, Unilever Research Vlaardingen, Vlaardingen, The Netherlands

John K. Lodge (121), Department of Molecular and Cell Biology, University of California, Berkeley, California 94720

Matilde Maiorino (103), Dipartimento di Chimica Biologica, Viale G. Colombo, 3.I-35121-Padova, Italy

Lucia Marcocci (481), Membrane Bioenergetics Group, Department of Molecular and Cell Biology, University of California, Berkeley, California 94720

Yasunari Masui (461), First Department of Medicine, Kyoto Prefectural University of Medicine, Kyoto 602-8566, Japan

Akitane Mori (471), Department of Molecular and Cell Biology, University of California, Berkeley, California 94720

Yuji Naito (461), First Department of Medicine, Kyoto Prefectural University of Medicine, Kyoto 602-8566, Japan

H. Nishino (231), Department of Biochemistry, Kyoto Prefectural University of Medicine, Kamigyoku, Kyoto 602, Japan

Yasuko Noda (323, 471), Department of Molecular and Cell Biology, University of California, Berkeley, California 94720

Tadashi Okamoto (165), Department of Biochemistry, Faculty of Pharmaceutical Sciences, Kobe Gakuin University, Kobe 651-2180, Japan

Takuo Okuda (393), Okayama University, Tsushima, Okayama 700-8530, Japan

Nesrin K. Özer (73), Department of Biochemistry, Faculty of Medicine, Marmara University, Istanbul, Turkey

Lester Packer (55, 111, 121, 135, 323, 343, 359, 471, 481), Department of Molecular and Cell Biology, University of California, Berkeley, California 94720

Catherine Pasquier (371), INSERM U479, CHU Xavier Bichat, Paris, France

Piergiorgio Pietta (283), ITBA-CNR, Milan, Italy

Juan José Poderoso (143), Laboratory of Oxygen Metabolism, University Hospital, University of Buenos Aires, Buenos Aires, Argentina

Catherine Rice-Evans (239), International Antioxidant Research Centre, Guy's, King's and St. Thomas' School of Biomedical Sciences, London SE1 9RT, United Kingdom

Sashwati Roy (111, 359), Membrane Bioenergetics Group, Department of Molecular and Cell Biology, University of California, Berkeley, California 94720

Hassan Sellak (371), INSERM U479, CHU Xavier Bichat, Paris, France

Chandan K. Sen (111, 359), Membrane Bioenergetics Group, Department of Molecular and Cell Biology, University of California, Berkeley, California 94720

Elena A. Serbinova (55), Department of Molecular and Cell Biology, University of California, Berkeley, California 94720

Victor S. Sharov (87), Institut für Physiologische Chemie I und Biologisch-Medizinisches Forschungszentrum, Heinrich-Heine-Universität Düsseldorf, D-40001 Düsseldorf, Germany

Helmut Sies (87, 183), Institut für Physiologische Chemie I und Biologisch-Medizinisches Forschungszentrum, Heinrich-Heine-Universität Düsseldorf, D-40001 Düsseldorf, Germany

Paolo Simonetti (283), diSTAM, University of Milan, Milan, Italy

Wilhelm Stahl (183), Institut für Physiologische Chemie I, Heinrich-Heine-Universität Düsseldorf, D-40001 Düsseldorf, Germany

Achim Stocker (73), Institut für Biochemie und Molekularbiologie, Universität Bern, 3012 Bern, Switzerland

Suzan Taha (73), Department of Biochemistry, Faculty of Medicine, Marmara University, Istanbul, Turkey

Takayuki Takahashi (165), Department of Biochemistry, Faculty of Pharmaceutical Sciences, Kobe Gakuin University, Kobe 651-21, Japan

Paul Talalay (203), Brassica Chemoprotection Laboratory, Department of Pharmacology and Molecular Science, Johns Hopkins University School of Medicine, Baltimore, Maryland 21205

Junji Terao (255), Department of Nutrition, School of Medicine, The University of Tokushima, Tokushima 770-8503, Japan

Lilian Tijburg (445), Unilever Nutrition Centre, Unilever Research Vlaardingen, Vlaardingen, The Netherlands

Maret G. Traber (55), Linus Pauling Institute, Oregon State University, Corvalis, Oregon 97331

F. Virgili (323), National Institute of Nutrition, Rome, Italy; Department of Molecular and Cell Biology, University of California, Berkeley, California 94720

Ute Weisgerber (445), Unilever Nutrition Centre, Unilever Research Vlaardingen, Vlaardingen, The Netherlands

Sheila Wiseman (445), Unilever Nutrition Centre, Unilever Research Vlaardingen, Vlaardingen, The Netherlands

Norimasa Yoshida (461), First Department of Medicine, Kyoto Prefectural University of Medicine, Kyoto 602-8566, Japan

Toshikazu Yoshikawa (461), First Department of Medicine, Kyoto Prefectural University of Medicine, Kyoto 602-8566, Japan

Preface

There has been increasing interest in recent years in healthy life styles and healthy aging; correspondingly, interest in antioxidants and food supplements has grown remarkably. However, it has only been relatively recently that basic science has matched this interest with research. The development of technologies for the study of free radicals and antioxidants has led to many new discoveries in the areas of oxygen and nitric oxide metabolism and pathophysiology, redox regulation of cell signaling, and in the identification of natural antioxidants and their mechanisms of action on free radicals and their role in health and disease.

This volume presents some of the results of this exciting new research. Contributions focus on two main areas: nutrient antioxidants (antioxidants that are either vitamins or commonly found in the diet) and natural source antioxidants (extracts and other products from various sources that are rich in a wide range of antioxidants).

Some nutrient antioxidants are essential: these include vitamins E and C and selenium. Vitamin E is perhaps the most well known of the nutrient antioxidants and was one of the first antioxidants for which epidemiological evidence indicated that health affects accrued when it was consumed in amounts greater than could be found in the diet (e.g., in protection against heart disease). Contributions to this volume examine the role of vitamin E in lung cancer prevention, the interaction between vitamins E and C, and the differing biological activities of the eight different forms of vitamin E (four tocopherols and four tocotrienols). In addition, selenium, an essential mineral found to be beneficial in supplemental amounts in the prevention of cancer, is examined for its role in selenoproteins, which reduce peroxynitrites, and in peroxidases in mammalian testis.

Other nutrient antioxidants, while not essential, are endogenous cellular components and are coming under increasing scrutiny as they offer numerous health benefits. α-Lipoic acid, found in α-ketoglutarate dehydrogenases, has been shown in numerous studies to have potent antioxidant function and potential in the prevention and therapy of numerous diseases, including diabetes. It is discussed here in relation to its role in the regulation of cell function, its natural sources, and the assay of protein-bound forms. Another antioxidant that is a natural cell component is coenzyme Q. The reduced form, ubiquinol, can react with nitric oxide; in addition, because coenzyme Q can cycle between reduced and oxidized (ubiquinone) forms, its redox cycle is examined for its antioxidant role.

Carotenoids and flavonoids represent a final class of nutrient antioxidants. Both of these groups of compounds contain hundreds of members and thus present a rich and complex area for antioxidant research (as well as the potential for seemingly conflicting results, as recent clinical trials with β-carotene have made clear). An overview of the carotenoids as well as their role in the prevention of cancer, are presented. Flavonoids are discussed in relation to methods for screening their antioxidant activity, their metabolic conversion and its effect on their antioxidant activity against lipid peroxidation, and their interaction with physiologic antioxidants.

Natural source antioxidants, comprising extracts, herbs, and other products, represent a vast and still little-studied area. These products can contain a bewildering array of compounds. Two of the most highly standardized and defined extracts are those of *Ginkgo biloba* (Egb 761) and pine bark (Pycnogenol). Contributions examine both Egb 761 and Pycnogenol for their broad array of antioxidant effects, as well as specifically reporting their effects on nitric oxide metabolism. Two of the most well-known "extracts" have been used as beverages since ancient times—wine and tea. Both are rich in polyphenols and both are discussed in terms of their health benefits. Tea is also discussed, particularly with reference to its effects on cardiovascular disease and on oral hygiene. Polyphenols are also found in various herbs and act as antioxidants, as discussed by Takuo Okuda, and mixed Japanese herbs are examined for their effects on age-related neuronal functions. Finally oyster extract, uyaku (a natural extract used in traditional medicine), and the fermentation product of *Carica pa-*

paya (Bio-Normalizer) are discussed in terms of their antioxidant properties and biological effects.

People everywhere have started to think more about health issues and have taken an interest in antioxidant food supplements. This book is the result of a symposium prompted by such interest: "Antioxidant Food Supplements in Human Health," held in Kaminoyama-city, Japan, October 12–16, 1997. The organizers are grateful to the sponsors and supporters of the symposium. Sponsors included the Society for Free Radical Research Asia, the Oxygen Club of California, and the UNESCO–MCBN (Global Network for Molecular and Cell Biology). Organizations supporting the symposium included A.O.A. Japan Co., Ltd., Kobe, Japan; Arsoa Ohsoa Corporation, Tokyo, Japan; ARZ Co., Ltd., Nara, Japan; BIO CORPORATION Inc., Kumamoto, Japan; Dainippon Pharmaceutical Co., Ltd., Tokyo, Japan; Daiwa Biological Research Institute, Kawasaki, Japan; Fujimi-Yohoen Co., Ltd. Saitama, Japan; Funakoshi Co., Ltd., Tokyo, Japan; Golden Neo-Life Diamite International, CA: Japan Clinic Company Ltd., Kyoto, Japan; Kenko Commerce & Co., Ltd., Tokyo, Japan; Medical Industries Co., Tokyo, Japan; Mitsui Nohrin Co., Ltd., Fujieda, Shizuoka, Japan; Nakachuou Hospital, Ibaragi, Japan; Osato Research Foundation, Inc., Gifu, Japan; Sanmei Co., Ltd., Miyagi, Japan; Sapporokousetsu Hospital, Sapporo, Japan; Saido Co., Ltd., Fukuoka, Japan; Sky Food Co., Ltd., Osaka, Japan; Sunstar Inc., Osaka, Japan; Suntory Limited, Osaka, Japan; the Shingu Hsu-Fu Association, Shingu, Wakayama, Japan; Tradepia Corporation, Saitama, Japan; Veritas Co., Tokyo, Japan; Wellness World Co., Ltd., Akita, Japan; Yamanouchi Pharmaceutical Co., Ltd., Tokyo, Japan; Yamagata Wine Association, Kaminoyama, Yamagata, Japan; and Zeria Pharmaceutical Co., Ltd., Tokyo, Japan.

Lester Packer
Midori Hiramatsu
Toshikazu Yoshikawa

I Health Effects of Antioxidant Nutrients

1 Vitamin E and Lung Cancer Prevention

Paul Knekt

National Public Health Institute
00300 Helsinki, Finland

INTRODUCTION

Tobacco smoking is the predominant cause of lung cancer, and it has been estimated that about 80% of lung cancers are attributed to smoking (Buiatti *et al.*, 1996). As a complement to smoking cessation in lung cancer prevention, considerable attention has been focused on possible protective factors in the diet. A diet rich in fruits and vegetables has rather consistently been reported to be associated with a reduced risk of lung cancer (Ziegler *et al.*, 1996). Several plausible mechanisms have been suggested, indicating that this reduction may be due to antioxidant micronutrients (Dorgan and Schatzkin, 1991). Thus far considerable attention has been directed to β-carotene, but intervention trials have not confirmed the presence of a protective effect (Albanes *et al.*, 1996; Hennekens *et al.*, 1996; Omenn *et al.*, 1996). Some evidence also suggests that the reduced risk of lung cancer associated with the intake of fruits and vegetables may be due to some other micronutrients, such as vitamin C (Block *et al.*, 1992), flavonoids (Knekt *et al.*, 1997a), and selenium (Clark *et al.*, 1996). The epidemiological evidence is, however, not yet persuasive for any of these, while the question of whether lung cancer can be prevented or slowed down by the antioxidant vitamin E has also been addressed in a number of studies (Knekt, 1994). The aim of this chapter is to review findings from epidemiological studies on the role of vitamin E in cancer prevention and to present some new results from the Finnish Mobile Clinic Health Examination Survey.

3

REVIEW OF STUDIES

Study Designs and Study Populations

Two intervention trials and 25 observational studies on the association between vitamin E status and lung cancer risk are considered here. In an intervention trial the investigator randomly assigns vitamin E or placebo to the study population and then waits for the outcome of lung cancer. Thus far the effect of vitamin E supplementation on the incidence of lung cancer has been studied in two large-scale double-blind intervention trials, one in Linxian, China (Blot *et al.*, 1993, 1994), and one in southwest Finland (The Alpha-Tocopherol, Beta Carotene Cancer Prevention Study Group, 1994; Albanes *et al.*, 1996; Table I). In an observational study the investigator looks for associations between vitamin E status and cancer occurrence, but makes no changes in the vitamin E level of the population. The observational studies considered are of cohort, nested case-control, or case-control design.

In a cohort study, a cancer-free population whose vitamin E status has been established is followed up over a period of time with respect to lung cancer occurrence. When a sufficient number of lung cancer cases have occurred, the incidence among individuals with high vitamin E status is compared with that among those with low vitamin E status. Thus far, the prospective Basel Study (Stähelin *et al.*, 1991; Eichholzer *et al.*, 1996), the Finnish Mobile Clinic Health Examination Survey (Knekt *et al.*, 1991b), a study carried out at a retirement community near Los Angeles, California (Shibata *et al.*, 1992), the Zutphen Study (Ocké *et al.*, 1997), and the NHANES I Epidemiologic Follow-up Study (Yong *et al.*, 1997) have been completed (Table II). Furthermore, the predictive value of circulating α-tocopherol at entry in the placebo group of the Alpha-Tocopherol, Beta-Carotene Cancer Prevention trial on subsequent occurrence of lung cancer has been reported (The Alpha-Tocopherol, Beta Carotene Cancer Prevention Study Group, 1994). A total of 37,008 individuals were monitored in these studies, and during a 5 to 22-year follow-up, a total of 859 lung cancer cases occurred.

A nested case-control study is a modified cohort study in which vitamin E status for lung cancer cases and controls is determined prediagnostically, and the controls are selected from the entire cohort at risk. The association between vitamin E status and cancer risk has been studied with nested case-control designs in eight different populations (Table III). The cohorts ranged from 6860 to 36,265 individuals, and the number of cancer cases occurring during the 2 to 10-year follow-up periods ranged from 17 to 144. A total of 599 lung cancer cases occurred. A total of two controls per can-

Table I Description of Intervention Trials on Vitamin E and Lung Cancer Risk

Study, country	Cohort size	Sex	Age	Starting year	Length of follow-up (years)	Reference
The Linxian Nutrition Intervention Trial, China	29,584	Both	40–69	1986	5	Blot *et al.* (1993)
The Alpha-Tocopherol, Beta Carotene Cancer Prevention Study, Finland	29,133	Men	50–69	1985	5–8	Albanes *et al.* (1996)

Table II   Description of Cohort Studies on Vitamin E and Lung Cancer Risk

Study, country	Cohort size	Sex	Age	Starting year	Length of follow-up (years)	Reference
Mobile Clinic Health Survey, Finland	4,538	Men	20–69	1966–1972	20	Knekt et al. (1991b)
Basel Study, Switzerland	2,974	Men	50 (9)[a]	1971–1973	12	Stähelin et al. (1991)
California, United States	11,580	Both	74 (7)	1981	9	Shibata et al. (1992)
The Alpha-Tocopherol, Beta Carotene Cancer Prevention Study, Finland	7,287	Men	50–69	1985	5–8	The Alpha-Tocopherol, Beta Carotene Cancer Prevention Study Group (1994)
The Zutphen Study, The Netherlands	561	Men	59 (5)[a]	1970	20	Ocké et al. (1997)
The NHANES I Study, United States	10,068	Both	25–74	1971–1975	17–22	Yong et al. (1997)

[a] Mean (SD).

Table III Description of Nested Case-Control Studies on Vitamin E and Lung Cancer

Study, country	Cohort size	Sex	Age (years)	Starting year	Length of follow-up (years)	Reference
Hypertension Detection and Follow-up Program, United States	4,480	Both	30–69	1973–1974	5	Willet et al. (1984)
Honolulu Heart Program, United States	6,680	Men	—	1971–1975	10	Nomura et al. (1985)
Washington County Study, United States	25,802	Both	≥18	1974	9	Menkes et al. (1986)
The Netherlands	10,532	Both	—	1975	9	Kok et al. (1987)
British United Provident Association, United Kingdom	22,000	Men	35–64	1975–1982	2–9	Wald et al. (1987)
Mobile Clinic Health Survey, Finland	36,265	Both	15–99	1968–1972	6–10	Knekt et al. (1988a, 1991a)
Multiple Risk Factor Intervention Trial, United States	12,866	Men	35–57	1973–1975	8–10	Connett et al. (1989)
Kaiser Permanente Medical Care Program, United States	—	Both	26–78	1969–1973	5–9	Orentreich et al. (1991)

cer case were selected among those individuals free from known cancer in these studies. The selection was, with one exception (Nomura *et al.*, 1985), performed by individual matching. The most common matching factors were age, sex, smoking status, and duration of serum sample storage.

In a case-control study, a group of lung cancer patients is compared with a group of controls with respect to vitamin E status. In all, 11 such studies were considered (Table IV). The number of cancer cases in the studies varied between 12 and 450. Most often the controls were patients from the same hospital as the cases. In only three of the studies were the controls selected from the general population (Byers *et al.*, 1987; Alavanja *et al.*, 1993; Mayne *et al.*, 1994). The controls were mostly selected by matching for sex and age.

Assessment of Vitamin E Status

Vitamin E exposure has been assessed as the daily dietary intake, supplement intake, or serum/plasma concentration in the studies considered. The dietary intake of vitamin E was assessed in seven observational studies. Various dietary recall methods were used, covering the total dietary intake or part of it, currently (Byers *et al.*, 1987; Hu *et al.*, 1997), during the previous 24 hr (Connett *et al.*, 1989; Yong *et al.*, 1997), during the preceding 6–12 months (Knekt *et al.*, 1991b; Ocké *et al.*, 1997), or over the preceding 4 years (Alavanja *et al.*, 1993). All studies estimated total vitamin E intake. One study also estimated the intake of α-, β-, γ-, and δ-tocopherol (Knekt *et al.*, 1991b).

The vitamin E supplement dose in the intervention trials consisted of 50 mg α-tocopherol alone or in combination with 20 mg β-carotene per day in the Finnish study (The Alpha-Tocopherol, Beta Carotene Cancer Prevention Study Group, 1994) and a combination of 30 mg α-tocopherol, 15 mg β-carotene, and 50 μg selenium in the Chinese study (Blot *et al.*, 1994), both in comparison to a placebo supplement. Three observational studies reported on voluntary self-supplementation; in one study supplement users were defined as individuals using a supplement at least once per week and compared with persons using less (Shibata *et al.*, 1992). The median daily dose among those using a supplement was 200 IU and more than 90% of the subjects who answered "yes" to the question had used the supplement for at least 1 year prior to entry into the study. Other studies evaluated the amount of daily consumption (Mayne *et al.*, 1994) and regularity of use as regular use (daily), irregular use (at least weekly but less than daily), and never (Yong *et al.*, 1997). In one study the individuals who had used supplements were placed in the category of highest intake level of vitamin E (Ocké *et al.*, 1997).

Table IV Description of Case-Control Studies on Vitamin E and Lung Cancer Risk

Study, country	Sex	Age	Starting year	Type of control	Reference
—	Both	46–82	—	Hospital	Atukorala et al. (1979)
New York, United States	Both	35–79	1980–1984	Population	Byers et al. (1987)
Hokkaido, Japan	Both	64 (11)[a]	1970–1982	Hospital	Miyamoto et al. (1987)
Bangkok, Thailand	Both	30–87	—	—	Skulchan and Ong-Ajyooth (1987)
New Orleans, United States	Both	58 (7)[a]	—	Hospital	Le Gardeur et al. (1990)
Oxford, United Kingdom	Men	59 (10)[a]	1979–1981	Hospital	Harris et al. (1991)
Tochigi, Japan	Both	44–78	—	Hospital	Tominaga et al. (1992)
Missouri, United States	Women	30–84	1986–1991	Population	Alavanja et al. (1993)
New York, United States	Both	20–80	1982–1985	Population	Mayne et al. (1994)
Ankara, Turkey	Both	—	—	—	Torun et al. (1995)
Heilongjiang Province, China	Both	52[a]	1985–1987	Hospital	Hu et al. (1997)

[a] Mean (SD).

The serum and plasma concentrations of α-tocopherol or vitamin E were determined from samples collected at the baseline survey. The samples were in general fresh in the cohort and case-control studies and were stored frozen (at -20 to $-75°C$) and thawed for analysis after the end of the follow-up period in the nested case-control studies. Determinations in the cohort and nested case-control studies were performed with high-pressure liquid chromatography (HPLC) and, in the case-control studies in general, by HPLC or spectrophotometry. In the nested case-control studies, serum samples of the case-control sets were performed simultaneously, and laboratory personnel were unaware of the case-control status within the sets. Coefficients of variation for the vitamin E determinations, reported in some of the studies, in general varied 1–5%. The mean serum vitamin E concentration of the case-control studies ranged from 7.0 to 17.4 mg/liter in the control samples. In the nested case-control studies the range for the control samples was 6.9–13.6 mg/liter, whereas the corresponding range for cohort studies was 6.3–15.4 mg/liter.

Statistical Methods

The strength of the association between vitamin E status and lung cancer occurrence was commonly expressed either as the relative risk (or odds ratio) of lung cancer at various concentrations of vitamin E or as the percentage mean vitamin E differences between cancer cases and controls. The relative risks were calculated by comparing the risk among persons between tertiles, quartiles, quintiles, around the median, or per standard unit of vitamin E distribution. In this chapter the lowest category is used as the reference category for which the risk is settled unity. The relative risk can be interpreted as the risk of developing lung cancer for exposed persons in comparison to unexposed individuals. A relative risk below 1.0 thus implies that vitamin E exposure is associated with a reduced risk of lung cancer. The percentage mean differences were estimated as [(case mean-control mean)/control mean] ×100.

Results

Intervention Trial

As part of the Nutrition Intervention Trial in Linxian, China, a combination of β-carotene (15 mg), α-tocopherol (30 mg), and selenium (50 μg) or placebo was randomly assigned to 29,584 men and women, 40–69 years of age, from 1986 to 1991 (Table I). During the 5-year follow-up, 31 individuals died of lung cancer (Blot *et al.*, 1994; Table V). The relative risk for

Table V   Results of Intervention Trials on Vitamin E and Lung Cancer Risk

Number of cases	Supplement	Relative risk (95% confidence interval)	Reference
31	30 mg α-tocopherol 15 mg β-carotene 50 μg selenium	0.55 (0.26–1.14)	Blot *et al.* (1994)
894	50 mg α-tocopherol	0.99 (0.87–1.13)	Albanes *et al.* (1996)

persons receiving the antioxidant supplement compared with those receiving placebo was 0.55 [95% confidence interval (CI) = 0.26–1.14].

In the Alpha-Tocopherol, Beta-Carotene Cancer Prevention study, undertaken in southwestern Finland, a total of 29,133 men 50–69 years of age who smoked five or more cigarettes daily were randomly assigned to receive α-tocopherol (50 mg), β-carotene (20 mg), α-tocopherol and β-carotene, or a placebo daily for 5–8 years (median 6.1 years) (The Alpha-Tocopherol, Beta Carotene Cancer Prevention Study Group, 1994; Albanes *et al.*, 1996; Table I). A total of 894 incident lung cancer cases were identified through the Finnish Cancer Registry (Table V). The relative risk of lung cancer between men receiving α-tocopherol and those receiving placebo was 0.99 (CI = 0.87–1.13). There was no evidence of an interaction between the two supplements in their effect on lung cancer. No significant effects were observed in strata of age, occupation, smoking, asbestos exposure, dietary intake of vitamin E, β-carotene, carotenoids, vitamin C, or retinol, alcohol consumption or serum α-tocopherol, β-carotene, or retinol level.

In summary, there was a suggestive protective effect of a combination of vitamin E, β-carotene, and selenium against lung cancer in a population with low lung cancer risk, whereas vitamin E supplementation alone or in combination with β-carotene provided no protection against lung cancer in a population of smokers with a high risk of lung cancer.

Cohort Studies

The relation between vitamin E intake and subsequent lung cancer risk was studied among 4538 initially cancer-free Finnish men 20–69 years of age (Knekt *et al.*, 1991b; Table II). During a 20-year follow-up, 117 lung cancer cases were diagnosed (Table VI). An inverse gradient was observed between vitamin E intake and the incidence of lung cancer among nonsmokers. For these individuals the age-adjusted relative risk of lung cancer in the highest tertile in comparison to the lowest was 0.33 (p for trend = 0.12). Adjustment for various potential confounding factors (geographic

Table VI Results of Cohort Studies on Vitamin E Status and Lung Cancer Risk

Number of cases	Vitamin E		Relative risk (95% confidence interval)	p value for trend	Control mean	Percentage case-control difference	Reference
	Source	Categories (highest vs lowest)					
117	Dietary vitamin E				8.5 mg	0	Knekt et al. (1991b)
	Nonsmoker	Tertile	0.33	0.12			
	Smoker	Tertile	1.25	0.58			
	Dietary α-tocopherol				7.2 mg	−1	
	Dietary β-tocopherol				0.7 mg	0	
	Dietary γ-tocopherol				2.8 mg	−21	
	Dietary δ-tocopherol				0.6 mg	−33	
68[a]	Plasma vitamin E	Quartile[b]	0.68 (0.41–1.15)	—	14.9 mg/liter	−2	Stähelin et al. (1991)
94 men	Vitamin E supplement	Yes/no	1.10 (0.73–1.65)	—	—	—	Shibata et al. (1992)
70 women			0.74 (0.46–1.18)				
208	Serum α-tocopherol	Quartile	0.74	—	11.4 mg/liter	—	The Alpha-Tocopherol Beta Carotene Cancer Prevention Study Group (1994)
	Dietary α-tocopherol	Quartile	0.66	—		—	
54	Dietary vitamin E	Tertile	0.68 (0.32–1.52)	—	15.4 mg	+5	Ocké et al. (1997)
248	Vitamin E intake	Quartile	0.88 (0.62–1.25)	0.30	6.3 mg	−4[c]	Yong et al. (1997)

[a] Bronchus cancer.
[b] Higher vs lowest quartile.
[c] $p < 0.05$.

area, social class, body mass index, height, and intakes of energy, fat, vitamin C, and β-carotene) did not notably alter the results. In addition, the association was apparently not due to preclinical cancer, as the relative risk after exclusion of the cancer cases diagnosed during the first 2 years of follow-up was 0.37 (CI = 0.10–1.43). The age-adjusted relative risk for smokers was 1.25 (p for trend = 0.58), suggesting the presence of no association. The age-adjusted relative risk for men with no vitamin E, β-carotene, or vitamin C in the lowest tertile of intake in comparison to those with all three micronutrients in the lowest tertile was 0.27 (CI = 0.06–1.11) in nonsmokers and 1.43 in smokers.

In the prospective Basel study, Switzerland, 2974 volunteer healthy male employees at three pharmaceutical companies were followed beginning in 1971 (Stähelin *et al.*, 1991; Table II). During a 12-year follow-up, a total of 68 men died of bronchial cancer (Table VI). No difference was observed in the plasma vitamin E levels between bronchial cancer cases and survivors; the lipid-adjusted mean difference was 3%. The relative risk of lung cancer between the three highest quartiles combined and the lowest quartile was 0.68 (CI = 0.41–1.15) after adjustment for age, lipid status, and smoking. Exclusion of lung cancer cases occurring during the first 2 years of follow-up did not materially alter the results. During a 17-year follow-up, 87 lung cancer cases occurred (Eichholzer *et al.*, 1996). Individuals with simultaneously high plasma levels of vitamin E and vitamin C showed a reduced lung cancer risk after adjustment for age, lipids, and smoking. This association persisted after exclusion of those cancer cases occurring during the first 2 years of follow-up. The relative risk for individuals occurring simultaneously in the higher quartiles of vitamin E and vitamin C status in comparison to those in the lowest quartiles of both micronutrients was 0.27 (CI = 0.12–0.62). Further adjustment for the joint effect of carotene and retinol did not materially alter the results.

The relationship between vitamin E supplement use and subsequent lung cancer occurrence was studied in a cohort of 11,580 residents of a retirement community initially free from cancer (Shibata *et al.*, 1992; Table II). During a 9-year follow-up period, a total of 94 lung cancer cases occurred among men and 70 cases among women (Table VI). No association was observed between vitamin E supplement use in men and a nonsignificant suggested association in women. The age- and smoking-adjusted relative risk of lung cancer between vitamin E supplement users and nonusers was 1.10 (CI = 0.73–1.65) in men and 0.74 (CI = 0.46–1.18) in women.

In the Alpha-Tocopherol, Beta Carotene Cancer Prevention study, 7287 male smokers, 50–69 years of age, from the placebo group were followed up for the predictive value of baseline serum β-carotene concentration (The Alpha-Tocopherol, Beta Carotene Cancer Prevention Study Group, 1994;

Table II). The incidence of lung cancer was lower among those subjects in the highest quartile of serum α-tocopherol intake than among those in the lowest, and the age- and smoking-adjusted relative risk between the two quartiles was 0.74 (Table VI). The corresponding relative risk for the dietary intake of α-tocopherol was 0.66.

The intake of vitamin E in relation to the incidence of lung cancer was studied among 561 men from the town of Zutphen, The Netherlands (Ocké *et al.*, 1997; Table II). During a 20-year follow-up, 54 new lung cancer cases were identified (Table VI). No significant relationship was observed between intake of vitamin E and lung cancer risk. The relative risk of lung cancer between the highest and the lowest quartiles of vitamin E intake, adjusted for age, smoking, and energy intake, was 0.68 (CI = 0.32–1.52). The baseline examination was carried out three times and, based on the results, the individuals were classified as having stable low or high intake. Comparison of the individuals with stable high intake and stable low intake of vitamin E revealed no significant reduction in risk. The relative risk was 0.65 (CI = 0.24–1.79).

The relation between dietary intake of vitamin E and lung cancer incidence was examined in the NHANES I Epidemiologic Followup Study cohort of 10,068 men and women, 25–74 years of age (Yong *et al.*, 1997; Table II). During a median follow-up time of 19 years, 248 individuals developed lung cancer (Table VI). The mean vitamin E level was significantly (4%) lower in lung cancer cases than in noncases (*p* for trend < 0.05). The relative risk of lung cancer between quartiles of vitamin E intake was 0.88 (CI = 0.62–1.25) after adjustment for sex, race, education, nonrecreational activity level, body mass index, family history, smoking status, total calorie intake, and alcohol intake. The corresponding values for carotenoids and vitamin C were 0.74 (CI = 0.52–1.06) and 0.66 (0.45–0.96), respectively. The effect of vitamin E and the other micronutrients was strengthened when they were combined. The relative risk when levels of all three nutrients were in the highest quartile in comparison to when levels of all nutrients were in the lowest quartile was 0.32 (CI = 0.14–0.74). The strength of association for vitamin E was similar for nonsmokers and current smokers. The relative risks were 0.87 (CI = 0.49–1.60) and 0.82 (CI = 0.50–1.37), respectively. The vitamin E intake–lung cancer relation was modified by the intensity of smoking with a significantly reduced risk confined to current smokers in the lowest tertile of pack–years of smoking (relative risk = 0.36, CI = 0.16–0.83). The corresponding relative risk for the highest tertile of pack–years was 1.65. No additional reduction associated with supplements of vitamin E beyond that provided through dietary intake was observed.

In summary, only one study found an inverse association between vitamin E status and lung cancer occurrence in the total population (The

Alpha-Tocopherol, Beta Carotene Cancer Prevention Study Group, 1994). Another study reported a suggestive association in persons with high vitamin C status (Eichholzer *et al.*, 1996), one reported such an association in women but not men (Shibata *et al.*, 1992), one in nonsmokers (Knekt *et al.*, 1991b), and one noted an association among light smokers (Yong *et al.*, 1997).

Nested Case-Control Studies

In the Hypertension Detection and Follow-up Program, a trial of hypertension treatment at 14 centers in the United States during 1973–1974, serum samples were collected from 4480 men and women 30–69 years of age with diastolic blood pressure of at least 90 mm Hg (Willett *et al.*, 1984; Table III). The serum samples were stored at −70°C. During a 5-year follow-up, 17 lung cancer cases occurred among the participants with no history of cancer (Table VII). Two controls per case were selected by matching for age, sex, race, smoking, time of blood collection, blood pressure, use of antihypertensive medication, random assignment, and (if female) parity and menopausal status. The mean serum vitamin E level in controls was 12.6 mg/liter. No significant association was present between lipid-adjusted serum α-tocopherol concentration and occurrence of lung cancer. On average, the lung cancer cases had 10% higher vitamin E levels than the controls.

A total of 6860 men of Japanese ancestry participated in the Honolulu Heart Program during 1971–1975 (Nomura *et al.*, 1985; Table III). Serum samples were stored at −75°C. During a 10-year follow-up, 74 lung cancer cases occurred (Table VII). An age-stratified random sample of 302 controls was selected. The coefficient of variation of the vitamin E determinations was very high (13%). The mean vitamin E level was 12.3 mg/liter in controls and the level in lung cancer cases was 4% higher; the difference was not significant.

In the Washington County study, serum samples from 25,802 volunteer men and women, 18 years of age or over, were collected in 1974 (Menkes *et al.*, 1986; Table III). The serum samples were stored at −73°C. During the follow-up times ending in 1983, 99 lung cancer cases occurred among originally cancer-free persons (Table VII). Two controls were drawn by matching for age, sex, race, time of blood collection, and smoking history from those subjects with no history of cancer. An association was present between serum vitamin E concentration and lung cancer occurrence (Menkes *et al.*, 1986). The mean vitamin E level in the lung cancer cases was 12% lower than in controls, and the relative risk of lung cancer between the highest and the lowest quartiles of the micronutrient was 0.40 (p for trend = 0.04). Adjustment for education, marital status, occupation, time since previous meal, treatment of hypertension, intake of vitamin sup-

Table VII Results of Nested Case-Control Studies on Vitamin E Status and Lung Cancer Risk

Number of cases/controls	Vitamin E		Relative risk (95% confidence interval)	p value for trend	Control mean (mg/liter)	Percentage case-control difference	Reference
	Source	Categories (highest vs lowest)					
17/28	Serum vitamin E	—	—	—	12.6	+10	Willett et al. (1984)
74/302	Serum vitamin E	—	—	—	12.3	+4	Nomura et al. (1985)
99/196	Serum vitamin E	Quintile	0.40	0.04	11.9	−12[b]	Menkes et al. (1986)
18/36	Serum vitamin E	—	—	—	8.5	−9	Kok et al. (1987)
50/99	Serum vitamin E	—	—	—	9.5	−4	Wald et al. (1987)
144/270 men	Serum α-tocopherol	Quintile	0.73	0.71	8.4	−4	Knekt et al. (1988a, 1991a)
8/16 women	Serum α-tocopherol	Quintile[a]	1.25	—	9.4	+2	
66/131	Serum α-tocopherol	—	—	—	13.6	−4	Connett et al. (1989)
	Dietary vitamin E	—	—	—		+1	
123/246	Serum α-tocopherol	Quintile	1.67	—	6.9	+4	Orentreich et al. (1991)

[a] Four highest vs lowest quintile.
[b] $p < 0.01$.

plements, serum β-carotene, retinol or selenium, pipe- or cigar-smoking history, hormone use (in women), and socioeconomic indicators did not alter the relation. No significant differences were present in the associations among cancer cell types. Study of the interaction between vitamin E and selenium showed that persons with high selenium levels and low vitamin E levels had an increased risk of lung cancer.

In a follow-up study of 10,532 subjects carried out in The Netherlands, 18 individuals died of lung cancer during a follow-up of 9 years (Kok *et al.,* 1987; Tables III and VII). Baseline serum vitamin E concentration was determined for these individuals and 36 controls matched for sex, age, and smoking status. Lung cancer cases had a nonsignificant (9%) lower mean vitamin E level than controls.

The British United Provident Association Medical Centre in London undertook a medical examination of 22,000 men, 35–64 years of age, during 1975–1982 (Wald *et al.,* 1987; Table III). Serum samples were stored at $-40°C$. During a 2- to 9-year follow-up period, 50 lung cancer cases occurred, and 99 controls were selected by matching for age, duration of sample storage, and smoking status (Table VII). No notable difference (4%) occurred among vitamin E concentrations in lung cancer cases and controls.

The Finnish Mobile Clinic undertook health examinations in various parts of Finland during 1968–1972 (Knekt *et al.,* 1988a, 1991a; Table III). A total of 36,265 men and women, 15 years of age or over, from 25 cohorts participated. Serum samples were taken and stored at $-20°C$. During a 6- to 10-year follow-up period, 152 lung cancer cases occurred among originally cancer-free individuals (Table VII). Two controls per case (total 286) were chosen by individual matching for sex, age, and municipality (which also matched for duration of storage and season). No association was noted between serum α-tocopherol level and lung cancer occurrence in the total population. The smoking-adjusted relative risk between the highest and the lowest quintiles of serum α-tocopherol concentration was 0.73 (p for trend $= 0.71$) for men (Knekt *et al.,* 1988a). Further adjustment for serum cholesterol, body mass index, hematocrit, and socioeconomic status weakened the association. Exclusion of the cancer cases occurring during the first 2 years of follow-up did not notably alter the result. The relative risk between tertiles in women was 1.25 (Knekt *et al.,* 1991a). Study of the association in men conducted separately for nonsmokers and smokers resulted in no association among current smokers but found an association among nonsmokers. The serum α-tocopherol level was 12% ($p = 0.19$) lower among nonsmoking lung cancer cases than among nonsmoking noncases (Knekt, 1993). The relative risk of lung cancer between the highest and the lowest tertiles of serum α-tocopherol were 1.25 (CI $= 0.63–2.50$) for current smokers and 0.15 (CI $= 0.03–0.77$) for nonsmokers.

The Multiple Risk Factor Intervention Trial (MRFIT) comprised 12,866 men, 35–57 years of age and at high risk of coronary heart disease, from several clinical centers in the United States (Connett *et al.*, 1989; Table III). A nested case-control study design used serum samples collected during 1973–1975 from randomized participants and were stored at −50 to −70°C. During a 10-year follow-up, 66 individuals died of lung cancer (Table VII). Two controls were selected from among the survivors through follow-up by individual matching for clinic, treatment group, age, smoking status, and date of randomization. No notable difference (4%) was present in mean serum α-tocopherol concentration between lung cancer cases and controls. A 24-hr dietary recall was also carried out and vitamin E intake was estimated. No notable differences were observed in the intake levels between cases and controls.

During 1969–1973, serum samples were received from men and women belonging to a study cohort of 143,574 persons, 26–78 years of age, participating in multiphasic health checkups at the Kaiser Permanente Medical Care Program in San Francisco (Friedman *et al.*, 1986; Table III). The serum samples were stored at −40°C. A total of 151 lung cancer cases occurred during a 5- to 9-year follow-up period among the cancer-free population (Table VII). A total of 302 controls were selected by individual matching for age, sex, skin color, smoking status, date of health checkup, and storage duration. Serum α-tocopherol concentrations were determined for 123 cases and 246 controls (Orentreich *et al.*, 1991). No notable differences (4%) were found between cases and controls in mean levels of serum α-tocopherol.

In summary, the nested case-control studies in general revealed no associations between the serum vitamin E concentration and the subsequent occurrence of lung cancer. The only exception was the study by Menkes *et al.* (1986), which found an inverse trend in the total population, and that by Knekt (1993), which found an elevated risk at low vitamin E levels among nonsmokers.

Case-Control Studies

Atukorala *et al.* (1979) studied 26 lung cancer cases 46–82 years of age, 10 controls with nonmalignant lung disease, and 11 controls with nonmalignant other diseases (Table IV); the controls were 47–75 years of age. The mean serum vitamin E concentration in the lung cancer patients was similar to that of the controls (Table VIII).

Byers *et al.* (1987) compared the dietary intake of vitamin E of 450 male and female lung cancer patients with that of 902 controls in three western New York counties (Table IV). The controls were matched for sex, age, and place of residence. No association was observed between vitamin

Table VIII Results of Case-Control Studies on Vitamin E Status and Lung Cancer Risk

Number of cases/controls	Vitamin E		Relative risk (95% confidence interval)	p value for trend	Control mean (mg/liter)	Percentage case-control difference	Reference
	Source	Categories (highest vs lowest)					
26/21	Serum vitamin E	—	—	—	9.3	−3	Atukorala et al. (1979)
296/587 men	Dietary vitamin E	Quartile	0.77	0.22	—	—	Byers et al. (1987)
154/315 women	Dietary vitamin E	Quartile	0.91	0.91	—	—	
37/56	Serum vitamin E	—	—	—	14.1	−22[c]	Miyamoto et al. (1987)
19/52	Serum vitamin E	—	—	—	17.4	−30[c]	Skulchan and Ong-Aiyooth (1987)
50/50	Serum vitamin E	—	—	—	12.2	−33[b]	Le Gardeur et al. (1990)
93/96	Serum vitamin E	—	—	—	7.0	−31	Harris et al. (1991)
31/31	Serum vitamin E	Tertile	0.12 (0.02–0.68)	—	15.0	−20[c]	Tominaga et al. (1992)
429/1021	Dietary vitamin E	Quintile	1.72	0.18	—	—	Alavanja et al. (1993)
413/413	Vitamin E supplement	Unit/day	0.55 (0.35–0.85)	—	—	—	Mayne et al. (1994)
12/156	Serum vitamin E	—	—	—	9.8	−41[a]	Torun et al. (1995)
227/227	Diet vitamin E	Quartile	1.1 (0.7–1.9)	0.92	—	—	Hu et al. (1997)

[a] $p < 0.05$.
[b] $p < 0.01$.
[c] $p < 0.001$.

E intake and occurrence of lung cancer (Table VIII). The smoking-adjusted relative risk of the disease between the highest and the lowest quartiles of vitamin E was 0.77 (p for trend = 0.22) in men and 0.91 (p for trend = 0.91) in women.

Miyamoto *et al.* (1987) compared the mean serum vitamin E levels among 37 lung cancer patients in Hokkaido, Japan (Table IV). The mean vitamin E level was 22% ($p < 0.001$) lower among the lung cancer patients than among the controls (Table VIII). In the same study, 115 healthy children of patients with lung cancer also had lower vitamin E levels than the controls. The mean difference was 9% ($p < 0.05$); thus familial factors, dietary or metabolic but independent of the disease process, may influence serum vitamin E levels. Significant differences were observed for both adenocarcinomas and squamous cell carcinomas. Because cigarette smoking did not correlate with the serum α-tocopherol level, it was concluded that the results are not confounded by smoking.

Skulchan and Ong-Ajyooth (1987) compared mean serum vitamin E levels among 19 lung cancer patients from Siraj Hospital in Bankok, Thailand, with those of 52 healthy individuals, matched for age group (Table IV). Mean serum vitamin E levels of lung cancer cases were significantly (30%) lower in lung cancer cases than in controls (Table VIII).

LeGardeur *et al.* (1990) studied 50 individuals in New Orleans with newly diagnosed lung cancer (Table IV). Controls were selected in a next-patient-encountered manner from the same hospital as the patients and matched by race, sex, age, and county of residence. A group of community controls with the same age, sex, and race as the hospitalized controls was also evaluated. Mean vitamin E levels were significantly lower (33%) in lung cancer cases than in hospital controls (Table VIII). Although the difference was reduced to 12% after adjustment for serum cholesterol, it was still significant ($p = 0.04$); results were not adjusted for smoking. Hospitalized controls showed 25% lower levels than community controls ($p = 0.013$).

Harris *et al.* (1991) determined serum vitamin E concentration for 93 men with lung cancer, 74 men with other epithelial cancers, and 96 hospital controls in Oxford, United Kingdom (Table IV). The controls were in-patients with nonmalignant conditions and showed a similar age distribution as the cancer cases. The mean vitamin E level for lung cancer cases was 31% lower than in the controls (Table VIII). No adjustment for smoking was made.

Tominaga *et al.* (1992) selected at random 31 clinically and histologically confirmed lung cancer patients from among 152 individuals who had been admitted to the Tochigi Cancer Center Hospital, Japan (Table IV). The controls were randomly selected from outpatients treated at the same hospital by individual matching for sex, age, smoking history, and month of blood collection. The relation between serum vitamin E level and risk of lung cancer was statistically significant, with a relative risk of 0.12 (CI = 0.02–0.68) between high and low concentrations (Table VIII). The relative risk for squamous cell carcinoma alone was 0.25 (CI = 0.04–1.45).

Alavanja *et al.* (1993) studied the effect of vitamin E intake on the risk of lung cancer in a population of nonsmoking white women 30–84 years of age (Table IV). A total of 429 women diagnosed with lung cancer and 1021 controls frequency matched for age were included. Dietary intake of vitamin E was not associated with the risk reduction of lung cancer (Table VIII). The relative risk of the disease between highest and lowest quintiles of vitamin E intake after adjustment for age, smoking history, prior lung disease, and energy intake was 1.72 (p for trend = 0.18). The corresponding value when supplement use was also included was 1.47.

In a population-based matched case-control study of lung cancer in nonsmokers in New York State in 1982–1985, Mayne *et al.* (1994) studied 413 individually matched case-control pairs of subjects (Table IV). The controls were matched for sex, age, and county of residence. Vitamin E supplement use was associated with a reduced risk of lung cancer. The relative risk of the disease was 0.55 (CI = 0.35–0.85) based on a case-control difference of one frequency unit per day (Table VIII). The magnitude of protective effect was approximately equal for men and women with relative risks of 0.59 (CI = 0.32–1.08) and 0.51 (CI = 0.27–0.97), respectively. Any use of vitamin E supplements was associated with reduced risk; more frequent users conferred no additional risk reduction in these data (test for linear trend not significant). Results were not confounded by income, education, or intake of fruits and vegetables.

Torun *et al.* (1995) compared 12 cancer patients in Turkey with 156 healthy subjects and found 41% lower serum vitamin E concentration in the lung cancer cases than in the controls ($p < 0.05$) (Tables IV and VIII); the results were not confounded by age. No adjustments were made for other potential confounding factors such as sex, body mass index, smoking, alcohol consumption, and diet. Smoking was not associated with the serum vitamin E level.

Hu *et al.* (1997) carried out a case-control study considering 227 lung cancer cases and 227 matched hospital controls in the Heilongjiang Province in northeast China (Table IV). The controls were individually matched for sex, age, and area of residence. No association was observed between vitamin E intake and lung cancer risk. The relative risk of lung cancer between the highest and the lowest quartiles of vitamin E intake was 1.1 (0.7–1.9) after adjustment for smoking and economic status (Table VIII). No significant interaction between smoking and vitamin E intake was observed. Relative risks of lung cancer between quartiles of vitamin E intake were 1.5 (CI = 0.7–3.1, p for trend = 0.43) and 0.9 (CI = 0.4–2.0, p for trend = 0.44) for smokers and nonsmokers, respectively, after adjustment for age, sex, economic status, and also for cigarettes per day and duration of smoking among smokers.

In summary, a strong inverse association was present between vitamin E status and lung cancer risk in 7 of the 11 case-control studies considered. Of these 7 studies, no adjustment for smoking was carried out in 5 (Miyamoto *et al.*, 1987; Skulchan and Ong-Ajyooth, 1987; LeGardeur *et al.*, 1990; Harris *et al.*, 1991; Torun *et al.*, 1995), and one study was carried out in a population of nonsmokers (Mayne *et al.*, 1994). A reduced risk of lung cancer was demonstrated in one study after allowance for smoking (Tominaga *et al.*, 1992).

Discussion

Causal Inference

In principle, an intervention trial can give a straightforward answer to the question of whether vitamin E provides protection against lung cancer. The two intervention trials published so far gave discrepant results, however (Blot *et al.* 1994; Albanes *et al.*, 1996). The main methodological difficulties possibly hiding the effect in such a study, which may be difficult to eliminate, are related to choice of study population and supplement, duration of supplementation, and length of follow-up (Knekt, 1997). Thus, conclusions about whether vitamin E provides protection against lung cancer must also rely on the information from observational studies. The main weakness of an observational study is that it can never occur under rigidly controlled conditions, thus leaving open the question of causality. To evaluate causality based on the studies considered, the criteria formulated by Hill (1971) were therefore applied. The criteria that should be fulfilled include consistency, strength, and dose–response of association, relation in time, specificity, biological plausability, and experimental evidence (van Poppel and Goldbohm, 1995).

In the case-control studies reviewed, vitamin E status was rather consistently inversely associated with the risk of lung cancer. Seven out of 11 studies gave an inverse association. Consistency may, in principle, be enhanced by publication bias caused by the tendency to publish studies indicating significant associations. The fact that, in general, the studies have also focused on variables other than vitamin E, however, diminishes this possibility. The nested case-control and the cohort studies did not give consistent results. Of the 6 cohort studies, only 2 reported a significant association in the total study population, but none of the studies definitely excludes the possibility of an association in some subpopulations. One of the 8 nested case-control studies revealed a significant association and 1 a nonsignificant association.

A strong association was present in almost all case-control studies published. The lung cancer patients had generally 20–40% lower vitamin E status than the controls. The difference was much smaller in cohort and nested case-control studies, being with two exceptions under 5%. The presence of a dose–response association between vitamin E status and cancer occurrence was evaluated in six studies, and a significant trend was observed in only one of these.

Vitamin E status should occur in a logical relationship in time to cancer occurrence in that the reduced level should precede the onset of disease. This condition is in principle met in cohort and nested case-control studies in which the serum samples are collected before the disease is diagnosed. In case-control studies, however, vitamin E intake or tissue concentration of vitamin E in the cancer patients may be altered by the cancer due to treatment or to changes in dietary habits or metabolism caused by the disease. In cohort and nested case-control studies the influence of the disease process on serum/plasma vitamin E status is not completely excluded but can be minimized by exclusion of cancer cases occurring during the first years of follow-up, as was done in the two nested case-control studies suggesting a risk reduction (Menkes *et al.*, 1986; Kok *et al.*, 1987). Of the case-control studies considered, only one has addressed this issue by demonstrating low vitamin E levels in family members of lung cancer patients also, thus supporting the hypothesis of a logical relationship in time (Miyamoto *et al.*, 1987).

A confounding factor that is a risk factor for cancer associated with vitamin E status may affect the association between vitamin E status and lung cancer occurrence by possibly hiding the true association or by resulting in an artifactual association. Various foodstuffs and nutrients, as well as dietary patterns, may be confounding factors. The intake of vitamin E is, for example, strongly associated with the intake of several potentially protective nutrients such as fiber, carotenoids, vitamin C, and flavonoids, both because they are included in the same foods and due to clustering dietary behavior. Smoking, the major risk factor for lung cancer, may also confound the results. The association observed can be regarded as specific for vitamin E only if the effects of confounding factors have been controlled. Smoking was adjusted for in the majority of cohort studies (Shibata *et al.*, 1992; The Alpha-Tocopherol, Beta Carotene Cancer Prevention Study Group, 1994; Eichholzer *et al.*, 1996; Ocké *et al.*, 1997) and in nested case-control studies (Menkes *et al.*, 1986; Wald *et al.*, 1987; Connett *et al.*, 1989; Knekt *et al.*, 1988a; Orentreich *et al.*, 1991), but only in one of the case-control studies (Tominaga *et al.*, 1992). Serum lipids (Nomura *et al.*, 1985; LeGardeur *et al.*, 1990; Eichholzer *et al.*, 1996) were adjusted for in only some of the studies. Adjustment for dietary factors in general has been incomplete.

Several possible biological mechanisms may occur, by which the hypothesis is plausible that vitamin E provides protection against cancer (Knekt, 1994). There is, however, insufficient evidence of a protective effect of vitamin E against lung cancer occurrence in intervention trials (Albanes *et al.*, 1996).

In summary, only partial evidence suggesting a causal connection between vitamin E status and lung cancer occurrence is provided by observational studies. The association can be regarded as consistent only in case-control studies, but the dose–response association has not been evaluated in any of these. Furthermore, control for confounding factors has been incomplete, and thus the association cannot be considered specific. There is also no convincing evidence that the relationship is logical in time. Although the association is biologically plausible, it has not been verified in an intervention trial. Thus, it cannot be concluded that there is a causal connection between vitamin E status and lung cancer occurrence.

Methodological Aspects

The lack of evidence for a causal connection between vitamin E and lung cancer in observational studies may be due to several sources of bias or variation that may hide the association (Knekt, 1997). The main sources of such bias and variation are the presence of selection bias, the low power of the study, vitamin E on an ineffective range, an inadequate choice of the statistical method, low reliability of the measurement method, and neglection of effect-modification.

Selection bias may occur when the controls chosen in case-control studies are not wholly representative of the source population from which the case originated. With the exception of three studies (Byers *et al.*, 1987; Alavanja *et al.*, 1993; Mayne *et al.*, 1994), the controls used in the case-control studies were hospital patients or normal values not necessarily resentative of the populations from which the cancer cases were selected.

A small number of cancer cases included, i.e., the low power of the study, may be one reason for the failure of some studies to detect significant associations (Kok *et al.*, 1987; Harris *et al.*, 1991; Stähelin *et al.*, 1991; Blot *et al.*, 1993; Ocké *et al.*, 1997). Another potential reason for the lack of association in some studies is that they have been conducted in populations with vitamin E at levels with no or constant effects on lung cancer development. Currently it is not known at what levels of vitamin E can a possible effect be present at different exposures. Because only four of the observational studies have reported on the effects of vitamin E supplementation, the current information regarding higher vitamin E levels is scarce (Shibata *et al.,* 1992; Mayne *et al.,* 1994). The choice of statistical method may also mask existing associations. The majority of the nested case-control studies have only reported the mean difference between cases and controls (Table VII), which may possibly not reveal associations present only at the ends of the distribution.

None of the studies on the dietary intake of vitamin E indicated any significant associations with lung cancer (Byers *et al.* 1987; Connett *et al.,* 1989; Knekt *et al.,* 1991b; Alavanja *et al.,* 1993; Hu *et al.,* 1997; Ocké *et al.,* 1997; Yong *et al.,* 1997). This may be due to methodological inadequacies associated with the measurement of vitamin E intake. Such issues derive from errors in a subject's estimates of the frequency or size of portions eaten, from inaccurate pictures of long-term intake, and from the low precision of the dietary transformation tables from which the vitamin E intake is estimated. Furthermore, the availability of vitamin E at the same intake level may vary among individuals due to interactions with dietary, lifestyle, and environmental factors.

Although some of the studies on circulating vitamin E reported an inverse association, it is difficult to interpret the results due to several methodological issues involved in the use of serum/plasma vitamin E for the estimation of vitamin E exposure. The serum/plasma concentration of vitamin E may not be representative of levels in the appropriate tissue. Circulating vitamin E concentration may not be a sufficient characteristic of long-term vitamin E exposure. One source of error, specific for nested case-control studies, is the possible degradation of vitamin E concentrations during storage of the serum samples. The blood vitamin E concentrations of samples stored at temperatures over $-70°C$ are especially susceptible to degradation (Comstock *et al.,* 1993). Furthermore, because the vitamin E status in all studies but one (Ocké *et al.,* 1997) was estimated based on a single measurement, conservative estimates of the strength of association were generally obtained.

It cannot be excluded that the strength of the possible effect of vitamin E on lung cancer risk is dependent on the relation between the amount of available vitamin E and the amount of exposure to possible effect-modifying factors. This means that the association would exist only at specific levels of these factors and that assessment of the association between vitamin E status and lung cancer risk in total populations may thus give conservative estimates. This suggestion is supported by the findings of a stronger association among persons with lower exposure to smoking than among persons with higher exposure in the majority of studies separating the effects among smokers and nonsmokers (Table IX). In studies on the

Table IX Relative Risk of Lung Cancer between High and Low Vitamin E Status at Different Exposures to Tobacco Smoke

| | Vitamin E | | | Relative risk (95% CI) by smoking status | | | | |
| | Source | Categories (highest vs lowest) | Non | | Current | | Reference |
Design			Relative risk	95% confidence internal	Relative risk	95% confidence interval	
Ca-Co	Diet	Quintile	1.72	—	—	—	Alavanja et al. (1993)
Nested	Serum	Tertile	0.15	(0.03–0.77)	1.25	(0.63–2.50)	Knekt et al. (1993)
Cohort	Diet	Tertile	0.30	(0.09–1.00)	1.25	(0.77–2.00)	
Cohort	Serum	Quartile	—	—	0.55[b]	—	The Alpha-Tocopherol, Beta Carotene Cancer Prevention Study Group (1994)
	Diet	Quartile	—	—	0.44[b]	—	
Trial	Supplement	Yes/no	—	—	0.99[b]	(0.87–0.13)	
Ca-Co	Supplement	Unit/day	0.55	(0.35–0.85)	—	—	Mayne et al. (1994)
Ca-Co	Diet	Quartile	0.9	(0.4–2.0)	1.5	(0.7–3.1)	Hu et al. (1997)
Cohort	Diet	Quartile	0.87	(0.49–1.60)	0.82	(0.50–1.37)	Yong et al. (1997)
			0.36[a]	(0.16–0.83)	1.65[b]	—	

[a] Light smoker.
[b] Heavy smoker.

effect of vitamin E in combination with other micronutrients, a stronger effect was found at lower vitamin C levels than at higher levels (Eichholzer *et al.,* 1996). Some studies also suggest that a combination of vitamin E with several other micronutrients (i.e., β-carotene, vitamin C, selenium, or retinol) may be more important than vitamin E alone (Blot *et al.,* 1994; Knekt, 1993; Yong *et al.,* 1997). However, contrary results were presented in one study suggesting that vitamin E does not need support from other micronutrients (Tominaga *et al.,* 1992). The associations were, with the previous exceptions, estimated in the total populations. Thus, there is a definite need for studies focusing on the effect of vitamin E in subgroups of exposure and in combination with other micronutrients.

THE MOBILE CLINIC HEALTH EXAMINATION SURVEY

Introduction

During 1968–1971 and 1972–1976 the Finnish Mobile Clinic carried out a two-stage health examination survey in various parts of Finland (Knekt, 1988a; Knekt *et al.,* 1997b). A serum sample bank was founded at the time of baseline examination and stored at $-20°C$. Cancer incidence information, collected by the nationwide Finnish Cancer Registry, has since been continuously linked to this population. The follow-up period currently covers the interval from the beginning of the study up to the end of 1994. Based on this population, a series of studies on vitamin E status and cancer risk has been carried out (Knekt, 1988a,b, 1993; Knekt *et al.,* 1988a,b, 1991a,b).

The aim of this section is to present findings from two new studies. The first describes the effect of vitamin E in combination with β-carotene and selenium on lung cancer occurrence, which was obtained from reanalysis of nested case-control study data published earlier and based on the first stage, between 1968 and 1971 (Knekt *et al.,* 1988a). The second study, which focuses on the interaction between vitamin E and selenium, is based on the second stage during 1972–1976 (Knekt *et al.,* 1997b).

Vitamin E in Combination with β-Carotene and Selenium

Introduction

The Linxian Nutrition Intervention trial suggested a protective effect of vitamin E in combination with β-carotene and selenium in a region with very low lung cancer incidence (Blot *et al.,* 1993, 1994). The relative risk of lung cancer among individuals receiving the supplement in comparison with those receiving placebo was 0.55 (CI = 0.26–1.14). In a reanalysis of the Mobile Clinic Health Examination data, the lung cancer incidence was compared among individuals at different serum levels of the same three micronutrients.

Population and Methods

The study population comprised 40,201 Finnish men and women, initially 15–99 years of age. The baseline examination was conducted in 1968–1972, and during a follow-up of 6–10 years 151 lung cancer cases were diagnosed. A total of 270 controls, matched for municipality, sex, and age, from among those who did not develop cancer during the follow-up, were selected. The serum α-tocopherol and β-carotene levels, measured from serum samples taken at the baseline examination and stored thereafter at $-20°C$, were simultaneously determined using liquid chromatography, and serum selenium was determined using a graphite furnace atomic absorption spectrometric method (Knekt, 1988a).

Results and Conclusions

The smoking-adjusted relative risks of lung cancer between the highest and the lowest quintiles were 0.87 (CI = 0.35–2.16, p for trend = 0.23) for α-tocopherol, 0.78 (CI = 0.32–1.86, p for trend = 0.05) for β-carotene, and 0.29 (CI = 0.12–0.70, p for trend = 0.002) for selenium. An antioxidant index was calculated as the product of the molar concentrations of α-tocopherol and β-carotene. The relative risk of lung cancer between the highest and the lowest tertiles of this index was 0.66 (CI = 0.21–2.08, p for trend = 0.03). When serum selenium was added to this index the corresponding number was 0.49 (CI = 0.19–1.22, p for trend = 0.01). Thus, in agreement with the suggestive finding from the Linxian trial, a reduced risk of lung cancer was observed for individuals at higher levels of the α-tocopherol, β-carotene, and selenium index. Because only selenium showed a significant inverse association with lung cancer risk, it is possible that the effect observed is mainly due to that micronutrient.

Vitamin E, Selenium, and Lung Cancer

Introduction

An interaction has been demonstrated between vitamins E and C on lung cancer incidence (Eichholzer *et al.*, 1996). The aim of the present nested case-control study was to examine whether a similar interaction occurs between α-tocopherol and selenium.

Population and Methods

Serum samples from 9101 cancer-free individuals were collected and stored at $-20°C$ by the Finnish Mobile Clinic in 1972–1976. During a follow-up of a maximum of 20 years between 1972 and the end of 1991, 95 lung cancer cases were diagnosed. Selenium and α-tocopherol concentrations were determined from the stored serum samples of these patients and 180 controls, individually matched for sex, age, and place of residence. The α-tocopherol concentration was determined from the stored serum samples by HPLC and the selenium concentrations by the graphite furnace technique.

Results and Conclusion

The mean level of serum selenium was 53.2 μg/liter in cancer cases and 57.8 μg/liter in controls. The corresponding results for α-tocopherol were 6.77 and 6.78 mg/liter, respectively. The relative risk of lung cancer between the highest and the lowest tertiles was 0.41 (CI = 0.17–0.94) for serum selenium and 0.83 (CI = 0.33–2.07) for α-tocopherol in a model including smoking, serum α-tocopherol, serum selenium, serum cholesterol, serum copper, serum orosomucoid, and body mass index. The risk was highest at low levels of both micronutrients. The relative risk of lung cancer for higher selenium and low α-tocopherol levels was 0.24 (CI = 0.07–0.85). For higher α-tocopherol and low selenium levels it was 0.41 (CI = 0.12–1.39). At higher levels of both micronutrients, the relative risk was 0.32 (CI = 0.10–0.98). Thus, the risk appears to be significantly elevated at low levels of both vitamin E and selenium, suggesting that selenium and α-tocopherol compensate for each other.

Conclusion

In earlier studies on serum or dietary vitamin E status and lung cancer occurrence, a significantly reduced risk of lung cancer at higher vitamin E levels was found only among nonsmokers (i.e., individuals with low exposure to tobacco smoke; Knekt *et al.*, 1991b; Knekt 1993). Knekt (1993) also showed that in both nonsmokers and smokers the inverse association between simultaneously high serum levels of α-tocopherol, β-carotene, retinol, and selenium and lung cancer risk was stronger than the association for the single nutrients (Knekt, 1993). The present studies demonstrated that selenium and α-tocopherol apparently compensated for each other. Furthermore, it was found that the reduced risk of lung cancer present at high levels of an antioxidant index combining α-tocopherol, β-carotene, and selenium apparently is mainly due to selenium.

CONCLUSIONS AND FURTHER RESEARCH

Evidence for a protective effect of vitamin E against lung cancer is not conclusive. This may be due partly to methodological issues involved in different study designs, and accordingly it cannot be excluded that vitamin E in some circumstances provides protection against the disease. Studies on the simultaneous effects of vitamin E and other antioxidant micronutrients, and of vitamin E in various categories of exposure to tobacco smoke, are in accordance with this hypothesis. To obtain more reliable information on the possible effect of vitamin E on lung cancer risk, and on the circumstances of the effect, meta-analyses combining previously published observational studies and new large observational studies, with special emphasis on interactions and other methodological issues, should be carried out. When that information becomes available it will be time to carry out new intervention trials.

REFERENCES

Alavanja, M. C. R., Brown, C. C., Swanson, C., and Brownson, R. C. (1993). Saturated fat intake and lung cancer risk among nonsmoking women in Missouri. *J. Natl. Cancer Inst.* **85**, 1906–1916.

Albanes, D., Heinonen, O. P., Taylor, P. R., Virtamo, J., Edwards, B. K., Rautalahti, M., Hartman, A. M., Palmgren, J., Freedman, L. S., Haapakoski, J., Barret, M. J., Pietinen, P., Malila, N., Tala, E., Liippo, K., Salomaa, E.-R., Tangrea, J. A., Teppo, L., Askin, F. B., Taskinen, E., Erozan, Y., Greenwald, P., and Huttunen, J. K. (1996). Alpha-tocopherol and beta-carotene supplements and lung cancer incidence in the Alpha-Tocopherol, Beta-Carotene Cancer Prevention Study: Effects of Base-line characteristics and study compliance *J. Natl. Cancer Inst.* **88**, 1560–1570.

Atukorala, S., Basu, T. K., Dickerson, J. W. T., Donaldson, D., and Sakula, A. (1979). Vitamin A, zinc and lung cancer. *Br. J. Cancer* **40**, 927–931.

Block, G., Patterson, B., and Subar, A. (1992). Fruit, vegetables, and cancer prevention: A review of the epidemiological evidence. *Nutr. Cancer* **18**, 1–29.

Blot, W. J., Li, J.-Y., Taylor, P. R., Guo, W., Dawsey, S., Wang, G.-Q., Yang, C. S., Zheng, S.-F., Gail, M., Li, G.-Y., Yu, Y., Liu, B., Tangrea, J., Sun, Y., Liu, F., Fraumeni, J. F., Jr., Zhang, Y.-H., and Li, B. (1993). Nutrition intervention trials in Linxian, China: Supplementation with specific vitamin/mineral combinations, cancer incidence, and disease-specific mortality in the general population. *J. Natl. Cancer Inst.* **85**, 1483–1492.

Blot, W. J., Li, J.-Y., Taylor, P. R., and Li, B. (1994). Lung cancer and vitamin supplementation. *N. Engl. J. Med.* **331**, 614.

Buiatti, E., Geddes, M., and Arniani, S. (1996). Epidemiology of lung cancer. *Ann. Ist. Super Sanita* **32**, 133–144.

Byers, T. E., Graham, S., Haughey, B. P., Marshall, J. R., and Swanson, M. K. (1987). Diet and lung cancer risk: Findings from the Western New York Diet Study. *Am. J. Epidemiol.* **125**, 351–363.

Clark, L. C., Combs, G. F., Jr., Turnbull, B. W., Slate, E. H., Chalker, D. K., Chow, J., Davis,

L. S., Glover, R. A., Graham, G. F., Gross, E. G., Krongrad, A., Lesher, J. L., Jr., Park, H. K., Sanders, B. B., Jr., Smith, C. L., and Taylor, J. R. (1996). Effects of selenium supplementation for cancer prevention in patients with carcinoma of the skin. *JAMA* **276**, 1957–1963.

Comstock, G. W., Alberg, A. J., and Helzlsouer, K. J. (1993). Reported effects of long-term freezer storage on concentrations of retinol, beta-carotene, and alpha-tocopherol in serum or plasma summarized. *Clin. Chem.* **39**, 1075–1078.

Connett, J. E., Kuller, L. H., Kjelsberg, M. O., Polk, B. F., Collins, G., Rider, A., and Hulley, S. B. (1989). Relationship between carotenoids and cancer: The Multiple Risk Factor Intervention Trial (MRFIT) Study. *Cancer* **64**, 126–134.

Dorgan, J. F., and Schatzkin, A. (1991). Antioxidant micronutrients in cancer prevention. *Nutr. Cancer* **5**, 43–68.

Eichholzer, M., Stähelin, H. B., Gey, K. F., Lüdin, E., and Bernasconi, F. (1996). Prediction of male cancer mortality by plasma levels of interacting vitamins: 17-year follow-up of the Prospective Basel Study. *Int. J. Cancer* **66**, 145–150.

Friedman, G. D., Blaner, W. S., Goodman, D. S., Vogelman, J. H., Brind, J. L., Hoover, R., Fireman, B. H., and Orentreich, N. (1986). Serum retinol and retinol-binding protein levels do not predict subsequent lung cancer. *Am. J. Epidemiol.* **123**, 781–789.

Harris, R. W. C., Key, T. J. A., Silcocks, P. B., Bull, D., and Wald, N. J. (1991). A case-control study of dietary carotene in men with lung cancer and in men with other epithelial cancers. *Nutr. Cancer* **15**, 63–68.

Hennekens, C. H., Buring, J. E., Manson, J. E., Stampfer, M., Rosner, B., Cook, N. R., Belanger, C., LaMotte, F., Gaziano, J. M., Ridker, P. M., Willet, W., and Peto, R. (1996). Lack of effect of long-term supplementation with beta carotene on the incidence of malignant neoplasms and cardiovascular disease. *N. Engl. J. Med.* **334**, 1145–1149.

Hill, A. B. (1971). "Principles of Medical Statistics," 9th ed. Oxford Univ. Press, New York.

Hu, J., Johnson, K. C., Mao, Y., Xu, T., Lin, Q., Wang, C., Zhao, F., Wang, G., Chen, Y., and Yang, Y. (1997). A case-control study of diet and lung cancer in northeast China. *Int. J. Cancer* **71**, 924–931.

Knekt, P. (1988a). Serum alpha-tocopherol and the risk of cancer: Publications of the Social Insurance Institution, ML:83. Helsinki, Finland.

Knekt, P. (1988b). Serum vitamin E level and risk of female cancers. *Int. J. Epidemiol.* **17**, 281–288.

Knekt, P. (1993). Vitamin E and smoking and the risk of lung cancer. *Ann. N. Y. Acad. Sci.* **686**, 280–288.

Knekt, P. (1994). Vitamin E and cancer prevention. *In* "Natural Antioxidants in Human Health and Disease" (B. Frei, ed.), pp. 199–237. Academic Press, New York.

Knekt, P. (1997). Vitamin E and cancer prevention: Methodological aspects of human studies. *In* "Food Factors for Cancer Prevention" (H. Ohigashi, T. Osawa, J. Terao, S. Watanabe, and T. Yoshikawa, eds.). Springer, Tokyo.

Knekt, P., Aromaa, A., Maatela, J., Aaran, R.-K., Nikkari, T., Hakama, M., Hakulinen, T., Peto, R., and Teppo, L. (1991a). Vitamin E and cancer prevention. *Am. J. Clin. Nutr.* **53**, 283s–286s.

Knekt, P., Aromaa, A., Maatela, J., Aaran, R.-K., Nikkari, T., Hakama, M., Hakulinen, T., Peto, R., Saxén, E., and Teppo, L. (1988a). Serum vitamin E and risk of cancer among Finnsh men during a 10-year follow-up. *Am. J. Epidemiol.* **127**, 28–41.

Knekt, P., Aromaa, A., Maatela, J., Alfthan, G., Aaran, R.-K., Teppo, L., and Hakama, M. (1988b). Serum vitamin E, serum selenium and the risk of gastrointestinal cancer. *Int. J. Cancer* **42**, 846–850.

Knekt, P., Järvinen, R., Seppänen, R., Rissanen, A., Aromaa, A., Heinonen, O. P., Albanes, D.,

Heinonen, M., Pukkala, E., and Teppo, L. (1991b). Dietary antioxidants and the risk of lung cancer. *Am. J. Epidemiol.* **134**, 471–479.

Knekt, P., Järvinen, R., Seppänen, R., Heliövaara, M., Teppo, L., and Aromaa, A. (1997a). Dietary flavonoids and the risk of lung cancer and other malignant neoplasms. *Am. J. Epidemiol.* **146**, 223–230.

Knekt, P., Marniemi, J., Teppo, L., Heliövaara, M., and Aromaa, A. (1997b). Is low selenium status a risk factor for lung cancer? *Am. J. Epidemiol.* (in press).

Kok, F. J., van Duijn, C. M., Hofman, A., Vermeeren, R., de Bruijn, A. M., and Valkenburg, H. A. (1987). Micronutrients and the risk of lung cancer. *N. Engl. J. Med.* **316**, 1416.

LeGardeur, B. Y., Lopez-S, A., and Johnson, W. D. (1990). A case-control study of serum vitamins A, E, and C in lung cancer patients. *Nutr. Cancer* **14**, 133–140.

Mayne, S. T., Janerich, D. T., Greenwald, P., Chorost, S., Tucci, C., Zaman, M. B., Melamed, M. R., Kiely, M., and McKneally, M. F. (1994). Dietary beta carotene and lung cancer risk in U. S. nonsmokers. *J. Natl. Cancer Inst.* **86**, 33–38.

Menkes, M. S., Comstock, G. W., Vuilleumier, J. P., Helsing, K. J., Rider, A. A., and Brookmeyer, R. (1986). Serum beta-carotene, vitamins A and E, selenium, and the risk of lung cancer. *N. Engl. J. Med.* **315**, 1250–1254.

Miyamoto, H., Araya, Y., Ito, M., Isobe, H., Dosaka, H., Shimizu, T., Kishi, F., Yamamoto, I., Honma, H., and Kawakami, Y. (1987). Serum selenium and vitamin E concentrations in families of lung cancer patients. *Cancer* **60**, 1159–1162.

Nomura, A. M. Y., Stemmermann, G. N., Heilbrun, L. K., Salkeld, R. M., and Vuilleumier, J. P. (1985). Serum vitamin levels and the risk of cancer of specific sites in men of Japanese ancestry in Hawaii. *Cancer Res.* **45**, 2369–2372.

Ocké, M. C., Bas Bueno-de-Mesquita, H., Feskens, E. J. M., van Staveren, W. A., and Kromhout, D. (1997). Repeated measurement of vegetables, fruits, beta-carotene, vitamins C and E in relation to lung cancer. *Am. J. Epidemiol.* **145**, 358–365.

Omenn, G. S., Goodman, G. E., Thornquist, M. D., Balmes, J., Cullen, M. R., Glass, A., Keogh, J. P., Meyskens, F. L., Valanis, B., Williams, J. H., Jr., Barnhart, S., and Hammar, S. (1996). Effects of a combination of beta carotene and vitamin A on lung cancer and cardiovascular disease. *N. Engl. J. Med.* **334**, 1150–1155.

Orentreich, N., Matias, J. R., Vogelman, H., Salkeld, R. M., Bhagavan, H., and Friedman, G. D. (1991). The predictive value of serum β-carotene for subsequent development of lung cancer. *Nutr. Cancer* **16**, 167–169.

Shibata, A., Paganini-Hill, A., Ross, R. K., and Henderson, B. E. (1992). Intake of vegetables, fruits, beta-carotene, vitamin C and vitamin supplements and cancer incidence among the elderly: A prospective study. *Br. J. Cancer* **66**, 673–679.

Skulchan, V., and Ong-Ajyooth, S. (1987). Serum vitamin E in Thai Cancer patients. *J. Med. Assoc. Thailand* **70**, 280–283.

Stähelin, H. B., Gey, K. F., Eichholzer, M., Lüdin, E., Bernasconi, F., Thurneysen, J., and Brubacher, G. (1991). Plasma antioxidant vitamins and subsequent cancer mortality in the 12-year follow-up of the Prospective Basel Study. *Am. J. Epidemiol.* **133**, 766–775.

The Alpha-Tocopherol, Beta Carotene Cancer Prevention Study Group. (1994). The effect of vitamin E and beta carotene on the incidence of lung cancer and other cancers in male smokers. *N. Engl. J. Med.* **330**, 1029–1035.

Tominaga, K., Saito, Y., Mori, K., Miyazawa, N., Yokoi, K., Koyama, Y., Shimamura, K., Imura, J., and Nagai, M. (1992). An evaluation of serum microelement concentrations in lung cancer and matched non-cancer patients to determine the risk of developing lung cancer: A preliminary study. *Jpn. J. Clin. Oncol.* **22**, 96–101.

Torun, M., Yardim, S., Gönenç, A., Sargin, H., Menevşe, A., and Şimşek, B. (1995). Serum β-

carotene, vitamin E, vitamin C and malondialdehyde levels in several types of cancer. *J. Clin. Pharm. Ther.* **20**, 259–263.

Van Poppel, G., and Goldbohm, R. A. (1995). Epidemiologic evidence for beta-carotene and cancer prevention. *Am. J. Clin. Nutr.* **62**(Suppl.), 1393s–1402s.

Wald, N. J., Thompson, S. G., Densem, J. W., Boreham, J., and Bailey, A. (1987). Serum vitamin E and subsequent risk of cancer. *Br. J. Cancer* **56**, 69–72.

Willett, W. C., Polk, B. F., Underwood, B. A., Stampfer, M. J., Pressel, S., Rosner, B., Taylor, J. O., Schneider, K., and Hames, C. G. (1984). Relation of serum vitamins A and E and carotenoids to the risk of cancer. *N. Engl. J. Med.* **310**, 430–434.

Yong, L-C., Brown, C. C., Schatzkin, A., Dresser, C. M., Slesinski, M. J., Cox, C. S., and Taylor, P. R. (1997). Intake of vitamins E, C, and A and risk of lung cancer: The NHANES I Epidemiologic Followup Study. *Am. J. Epidemiol.* **146**, 231–243.

Ziegler, R. G., Mayne, S. T., and Swanson, C. A. (1996). Nutrition and lung cancer. *Cancer Causes Control* **7**, 157–177.

2 Natural Antioxidants in the Protection against Cigarette Smoke Injury

Garry G. Duthie
Rowett Research Institute
Scotland, United Kingdom

INTRODUCTION

There are currently 3 million tobacco-related deaths in the world each year, and in general smokers can expect to live 7 years less than nonsmokers (World Health Organisation, 1977). This premature mortality is because habitual smoking is associated with an increased risk of developing many diseases, including coronary heart disease, lung cancer, stroke, and emphysema (Health Education Authority, 1992). Paradoxically, analysis across countries reveals little relationship between smoking levels and mortality from diseases such as coronary heart disease and cancer. For example, Japan has one of the lowest incidences of lung cancer in the world despite having one of the highest per capita consumption of cigarettes (Diana, 1993). Similarly, coronary heart disease rates in countries such as Greece and Spain are low despite very high cigarette usage (Fig. 1). This suggests that indigenous factors within countries such as diet may modify the risk of developing smoking-related diseases. Many of the clinical conditions implicated with smoking are also associated with increased indices of free radical-mediated damage to proteins, lipids, and DNA (Duthie and Arthur, 1994), indicating that smoking may exacerbate the initiation and propagation of oxidative stresses, which are potential underlying processes in the pathogenesis of many diseases (Diplock, 1994). Consequently, one expla-

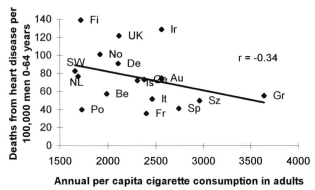

Figure 1   The lack of a positive and strong cross-country relationship between standardized mortality rates for coronary heart disease calculated as an average from 1985 to 1987 and total adult cigarette consumption in adults over 15 years of age in Europe. Au, Austria; Be, Belgium; De, Denmark; Ge, Germany (former West); Fi, Finland; Fr, France; Gr, Greece; Ir, Ireland; Is, Israel; It, Italy; NL, Netherlands; No, Norway; Po, Portugal; Sp, Spain; Sz, Switzerland; UK, United Kingdom. Data courtesy of Mary Bellizzi, Rowett Research Institute. Similar data including non-European countries can be found in Duthie and Arthur (1994).

nation for the poor cross-country association between smoking intensity and disease may be due to the variation in antioxidant intakes, which modify smoking-induced oxidative stress. Therefore, the aims of this chapter are to review how smoking challenges the antioxidant defence system and to discuss whether increased intakes of natural antioxidants can influence smoking-induced oxidative stress and disease morbidity.

SMOKING AS AN OXIDATIVE STRESS

Smokers inhale large amounts of reactive free radicals arising from the combustion of tobacco. The tar in cigarettes contains more than 10^{17} stable long-lived quinone–semiquinone radicals per gram, which are generated by the oxidation of polycyclic aromatic hydrocarbons during the combustion process. These can reduce oxygen to superoxide and hydrogen peroxide and result in the production of the highly reactive hydroxyl radical. The gas phase smoke contains more than $\sim 10^{15}$ free radicals per puff of short-lived, reactive carbon- and oxygen-centered peroxy species. These can achieve a steady state between production and destruction as a result of the slow oxidation of nitric oxide in cigarette smoke to nitrogen dioxide, which then reacts with aldehydes and olefins to continually produce peroxy radicals (see review by Pryor, 1997).

One consequence of the production of free radicals in cigarette smoke is marked oxidizing reactions in biological fluids. For example, exposure of plasma to gas phase cigarette smoke causes depletion in antioxidants such as vitamin C, vitamin E, urate, ubiquinol-10, and β-carotene and increased products of free radical-mediated damage to lipids and proteins (Eiserich *et al.*, 1995; Handelman *et al.*, 1996). Moreover, as a result of this sustained oxidant load, many indices of oxidative damage to biomolecules are elevated in smokers (Table I). For example, plasma of smokers has a twofold higher concentration of F_2-isoprostanes than that of nonsmokers, which is indicative of the enhanced peroxidation of arachidonic acid (Morrow *et al.*, 1995). In addition, increased hydrocarbon expiration in smokers compared with nonsmokers suggests that smoking stimulates the peroxidation of endogenous fatty acids *in vivo* (Hoshina *et al.*, 1990). Moreover, smoking increases DNA damage *in vivo* as estimated by the enhanced detection of mutations in circulating T lymphocytes and 8-hydroxydeoxyguanosine in the urine and circulating leukocytes of smokers (Duthie *et al.*, 1995; Loft *et al.*, 1992).

Concentrations of natural antioxidants in blood and tissue of smokers tend to be lower than in nonsmokers (Table II). This may be partly due to

Table I   Studies Indicating Increased Indices of Oxidative Stress in Smokers

Author	Comments
Sakamoto (1985)	Smokers have enhanced ethane expiration, indicating higher peroxidation of *n*-3 fatty acids *in vivo*
Duthie *et al.* (1989, 1991)	Smokers have increased plasma concentrations of conjugated dienes and dehydroascorbate
Hoshino *et al.* (1990)	Increased pentane expiration by smokers indicates enhanced peroxidation of *n*-6 fatty acids *in vivo*
Loft *et al.* (1992)	Smoking increases oxidative DNA damage *in vivo* by approximately 50% as indicated by increased urinary levels of the DNA repair product, 8-hydroxydeoxyguanosine
Brown *et al.* (1994)	Plasma concentrations of lipid peroxides, thiobarbituric acid-reactive substances, and conjugated dienes elevated in smokers suggesting increased lipid peroxidation
Duthie *et al.* (1995)	Increased DNA damage in smokers indicated by positive relationship between *hprt* mutant frequency of peripheral T lymphocytes and reported smoking intensity
Morrow *et al.* (1995)	Increased circulating levels of F_2 isoprostanes, a measure of lipid peroxidation, in smokers compared with nonsmokers
Reilly *et al.* (1996)	Increased levels of 8-epi-prostoglandin F_2-α, a stable product of lipid peroxidation, in urine of smokers
Miller *et al.* (1997)	Increased expired ethane and plasma thiobarbituric acid-reactive substances in smokers compared with nonsmokers

Table II   Example of the Differences in Plasma Concentrations of Dietary Antioxidants between 50 Scottish Smokers and 50 Age-Matched Nonsmokers from the Same Population[a]

Parameter	Smokers	Nonsmokers	p
α-Tocopherol (μg/ml)	11.38 ± 0.38	11.32 ± 0.33	0.574
γ-Tocopherol (μg/ml)	0.88 ± 0.94	0.85 ± 0.07	0.663
Vitamin C (μM)	25.6 ± 3.3	37.6 ± 2.4	**0.003**
α-Carotene (μg/ml)	0.042 ± 0.003	0.061 ± 0.004	**0.0006**
β-Carotene (μg/ml)	0.2 ± 0.02	0.32 ± 0.02	**0.026**
β-Cryptoxanthin (μg/ml)	0.042 ± 0.006	0.063 ± 0.007	**0.041**
Lycopene (μg/ml)	0.27 ± 0.02	0.28 ± 0.02	0.879
Lutein/zeaxanthin (μg/ml)	0.24 ± 0.02	0.29 ± 0.02	0.052

[a] Data adapted from Ross *et al.* (1995) and compared using students t test with $p < 0.05$ being taken as indicating statistical significance. Significant differences indicated by bold type.

different dietary habits as intakes of fruit and fresh vegetables, which are rich sources of vitamin C, carotenoids, and polyphenols, tend to be lower in smokers than in nonsmokers (Whichelow *et al.*, 1988). However, low concentrations of antioxidant nutrients may also result from nondietary factors responding to the enhanced oxidative stress due to smoking. Metabolic studies and dietary surveys indicate that smokers have a nicotine-induced decrease in intestinal absorption efficiency, a decreased urinary excretion rate, and a higher turnover of vitamin C (reviewed by Duthie and Arthur, 1994).

ANTIOXIDANT SUPPLEMENT OF SMOKERS

Supplementation with antioxidants can decrease the elevated indices of oxidative damage associated with smoking. For example, increased pentane expiration by smokers can be reduced by supplementation with 800 mg vitamin E/day for 2 weeks (Hoshina *et al.*, 1990), and the increased susceptibility of washed erythrocytes of smokers to hydrogen peroxide-induced peroxidation can be abolished by supplementing with 1000 mg vitamin E per day for 2 weeks (Duthie *et al.*, 1989). Brown *et al.* (1994) initially showed that erythrocytes of male smokers from a Scottish population with a habitually low vitamin E intake were more susceptible to hydrogen peroxide-stimulated peroxidation than those from nonsmokers. Plasma concentrations of lipid peroxides, thiobarbituric acid-reactive substances, and conjugated dienes were also elevated in smokers compared with nonsmokers. These indices of oxidative stress were significantly decreased in smokers following consumption of 280 mg dl-α-tocopherol acetate/day for 10 weeks. Lower intakes of vitamin E may also be effective as in a second study (Brown *et al.*, 1997), erythrocyte vitamin E concentrations of smokers increased in a dose-dependent manner during 20 weeks of supplementation with 70, 140, 560, or 1050 mg vitamin E/day. However, the lower dose was equally as effective as the higher doses in suppressing indices of lipid peroxidation. Markers of DNA damage, which are elevated in smokers, also respond to antioxidants. For example, there is a very strong protective effect of supplementation of smokers with a cocktail containing vitamin E, vitamin C, and β-carotene against endogenous oxidation of pyrimidines in DNA (Duthie *et al.*, 1996). It should be noted, however, that such effects on indices of oxidative damage do not necessarily imply that increased intakes of antioxidants will prevent the pathogenesis of major clinical conditions.

NATURAL PHENOLICS AND SMOKING

Dietary intakes of some plant-derived phenolic compounds amount to approximately 7–40 mg/day in the United States (Rimm *et al.*, 1996) and may be considerably higher in other countries with more plant-based diets. Many phenolics can act chemically as antioxidants as their extensive conjugated π-electron systems allow ready donation of electrons or hydrogen atoms from the hydroxyl moieties to free radicals. As yet few, if any, studies have assessed whether intakes or plasma and tissue levels of compounds such as flavonoids, catechins, coumarins, and anthocyanins are less in smokers compared with nonsmokers. However, the relatively low incidence of coronary heart disease and lung cancer in Japan, despite a high *per capita* smoking incidence, has been ascribed to the high intake of polyphenolic compounds. In particular, polyphenols in tea may have important functions as nutritional antioxidants as epidemiological studies have found that tea consumption is inversely related to risk of strokes, coronary heart disease, certain cancers, and liver disorders (Hertog *et al.*, 1993; Imai and Nakachi, 1995; Knekt *et al.*, 1996; Yang and Wang, 1993).

Green tea leaves can contain 35% of their dry weight as polyphenols (Balentine, 1992); these are mainly flavanols such as catechin, epicatechin, epigallocatechin, epicatechin gallate, and epigallocatechin gallate (Fig. 2). These compounds have marked antioxidant activity in chemical systems. For example, electron spin resonance methodology to assess the hydrogen-donating (antioxidant) ability of catechins and extracts of green and black tea to Fremy's radical and galvinoxyl radical in aqueous and organic solutions indicates that catechin flavanols account for 77–82% of the total antioxidant activity of green tea and 47–58% of that of black tea (Gardner *et*

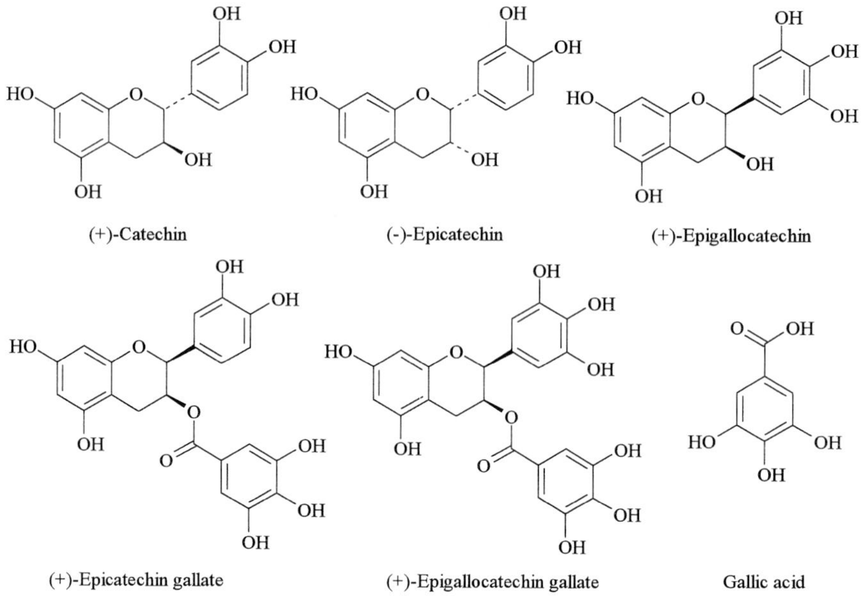

Figure 2   Chemical structures of the catechins used in the study and of gallic acid.

al., 1998). Epigallocatechin gallate was the most effective at reducing the two radical species, whereas epigallocatechin was least effective in reducing the galvinoxyl/ethanol system and catechin was the least effective in the Fremy's radical/water system. Possession of a galloyl ester moiety at the three-position on the C ring of the catechin appears important in conferring antioxidant activity (Duthie *et al.*, 1997). However, the significance of catechins and other polyphenolic products of the phenylpropanoid biosynthetic pathway such as anthocyanidins, flavanols, transresveratrol, and hydroxycinnamic and hydroxybenzoic acids as biological antioxidants, which benefit smokers, will remain unclear until the bioavailability, mechanism of uptakes, and consequences of biotransformation by intestinal flora of these compounds are established.

CONCLUSIONS

The inhalation of tobacco smoke results in a large intake of long- and short-lived free radicals. This sustained oxidative burden incurred by smokers results in elevated indices of free radical-mediated damage to lipids and DNA and decreased blood and tissue concentrations of some antioxidant nutrients such as vitamin E, vitamin C, and certain carotenoids. Increased intakes of these antioxidants by supplementation can moderate indices of oxidative stress in smokers, although it is not clear whether the risk of developing smoking-related diseases is also decreased. Although some evidence exists to suggest that variation in dietary intakes of natural plant-based antioxidants may explain the lack of association between smoking and disease on a cross-country basis, it is still a very good idea to stop smoking, particularly as within countries smokers have an increased incidence of disease.

ACKNOWLEDGMENTS

The author is funded by the Scottish Office Agriculture, Environment and Fisheries Department (SOAEFD) and gratefully acknowledges financial support from the Ministry of Agriculture, Fisheries and Food, the World Cancer Research Fund, the Association for International Cancer Research, and the EC.

REFERENCES

Balentine, A. D. (1992). Manufacturing and chemistry of tea. *In* "Phenolic Compounds in Food and Their Effects on Health" (H. Chi-Tang, Y. L. Chang, and H. Mou-Tuan, eds.), pp. 103–117. ACS, Washington, DC.

Brown, K. M., Morrice, P. C., and Duthie, G. G. (1994). Vitamin E supplementation suppresses indices of lipid peroxidation and platelet counts in blood of smokers and non-smokers but plasma lipoprotein concentrations remain unchanged. *Am. J. Clin. Nutr.* **60**, 383–387.

Brown, K. M., Morrice, P. C., and Duthie, G. G. (1997). Erythrocyte vitamin E and plasma ascorbate concentrations in relation to erythrocyte peroxidation in smokers and non-smokers: Dose response of vitamin E supplementation. *Am. J. Clin. Nutr.* **65**, 496–502.

Diana, J. N. (1993). Tobacco smoking and nutrition: Overview. *Ann. N.Y. Acad. Sci.* **686**, 1–11.

Diplock, A. T. (1994). Antioxidants and disease prevention. *Mol. Aspects Med.* **15**, 293–376.

Duthie, G. G., and Arthur, J. R. (1994). Cigarette smoking as an inducer of oxidative stress. *In* "Exercise and Oxygen Toxicity" (C. Sen and L. Packer, eds.), pp. 297–317. Elsevier Science, B.V.

Duthie, G. G., Arthur, J. R., and James, W. P. T. (1991). Effects of smoking and vitamin E on blood antioxidant status. *Am. J. Clin. Nutr.* **53**, 1061–1063S.

Duthie, G. G., Arthur, J. R., James, W. P. T., and Vint H. M. (1989). Antioxidant status of smokers and non-smokers: Effects of vitamin E supplementation. *Ann. N.Y. Acad. Sci.* **570**, 435–438.

Duthie, G. G., Gardner, P. T., Morrice, P. C., Crozier, A., and McPhail, D. B. (1997). Antioxidant efficacy of plant polyphenols in chemical and biological systems. *In* COST-916 Action "Polyphenols in Foods," (R. Armado, ed.), pp 117–122. Office for Official Publications of the European Communities, Luxembourg.

Duthie, S. J., Ma, A.-G., Ross, M. A., and Collins, A. R. (1996). Antioxidant supplementation decreases oxidative DNA damage in human lymphocytes. *Cancer Res.* **56,** 1291–1295.

Duthie, S. J., Ross, M., and Collins, A. (1995). The influence of smoking and diet on the hypoxanthine phosphoribosyltransferase (hprt) mutant frequency in circulating T lymphocytes from a normal human population. *Mutat. Res.* **331,** 55–64.

Eiserich, J. J., van der Vliet, A., Handelman G. J., Halliwell, B., and Cross, C. E. (1995). Dietary antioxidants and cigarette smoke-induced biomolecular damage: A complex interaction. *Am. J. Clin. Nutr.* **62,** 1490S–1500S.

Gardner, P. T., McPhail, D. B., and Duthie G. G. (1997). Electron spin resonance spectroscopic assessment of the antioxidant potential of teas in aqueous and organic media. *J. Sci. Food Agri.* **76,** 257–262.

Handelman, G. J., Packer, L., and Cross C. E. (1996). Destruction of tocopherols, carotenoids, and retinol in human plasma by cigarette smoke. *Am. J. Clin. Nutr.* **63,** 559–565.

Health Education Authority. (1992). "The Smoking Epidemic," Vol. A, pp. 81–98. Martins of Berwick, UK.

Hertog, M. G. L., Feskens, E. J. M., Hollman, P. C. H., Katan, M. B., and Kromhout, D. (1993). Dietary antioxidant flavonoids and the risk of coronary heart disease: The Zuphthen elderly study. *Lancet* **342,** 1007–1011.

Hoshina, E. R., Shariff, A., Van Gossum, A., Allard, J. P., Pichard, C., Kurian, R., and Jeejebhoy, K. N. (1990). Vitamin E suppresses increased lipid peroxidation in cigarette smokers. *J. Parenter. Enteral. Nutr.* **14,** 300–305.

Imai, K., and Nakachi, K., (1995). Cross sectional study of the effects of drinking green tea on cardiovascular and liver diseases. *Br. Med. J.* **310,** 693–696.

Knekt, P., Järvine, R., Reunanan, A., and Maatela, J. (1996). Flavonoid intake and coronary mortality in Finland: A cohort study. *Br. Med. J.* **312,** 478–481.

Loft, S., Virtisen K, Ewertz, M., Tjønnland A., Overvad, K., and Poulsen, H. E. (1992). Oxidative DNA damage estimated by 8-hydroxydeoxyguanosine excretion in humans: Influence of smoking, gender and body mass index. *Carcinogenesis* **13,** 2241–2247.

Miller, E. R., Appel, L. J., Jiang, L., and Risby, T. H. (1997). Association between cigarette smoking and lipid peroxidation in a controlled feeding study. *Circulation* **96,** 1097–1101.

Morrow, J. D., Frei, B., Longmire, W., Gaziano , M., Lynch S. M., Shyr, Y., Strauss, W., Oates, J. A., and Roberts, L. J. (1995). Increase in circulating products of lipid peroxidation (F2-isoprostanes) in smokers. *N. Engl. J. Med.* **64,** 1198–1203.

Pryor, W. A., (1997). Cigarette smoke radicals and the role of free radicals in chemical carcinogenicity. *Environ. Health Persp.* **105,** 875–882.

Reilly, M., Delanty, N., Lawson, J. A., and FitzGerald, G. G. (1996). Modulation of oxidant stress *in vivo* in chronic cigarette smokers. *Circulation* **94,** 19–25.

Rimm, E., Katan, M. B., Ascherio, A. B., Stampfer, M. J., and Willett, W. C. (1996). Relation between intake of flavonoids and risk for coronary heart disease in male health professionals. *Ann. Intern. Med.* **125,** 384–389.

Ross, M., Crosley, K. M., Brown, K. M., Duthie, S. J., Arthur, J. R., and Duthie, G. G. (1995). Plasma concentrations of carotenoids and antioxidant vitamins in Scottish males: Influences of smoking. *Eur. J. Clin. Nutr.* **49,** 861–865.

Sakamoto, M. (1985). Ethane expiration among smokers and non-smokers. *Jpn. J. Hyg.* **40,** 835–840.

World Health Organisation. (1997). "World Health Organisation Report of the Director General." WHO, Geneva.

Whichelow, M. J., Golding, J. F., and Treasure, F. P. (1988). Comparison of some dietary habits of smokers and non-smokers. *Br. J. Addict.* **83,** 295–304.

Yang, C. S., and Wang, Z.-Y. (1993). Tea and cancer. *J. Natl. Cancer Inst.* **85,** 1038–1049.

II Nutrients

3 Emerging Role of Nutrition in Chronic Disease Prevention: A Look at the Data, with an Emphasis on Vitamin C

Gladys Block
Department of Epidemiology and Public Health Nutrition
University of California
Berkeley, California 94720

INTRODUCTION

We are at the threshold of the second revolution in our understanding of the role of nutrition in disease and health. The first revolution took place early in this century, with the discovery of the frank nutrient deficiency diseases and their causes. As a result of that research in the nutritional sciences, we essentially eliminated beriberi, pellagra, rickets, and goiter. We did so, incidentally, not by education but by fortification.

Today we are at the threshold of an even greater revolution, and it involves the antioxidant nutrients, including vitamin C, vitamin E, and the carotenoids. Increasingly, research suggests that these nutrients are of great importance in reducing the risk of cancer and heart disease, the two major killers in Western society. However, beyond these diseases, it is increasingly apparent that antioxidants may be important in most of the diseases of aging, including age-related eye diseases such as cataracts, and impaired immune function resulting in increased susceptibility to infection.

Evidence for an important role for antioxidant nutrients comes from the complete spectrum of biomedical research fields, from biochemical research, animal studies, epidemiologic data, and clinical trials. Any one of these alone would be insufficient as a basis for public policy. However, in

Antioxidant Food Supplements in Human Health

fact, all of these types of research are reaching the same conclusion, that antioxidant nutrients may be of great importance in preserving health and vigor and in preventing disease. This chapter reviews some of that evidence, with a special emphasis on vitamin C.

LABORATORY AND ANIMAL DATA ON OXIDATION AND ANTIOXIDANTS

Oxidation

Oxidation is the transfer of electrons from one atom to another. It is an essential part of normal metabolism. The process of extracting energy from food involves the transfer of electrons, with release of energy at each step, through a series of electron acceptors until finally carbon dioxide and water result. Plants capture the energy of light through a series of electron transfer reactions in the chloroplast. Numerous other biochemical processes involve oxidation and the transfer of electrons. Serious problems arise, however, when electrons are transferred not in pairs, but escape the process as unpaired single electrons. Atoms or molecules that have unpaired electrons are called free radicals and are extremely damaging to the molecules in nearby cells and tissues. Free radicals are highly reactive and "seek" to capture another electron to complete the pair; in doing so, they damage or destroy the function of the other molecule. Molecules thus damaged might be the lipids in cell membranes, the proteins in tissue or enzymes, or the DNA, the basic instruction manual for the body's functions. Oxidant damage to DNA, protein, and other macromolecules appears to be a major contributing factor to aging and the many degenerative processes associated with it, including cancer, heart disease, cataracts, and cognitive dysfunction (1–8).

Evidence for radical mediation of many events in carcinogenesis is very strong. Humans are exposed to a sea of oxidative and radical agents producing damage to lipids, proteins, membranes, and DNA. The sources are both external and internal. External ones include polynuclear aromatic hydrocarbons (PAH) in food, polluted air, and cigarette smoke; heterocyclic amines in food and cigarette smoke; background radiation; and ozone. Internal sources include oxidative bursts from activated neutrophils; oxidative transformations in prostaglandin synthesis; redox cycling of quinones; and oxyradical attack on DNA. Oxidative damage to DNA is extensive. It is possible to measure the amount of this damage by measuring the fragments that result from repair of that oxidative DNA damage. Based on the urinary excretion of these DNA fragments, oxidative damage to DNA occurs at an estimated rate of 10^5 hits per cell per day in the rat (5) and about 10^4 hits per cell per day in the human (6). These oxidative lesions are effectively but not perfectly repaired; the normal steady-state level of oxidative DNA lesions is about 10^6 per cell in the young rat and about twice this in the old rat (5, 7). Thus, the burden of oxidatively damaged DNA builds up as we age. The extensive oxidative damage is also seen in human sperm (8), and ascorbic acid has been shown to protect human sperm from endogenous oxidative DNA damage.

Lipids are the major components of all cell membranes and play an important role in nerve function. Oxidation of lipids impairs cell-to-cell communication and nerve transmission, and oxidized lipids are the main culprits in atherosclerosis. Vitamin E is the major lipid-soluble antioxidant, but vitamin C plays an important role in conjunction with vitamin E in scavenging oxygen radicals and in protecting cell membranes (9, 10). In laboratory studies by Frei *et al.* (11, 12), a 50 μmol/liter concentration of vitamin C was shown to fully prevent the free radical damage to lipids caused by cigarette smoke (11, 12). Only when all plasma ascorbate was depleted did lipid peroxidation begin.

Nitrosation

Most nitrosamines and nitrosamides are well-established animal carcinogens. They are formed in the stomach and other human tissues and in foods and cigarette smoke, when nitrites react with amines and amides. Biochemical and experimental animal evidence going back many years has established conclusively that vitamin C can inhibit nitrosation reactions, thereby acting as effective anticarcinogens under experimental conditions (13–16).

Although blocking of nitrosation is a well-known effect of the reducing capacity of vitamin C, other chemical carcinogens are also blocked by this antioxidant action. Warren (17) found that anthracene and 3,4-benzpyrene are converted in the presence of ascorbic acid to less harmful compounds. Detoxification of organochlorine pesticides such as DDT, dieldrin, and lindane by ascorbate has also been shown (18).

Numerous studies have shown an antitumorigenic role of vitamin C in animal models (19). The administration of ascorbic acid was found to inhibit the development of cancerous lesions of the respiratory tract of mice exposed to fiberglass dust (20) and of rats exposed to plutonium dioxide particles (21) and to protect against abnormal growth and malignant transformation of hamster lung cultures that were subjected repeatedly to cigarette smoke (22). Sodium ascorbate, fed *ad libitum* in the diet, completely prevented 1,2-dimethylhydrazine(DMH)-induced colon tumors in the rat and reduced DMH-induced kidney tumors (23). Pretreatment with vitamin C reduced estradiol-induced renal tumors by 50% (24). Although an effect is not seen for every nutrient in every animal model and dosage regimen, data quite consistently show a strong and significant effect.

EPIDEMIOLOGIC DATA ON ANTIOXIDANTS AND CANCER

An extensive review was conducted of epidemiologic data on fruits and vegetables and cancer (25) and revealed extremely consistent protective effects of high fruit and vegetable intake. Of 156 dietary studies, 128 found a statistically significant protective effect of fruit and vegetable consumption. For most cancer sites, the one-fourth of the population with the lowest fruit and vegetable intake experience about twice the risk of cancer compared with those with high intake, even after control for potentially confounding factors.

Most of the epidemiologic studies of cancer focused on β-carotene. Many used dietary questionnaires that were incapable of measuring vitamin C well; most completely omitted the measurement of intake of foods fortified with vitamin C or of vitamin supplements; and of those that did measure vitamin supplement use, most analyzed their data in a way that would make it difficult to detect a role for vitamin C (26). As noted in an extensive review (25), almost all of these epidemiologic studies found a statistically significant protective role of antioxidant-containing fruits and vegetables. For the most part this was interpreted as an effect of β-carotene. Subsequent intervention trials (27, 28) have called this conclusion into question. Interpretation after these intervention trials has pointed out that the β-carotene effect observed in earlier studies may represent a marker of something else in the foods (29).

Numerous food components have been proposed, including indoles, isothiocyanates, elagic acid, and similar fairly obscure components. Vitamin C has seemed to evoke less interest. Nevertheless, vitamin C has also been found to have a significant cancer-reducing role in numerous epidemiologic studies, despite the methodologic flaws mentioned earlier that make it difficult to detect a vitamin C effect (30). In epidemiologic studies of lung cancer, 9 of 11 investigators who studied the role of vitamin C have found reduced risk with high intake, even after control for smoking. All 8 studies of vitamin C and cancers of the esophagus and oral cavity found a statistically significant reduced risk with high intake. All 7 studies that reported on dietary vitamin C intake and stomach cancer have found statistically significant reduced risks with higher intake. Similarly consistent results for vitamin C have been found for cervical cancer and pancreatic cancer. Finally, a meta-analysis (31) of 12 major breast cancer studies found an association with vitamin C intake that was as strong and significant in the protective direction as saturated fat was in the harmful direction.

Most studies of fruit/vegetable intake and cancer risk have found a dose–response relationship, with approximately a twofold increased risk for those in the lower end of the distribution. These at-risk groups are not tiny fractions of the population, but represent the lower fifth to lower third of the population, or even more in populations or subgroups with a low overall fruit/vegetable intake. Bjelke (32) suggests that for colorectal cancer, the *majority* of the population may be at substantial increased risk compared to a small minority whose fruit/vegetable consumption is high. Few, if any, risk factors besides smoking confer risks of this magnitude, to populations of this magnitude, with evidence as consistent.

Heart Disease

A role for antioxidants in the prevention and treatment of heart disease has gained widespread support. Oxidized low-density lipoprotein (LDL) has been implicated in the initiation of arterial plaques, and antioxidants including vitamin C have been shown in animal studies to prevent that process (33, 34). The potential role of antioxidants, including vitamin C, in the prevention of heart disease has been examined in several reviews (35–38). Large observational studies have suggested the possibility that vitamin C specifically may have a beneficial effect on heart disease risk. For example, a follow-up study of a large national cohort of over 10,000 persons followed for approximately 10 years obtained data on dietary vitamin C intake and vitamin supplement use (39). Respondents were divided into three groups according to their baseline vitamin C intake: (1) persons with dietary intake below 50 mg/day, (2) persons with dietary intake above 50 mg/day and not consuming supplements containing vitamin C, and (3) persons with dietary intake above 50 mg/day and consuming vitamin C supplements regularly. Total mortality and mortality from heart disease and cancer decreased notably across these three groups, with strong and statistically significant reductions in heart disease and total mortality among those in the third group. Results were controlled for smoking and other socioeconomic factors.

In addition, cross-sectional data also suggest a role for vitamin C in lowering blood pressure (40) and in improving serum lipid profiles (41, 42), two of the major risk factors for heart disease. However, cross-sectional data, while suggestive, cannot rule out the possibility that other factors highly correlated with vitamin C are the actual effective agents. A controlled diet feeding study (43), however, strongly suggests that observed effects associated with blood pressure and serum lipids may be specific to vitamin C (G. Block, unpublished data). After depletion of plasma vitamin C status for 1 month, a highly significant inverse relationship was found between ascorbic acid plasma levels and blood pressure and serum lipids measured 1 month later. Only well-designed clinical trials in appropriate populations will ultimately be conclusive.

Eye Disease, General Well-Being

The antioxidant nutrients that are found in fruits and vegetables also appear to have important protective roles in age-related eye diseases such as cataracts and macular degeneration (44–47). In the presence of oxygen, ultraviolet light generates free radicals or precursors, and oxidative damage to lens proteins leads to increasing opacity. In the young, repair mechanisms exist, but they become progressively less effective as we age. It has been hypothesized that oxidation may be involved in most or all types of cataract and of age-related macular degeneration. This hypothesis is supported by animal as well as human studies. A number of epidemiologic studies have found that persons wtih age-related eye diseases consumed significantly less of the antioxidant nutrients, particularly vitamin C, than did persons who did not develop those eye conditions.

INTAKES AND BLOOD LEVELS

Given this substantial body of evidence suggesting an important role of vitamin C in preventing the major chronic diseases of Western nations, it is important to examine the intake and the nutritional status of the U.S. population with regard to this nutrient. The mean vitamin C intake in the United States is approximately 100 mg per day. Because the recommended dietary allowance (RDA) is 60 mg, it is often assumed that there is no problem with inadequate intakes of this vitamin. The mean, however, is influenced by high intakes of a relatively few individuals and obscures the fact that substantial segments of the population consume considerably less than this amount. Examinations of the distribution of intake consistently find, in large national data sets (48), that the median intake (at which 50% are at that level or lower) is approximately 65–70 mg/day. Among those below poverty the figures are startling. Among women below poverty in the 25- to 34- and 35- to 44-year age groups, the *median* intake was only 36 and 32 mg, respectively.

Serum data from the Second National Health and Nutrition Examination Survey (NHANES II) confirm that these apparently low intakes are indeed reflected in low serum levels of ascorbic acid. Approximately 10–15% of white men had plasma levels at or below 0.3 mg/dl. Among African American men, however, the proportion with low serum ascorbate levels was shocking. Approximately one of every four to five African American males had serum ascorbate levels at or below 0.3 mg/dl. As a frame of reference for these serum levels, well-nourished persons tend to have serum ascorbate levels between about 0.8 and 1.2 mg/dl, and in a study conducted by the author, healthy volunteers on an ascorbate-restricted diet of 10 mg/day fell to a blood level of about 0.3 mg/dl total ascorbate. Most investigators would agree that 0.3 mg/dl is a low blood level, which, if maintained for any length of time, could be associated with symptoms of ascorbate deficiency. Levels such as this confer an increased risk even of the acute symptoms of low ascorbate status, such as increased fatigue and irritability and decreased resistance to infections. Given the growing body of data on increased risk of chronic diseases, such levels may confer an increased risk of cancer, cardiovascular disease, and various chronic and age-related conditions of the lung, eye, and other organs.

DISCUSSION

Experimental and epidemiologic data are consistent in finding a substantial reduced risk of cancer, heart disease, cataracts, and other major diseases among those with a high antioxidant status. Reduced total mortality, decreased susceptibility to infection, improved wound healing and operative outcome, reduced hospital stays attributable to bed sores, and improved mental acuity have all been seen. Almost without exception, those studies have used levels or have observed protection at levels well above the RDA standards. If these data are correct, an increase in antioxidant levels in the U.S. population could substantially reduce disease rates, improve the outcomes of treatment, and reduce hospital stays. In other words, it could have a substantial impact on health care costs. Yet national data indicate that most Americans do not even get those minimal RDA levels. What are the implications for public health policy?

The Department of Health and Human Services and the U.S. Department of Agriculture have both called for dietary improvement, emphasizing reduced fat intake, and to a lesser extent increases in fiber and/or fruits and vegetables. The National Cancer Institute has supported a national program to increase fruit and vegetable consumption to five servings per day. While these are laudable goals, there is little cause for optimism that any of these goals will be achieved in any substantial way in the foreseeable future. Dietary habits change, but change is driven in large part by life-style and economics. America's life-style today is one of haste, of working mothers, and of the enormous availability and variety of fast foods and snack foods. There is little reason to suspect, based on national data, that any but a small minority of the U.S. public will give up the pleasure and convenience of fatty foods and snack foods and greatly increase their fruit and vegetable consumption to the recommended five to nine servings per day.

Educational efforts are desirable and should continue. However, it appears to this observer that a widespread and substantial change in antioxidant levels in the population will come about only through fortification and through the encouragement of supplementation. Eradication of beriberi, pellagra, rickets, and goiter was not achieved through educational efforts to encourage the population to eat better. Eradication of many communicable diseases did not come about through education of the population, but through the commitment and reimbursement of the medical community to practice inoculation.

It is understandable that at this point public policy does not yet support systematic fortification with vitamin C or encouragement of supplementation. Data supporting a benefit are as yet insufficient. However, it is critical that public and scientific policy be directed toward supporting and conducting well-designed studies, in appropriate populations, to obtain sufficient data. Laboratory studies have focused on more interesting or profitable agents. Animal studies have been conducted on vitamin C-synthesizing species, making their relevance to the human condition questionable. Epidemiologic studies in the past have been designed and analyzed too poorly to provide much useful information on the role of vitamin C. Virtually no clinical trials are examining the possible effect of vitamin C alone. At all of these levels of research, more well-designed studies are needed in order to determine whether public policy initiatives in the areas of fortification and supplementation could have the disease-preventive benefits suggested by existing data.

REFERENCES

1. Ames, B. N. (1983). Dietary carcinogens and anticarcinogens: Oxygen radicals and degenerative diseases. *Science* **221**, 1256–1264.
2. Ames, B. N. (1989). Endogenous oxidative DNA damage, aging, and cancer. *Free Radic. Res. Commun.* 7, 121–128.
3. Halliwell, B., and Gutteridge, J. M. C. (1990). "Free Radicals in Biology and Medicine," 2nd ed. Clarendon, Oxford.
4. Carney, J. M., Starke-Reed, P. E., Oliver, C. N., et al. (1991). Reversal of age-related increase in brain protein oxidation, decrease in enzyme activity, and loss in temporal and spatial memory by chronic administration of the spin-trapping compound N-*tert*-butyl-a-phenylnitrone. *Proc. Natl. Acad. Sci. USA* 88, 3633–3636.
5. Fraga, C. G., Shigenaga, M. K., Park, J.-W., Degan, P., and Ames, B. N. (1990). Oxidative damage to DNA during aging: 8-hydroxy-2′-deoxyguanosine in rat organ DNA and urine. *Proc. Natl. Acad. Sci. USA* 87, 4533–4537.
6. Cathcart, R., Schwiers, E., Saul, R. L., and Ames, B. N. (1984). Thymine glycol and thymidine glycol in human and rat urine: A possible assay for oxidative DNA damage. *Proc. Natl. Acad. Sci. USA* **81**, 5633–5637.
7. Richter, D., Park, J.-W., and Ames, B. N. (1988). Normal oxidative damage to mitochondrial and nuclear DNA is extensive. *Proc. Natl. Acad. Sci. USA* **85**, 6465–6467.
8. Fraga, C. G., Motchnik, P. A., Shigenaga, M. K., Helbock, H. J., Jacob, R. A., and Ames,

B. N. (1991). Ascorbic acid protects against endogenous oxidative DNA damage in human sperm. *Proc. Natl. Acad. Sci. USA* 24, 1–4.

9. Niki, E. (1991). Action of ascorbic acid as a scavenger of active and stable oxygen radicals. *Am. J. Clin. Nutr.* 54, 1119S–1124S.
10. Niki, E., Yamamoto, Y., Komuro, E., and Sato, K. (1991). Membrane damage due to lipid oxidation. *Am. J. Clin. Nutr.* 53, 201S–205S.
11. Frei, B., England, L., and Ames, B. N. (1989). Ascorbate is an outstanding antioxidant in human blood plasma. *Proc. Natl. Acad. Sci. USA* 86, 6377–6381.
12. Frei, B., Stocker, R., and Ames, B. N. (1988). Antioxidant defenses and lipid peroxidation in human blood plasma. *Proc. Natl. Acad. Sci. USA* 85, 9748–9752.
13. Ranieri, R., and Weisburger, J. H. (1975). Reduction of carcinogens with ascorbic acid. *Ann. N.Y. Acad. Sci.* 258, 181.
14. Mirvish, S. S., Cardesa, A., Wallcave, L., and Shubik, P. (1973). Effect of sodium ascorbate on lung adenoma induction by amines plus nitrite. *Proc. Am. Assoc. Cancer Res.* 14, 102. [Abstract]
15. Mirvish, S. S., Cardesa, A., Wallcave, L., and Shubik, P. (1975). Induction of mouse lung adenomas by amines or ureas plus nitrite and by N-nitroso compounds: Effect of ascorbate, gallic acid, thiocyanate, and caffeine. *J. Natl. Cancer Inst.* 55, 633–636.
16. Pipkin, G. E., Schlegel, J. U., Nishimura, R., and Shultz, G. N. (1969). Inhibitory effect of L-ascorbate on tumor formation in urinary bladders implanted with 3-hydroxyanthranilic acid. *Proc. Soc. Exp. Biol. Med.* 131, 522–524.
17. Warren, F. L. (1943). Aerobic oxidation of aromatic hydrocarbons in the presence of ascorbic acid. *Biochem. J.* 37, 338.
18. Street, J. C., and Chadwick, W. R. (1975). Ascorbic acid requirements and metabolism in relation to organochlorine pesticides. *Ann. N.Y. Acad. Sci.* 258, 132.
19. Block, G., and Schwarz, R. (1994). Ascorbic acid and cancer: Animal and cell culture data. *In* "Natural Antioxidants in Human Health and Disease" (B. Frei, ed.), pp. 129–155. Academic Press, San Diego.
20. Morrison, D. G., Daniel, J., Lynd, F. T., *et al.* (1981). Retinyl palmitate and ascorbic acid inhibit pulmonary neoplasms in mice exposed to fiberglass dust. *Nutr. Cancer* 3, 81–85.
21. Sanders, C. L., and Mahaffey, J. A. (1983). Action of vitamin C on pulmonary carcinogenesis from inhaled 239PuO2. *Health Phys.* 43, 794.
22. Leuchtenberger, C., and Leuchtenberger, R. (1977). Protection of hamster lung cultures by L-cysteine or vitamin C against carcinogenic effects of fresh smoke from tobacco or marihuana cigarettes. *Br. J. Exp. Pathol.* 58, 625.
23. Reddy, B. S., and Hirota, N. (1979). Effect of dietary ascorbic acid on 1,2-dimethylhydrazine-induced colon cancer in rats. *Fed. Proc.* 38, 714. [Abstract]
24. Liehr, J. G. (1991). Vitamin C reduces the incidence and severity of renal tumors induced by estradiol and diethylstilbestrol. *Am. J. Clin. Nutr.* 54, 1256S–1260S.
25. Block, G., Patterson, B., and Subar, A. (1992). Fruit, vegetables, and cancer prevention: A review of the epidemiologic evidence. *Nutr. Cancer* 18, 1–29.
26. Block, G., Sinha, R., and Gridley, G. (1994). Collection of dietary-supplement data and implications for analysis. *Am. J. Clin. Nutr.* 59, 232S–239S.
27. The Alpha-Tocopherol BcCPS. (1994). The effect of vitamin E and beta carotene on the incidence of lung cancer and other cancers in male smokers. *N. Engl. J. Med.* 330, 1029–1035.
28. Omenn, G. S., Goodman, G. E., Thornquist, M. D., *et al.* (1996). Effects of a combination of beta carotene and vitamin A on lung cancer and cardiovascular disease. *N. Engl. J. Med.* 334, 1150–1155.
29. Hennekens, C. H., Buring, J. E., Manson, J. E., *et al.* (1996). Lack of effect of long-term

supplementation with beta carotene on the incidence of malignant neoplasms and cardio-vascular disease. *N. Engl. J. Med.* **334**, 1145–1149.

30. Block, G. (1991). Vitamin C and cancer prevention: The epidemiologic evidence. *Am. J. Clin. Nutr.* **53**, 270S–282S.
31. Howe, G. R., Hirohata, T., Hislop, T. G., *et al.* (1990). Dietary factors and risk of breast cancer: Combined analysis of 12 case-control studies. *J. Natl. Cancer Inst.* **82**, 561–569.
32. Bjelke, E. (1973). "Epidemiologic Studies of Cancer of the Stomach, Colon, and Rectum; with Special Emphasis on the Role of Diet," Vols. I–IV. Doctoral dissertation, Ann Arbor, Michigan.
33. Verlangieri, A. J., and Bush, M. J. (1992). Prevention and regression of primate athero-sclerosis by alpha-tocopherol. *J. Am. Coll. Nutr.* **11**(2), 131–138.
34. Bjokhem, I., Henriksson-Freyschuss, A., Breuer, O., Diczfalusy, U., Berglund, L., and Hen-riksson, P. (1991). The antioxidant butylated hydroxytoluene protects against atheroscle-rosis. *Arteriosclero. Thromb.* **11**, 15–22.
35. Gey, K. F. (1995). 10-year retrospective on the antioxidant hypothesis of arteriosclerosis: Threshold plasma levels of antioxidant micronutrients related to minimum cardiovascular risk. *J. Nutr. Biochem.* **6**, 206–236.
36. Todd, S., Woodward, M., Bolton-Smith, C., and Tunstall-Pedoe, H. (1995). An investiga-tion of the relationship between antioxidant vitamin intake and coronary heart disease in men and women using discriminant analysis. *J. Clin. Epidemiol.* **48**, 297–305.
37. Manson, J. E., Gaziano, J. M., Jonas, M. A., and Hennekens, C. H. (1993). Antioxidants and cardiovascular disease: A review. *J. Am. Coll. Nutr.* **12**, 426–432.
38. Simon, J. A. (1992). Vitamin C and cardiovascular disease: A review. *Am. J. Clin. Nutr.* **11**, 107–125.
39. Enstrom, J. E., Kanim, L. E., and Klein, M. A. (1992). Vitamin C intake and mortality among a sample of the United States population. *Epidemiology* **3**, 194–202.
40. Salonen, J. T., Salonen, R., Ihanainen, M., *et al.* (1988). Blood pressure, dietary fats, and antioxidants. *Am. J. Clin. Nutr.* **48**, 1226–1232.
41. Jacques, P. F., Hartz, S. C., McGandy, R. B., Jacob, R. A., and Russell, R. M. (1987). Vi-tamin C and blood lipoproteins in an elderly population. *Ann. N.Y. Acad. Sci.* **498**, 100–109.
42. Dallal, G. E., Choi, E., Jacques, P., Schaefer, E. J., and Jacob, R. A. (1989). Ascorbic acid, HDL cholesterol, and apolipoprotein A-I in an elderly Chinese population in Boston. *J. Am. Coll. Nutr.* **8**, 69–74.
43. Mangels, A. R., Block, G., Frey, C. M., *et al.* (1993). The bioavailability to humans of ascorbic acid from oranges, orange juice and cooked broccoli is similar to that of synthetic ascorbic acid. *J. Nutr.* **123**, 1054–1061.
44. Jacques, P. F., Chylack, L. T., McGandy, R. B., and Hartz, S. C. (1988). Antioxidant sta-tus in persons with and without senile cataract. *Arch. Ophthalmol.* **106**, 337–340.
45. Jacques, P. F., Hartz, S. C., Chylack, L. T., McGandy, R. B., and Sadowski, J. A. (1988). Nutritional status in persons with and without senile cataract: Blood vitamin and mineral levels. *Am. J. Clin. Nutr.* **48**, 152–158.
46. Robertson, J. M., Donner, A. P., and Trivithick, J. R. (1989). Vitamin E intake and risk of cataracts in humans. *Ann. N.Y. Acad. Sci.* **570**, 372–382.
47. Leske, M. C., Chylack, L. T., Jr., Wu, S., The Lens Opacities Case-Control Study Group. (1991). The Lens Opacities Case-Control Study: Risk factors for cataract. *Arch. Oph-thalmol.* **109**, 244–251.
48. Block, G., and Sorenson, A. (1987). Vitamin C intake and dietary sources by demographic characteristics. *Nutr. Cancer* **10**, 53–65.

4 Biological Activities of Tocotrienols and Tocopherols

Maret G. Traber*, Elena A. Serbinova†, and Lester Packer†

* Linus Pauling Institute
Oregon State University
Corvalis, Oregon 97331

†Department of Molecular and Cell Biology
University of California
Berkeley, California 94720

INTRODUCTION

Vitamin E deficiency was first described at the University of California in Berkeley in 1922 by Evans and Bishop during their investigations of infertility in rancid lard-fed rats. In 1936, Evans and colleagues isolated a factor that prevented vitamin E deficiency symptoms and named it α-tocopherol (structures are shown in Fig. 1). In the subsequent year, two other tocopherols, β and γ, were isolated from vegetable oils, but these had lower biologic activities than α-tocopherol (Emerson *et al.*, 1937). This was the first description that different naturally occurring forms of virtamin E exist and that α-tocopherol is the most effective form in preventing vitamin E deficiency symptoms.

The existence of eight different naturally occurring forms of vitamin E, all with relatively potent antioxidant activities, has provoked interest as to the function of this vitamin. If α-tocopherol is the most biologically potent form of vitamin E, why do the other forms exist? This chapter attempts to answer this question by evaluating the available information on the func-

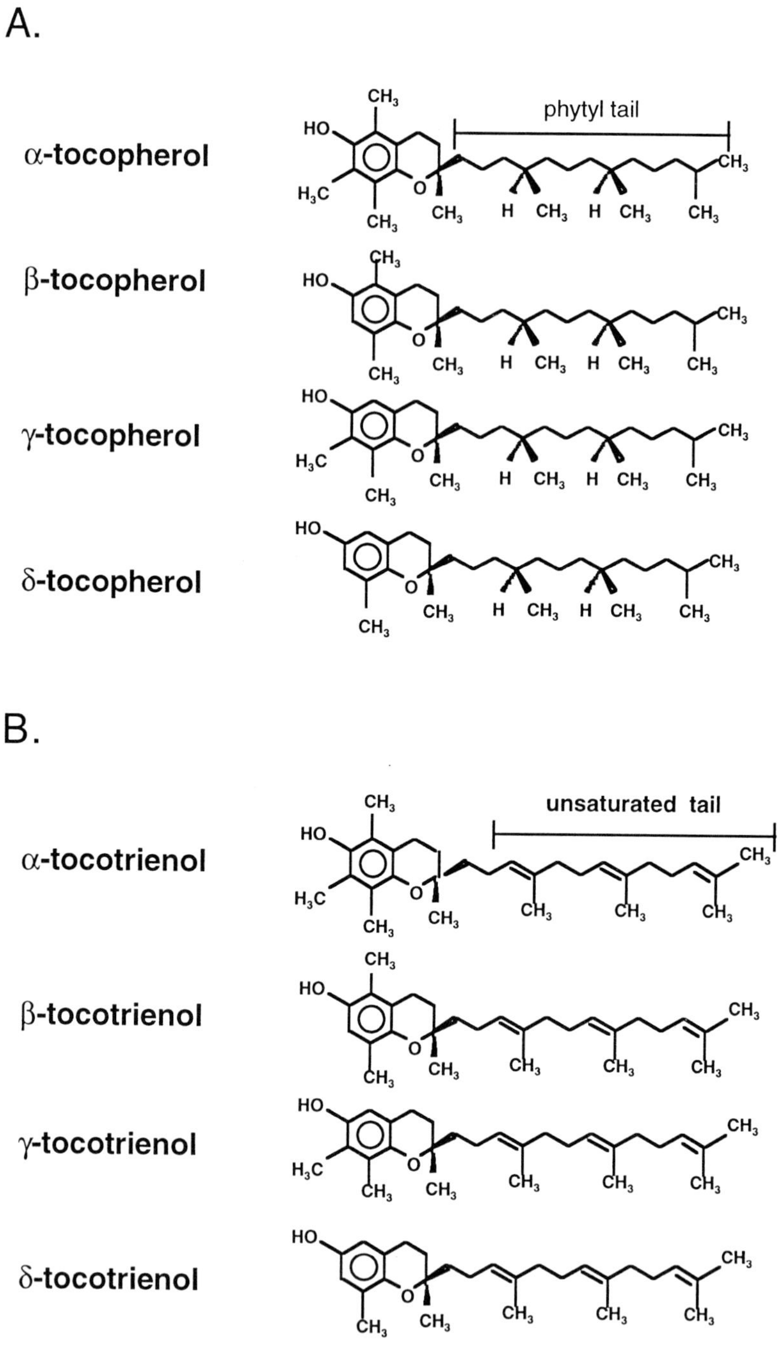

A.

α-tocopherol

β-tocopherol

γ-tocopherol

δ-tocopherol

B.

α-tocotrienol

β-tocotrienol

γ-tocotrienol

δ-tocotrienol

Figure 1 Vitamin E homologs. (A) Four naturally occurring tocopherols and (B) four naturally occurring tocotrienols.

tional and physiological roles of the various tocopherol and tocotrienol homologs of vitamin E.

VITAMIN E CHEMISTRY AND BIOLOGIC ACTIVITY

Vitamin E Homologs

Vitamin E occurs in nature in eight different forms: α-, and β-, γ-, and δ-tocopherols and α-, β-, γ-, and δ-tocotrienols (Fig. 1). Tocotrienols differ from tocopherols in that tocotrienols have an unsaturated side chain, whereas tocopherols have a phytyl tail with three chiral centers (Fig. 1A: 2, 4′ and 8′).

Chemically synthesized a-tocopherol is not identical to the naturally occurring α-tocopherol because the synthetic contains eight different stereoisomers arising from these three chiral centers. RRR-α-Tocopherol, the naturally occurring form, is only one of the eight stereoisomers present in *all-rac-α*-tocopherol.

Historically, an international unit (IU) of vitamin E activity was defined as 1 mg of *all-rac-α*-tocopheryl acetate, whereas 1 mg RRR-α-tocopherol equaled 1.49 IU. A currently accepted measure of vitamin E activity in food is the conversion of the amounts of each homolog into α-tocopherol equivalents (α-TE) (Eitenmiller and Landen, 1995). One α-TE is the activity of 1 mg of RRR-α-tocopherol (Food and Nutrition Board, 1989).

The conversion factors to calculate α-TE are based on the biologic activities of the various vitamin E forms. As defined by Machlin (1991), the biologic activity of a vitamin E homolog is dependent on its ability to prevent or reverse specific vitamin E deficiency symptoms (e.g., fetal resorption, muscular dystrophy, and encephalomalacia). Factors for the conversion of tocopherols and tocotrienols to α-TE units are α-tocopherol, mg $\times$ 1.0; β-tocopherol, mg $\times$ 0.5; γ-tocopherol, mg $\times$ 0.1; δ-tocopherol, mg $\times$ 0.03; α-tocotrienol, mg $\times$ 0.3; and β-tocotrienol, mg $\times$ 0.05. The factors for γ- and δ-tocotrienol are unknown. For synthetic *all-rac-α*-tocopherol, the conversion factor is mg $\times$ 0.74.

Antioxidant Activity

Vitamin E is a potent peroxyl radical scavenger (Burton and Ingold, 1986) and can protect polyunsaturated fatty acids (PUFA) within phospholipids of biological membranes (Burton *et al.*, 1983b) and in plasma lipoproteins (Jialal *et al.*, 1995). When vitamin E reacts with a peroxyl radical, it forms a tocopheroxyl radical. In this regard, α-tocotrienol is a more potent antioxidant than α-tocopherol (Serbinova *et al.*, 1991). Serbinova *et al.* (1991) suggested that the higher antioxidant potency of α-tocotrienol is due to the combined effects of (1) its higher recycling efficiency from chromanoxyl radical, (2) its more uniform distribution in membrane bilayers, and (3) its stronger disordering of membrane lipids allowing interaction of the chromanol nucleus with lipid radicals. Kamal-Eldin and Appelqvist (1996) reviewed in detail the various antioxidant reactions of tocopherols and tocotriencols. Packer (1995) contrasted the antioxidant properties of tocotrienols and tocopherols in membrane systems including model membranes, microsomes, and low-density lipoproteins. The major reasons why tocotrienols are better antioxidants are listed in Table I.

Biologic Activity

Biologic activities of tocopherols and tocotrienols are unlikely to be dependent solely on their antioxidant activities. α-Tocotrienol has only one-third the biological activity of α-tocopherol (Bunyan *et al.*, 1961; Weimann and Weiser, 1991), yet it has higher (Kamat *et al.*, 1997; Serbinova *et al.*, 1993) or equivalent (Suarna *et al.*, 1993a) antioxidant activity. A vitamin E analog [2,4,6,7-tetramethyl-2-(4', 8', 12'-trimethyltridecyl)-5-hydroxy-3,4-dihydrobenzofuran] with equivalent biological activity to RRR-α-tocopherol (Ingold *et al.*, 1990) has 1.5 times the antioxidant activity (Burton *et al.*, 1983a). Furthermore, the eight different stereoisomers of synthetic vitamin E (*all-rac-α*-tocopherol) have equivalent antioxidant activity, but different biological activities (Weiser *et al.*, 1986, 1996).

Overall, the highest biological activity is found in molecules with three methyl groups and a free hydroxyl group on the chromanol ring with the phytyl tail meeting the ring in the R orientation (Fig. 1). This specific requirement for biological, but not antioxidant, activity can best be rationalized by the preferential interactions of RRR-α-tocopherol with some stereospecific ligands in cells. However, this also raises the question of why there are eight different naturally occurring forms of vitamin E. Do these various forms have specific functions or characteristics that yield preferential tissue localization?

Table I   Antioxidant Activity of α-Tocotrienol as Compared with α-Tocopherol

---

Greater in α-tocotrienol due to
    More uniform distribution in membrane bilayer
    Stronger disordering of membrane lipids
    More effective collision with radicals
    Greater recycling activity of chromanoxyl radicals
    Recycling activity correlates with inhibition of lipid peroxidation

---

ABSORPTION, TRANSPORT, AND METABOLISM

Despite the variety of dietary vitamin E forms, human plasma contains 5 to 10 times higher concentrations of α-tocopherol than γ-tocopherol (Baker *et al.*, 1986; Handelman *et al.*, 1985; Hayes *et al.*, 1993; Traber and Kayden, 1989) and virtually no tocotrienols (Hayes *et al.*, 1993). Furthermore, supplementation with α-tocopherol not only increases α-tocopherol concentrations, it also reduces plasma γ-tocopherol (Baker *et al.*, 1986; Handelman *et al.*, 1985) and adipose tissue γ-tocopherol (Handelman *et al.*, 1994). The following section discusses how absorption, transport, and metabolism of vitamin E interact to result in this preference for α-tocopherol in the plasma and how tissues contain other forms of vitamin E; notably skin is preferentially enriched in tocotrienols (Podda *et al.*, 1996).

Intestinal Absorption

The absorption of vitamin E from the intestinal lumen is dependent on processes necessary for fat digestion and uptake into enterocytes. Bile acids and free fatty acids are important components for the formation of mixed micelles, which contain products of lipolysis and deliver these along with vitamin E to enterocytes (Traber *et al.*, 1990a). Indeed, bile acids are essential for vitamin E absorption (Sokol *et al.*, 1983), and vitamin E deficiency occurs in humans with cholestatic liver disease because they lack the ability to secrete bile acids (Sokol, 1993).

Pancreatic esterases, required for release of free fatty acids from dietary triglycerides, are also required for the hydrolytic cleavage of tocopheryl esters (Nakamura *et al.*, 1975), a common synthetic form of vitamin E in dietary supplements. Generally, these esterases are quite effective; the apparent absorption of *RRR*-α-tocopherol in humans was studied by Cheesmen *et al.* (1995) using deuterated forms of vitamin E and was found to be similar whether administered to α-tocopherol, α-tocopheryl acetate, or α-tocopheryl succinate.

Importantly, Ikeda *et al.* (1996) have reported that α-tocotrienol absorption in rats is enhanced compared with α-tocopherol or γ- or δ-tocotrienols. This occurred whether the vitamin E homolog were administered simultaneously or individually. In contrast, Hayes *et al.* (1993) found that α-tocopherol in hamsters was absorbed and secreted preferentially in lymph when the animals were fed a diet containing a mixture of vitamin E homologs obtained as the tocotrienol-rich fraction from palm oil (TRF). Hayes *et al.* (1993) studied humans and found that tocotrienols could be detected in postprandial plasma.

Vitamin E absorption requires chylomicron synthesis and secretion (Kayden and Traber, 1993). The movement of vitamin E through the absorptive cells is not well understood. Although of interest, no intestinal to-copherol transfer proteins have been described and therefore this is an important area for future study.

Plasma Transport and Distribution to Tissues

During chylomicron catabolism in the circulation, some of the newly absorbed vitamin E is transferred to circulating lipoproteins and some remains with the chylomicron remnants (Kayden and Traber, 1993). During the delipidation of chylomicrons by lipoprotein lipase, vitamin E is distributed to all of the circulating lipoproteins. Vitamin E can transfer spontaneously between lipoproteins, but its transfer is enhanced by the plasma phospholipid transfer protein (Kostner *et al.*, 1995).

Suarna *et al.* (1993) demonstrated that after feeding TRF to rats, all lipoprotein classes contained tocotrienols. Although Hayes *et al.* (1993) were unable to detect tocotrienol-fasting human plasma using an HPLC method with fluorescence detection, they did find all of the vitamin E forms in the 2-hr postprandial plasma of humans fed TRF. Similarly, E. A. Serbinova and L. Packer (unpublished observations) found tocotrienols in the nonfasting plasma of subjects supplemented with TRF (Fig. 2). They fed TRF (120 mg α-tocopherol and 240 mg total tocotriencols daily) to seven healthy volunteers for 8 weeks. Plasma vitamin E concentrations demonstrated an increase in α-tocopherol, a decrease in γ-tocopherol, and total tocotrienols were increased from 0 to approximately 2.1 μM. Although TRF supplementation raised plasma tocotrienol concentrations, feeding palmolein-fried potato crisps did not increase them (Choudhury *et al.*, 1997).

Chylomicron remnants are taken up by the liver, following partial delipidation by lipoprotein lipase and acquisition of apo(E). In the liver, dietary fat and vitamin E are repackaged and secreted into the plasma in very low density lipoproteins (VLDL). Remarkably, only one form of vitamin E, *RRR*-α-tocopherol, is apparently secreted in a preferential manner by the liver (Traber *et al.*, 1990b). Thus, the liver, not the intestine, discriminates between tocopherols and tocotrienols.

Hepatic α-Tocopherol Transfer Protein

α-TTP ($\sim$32 kDa) has been purified to homogeneity from rat and human liver (Arita *et al.*, 1995; Sato *et al.*, 1991, 1993). Defects in the gene for α-TTP result in vitamin E deficiency in humans (Gotoda *et al.*, 1995; Hentati *et al.*, 1996; Ouahchi *et al.*, 1995; Yokota *et al.*, 1996). Biologic activities of the vitamin E forms have been suggested to be dependent on the function of α-TTP (Hosomi *et al.*, 1997; Traber, 1994).

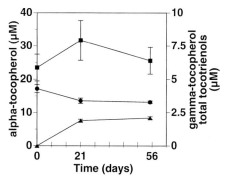

Figure 2   Distribution of plasma vitamin E concentration in seven normal healthy subjects following TRF supplementation. Seven normal, healthy subjects consumed TRF (120 mg tocopherol and 240 mg total tocotrienols) daily for 8 weeks. Plasma samples were obtained 6 hr after supplementation on the indicated day. Plasma α-tocopherol (■), α-tocopherol (●) and total tocotrienols (▲) are shown. Plasma α-tocopherol and total tocotrienol-increased concentrations were statistically significant ($p < 0.005$).

In vitro, purified α-TTP transfers *RRR-α*-tocopherol between liposomes and microsomes (Arita *et al.,* 1995; Hosomi *et al.,* 1997; Sato *et al.,* 1991) and distinguishes between *RRR-* and *SRR-α*-tocopherols (natural and synthetic stereoisomers) and between α- and γ-tocopherols (Hosomi *et al.,* 1997). This ability to recognize α-tocopherol and to transfer it makes that α-TTP a likely candidate for the enrichment of VLDL with vitamin E by the liver, as reviewed (Traber, 1994; Traber and Sies, 1996). However, Arita *et al.* (1997) demonstrated in McArH7777 cells overexpressing α-TTP that triglyceride secretion and α-tocopherol secretion are not linked. Because triglycerides are secreted from hepatocytes in VLDL, their data suggest that α-tocopherol is secreted independently of lipoproteins from hepatocytes and that VLDL secretion is not a prerequisite for α-TTP-mediated α-tocopherol secretion. Thus, the specific mechanism by which α-TTP functions remains to be elucidated.

Distribution to Tissues

Vitamin E is transported in plasma lipoproteins in a nonspecific manner. No plasma-specific vitamin E transport proteins have been described. It is likely that the mechanisms of lipoprotein metabolism determine the delivery of vitamin E to tissues (Kayden and Traber, 1993). There are at least two major routes by which tissues likely acquire vitamin E: (1) via lipoprotein lipase-mediated lipoprotein catabolism and (2) via uptake of LDL by the LDL receptor. In addition, vitamin E exchanges rapidly between liposomes (Kagan *et al.,* 1990), between lipoproteins, and between lipoproteins and membranes and may enrich membranes with vitamin E (Traber *et al.,*

1992). This process of exchange is enhanced by the plasma phospholipid transfer protein (Kostner *et al.*, 1995).

Metabolism and Excretion

Due to its low intestinal absorption, the major route of excretion of ingested vitamin E is fecal elimination.

The chromanoxyl radical of vitamin E (one electron oxidation) can be reduced back to tocopherol by ascorbate, semiascorbyl radical, and other reducing agents (Packer, 1994). Therefore, continuous oxidation and reduction of α-tocopherol may be greater than the complete oxidation of vitamin E to α-tocopheryl quinone, the nonreversible oxidation product of α-tocopherol (two electron oxidation). The quinone can be reduced to the hydroquinone, then conjugated to yield the glucuronate. The glucuronate can be excreted into bile or further degraded in the kidneys to tocopheronic acid and processed for urinary excretion (Drevon, 1991). Other vitamin E oxidation products, including dimers and trimers and other adducts, have been described (Kamal-Eldin and Appelqvist, 1996).

Studies have demonstrated that urinary vitamin E metabolites arise from shortening the side chain without oxidation of the chromanol nucleus. Metabolites found in human urine include both the metabolite of α-tocopherol [2,5,7,8-tetramethyl-2(2'carboxyethyl)-6-hydroxychroman, α-CEHC] and that of γ-tocopherol [2,7,8-trimethyl-2-(2'carboxyethyl-6-hydroxychroman, LLUα]. Chiku *et al.* (1984) reported a similar urinary metabolite of δ-tocopherol. Presumably, the vitamin E metabolites result from β-oxidation of the side chain. No metabolites of the tocotrienols have thus far been reported. It has been suggested that only high levels of vitamin E intake result in metabolite formation and excretion (Schultz *et al.*, 1995).

Surprisingly, LLUα (the γ-tocopherol metabolite) has a unique role in regulating kidney function, in that it is a natriuretic factor (Wechter *et al.*, 1996). LLUα is a highly specific reversible inhibitor of the 70 pS K+ channel in TAL cells with an ED_{50} of 3 nM, making it the most potent inhibitor of K+ recycling yet to be reported (Murray *et al.*, 1997). α-CEHC, the analogous metabolite of α-tocopherol, is neither natriuretic nor inhibitory of the 70 pS K+ channel (Murray *et al.*, 1997). The importance of LLUα in human physiology remains to be elucidated.

The relationship between plasma vitamin E and urinary metabolites is unclear. α-CEHC increases in the urine with increasing vitamin E supplements. It is unknown what happens to LLUα. Vitamin E supplements (α-tocopheryl acetate) decrease plasma concentrations of γ-tocopherol (Baker *et al.*, 1986; Handelman *et al.*, 1985), which may increase liver γ-

tocopherol concentrations, thereby increasing the production of LLUα. γ-Tocotrienol could also be converted to LLUα, but this has not been demonstrated directly.

In vitro studies have shown that γ- β-, or δ-tocopherols, unlike α-tocopherol, can be nitrated (Christen *et al.*, 1997; Cooney *et al.*, 1993; Green *et al.*, 1959), a potentially important reaction with nitric oxide or cigarette smoke-derived components (Cross *et al.*, 1997). If smokers have a higher nitration of γ-tocopherol, then less γ-tocopherol would be available for conversion to LLU-α. This suggests that high blood pressure in smokers could be a consequence of γ-tocopherol depletion.

TISSUE VITAMIN E

The tissue-specific distribution of the various vitamin E forms, tocotrienols and tocopherols, suggests that they have unique roles in cellular functions. Podda *et al.* (1996) described in chow-fed mice that of the vitamin E forms, brain contained virtually only α-tocopherol (5.4 ± 0.1 nmol/g; 99.8%), whereas skin contained nearly 15% tocotrienols and 1% γ-tocopherol. In other tissues, the α-tocopherol content was higher (20 nmol/g), while each of the other forms represented about 1% of the total (γ-tocopherol, 0.2 to 0.4 nmol/g; α-tocotrienol, 0.1; γ-tocotrienol, 0.2). The unique distribution of these various forms in the tissues measured suggests that vitamin E levels may be regulated independently in each tissue.

Of mouse tissues, skin is unique in containing appreciable concentrations of tocotrienols, whereas other mouse tissues contain substantially smaller tocotrienol concentrations (Podda *et al.*, 1996). Hayes *et al.* (1993) examined the vitamin E contents of tissues from hamsters fed tocotrienol-enriched diets and also found tocotrienols in all tissues, except brain. Serbinova and Packer (1994) fed vitamin E-deficient rats diets that contained either α-tocopherol or α-tocotrienol (3 g/kg diet). Only heart and liver contained detectable amounts of α-tocotrienol. Furthermore, the rate of α-tocotrienol accumulation was slower and its final concentration was lower than that of α-tocopherol (Serbinova and Packer, 1994). Interestingly, when the experiment was designed such that the rat heart tissue contained approximately 0.5 nmol of either α-tocopherol or α-tocotrienol/mg protein (rats sacrificed on day 10 for α-tocopherol-fed and day 21 for α-tocotrienol-fed rats) and the heart tissue homogenates were tested for their susceptibility to lipid peroxidation *in vitro*, the rate of lipid peroxidation was lower in the hearts from the α-tocotrienol-fed rats (Serbinova and Packer, 1994), thereby demonstrating increased antioxidant protection by α-tocotrienols.

To evaluate differences in vitamin E structure on plasma and tissue α-tocopherol concentrations, equimolar amounts of *RRR*-α-tocopheryl acetate and *all-rac*-α-tocopheryl acetate, labeled with deuterium (d_3 and d_6, respectively), were consumed by (1) normal humans, (2) 26 patients undergoing elective surgery, and (3) 2 terminally ill patients (Burton *et al.*, 1998). On the morning following either a 30- or a 300-mg dose (d_3RRR- and d_6all-*rac*-α-tocopheryl acetate) consumed with an evening meal, the ratio of the natural to the synthetic stereoisomers (*RRR/rac* ratio) in 5 subjects was 1.51 ± 0.12 and 1.57 ± 0.09, respectively; the % deuterated of the plasma α-T was 10 and 55%, respectively. Following eight daily doses of 30 or 300 mg to 5 other subjects, the ratio was 1.77 ± 0.08 and 1.51 ± 0.02, respectively; the deuterated α-tocopherols represented 35 and 80%, respectively. When dosing ceased, *RRR/rac* increased to ~ 2 and % deuterated declined rapidly. In the elective surgery study, patients consumed the 150-mg dose up to 78 days prior to surgery; *RRR/rac* ratios in skin (1.48 ± 0.18, $n = 4$) and adipose tissue (1.58 ± 0.28, $n = 18$) were significantly lower than in their respective plasma samples (1.74 ± 0.06 and 1.77 ± 0.09). Except for liver and gall bladder, tissue % deuterated α-tocopherol were significantly lower than in plasma. In the terminally ill subject given 30 mg for 361 days, plasma and tissue *RRR/rac* ratios at autopsy were 2.0 and 1.71 ± 0.24, respectively, and % deuterated α-tocopherol were 6.3 and $5.8\% \pm 2.2\%$, respectively. In the patient given 300 mg for 615 days, plasma and tissue *RRR/rac* were 2.11 and 2.01 ± 0.17, respectively, and % deuterated α-tocopherol were 68 and $65\% \pm 10\%$, respectively. Remarkably, a 10-fold increase in dose and 250 days of additional dosing increased % deuterated α-tocopherol 10-fold with an apparent doubling of plasma and tissue α-tocopherol. The effect of vitamin E supplementation with 300 mg was to increase plasma concentrations of total α-tocopherol 3-fold and to at least double most tissue α-tocopherol concentrations, whereas supplementation with 30 mg had little effect on increasing concentrations of α-tocopherol in either plasma or tissues.

Brain

Brain vitamin E content is especially noteworthy because brain has been found to contain only α-tocopherol (Hayes *et al.*, 1993; Podda *et al.*, 1996). Furthermore, brain vitamin E is spared in response to a vitamin E-deficient diet (Bourre and Clement, 1991; Pillai *et al.*, 1994; Podda *et al.*, 1995; Southam *et al.*, 1991; Vatassery *et al.*, 1984a,b) and does not increase markedly in response to a vitamin E-supplemented diet (1000 IU/kg) (Vatassery *et al.*, 1988). Furthermore, feeding a diet deficient in α-tocopherol, but supplemented with γ-tocopherol, did not enrich markedly the brain with γ-tocopherol (Clement *et al.*, 1995) or supplemented with α-tocotrienol did not enrich the brain with α-tocotrienol (Serbinova and Packer, 1994). Taken together, these data suggest that there may be a specific transport mechanism for α-tocopherol through the blood/brain barrier.

Adipose Tissue

The bulk of vitamin E in the human body is localized in the adipose tissue (Traber and Kayden, 1987). More than 90% of the human body pool of α-tocopherol is located in the adipose tissue, and more than 90% of adipose tissue α-tocopherol is in fat droplets, not membranes (Traber and Kayden, 1987). Handelman *et al.* (1994) estimated that ≥2 years are required for ratios of α-/γ-tocopherols to reach new steady-state levels in response to changes in dietary intake. Thus, the analysis of adipose tissue α-tocopherol content is a useful estimate of the long-term vitamin E status (Handelman *et al.*, 1988; Kayden *et al.*, 1983). Hayes *et al.* (1993b) measured the vitamin E contents of adipose tissue from hamsters fed a diet containing both tocopherols and tocotrienols and found that the adipose tissue was enriched in tocotrienols.

Skin

The presence of α- and γ-tocotrienols in untreated mouse skin reported by Podda *et al.* (1996) was unexpected because (1) the mouse diets were not specially enriched with tocotrienols and (2) the liver discriminates against tocotrienols in favor of α-tocopherol during the repackaging of dietary fats into VLDL for secretion into the circulation (Traber, 1994; Traber and Sies, 1996). Apparently, the transfer of tocotrienols to mouse skin must take place following the absorption and transport of dietary vitamin E in chylomicrons during postprandial chylomicron clearance and during the delivery of tocotrienol-containing lipoproteins (Traber, 1994; Traber and Sies, 1996).

Skin has been suggested to be an important storage site for vitamin E and a major excretory organ for this vitamin (Shiratori, 1974). However, skin vitamin E could also have a regulatory role in maintaining barrier function. Skin contains a high proportion of tocotrienols, which could possibly inhibit cholesterol synthesis (Parker *et al.*, 1993) and destroy barrier function because cholesterol is a key component of the lipid barrier of the stratum corneum (Mao-Qiang *et al.*, 1993). In addition, vitamin E may enhance penetration and resorption of skin lipids (Trivedi *et al.*, 1995).

The rationale for examining protective effects of applied vitamin E to

Table II   Rationale for Testing Antioxidant Protective Effects of Tocotrienols

α- and γ-tocotrienols
    Unsaturated side chain
    Greater fluidity
    May interact more effectively with lipids and peroxyl radicals
    Higher antioxidant activity than other forms of vitamin E
    May decrease LDL susceptibility to oxidation

skin are given in Table II. Application of γ-tocotrienol, α-tocotrienol, or α-tocopherol to murine skin resulted in a major increase in murine stratum corneum vitamin E (Traber *et al.*, 1998). Vitamin E homologs penetrated through the entire skin to the subcutaneous fat layer within the first 0.5 hr. It is not clear whether this rapid penetration is into skin cells (keratinocytes or fibroblasts), around the cells in the skin lipids, or perhaps down the hair follicles into the deepest layers. However, it is clear from these studies that

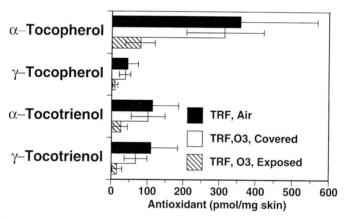

Figure 3   Vitamin E content of murine skin after 2 hr of ozone exposure. A 5% (w/v) solution of a tocotrienol rich fraction (TRF) was prepared in polyethylene glycol 400 (PEG). Mice were anesthetized, and four polypropylene plastic rings (1 cm^2) were glued onto the animals' backs with TRF solution (20 μl) applied to two rings and PEG to the other two rings. After 2 hr, the treated areas were washed, half of the sites were sealed with cellophane tape, and mice were exposed either to air ($n = 3$, control) or to 10 ppm ozone ($n = 4$). After air or ozone exposure, the skin was excised for vitamin E analysis. Concentrations of α-tocopherol, γ-tocopherol, α-tocotrienol, and γ-tocotrienol from air-exposed or from ozone-exposed covered or uncovered sites, are shown. Significant decreases in each homolog concentration were observed after ozone exposure in TRF-treated skin compared to covered sites.

papillary dermis has a special affinity for vitamin E; the mechanism for this affinity remains to be determined. If skin is exposed to oxidative stress following the topical application of vitamin E, the increased antioxidant content is sufficient to provide protection, as has been described in studies with ozone exposure (Fig. 3) (Thiele *et al.*, 1997) or ultraviolet irradiation (Weber *et al.*, 1997).

CONCLUSIONS AND IMPLICATIONS

Vitamin E was discovered in 1922 by Evans and Bishop, yet we are still now discovering new aspects about this vitamin. What remains to be investigated are the cellular regulatory effects of the vitamin E homologs. There may be many pathways by which vitamin E homologs may modulate signal transduction pathways and control cellular proliferation. Importantly, the pathways for apoptosis and necrosis are modulated by redox-sensitive signals and these too could be dependent on vitamin E.

ACKNOWLEDGMENTS

The authors gratefully acknowledge the support of the Palm Oil Research Institute of Malaysia. Special thanks also to our collaborators without whom the studies described herein could not have been carried out.

REFERENCES

Arita, M., Nomura, K., Arai, H., and Inoue, K. (1997). α-Tocopherol transfer protein stimulates the secretion of α-tocopherol from a cultured liver cell line through a brefeldin A-insensitive pathway. *Proc. Natl. Acad. Sci. USA* **94**, 12437–12441.

Arita, M., Sato, Y., Miyata, A., Tanabe, T., Takahashi, E., Kayden, H., Arai, H., and Inoue, K. (1995). Human alpha-tocopherol transfer protein: cDNA cloning, expression and chromosomal localization. *Biochem. J.* **306**, 437–443.

Baker, H., Handelman, G. J., Short, S., Machlin, L. J., Bhagavan, H. N., Dratz, E. A., and Frank, O. (1986). Comparison of plasma α- and γ-tocopherol levels following chronic oral administration of either all rac-α-tocopheryl acetate or RRR-α-tocopheryl acetate in normal adult male subjects. *Am. J. Clin. Nutr.* **43**, 382–387.

Bourre, J., and Clement, M. (1991). Kinetics of rat peripheral nerve, forebrain and cerebellum α-tocopherol depletion: Comparison with different organs. *J. Nutr.* **121**, 1204–1207.

Bunyan, J., McHale, D., Green, J., and Marcinkiewicz, S. (1961). Biological potencies of ϵ- and ζ_1-tocopherol and 5-methyltocol. *Br. J. Nutr.* **15**, 253–257.

Burton, G. W., Hughes, L., and Ingold, K. U. (1983a). Antioxidant activity of phenols related to vitamin E: Are there chain-breaking antioxidants better than α-tocopherol? *J. Am. Chem. Soc.* **105**, 5950–5951.

Burton, G. W., and Ingold, K. U. (1986). Vitamin E: Application of the principles of physical organic chemistry to the exploration of its structure and function. *Acc. Chem. Res.* **19**, 194–201.

Burton, G. W., Joyce, A., and Ingold, K. U. (1983b). Is vitamin E the only lipid-soluble, chain-breaking antioxidant in human blood plasma and erythrocyte membranes? *Arch. Biochem. Biophys.* **221**, 281–290.

Burton, G. W., Traber, M. G., Acuff, R. V., Walters, D. N., Kayden, H., Hughes, L., and Ingold, K. (1998). Human plasma and tissue α-tocopherol concentrations in response to supplementation with deuterated natural and synthetic vitamin E. *Am. J. Clin. Nutr.* **67**, 669–684.

Cheesemen, K. H., Holley, A. E., Kelly, F. J., Wasil, M., Hughes, L. and Burton, G. (1995). Biokinetics in humans of *RRR*-α-tocopherol: The free phenol, acetate ester, and succinate ester forms of vitamin E. *Free Radic. Biol. Med.* **19**, 591–598.

Chiku, S., Hamamura, K., and Nakamura, T. (1984). Novel urinary metabolite of δ-delta-tocopherol in rats. *J. Lipid Res.* **25**, 40–48.

Choudhury, N., Truswell, A. S., and McNeil, Y. (1997). Comparison of plasma lipids and vitamin E in young and middle-aged subjects on potato crisps fried in palmolein and highly oleic sunflower oil. *Ann. Nutr. Metabol.* **41**, 137–148.

Christen, S., Woodall, A. A., Shigenaga, M. K., Southwell-Keely, P. T., Duncan, M. W., and Ames, B. N. (1997). γ-Tocopherol traps mutagenic electrophiles such as NO_x and complements α-tocopherol: Physiological implications. *Proc. Natl. Acad. Sci. USA* **94**, 3217–3222.

Clement, M., Dinh, L., and Bourre, J. (1995). Uptake of dietary *RRR*-alpha- and *RRR*-gamma-tocopherol by nervous tissues, liver and muscle in vitamin-E-deficient rats. *Biochim. Biophys. Acta* **1256**, 175–180.

Cooney, R. W., France, A. A., Harwood, P. J., Hatch-Pigott, V., Custer, L. J., and Mordan, L. J. (1993). γ-Tocopherol detoxification of nitrogen dioxide: Superiority to α-tocopherol. *Proc. Natl. Acad. Sci. USA* **90**, 1771–1775.

Cross, C. E., Eiserich, J. P., and Halliwell, B. (1997). General biological consequences of inhaled environmental toxicants. *In* "The Lung" (R. G. Crystal and J. B. West, eds). Lippincott-Raven Publishers, Philadelphia.

Drevon, C. A. (1991). Absorption, transport and metabolism of vitamin E. *Free Radic. Res. Commun.* **14**, 229–246.

Eitenmiller, R. R., and Landen, W. O., Jr. (1995). Vitamins. *In* "Analyzing Food for Nutrition Labeling and Hazardous Contaminants" (I. J. Jeon and W. G. Ikins, eds). Dekker, New York.

Emerson, O. H., Emerson, G. A., Mohammed, A., and Evans, H. M. (1937). The chemistry of vitamin E: Tocopherols from various sources. *J. Biol. Chem.* **122**, 99–107.

Evans, H. M., and Bishop, K. S. (1922). On the existence of a hitherto unrecognized dietary factor essential for reproduction. *Science* **56**, 650–651.

Evans, H. M., Emerson, O. H., and Emerson, G. A. (1936). The isolation from wheat germ oil of an alcohol, alpha-tocopherol, having the properties of vitamin E. *J. Biol. Chem.* **113**, 319–332.

Food and Nutrition Board, N. R. C. (1989). "Recommended Dietary Allowances." National Academy of Sciences Press, Washington, DC.

Gotoda, T., Arita, M., Arai, H., Inoue, K., Yokota, T., Fukuo, Y., Yazaki, Y., and Yamada, N. (1995). Adult-onset spinocerebellar dysfunction caused by a mutation in the gene for the alpha-tocopherol-transfer protein. *N. Engl. J. Med.* **333**, 1313–1318.

Green, J., McHale, D., Marcinkiewwicz, S., Hamalis, P., and Watt, P. R. (1959). Tocopherols. V. Structural studies on ε- and ζ-tocopherol. *J. Chem. Soc.* 3362–3373.

Handelman, G. J., Epstein, W. L., Machlin, L. J., van Kujik, F. J. G. M., and Dratz, E. A. (1988). Biopsy method for human adipose with vitamin E and lipid measurements. *Lipids* **23**, 598–604.

Handelman, G. J., Epstein, W. L., Peerson, J., Spiegelman, D., and Machlin, L. J. (1994). Human adipose α-tocopherol and γ-tocopherol kinetics during and after 1 year of α-tocopherol supplementation. *Am. J. Clin. Nutr.* **59**, 1025–1032.

Handelman, G. J., Machlin, L. J., Fitch, K., Weiter, J. J., and Dratz, E. A. (1985). Oral α-tocopherol supplements decrease plasma γ-tocopherol levels in humans. *J. Nutr.* **115**, 807–813.

Hayes, K. C., Pronczuk, A., and Liang, J. S. (1993). Differences in the plasma transport and tissue concentrations of tocopherols and tocotrienols: Observations in humans and hamsters. *Proc. Soc. Exp. Biol. Med.* **202**, 353–359.

Hentati, A., Deng, H.-X., Hung, W.-Y., Nayer, M., Ahmed, M. S., He, X., Tim, R., Stumpf, D. A., and Siddique, T. (1996). Gene structure and mutations in familial vitamin E deficiency. *Ann. Neurol.* **39**, 295–300.

Hosomi, A., Arita, M., Sato, Y., Kiyose, C., Ueda, T., Igarashi, O., Arai, H., and Inoue, K. (1997). Affinity for alpha-tocopherol transfer protein as a determinant of the biological activities of vitamin E analogs. *FEBS Lett.* **409**, 105–108.

Ikeda, I., Imasato, Y., Sasaki, E., and Sugano, M. (1996). Lymphatic transport of alpha-, gamma- and delta-tocotrienols and alpha-tocopherol in rats. *Int. J. Vitam. Nutr. Res.* **66**, 217–221.

Ingold, K. U., Burton, G. W., Foster, D. O., and Hughes, L. (1990). Further studies of a new vitamin E analogue more active than alpha-tocopherol in the rat curative myopathy bioassay. *FEBS Lett.* **267**, 63–65.

Jialal, I., Fuller, C. J., and Huet, B. A. (1995). The effect of α-tocopherol supplementation on LDL oxidation: A dose-response study. *Arterioscler. Thromb. Vasc. Biol.* **15**, 190–198.

Kagan, V. E., Bakalova, R. A., Zhelev, Z. Z., Rangelova, D. S., Serbinova, E. A., Tyurin, V. A., Denisova, N. K., and Packer, L. (1990). Intermembrane transfer and antioxidant action of alpha-tocopherol in liposomes. *Arch. Biochem. Biophys.* **280**, 147–152.

Kamal-Eldin, A., and Appelqvist, L. A. (1996). The chemistry and antioxidant properties of tocopherols and tocotrienols. *Lipids* **31**, 671–701.

Kamat, J. P., Sarma, H. D., Devasagayam, T. P., Nesaretnam, K., and Basiron, Y. (1997). Tocotriencols from palm oil as effective inhibitors of protein oxidation and lipid peroxidation in rat liver microsomes. *Mol. Cell. Biochem.* **170**, 131–137.

Kayden, H. J., Hatam, L. J., and Traber, M. G. (1983). The measurement of nanograms of tocopherol from needle aspiration biopsies of adipose tissue: Normal and abetalipoproteinemic subjects. *J. Lipid Res.* **24**, 652–656.

Kayden, H. J., and Traber, M. G. (1993). Absorption, lipoprotein transport and regulation of plasma concentrations of vitamin E in humans. *J. Lipid Res.* **34**, 343–358.

Kostner, G. M., Oettl, K., Jauhiainen, M., Ehnholm, C., Esterbauer, H., and Dieplinger, H. (1995). Human plasma phospholipid transfer protein accelerates exchange/transfer of alpha-tocopherol between lipoproteins and cells. *Biochem. J.* **305**, 659–667.

Machlin, L. J. (1991). Vitamin E. *In* "Handbook of Vitamins" (L. J. Machlin, eds.), 2nd ed., pp. 99–144. Dekker, New York.

Mao-Qiang, M., Feingold, K. R., and Elias, P. M. (1993). Inhibition of cholesterol and sphingolipid synthesis causes paradoxical effects on permeability barrier homeostasis. *J. Invest. Dermatol.* **101**, 185–190.

Murray, E. D. J., Wechter, W. J., Kantoci, D., Wang, W. H., Phan, T., Quiggle, D. D., Givson, K. M., Leipold, D., and Anner, B. (1997). Endogenous natriuretic factors 7: Biospecificity of a natriuretic γ-tocopherol metabolite LLU-α. *J. Exp. Pharm. Ther.*

Nakamura, T., Aoyama, Y., Fujita, T., and Katsui, G. (1975). Studies on tocopherol derivatives. V. Intestinal absorption of several d,1-3,4-3H2-alpha-tocopheryl esters in the rat. *Lipids* 10, 627–633.

Ouahchi, K., Arita, M., Kayden, H., Hentati, F., Ben Hamida, M., Sokol, R., Arai, H., Inoue, K., Mandel, J.-L., and Koenig, M. (1995). Ataxia with isolated vitamin E deficiency is caused by mutations in the α-tocopherol transfer protein. *Nature Genet.* 9, 141–145.

Packer, L. (1994). Vitamin E is nature's master antioxidant. *Sci. Am. Sci. Med.* 1, 54–63.

Packer, L. (1995). Nutrition and biochemistry of the lipophilic antioxidants vitamin E and carotenoids. *In* "Nutrition, Lipids, Health and Disease" (A. S. H. Ong, E. Niki, and L. Packer, eds.), pp. 8–35. AOCS Press, Champaign, IL.

Parker, R. A., Pearce, B. C., Clark, R. W., Gordon, D. A., and Wright, J. J. (1993). Tocotrienols regulate cholesterol production in mammalian cells by post-transcriptional suppression of 3-hydroxy-3-methylglutaryl-coenzyme A reductase. *J. Biol. Chem.* 268, 11230–11238.

Pillai, S. R., Traber, M. G., Kayden, H. J., Cox, N. R., Toivio-Kinnucan, M., Wright, J. C., and Steiss, J. E. (1994). Concomitant brain stem axonal dystrophy and necrotizing myopathy in vitamin E-deficient rats. *J. Neuro. Sci.* 123, 64–73.

Podda, M., Descans, B., Traber, M. G., and Packer, L. (1995). Vitamin E deficiency symptoms are prevented by alpha-lipoic acid. *FASEB J.* 9, A473.

Podda, M., Weber, C., Traber, M. G., and Packer, L. (1996). Simultaneous determination of tissue tocopherols, tocotrienols, ubiquinols and ubiquinones. *J. Lipid Res.* 37, 893–901.

Sato, Y., Arai, H., Miyata, A., Tokita, S., Yamamoto, K., Tanabe, T., and Inoue, K. (1993). Primary structure of alpha-tocopherol transfer protein from rat liver: Homology with cellular retinaldehyde-binding protein. *J. Biol. Chem.* 268, 17705–17710.

Sato, Y., Hagiwara, K., Arai, H., and Inoue, K. (1991). Purification and characterization of the α-tocopherol transfer protein from rat liver. *FEBS Lett.* 288, 41–45.

Schultz, M., Leist, M., Petrzika, M., Gassmann, B., and Brigelius-Flohé, R. (1995). A novel urinary metabolite of α-tocopherol, 2,5,7,8-tetramethyl-2(2'carboxyethyl)-6-hydroxychroman (α-CEHC) as an indicator of an adequate vitamin E supply? *Am. J. Clin. Nutr.* 62(Suppl.), 1527S–1534S.

Serbinova, E., Kagan, V., Han, D., and Packer, L. (1991). Free radical recycling and intramembrane mobility in the antioxidant properties of alpha-tocopherol and alpha-tocotrienol. *Free Radic. Biol. Med.* 10, 263–275.

Serbinova, E. A., and Packer, L. (1994). Antioxidant and biological activities of palm oil vitamin E. *Food Nutr. Bull.* 15, 138–143.

Serbinova, E. A., Tsuchiya, M., Goth, S., Kagan, V. E., and Packer, L. (1993). Antioxidant action of α-tocopherol and α-tocotrienol in membranes. *In* "Vitamin E in Health and Disease" (L. Packer and J. Fuchs, eds.), pp. 235–243. Dekker, New York.

Shiratori, T. (1974). Uptake, storage and excretion of chylomicra-bound 3H-alpha-tocopherol by the skin of the rat. *Life Sci.* 14, 929–935.

Sokol, R. J. (1993). Vitamin E deficiency and neurological disorders. *In* "Vitamin E in Health and Disease" (L. Packer and J. Fuchs, eds.), pp. 815–849. Dekker, New York.

Sokol, R. J., Heubi, J. E., Iannaccone, S., Bove, K. E., Harris, R. E., and Balistreri, W. F. (1983). The mechanism causing vitamin E deficiency during chronic childhood cholestasis. *Gastroenterology* 85, 1172–1182.

Southam, E., Thomas, P. K., King, R. H. M., Goss-Sampson, M. A., and Muller, D. P. R. (1991). Experimental vitamin E deficiency in rats: Morphological and functional evidence of abnormal axonal transport secondary to free radical damage. *Brain* 114, 915–936.

Suarna, C., Food, R. L., Dean, R. T., and Stocker, R. (1993). Comparative antioxidant activity of tocotrienols and other natural lipid-soluble antioxidants in a homogeneous system, and in rat and human lipoproteins. *Biochim. Biophys. Acta* 1166, 163–170.

Thiele, J. J., Traber, M. G., Podda, M., Tsang, K., Cross, C., and Packer, L. (1997). Ozone depletes tocopherols and tocotrienols topically applied to murine skin. *FEBS Lett.* **401,** 167–170.

Traber, M. G. (1994). Determinants of plasma vitamin E concentrations. *Free Radic Biol. Med.* **16,** 229–239.

Traber, M. G., Goldberg, I., Davidson, E., Lagmay, N., and Kayden, H. J. (1990a). Vitamin E uptake by human intestinal cells during lipolysis in vitro. *Gastroenterology* **98,** 96–103.

Traber, M. G., and Kayden, H. J. (1987). Tocopherol distribution and intracellular localization in human adipose tissue. *Am. J. Clin. Nutr.* **46,** 488–495.

Traber, M. G., and Kayden, H. J. (1989). Preferential incorporation of α-tocopherol vs γ-tocopherol in human lipoproteins. *Am. J. Clin. Nutr.* **49,** 517–526.

Traber, M. G., Lane, J. C., Lagmay, N., and Kayden, H. J. (1992). Studies on the transfer of tocopherol between lipoproteins. *Lipids* **27,** 657–663.

Traber, M. G., Rallis, M., Podda, M., Weber, C., Maibach, H. I., and Packer, L. (1998). Penetration and distribution of α-tocopherol, α- or γ-tocotrienols applied individually onto murine skin. *Lipids* **33,** 87–91.

Traber, M. G., Rudel, L. L., Burton, G. W., Hughes, L., Ingold, K. U., and Kayden, H. J. (1990b). Nascent VLDL from liver perfusions of cynomolgus monkeys are preferentially enriched in *RRR*- compared with *SRR-α* tocopherol: Studies using deuterated tocopherols. *J. Lipid Res.* **31,** 687–694.

Traber, M. G., and Sies, H. (1996). Vitamin E in humans: Demand and delivery. *Annu. Rev. Nutr.* **16,** 321–347.

Trivedi, J. S., Krill, S. L., and Fort, J. J. (1995). Vitamin E as a human skin penetration enhancer. *Eur. J. Pharm. Sci.* **3,** 241–243.

Vatassery, G. T., Angerhofer, C. K., Knox, C. A., and Deshmukh, D. S. (1984a). Concentrations of vitamin E in various neuroanatomical regions and subcellular fractions, and the uptake of vitamin E by specific areas, of rat brain. *Biochim. Biophys. Acta* **792,** 118–122.

Vatassery, G. T., Angerhofer, C. K., and Peterson, F. J. (1984b). Vitamin E concentrations in the brains and some selected peripheral tissues of selenium-deficient and vitamin E-deficient mice. *J. Neurochem.* **42,** 554–558.

Vatassery, G. T., Brin, M. F., Fahn, S., Kayden, H. J., and Traber, M. G. (1988). Effect of high doses of dietary vitamin E upon the concentrations of vitamin E in several brain regions, plasma, liver and adipose tissue of rats. *J. Neurochem.* **51,** 621–623.

Weber, C., Podda, M., Rallis, M., Traber, M. G., and Packer, L. (1997). Efficacy of topical application of tocopherols and tocotrienols in protection of murine skin from oxidative damage induced by UV-irradiation. *Free Radic. Biol. Med.* **22,** 761–769.

Wechter, W. J., Kantoci, D., Murray, E. D. J., D'Amico, D. C., Jung, M. E., and Wang, W.-H. (1996). A new endogenous natriuretic factor: LLU-alpha. *Proc. Natl. Acad. Sci. USA* **93,** 6002–6007.

Weimann, B. J., and Weiser, H. (1991). Functions of vitamin E in reproduction and in prostacyclin and immunoglobulin synthesis in rats. *Am. J. Clin. Nutr.* **53,** 1056S–1060S.

Weiser, H., Riss, G., and Kormann, A. W. (1996). Biodiscrimination of the eight alpha-tocopherol stereoisomers results in preferential accumulation of the four 2R forms in tissues and plasma of rats. *J. Nutr.* **126,** 2539–2549.

Weiser, H., Vecchi, M., and Schlachter, M. (1986). Stereoisomers α-tocopheryl acetate. IV. USP units and α-tocopherol equivalents of *all-rac-, 2-ambo-* and *RRR-α*-tocopherol evaluated by simultaneous determination of resorption-gestation, myopathy and liver storage capacity in rats. *Int. J. Vit. Nutr. Res.* **56,** 45–56.

Yokota, T., Shiojiri, T., Gotoda, T., and Arai, H. (1996). Retinitis pigmentosa and ataxia caused by a mutation in the gene for the α-tocopherol-transfer protein. *N. Engl. J. Med.* **335,** 1769–1770.

5 Effects of High Cholesterol, Vitamin E, and Probucol on Protein Kinase C Activity and Proliferation of Smooth Muscle Cells

Nesrin K. Özer, * *Suzan Taha,* * *Achim Stocker,†*
and Angelo Azzi †

*Department of Biochemistry
Faculty of Medicine
Marmara University
Istanbul, Turkey

†Institut für Biochemie und Molekularbiologie
Universität Bern
3012 Bern, Switzerland

α-Tocopherol, but not β-tocopherol, decreases in a concentration-dependent way proliferation of smooth muscle cells. At the same concentrations (10–50 μM) it induces the inhibition of protein kinase C (PKC) activity. Proliferation and protein kinase C inhibition by α-tocopherol, the lack of inhibition by β-tocopherol, and the prevention by β-tocopherol indicate that the mechanism involved is not related to the radical-scavenging properties of these two molecules, which are essentially equal. Probucol (10–50 μM), a potent lipophilic antioxidant, does not inhibit smooth muscle cell proliferation and protein kinase C activity. In rabbit studies, atherosclerosis was induced by a 2% cholesterol-containing, vitamin E-poor diet. Six different groups of rabbits each received vitamin E, probucol, and probucol *plus* vitamin E. After 4 weeks, aortas were analyzed for PKC activity. Their media smooth muscle cells exhibited an increase in protein kinase C activity. Vitamin E fully prevented cholesterol-induced atherosclerotic lesions and the induction of protein kinase C activity. Probucol was not effective in preventing both cholesterol-induced atherosclerotic lesions and the induction of PKC activity. These results show that the protective effect of vitamin E against hypercholesterolemic atherosclerosis is not produced by another antioxidant such as probucol and therefore may not be linked to the antioxidant properties of this vitamin. The effects observed at the level of smooth muscle cells *ex vivo* and cell culture suggest an involvement of signal transduction events on the onset of atherosclerosis. It is at this level that the protective effect of vitamin E against atherosclerosis is exerted.

INTRODUCTION

Atherosclerosis (Ross, 1993) initiates with the migration and proliferation of media smooth muscle cells to the intima in the arterial wall (Raines and Ross, 1993, Bornfeldt *et al.*, 1994). Antioxidant vitamins, especially vitamin E, have been shown to protect against the progress of atherosclerosis, as documented by epidemiological and intervention studies (Gey, 1990; Rimm *et al.*, 1993; Stampfer *et al.*, 1993). The Cambridge Heart Antioxidant Study (CHAOS) reported a strong protection by high vitamin E doses against the risk of fatal and nonfatal myocardial infarction (Stephens *et al.*, 1996). One of the mechanisms of α-tocopherol action is to protect low-density lipoproteins from oxidation by peroxiradicals (Esterbauer *et al.*, 1992; Carew *et al.*, 1987). Other functions of α-tocopherol are independent of its antioxidant characteristics (Boscoboinik *et al.*, 1991b, 1995; Azzi *et al.*, 1993; Azzi *et al.*, 1995), such as that of moderating vascular smooth muscle cell proliferation. Smooth muscle cell proliferation is controlled by growth factor receptors, which, via a cascade of kinases and phosphatases, produce activation and expression of proteins necessary for the progression and completion of the cell cycle (Muller *et al.*, 1993). Part of the main signal transduction path initiated by growth factors or hormones (reviewed in Azzi *et al.*, 1992), protein kinase C has a complex activation mechanism and interacts to permit phosphorylation (Dutil *et al.*, 1994; Newton, 1995; Borner *et al.*, 1989; Bornancin and Parker, 1996).

Although epidemiological data show a reduced coronary heart disease risk, cell culture studies reveal inhibition by α-tocopherol of smooth muscle cell proliferation. The latter event takes place via the inhibition of protein kinase C activity. α-Tocopherol also inhibits low-density, lipoprotein-stimulated smooth muscle cell proliferation and PKC activity. Because β-tocopherol and other potent antioxidants are not inhibitory, the effect of

α-tocopherol is considered to be due to a nonoxidant mechanism (Özer *et al.*, 1993, 1995; Azzi *et al.*, 1993). Moreover, increased protein kinase C activity and expression in atherosclerotic rabbits were shown (Sirikci *et al.*, 1996).

The following cell culture and animal studies were carried out to investigate the mechanism of α-tocopherol on smooth muscle cell proliferation, its effect in protection against atherosclerosis, and the effect of α-tocopherol as compared with β-tocopherol and probucol.

MATERIALS AND METHODS

Growth media and serum for cell culture were obtained from Gibco Laboratories. A7r5 rat aortic smooth muscle cells (SMC) were obtained from the American Type Culture Collection (Rockville, MD). Phorbol 12-myristate 13-acetate (PMA) and streptolysin-O (25,000 units) were from Sigma (Buchs, Switzerland). [γ-^{32}P] ATP (30 Ci/mmol) and [*methyl*-^{3}H]thymidine (25 Ci/mmol) were purchased from Amersham International and α-tocopherol and β-tocopherol from Merck (Darmstadt, Germany). Vitamin E-free rabbit diet and vitamin E (Ephynal) were kindly donated by Dr. U. Moser from Hoffman La Roche (Basel, Switzerland and Istanbul, Turkey) and probucol was a gift from Marion Merrell Dow Pharmaceuticals Inc., Switzerland. The peptide PLSRTLSVAAKK used as a substrate for the protein kinase C assay was synthesized by Dr. Servis, Epalinges (Switzerland). The kit for the measurement of PKC activity was obtained from Upstate Biotechnology Inc., Lake Placid, New York. All other chemicals used were of the purest grade commercially available.

Cell Culture

A7r5 cells are maintained in Dulbecco's modified Eagle's medium (DMEM) supplemented with 10% (v/v) fetal calf serum (FCS). Cells in a subconfluent state are made quiescent by incubation in DMEM containing 0.2% FCS for 48 hr. Cells are then washed with phosphate-buffered saline (PBS) and treated as described in figure legends 1–4. Viability is determined by the Trypan blue dye method.

Determination of Cell Number and [^{3}H]Thymidine Incorporation

Quiscent A7r5 cells were restimulated to grow by the addition of 10% FCS. Tocopherols and probucol were added to cells at the indicated concentrations. The cell number was determined 48 hr later using an hemocytometer.

To measure DNA synthesis, cells were pulsed for 1 hr with [³H]thymidine (1 μCi/well) at the indicated times following restimulation with FCS. After labeling, cells were washed twice with PBS, fixed for 20 min with ice-cold 5% trichloroacetic acid, and solubilized in 0.1 M NaOH/2% Na_2CO_3/1% SDS. The radioactivity incorporated into the acid-insoluble material was determined in a liquid scintillation analyzer.

Determination of Protein Kinase C Activity in Permeabilized Cells

The activity of PKC in permeabilized smooth muscle cells was determined using the procedure of Alexander *et al.* (1990) with minor modifications. Quiescent A7r5 cells were subjected to different treatments as indicated. During the last hour of the preincubation period, cells were treated with 100 nM phorbol 12-myristate 13-acetate. Cells were then washed twice with PBS, resuspended in intracellular buffer (5.2m M $MgCl_2$, 94 mM KCl, 12.5 mM HEPES, 12.5 mM EGTA, 8.2 mM $CaCl_2$, pH 7.4), and divided into 220-μl portions (1.5 × 10⁵ cells). Assays were started by adding [γ-³²P] ATP (9cpm/pmol, final concentration 250 μM), peptide substrate (final concentration 70 μM), and streptolysin-O (0.3 IU). The reaction mixtures were incubated at 37°C for 10 min, and the reaction was stopped by the addition of 100 μl of 25% (w/v) trichloroacetic acid in 2 M acetic acid. After 10 min on ice, 100-μl aliquots were spotted onto 4.5 × 4.5 cm P81 ion-exchange chromatographic paper within predrawn squares (Whatman International). The paper was then washed twice for 10 min with 30% (v/v) acetic acid containing 1% phosphoric acid and once with ethanol. The amount of the two washes was 10 ml each per spotted square. The P81 paper was dried, the squares were cut, and radioactivity was counted in a liquid scintillation analyzer. Basal phosphorylation in the absence of the peptide was subtracted from experimental data to determine the specific activity.

Animal Experiments

Thirty male New Zealand albino rabbits 2–4 months of age were assigned randomly to one of the following six groups. All rabbits were fed 100 g per day of a vitamin E-poor diet. Probucol and cholesterol were added to the diet as a diethyl ether solution. The control diet was treated with the same amount of pure solvent. All diets were dried of the solvent before use. Concentrations of cholesterol, vitamin E, and probucol used were based on previous literature reports (Stein *et al.*, 1989; Ferns *et al.*, 1992; Bocan *et al.*, 1993; Mantha *et al.*, 1993; Konneh *et al.*, 1995).

One group of rabbits was fed with the diet only, without additions or treatments. The second group received injections of 50 mg/kg of vitamin E intramuscularly on alternate rear legs, once daily. The diet of the third group contained 2% cholesterol. The diet of the fourth group contained 2% cholesterol and the rabbits received injections of 50 mg/kg of vitamin E per day. The fifth group of rabbits received 1% probucol in addition to 2% cholesterol in the daily diet. The sixth group had 2% cholesterol and 1% probucol in the diet and were injected with 50 mg/kg of vitamin E per day.

After 4 weeks, plasma samples were taken and thoracic aortas were removed. Samples from thoracic aorta media were taken to measure smooth muscle cell protein kinase C activity by using a nonradioactive kit (Upstate Biotechnology Inc.). Plasma cholesterol levels were determined using an automated enzymatic technique. Vitamin E levels were measured by reverse phase high-pressure liquid chromatography (Nierenberg and Nann, 1992).

RESULTS

The effect of α-tocopherol on smooth muscle cell proliferation and on PKC is shown in Fig. 1. A parallel inhibition of protein kinase C activity and of proliferation is observed to occur at concentrations of α-tocopherol close to those measured in healthy adults (Gey, 1990; Riemersma *et al.*, 1991). The following experiments were designed to establish if the previously described effects of α-tocopherol were related to its antioxidant properties, if a direct interaction was responsible for the inhibition, and which

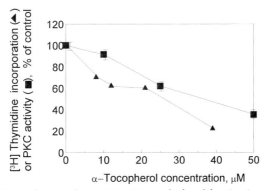

Figure 1   Inhibition of protein kinase C activity and of proliferation in smooth muscle cell. Serum-deprived quiescent cells were stimulated to grow by FCS in the presence or absence of α-tocopherol. After 7 hr, PKC was measured for DNA synthesis determination, and a 1-hr pulse of 1 μCi/well [^{3}H]thymidine was given to the cells 15 hr after restimulation with FCS.

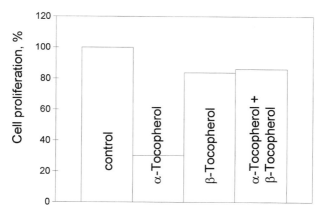

Figure 2   Differential effects of α-tocopherol and β-tocopherol on smooth muscle cell proliferation. Quiescent cells were restimulated to grow with FCS (10%) in the presence of α-tocopherol and/or β-tocopherol (50 μM). After 48 hr of restimulation, cells were counted with a hemocytometer. Viability was greater than 95%.

of the two events, protein kinase C activity or proliferation inhibition, was the cause of the other.

α-Tocopherol at concentrations of 50 μM inhibits rat A7r5 smooth muscle cell proliferation (Fig. 2), whereas β-tocopherol is ineffective. When α-tocopherol and β-tocopherol are added together, no inhibition of cell growth is seen. Both compounds are transported equally in cells and do not compete with each other for uptake (Azzi *et al.*, 1995). The prevention by β-tocopherol of the proliferation inhibition caused by α-tocopherol suggests a site-directed event at the basis of α-tocopherol inhibition rather than a general radical-scavenging reaction. The oxidized product of α-tocopherol, α-tocopherylquinone, is not inhibitory, indicating that the effects of α-tocopherol are not related to its antioxidant properties (Azzi *et al.*, 1995). Inhibitory effects of α-tocopherol are also observed in primary human aortic smooth muscle cells (hAOMSC from Clonetics Corp., San Diego, CA, and in four cell strains provided by Dr. T. Resink, Kantonspital, Basel, Switzerland).

In smooth muscle cells permeabilized with streptolysin-O, to permit the entry of a peptide substrate to measure protein kinase C activity (Fig. 3), α-tocopherol inhibits protein kinase C activity, whereas β-tocopherol is ineffective. When both are present, β-tocopherol prevents the inhibitory effect of α-tocopherol. The inhibition by α-tocopherol and the lack of inhibition by β-tocopherol of cell proliferation and protein kinase C activity shows that the mechanism involved is not related to the radical scavenging

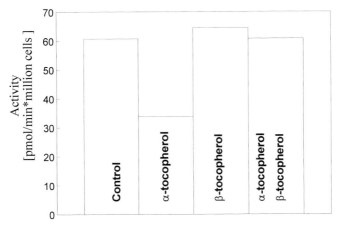

Figure 3   Effect of α-tocopherol and β-tocopherol on smooth muscle cell protein kinase C activity. Quiescent cells were restimulated to grow with FCS (10%) in the presence of α-tocopherol and/or β-tocopherol (50 μM). After 7 hr of restimulation, cells were permeabilized and protein kinase C was measured as described in Materials and Methods. Phorbol myristate acetate (100 nM) was added 60 min before assaying activity. The basal kinase activity was subtracted in all samples and only the PMA-stimulated activity is shown.

properties of these two molecules, which are essentially equal (Pryor *et al.*, 1993).

Figure 4 addresses the question as to whether probucol, a potent lipophilic antioxidant, exerts an effect on smooth muscle cell proliferation and protein kinase C activity. Probucol (10–50 μM) does not inhibit smooth muscle cell proliferation and protein kinase C activity.

The following animal experiments address the question of whether vitamin E can prevent the development of atherosclerosis *in vivo* and whether probucol can mimic the effect of α-tocopherol. Cholesterol and vitamin E plasma concentrations of the six animal groups are shown in Table I. Two percent cholesterol diet supplementation for 4 weeks resulted in an approximately 20-fold increase of plasma cholesterol. After additional supplementation with vitamin E or probucol, plasma cholesterol increased 12- and 13-fold, respectively, relative to the controls. Plasma vitamin E concentrations were higher in the cholesterol-fed rabbits in agreement with literature data (Godfried *et al.*, 1989; Wilson *et al.*, 1978; Bitman *et al.*, 1976; Bjorkhem *et al.*, 1991; Morel *et al.*, 1994), but the values corrected for the plasma cholesterol concentrations were of a similar order of magnitude.

Previous results have shown that protein kinase C activity and expression were upregulated by cholesterol, whereas vitamin E counteraction, al-

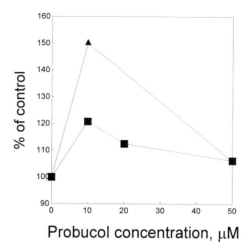

Probucol concentration, μM

Figure 4 Effect of probucol on smooth muscle cell proliferation (■) and protein kinase C activity(▲). Quiescent cells were restimulated to grow with FCS (10%) in the presence of indicated amounts of probucol. Cell proliferation and protein kinase C activity were measured as indicated under Materials and Methods.

though visible as a tendency, was not statistically significant (Sirikci *et al.,* 1996). In this study (Table II), in the total absence of vitamin E in the diet, the activity of smooth muscle cell protein kinase C was 8.4 Δ-Absorbance units/min/mg protein. After vitamin E treatment, an approximately 50% reduction of protein kinase C activity was seen. With cholesterol supplemen-

Table I Effect of Cholesterol Probucol and Vitamin E Treatment on Plasma Levels in Rabbits

Group	Cholesterol (mg/dl)	Vitamin E (μg/ml)
Control	37.4 ± 22.0	2.30 ± 1.04
Vitamin E	35.2 ± 3.6	21.80 ± 4.38[a]
Cholesterol	815.6 ± 416.7[a]	6.30 ± 2.80[b]
Cholesterol + vitamin E	480.8 ± 238.8[a]	93.22 ± 30.26[a]
Cholesterol + probucol	527.8 ± 258.0[a]	5.30 ± 1.96[b]
Cholesterol + probucol + vitamin E	448.8 ± 137.2[a]	109.80 ± 27.60[a]

The plasma levels of cholesterol and vitamin E have been measured in all five rabbits of the six diet groups. The numbers (mean ± SD) represent the plasma values measured after a 1-month diet.
[a] $p < 0.01$.
[b] Not statistically significant.

Table II   Protein Kinase C Activity from Smooth Muscle Cell Homogenates Obtained from Differently Treated Rabbits

Group	Total protein kinase C activity Δ-Absorbance/min/mg protein	
Control	8.4 ± 1.1	
Vitamin E	4.5 ± 2.5	$p < 0.01^a$
Cholesterol	10.2 ± 2.4	
Cholesterol + vitamin E	4.5 ± 1.0	$p < 0.02^b$
Cholesterol + probucol	6.8 ± 1.8	NS^b
Cholesterol + probucol + vitamin E	5.2 ± 1.4	$p < 0.04^b$

Aortic medias were minced, homogenized, and nuclei sedimented by centrifugation. Supernatants were centrifuged again at 100,000 g to obtain cytosolic fractions. Pellets were used for the preparation of membrane fractions. Protein kinase C activity was measured in both fractions. Because they did not show significant membrane/cytosol distribution changes, only the values of total protein kinase C activity are reported. Results are expressed as mean $\pm$ SD. ($n = 5$).
^{a} Control group.
^{b} Cholesterol group. NS, not statistically significant.

tation the activity of protein kinase C increased to 10.2 Δ-Absorbance units/min/mg protein. Vitamin E treatment was able to reduce in this case protein kinase C activity values to those measured in the absence of cholesterol (4.5 Δ-Absorbance units/min/mg protein). Smooth muscle cells from probucol *plus* cholesterol-treated rabbits showed protein kinase C activity lower than that in the presence of cholesterol alone, although data were not statistically significant. Instead, when the rabbits were also treated with vitamin E the protein kinase C value decreased to 5.2 Δ-Absorbance units/min/mg protein and reached significance relative to the cholesterol-fed rabbit group.

DISCUSSION

The proliferation of smooth muscle cells is fundamental in the onset of accelerated atherosclerosis, after cardiac transplantation and restenosis. Multiple pathways are involved in the proliferative response of smooth muscle cell to an extracellular stimulus, which has rendered is difficult to identify precise targets of antiproliferative agents. One of the important elements in signal transduction cascades is protein kinase C, and its inhibition appears to be sufficient to cause the inhibition of smooth muscle cell proliferation *in vitro*. In this study, a specific inhibition of proliferation *in vitro* of smooth muscle cells by α-tocopherol at physiological concentrations was observed. Of particular relevance was the finding of an antiproliferative effect of α-tocopherol on rat and human smooth muscle cells.

The inhibition of smooth muscle cell proliferation *in vitro* by α-tocopherol at physiological concentrations may explain the finding the *in vivo* smooth muscle cell only proliferate under stress situations (Raines and Ross, 1993; Clowes and Schwartz, 1985). A local or generalized diminution of α-tocopherol concentration, caused by dietary or oxidative factors, can lead to cell growth stimulation and atherosclerosis progress. β-Tocopherol, an antioxidant almost as potent as α-tocopherol, not only does not show any effect at the level of cell proliferation or protein kinase C activity, but also prevents the effects of α-tocopherol. Thus, it is legitimate to conclude that the mechanism of action of α-tocopherol, as a regulator of smooth muscle cell proliferation, is not due to its antioxidant properties. Data discussed earlier speak for the existence of an α-tocopherol-binding protein, binding α-tocopherol as an agonist and β-tocopherol as an antagonist.

β-Tocopherol shows only a minor effect at the level of cell proliferation or protein kinase C activity. Because β-tocopherol prevents the effects of α-tocopherol, it seems justifiable to conclude that the mechanism of action of α-tocopherol as a regulator of smooth muscle cell proliferation cannot be associated with its antioxidant properties. Moreover, these results can be interpreted in terms of the existence of a common intermediate, a putative α-tocopherol-binding protein, able to bind α- and β-tocopherol with similar affinity, with α-tocopherol acting as an agonist and β-tocopherol as an antagonist.

By comparing α-tocopherol with analogous compounds exhibiting similar antioxidant properties, such as β-tocopherol or probucol, it was concluded that α-tocopherol exerts its action independently of its free radical scavenger capacity, most probably by interacting with an as yet not characterized receptor molecule in smooth muscle cells (Boscoboinik *et al.*, 1991b, 1994, 1995).

Several groups have reported that antioxidant vitamins, especially vitamin E, have an important antiatherogenic role (Rimm *et al.*, 1993; Stampfer *et al.*, 1993; Gey *et al.*, 1993; Stahelin *et al.*, 1992; Steinberg, 1995). The question posed in this study has been whether molecular events in signal transduction can be regulated by cholesterol and vitamin E. Because the importance of protein kinase C in smooth muscle cell proliferation has been reported in many studies (Castellot, *et al.*, 1989; Matsumoto and Sasake, 1989; Chatelain *et al.*, 1993; Newby *et al.*, 1995) and because its proliferation is inhibited by α-tocopherol (Boscoboinik *et al.*, 1991a,b, 1995), the activity in rabbit aorta smooth muscle cells was measured. The present *in vivo* data of cholesterol stimulation and vitamin E inhibition of protein kinase C activity agree with previous results obtained both *in vitro* (Boscoboinik *et al.*, 1991a,b, 1995; Özer *et al.*, 1993, 1995) and *in vivo* (Sirikci *et al.*, 1996). It appears that probucol does not act as a protective agent against atherosclerotic plaque formation in the absence of vitamin E (Özer *et al.*, 1997). It also appears that among the molecular events that may be responsible for the protection by vitamin E against atherosclerotic plaque formation, inhibition of smooth muscle protein kinase C may play a crucial role.

In conclusion, α-tocopherol effect are indicative of a site-directed recognition mechanism involving a series of events, including the binding of α-tocopherol to a "receptor protein," activation (or expression) of a protein phosphatase, dephosphorylation of protein kinase, and inhibition of protein kinase C activity, along with affecting gene transcription inhibition of cell proliferation. These studies provide a molecular interpretation to the epidemiological information linking a decrease of plasma α-tocopherol with an increased risk of ischemic heart disease.

ACKNOWLEDGMENTS

This study was supported by the Marmara University Research Foundation, F. Hoffmann-La Roche A. G., Swiss National Science Foundation, and by a UNESCO-MCBN grant.

REFERENCES

Alexander, D. R., Graves, J. D., Lucas, S. C., Cantrell, D. A., and Crumpton, M. J. (1990). A method for measuring protein kinase C activity in permeabilized T lymphocytes by using peptide substrates: Evidence for multiple pathways of kinase activation. *Biochem. J.* **268**, 303–308.

Azzi, A., Boscoboinik, D., Chatelain, E., Özer, N. K., and Stäuble, B. (1993). d-alpha-tocopherol control of cell proliferation. *Mol. Aspects Med.* **14**, 265–271.

Azzi, A., Boscoboinik, D., and Hensey, C. (1992). The protein kinase C family. *Eur. J. Biochem.* **208**, 547–557.

Azzi, A., Boscoboinik, D., Marilley, D., Özer, N. K., and Stäuble, B., and Tasinato, A. (1995). Vitamin E: A sensor and an information transducer of the cell oxidation state. *Am. J. Clin. Nutr.* **62**(Suppl.), 1337S–1346S.

Bitman, J., Weyant, J., Wood, D. L., and Wrenn, T. R. (1976). Vitamin E, cholesterol, and lipids during atherogenesis in rabbits. *Lipids* **11**, 449–461.

Bjorkhem, I., Henriksson Freyschuss, A., Breuer, O., Diczfalusy, U., Berglund, L., and Henriksson, P. (1991). The antioxidant butylated hydroxytoluene protects against atherosclerosis. *Arterioscler. Thromb.* **11**, 15–22.

Bocan, T. M., Mueller, S. B., Mazur, M. J., Uhlendorf, P. D., Brown, E. Q., and Kieft, K. A.

(1993). The relationship between the degree of dietary-induced hypercholesterolemia in the rabbit and atherosclerotic lesion formation. *Atherosclerosis* **102**, 9–22.

Bornancin, F., and Parker, P. J. (1996). Phosphorylation of threonine 638 critically controls the dephosphorylation and inactivation of protein kinase Cα. *Curr. Biol.* **6**, 1114–1123.

Borner, C., Filipuzzi, I., Wartmann, M., Eppenberger, U., and Fabbro, D. (1989). Biosynthesis and posttranslational modifications of protein kinase C in human breast cancer cells. *J. Biol. Chem.* **264**, 13902–13909.

Bornfeldt, K. E., Raines, E. W., Nakano, T., Graves, L. M., Krebs, E. G., and Ross, R. (1994). Insulin-like growth factor-I and platelet-derived growth factor-BB induce directed migration of human arterial smooth muscle cells via signaling pathways that are distinct from those proliferation *J. Clin. Invest.* **93**, 1266–1274.

Boscoboinik, D., Özer, N. K., Moser, U., and Azzi, A. (1995). Tocopherols and 6-hydroxy-chroman-2-carbonitrile derivatives inhibit vascular smooth muscle cell proliferation by a nonantioxidant mechanism. *Arch. Biochem. Biophys.* **318**, 241–246.

Boscoboinik, D., Szewczyk, A., and Azzi, A. (1991a). Alpha-tocopherol (vitamin E) regulates vascular smooth muscle cell proliferation and protein kinase C activity. *Arch. Biochem. Biophys.* **286**, 264–269.

Boscoboinik, D., Szewczyk, A., Hensey, C., and Azzi, A. (1991b). Inhibition of cell proliferation by alpha-tocopherol: Role of protein kinase C. *J. Biol. Chem.* **266**, 6188–6194.

Boscoboinik, D. E., Chatelain, E., Bartoli, G. M., Stäuble, B., and Azzi, A. (1994). Inhibition of protein kinase C activity and vascular smooth muscle cell growth by d-alpha-tocopherol. *Biochim. Biophys. Acta* **1224**, 418–426.

Carew, T. E., Schwenke, D. C., and Steinberg, D. (1987). Antiatherogenic effect of probucol unrelated to its hypocholesterolemic effect: Evidence that antioxidants in vivo can selectively inhibit low density lipoprotein degradation in macrophage-rich fatty streaks and slow the progression of atherosclerosis in the Watanabe heritable hyperlipidemic rabbit. *Proc. Natl. Acad. Sci. USA* **84**, 7725–7729.

Castellot, J. J., Jr., Pukac, L. A., Caleb, B. L., Wright, T. C., Jr., and Karnovsky, M. J. (1989). Heparin selectively inhibits a protein kinase C-dependent mechanism of cell cycle progression in calf aortic smooth muscle cells [published erratum appears in *J. Cell Biol.* 1990 Mar; **110**(3), 863]. *J. Cell Biol.* **109**, 3147–3155.

Chatelain, E., Boscoboinik, D. O., Bartoli, G. M., Kagan, V. E., Gey, F. K., Packer, L., and Azzi, A. (1993). Inhibition of smooth muscle cell proliferation and protein kinase C activity by tocopherols and tocotrienols. *Biochim. Biophys. Acta* **1176**, 83–89.

Clowes, A. W., and Schwartz, S. M. (1985). Significance of quiescent smooth muscle migration in the injured rat carotid artery. *Circ. Res.* **56**, 139–145.

Dutil, E. M., Keranen, L. M., DePaoli-Roach, A. A., and Newton, A. C. (1994). *In vivo* regulation of protein kinase C by trans-phosporylation followed by autophosphorylation. *J. Biol. Chem.* **269**, 29359–29362.

Esterbauer, H., Waeg, G., Puhl, H., Dieber Rotheneder, M., and Tatzber, F. (1992). Inhibition of LDL oxidation by antioxidants. *EXS* **62**, 145–157.

Ferns, G. A., Forster, L., Stewart Lee, A., Konneh, M., Nourooz Zadeh, J., and Anggard, E. E., (1992). Probucol inhibits neointimal thickening and macrophage accumulation after balloon injury in the cholesterol-fed rabbit. *Proc. Natl. Acad. Sci. USA* **89**, 11312–11316.

Gey, K. F., (1990). The antioxidant hypothesis of cardiovascular disease: Epidemiology and mechanisms. *Biochem. Soc. Trans.* **18**, 1041–1045.

Gey, K. F., Moser, U. K., Jordan, P., Stahelin, H. B., Eichholzer, M., and Ludin, E. (1993). Increased risk of cardiovascular disease at suboptimal plasma concentrations of essential antioxidants: An epidemiological update with special attention to carotene and vitamin C. *Am. J. Clin. Nutr.* **57**, 787S–797S.

Godfried, S. L., Combs, G. F., Jr., Saroka, J. M., and Dillingham, L. A. (1989). Potentiation of atherosclerotic lesions in rabbits by a high dietary level of vitamin E. *Br. J. Nutr.* **61**, 607–617.

Konneh, M. K., Rutherford, C., Li, S.-R., Änggård, E. E., and Ferns, G. A. A. (1995). Vitamin E inhibits the intimal response to balloon catheter injury in the carotid artery of the cholesterol-fed rat. *Atherosclerosis* **113**, 29–39.

Mantha, S. V., Prasad, M., Kalra, J., and Prasad, K. (1993). Antioxidant enzymes in hypercholesterolemia and effects of vitamin E in rabbits. *Atherosclerosis* **191**, 135–144.

Matsumoto, H., and Sasaki, Y. (1989). Staurosporine, a protein kinase C inhibitor interferes with proliferation of arterial smooth muscle cells. *Biochem. Biophys. Res. Commun.* **158**, 105–109.

Morel, D. W., De la Llera-Moya, M., and Friday, K. E. (1994). Treatment of cholesterol-fed rabbits with dietary vitamins E and C inhibits lipoprotein oxidation but not development of atherosclerosis *J. Nutr.* **124**, 2123–2130.

Muller, R., Mumberg, D., and Lucibello, F. C. (1993). Signals and genes in the control of cell-cycle progression. *Biochim. Biophys. Acta* **1155**, 151–179.

Newby, A. C., Lim, K., Evans, M. A., Brindle, N. P., and Booth, R. F. (1995). Inhibition of rabbit aortic smooth muscle cell proliferation by selective inhibitors of protein kinase C. *Br. J. Pharmacol.* **114**, 1652–1656.

Newton, A. C. (1995). Protein kinase C: Structure, function, and regulation. *J. Biol. Chem.* **270**, 28495–28498.

Nierenberg, D. W., and Nann, S. L. (1992). A method for determining concentrations of retinol, tocopherol, and five carotenoids in human plasma and tissue samples. *Am. J. Clin. Nutr.* **56**, 417–426.

Özer, N. K., Boscoboinik, D., and Azzi, A. (1995). New roles of low density lipoproteins and vitamin E in the pathogenesis of atherosclerosis. *Biochem. Mol. Biol. Int.* **35**, 117–124.

Özer, N. K., Palozza, P., Boscoboinik, D., and Azzi, A. (1993). d-alpha-Tocopherol inhibits how density lipoprotein induced proliferation and protein kinase C activity in vascular smooth muscle cells. *FEBS Lett.* **322**, 307–310.

Özer, N. K., Sirikci, Ö., Taha, S., San, T., Moser, U., and Azzi, A. (1997). Effect of vitamin E and probucol on dietary cholesterol-induced atherosclerosis in rabbits. *Free Radic. Biol. Med.* **24**, 226–233.

Pryor, A. W., Cornicelli, J. A., Devall, L. J., Tait, B., Trivedi, B. K., Witiak, D. T., and Wu, M. (1993). A rapid screening test to determine the antioxidant potencies of natural and synthetic antioxidants. *J. Org. Chem.* **58**, 3521–3532.

Raines, E. W., and Ross, R. (1993). Smooth muscle cells and the pathogenesis of the lesions of atherosclerosis. *Br. Heart J.* **69**, S30–S37.

Riemersma, R. A., Wood, D. A., Macintyre, C. C., Elton, R. A., Gey, K. F., and Oliver, M. (1991). Risk of angina pectoris and plasma concentrations of vitamins A, C, and E and carotene (see comments). *Lancet* **337**, 1–5.

Rimm, E. B., Stampfer, M. J., Ascherio, A., Giovannucci, E., Colditz, G. A., and Willett, W. C. (1993). Vitamin E consumption and the risk of coronary heart disease in men (see comments). *N. Engl. J. Med.* **328**, 1450–1456.

Ross, R. (1993). The pathogenesis of atherosclerosis: A perspective for the 1990s. *Nature* **362**, 801–809.

Sirikci, Ö., Özer, N. K., and Azzi, A. (1996). Dietary cholesterol-induced changes of protein kinase C and the effect of vitamin E in rabbit aortic smooth muscle cells. *Atherosclerosis* **126**, 253–263.

Stahelin, H. B., Eichholzer, M., and Gey, K. F. (1992). Nutritional factors correlating with cardiovascular disease: Results of the Basel study. *Bibl. Nutr. Diet.* **49**, 24–35.

Stampfer, M. J., Hennekens, C. H., Manson, J. E., Colditz, G. A., Rosner, B., and Willett, W. C. (1993). Vitamin E consumption and the risk of coronary disease in women. *N. Engl. J. Med.* **328**, 1444–1449.

Stein, Y., Stein, O., Delplanque, B., Fesmire, J. D., Lee, D. M., and Alaupovic, P. (1989). Lack of effect of probucol on atheroma formation in cholesterol-fed rabbits kept at comparable plasma cholesterol levels. *Atherosclerosis* **75**, 145–155.

Steinberg, D. (1995). Clinical trials of antioxidants in atherosclerosis: Are we doing the right thing? *Lancet* **346**, 36–38.

Stephens, N. G., Parsons, A., Schofield, P. M., Kelly F., Cheeseman, K., Mitchinson, M. J., and Brown, M. J. (1996). Randomised controlled trial of vitamin E in patients with coronary disease: Cambridge Heart Antioxidant Study (CHAOS). *Lancet* **347**, 781–786.

Wilson, R. B., Middleton, C. C., and Sun, G. Y. (1978). Vitamin E, antioxidants and lipid peroxidation in experimental atherosclerosis of rabbits. *J. Nutr.* **198**, 1858–1867.

6 A New Function for Selenoproteins: Peroxynitrite Reduction

Helmut Sies, Lars-Oliver Klotz, Victor S. Sharov, Annika Assmann, and Karlis Briviba

Institut für Physiologische Chemie I und
Biologisch-Medizinisches Forschungszentrum
Heinrich-Heine-Universität Düsseldorf
D-40001 Düsseldorf, Germany

Cellular defense against excessive peroxynitrite generation is required to protect against DNA strand breaks and mutations and against interference with protein tyrosine-based signaling and other protein functions due to the formation of 3-nitrotyrosine. A role of selenium-containing enzymes catalyzing peroxynitrite reduction has been demonstrated. Glutathione peroxidase (GPx) protected against the oxidation of dihydrorhodamine 123 (DHR) by peroxynitrite more effectively than ebselen [2-phenyl-1,2-benzisoselenazol-3(2H)-one], a selenoorganic compound exhibiting a high second-order rate constant for the reaction with peroxynitrite, $2 \times 10^6 \ M^{-1}sec^{-1}$. The maintenance of protection by GPx against peroxynitrite requires GSH as the reductant. Similarly, selenomethionine but not selenomethionine oxide exhibited inhibition of rhodamine 123 formation from DHR caused by peroxynitrite.

In steady-state experiments, in which peroxynitrite was infused to maintain a 0.2 μM concentration, GPx in the presence of GSH, but neither GPx nor GSH alone, effectively inhibited the hydroxylation of benzoate by peroxynitrite. Under these steady-state conditions, peroxynitrite did not cause loss of "classical" GPx activity. GPx, like selenomethionine, protected against protein 3-nitrotyrosine formation in human fibroblast lysates shown in Western blots. The formation of nitrite rather than nitrate from peroxynitrite was enhanced by GPx, ebselen, or selenomethionine. Selenoxides can be reduced effectively by glutathione, establishing a biological line of defense against peroxynitrite.

The novel function of GPx as a peroxynitrite reductase may extend to other selenoproteins containing selenocysteine or selenomethionine.

Work on organotellurium compounds has revealed peroxynitrite reductase activity as well. Inhibition of dihydrorhodamine 123 oxidation correlated well with the GPx-like activity of a variety of diaryl tellurides.

INTRODUCTION

Peroxynitrite is a biological oxidant generated by endothelial cells, Kupffer cells, neutrophils, and macrophages (for a review see Beckman, 1996). Peroxynitrite is a mediator of toxicity in inflammatory processes with strong oxidizing properties toward biological molecules, including sulfhydryls, ascorbate, lipids, amino acids, and nucleotides, and it can cause strand breaks in DNA. Free or protein-bound tyrosine residues and other phenolics can be nitrated by peroxynitrite. Protein tyrosine nitration may interfere with phosphorylation/dephosphorylation signaling, and the *in vivo* occurrence of protein nitration in the human has been demonstrated (Beckman *et al.,* 1994; MacMillan-Crow *et al.,* 1996).

Peroxynitrite ($ONOO^-$) is a relatively stable species as compared with free radicals, but peroxynitrous acid (ONOOH) decays with a rate constant of $1.3 \ sec^{-1}$. The selenium-containing compound, ebselen, (Masumoto and Sies, 1996a) and its main metabolite *in vivo*, 2-(methylseleno)benzanilide (Masumoto and Sies, 1996b), react with peroxynitrite very efficiently. Ebselen, selenocystine, and selenomethionine protected DNA from single-strand break formation caused by peroxynitrite more effectively than their sulfur-containing analogs (Roussyn *et al.,* 1996). Furthermore, these selenocompounds were protective in model oxidation and nitration reactions mediated by peroxynitrite (Briviba *et al.,* 1996). Ebselen is known as a mimic of the GPx reaction. It has been hypothesized that its newly found reactivity with peroxynitrite mimics a so far undescribed peroxynitrite reductase activity of selenoproteins (Sies and Masumoto, 1997). Evidence has established a protective function for GPx, ebselen, and selenomethionine against peroxynitrite (Assmann *et al.,* 1998; Sies *et al.,* 1997b).

MATERIALS AND METHODS

Reagents

Glutathione peroxidase from bovine erythrocytes was from Calbiochem (La Jolla, CA). Seleno-*dl*-methionine, glutathione, and dithiothreitol were obtained from Sigma (Deisenhofen, Germany). Ebselen, 2-phenyl-1,2-benzisoselenazol-3*(2H)*-one, and its derivatives, 2-(methylseleno)benzanilide, and ebsulfur, 2-phenyl-1,2-benzisothiazol-3*(2H)*-one, were kindly provided by Rhône-Poulenc-Rorer (Cologne, Germany). Organotellurides were a kind gift from Dr. Lars Engman (Uppsala, Sweden). Dihydrorhodamine 123 was from Molecular Probes (Eugene, OR), and rhodamine 123 was from ICN Biomedicals (Aurora, OH). Peroxynitrite was synthesized from potassium superoxide and nitric oxide (Koppenol *et al.*, 1996), and H_2O_2 was eliminated by passage of the peroxynitrite solution over MnO_2 powder. The peroxynitrite concentration was determined spectrophotometrically at 302 nm $\epsilon = 1670\ M^{-1}$ cm^{-1}).

GSH Peroxidase Assay

GPx activity was followed spectrophotometrically at 340 nm. The test mixture contained GSH (1 mM), DTPA (1 mM), glutathione disulfide reductase (0.6 U/ml), and NADPH (0.1 mM) in 0.1 M sodium phosphate, pH 7.3. GPx samples were added to the test mixture at room temperature, and the NADPH oxidation rate was recorded for 1 min. The reaction was started by the addition of *tert*-butyl hydroperoxide (1.2 mM). Activity was calculated from the rate of NADPH oxidation. Carboxymethylation of the selenol in selenocysteine of GPx was carried out as described (Sies *et al.*, 1997b).

Assay of Peroxynitrite-Mediated Oxidation of Dihydrorhodamine 123

The peroxynitrite-mediated oxidation of dihydrorhodamine 123 was followed as described (Kooy *et al.*, 1994; Briviba *et al.*, 1996) using a fluorescence spectrophotometer LS-5 (Perkin-Elmer Co., Norwalk, CT) with excitation and emission wavelengths of 500 and 536 nm, respectively, at room temperature. The fluorescence intensity was related linearly to a rhodamine 123 concentration between 0 and 400 nM. Results are reported as means $\pm$SD ($n = 3$–6) for the final fluorescence intensity minus background fluorescence.

Hydroxylation of Benzoate Caused by Steady-State Infusion of Peroxynitrite

The peroxynitrite-mediated hydroxylation of benzoate was measured as described (Szabo *et al.*, 1997). Peroxynitrite was infused with a micropump at a rate of 175 μl/min from a stock solution of 50 μM under constant mixing with a magnetic stirrer at room temperature into a mixture (1.5 ml) containing benzoate (10 mM) and DTPA (0.1 mM) in 0.5 M potassium phosphate buffer (pH 7.4) For further details see Sies *et al.* (1997b). In control experiments, the peroxynitrite solution was incubated with phosphate buffer at pH 7.4 for 10 min at room temperature to decompose the peroxynitrite before infusion into the reaction mixture.

Western Blot Analysis

After lysis and homogenization by sonication of human skin fibroblasts grown to near confluency and the separation of proteins by SDS–PAGE, Western blots using a mouse monoclonal antinitrotyrosine antibody (kindly provided by J. S. Beckman, Birmingham, AL) were performed essentially as described (MacMillan-Crow *et al.*, 1996). The exposure to peroxynitrite (200 μM) was by injection into the cell lysate (1 mg protein/ml) under vortexing. GPx, selenomethionine, ebselen, or bis(4-aminophenyl)telluride was present from the beginning. After the usual processing and incubation with a secondary goat antimouse antibody coupled to alkaline phosphatase and appropriate washings, nitrated proteins were detected using a chemiluminescent substrate (Starlight, ICN, Costa Mesa, CA). The reaction was abolished in the presence of 10 mM 3-nitrotyrosine as ascertained in dot blots.

RESULTS

Glutathione Peroxidase Protects Against Peroxynitrite

Dihydrorhodamine 123 Oxidation by Peroxynitrite

The peroxynitrite-mediated oxidation of dihydrorhodamine 123 to fluorescent rhodamine 123 is an efficient and selective probe of peroxynitrite production in model systems (Kooy *et al.*, 1994). When peroxynitrite (100 nM) was added to 500 nM dihydrorhodamine 123, about 10 nM rhodamine 123 was formed. In the experiments shown in Fig. 1, this value is set to 100%. As shown in Fig. 1, the addition of a GPx preparation from bovine erythrocytes up to 200 nM had no effect on rhodamine 123 formation (open circles). However, in the presence of GSH at the low concentration of 1 μM, GPx exhibited a pronounced inhibition of rhodamine 123 formation

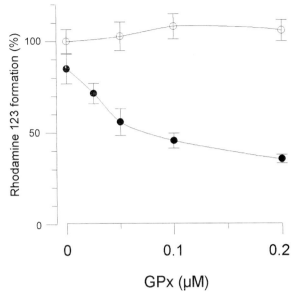

Figure 1   Protection by GSH peroxidase (GPx) against dihydrorhodamine 123 oxidation caused by peroxynitrite. Peroxynitrite (100 nM) was added to 0.5 μM dihydrorhodamine 123 and different concentrations of GPx without (○) and with 1 μM GSH (●) in 0.1 M phosphate buffer, 0.1 mM DTPA, pH 7.3, under intense stirring at room temperature. Modified from Sies *et al.* (1997b).

(solid circles). It can be noted at the *y* axis in Fig. 1 that the addition of 1 μM GSH alone, without GPx, led to a 15% loss of rhodamine 123 production. The half-maximal inhibitory concentration of GPx is 150 nM (Table I).

When GPx was carboxymethylated by iodoacetate (solid diamonds), the glutathione peroxidase activity was lost (Fig. 2A), whereas its protective activity against the peroxynitrite-mediated oxidation of dihydrorhodamine 123 was retained or even slightly enhanced (Fig. 2B).

Hydroxylation of Benzoate under Steady-State Infusion of Peroxynitrite

A suitable detector system for examining steady-state conditions is given by the hydroxylation of benzoate (Szabo *et al.*, 1997). In the experiments shown in Fig. 3, peroxynitrite was infused with a micropump to give a steady-state concentration of 0.2 μM over 3 min. The cumulative peroxynitrite concentration was 13 μM. The hydroxylation of benzoate in the controls is shown as solid squares in Fig. 3; 330 nM GPx (not shown), 10 μM GSH (open squares), or 20 μM GSH (not shown) alone had only a small protective effect. However, in the presence of 10 μM GSH (solid

Table I   Half-Maximal Inhibitory Concentrations of Selenoproteins and Some Selenocompounds in Peroxynitrite-Mediated Oxidation of Dihydrorhodamine 123[a]

Compound	Half-maximal inhibitory concentration (μM)[b]
PHGPx[c]	0.05
Carboxymethylated GPx	0.1
GPx	0.15
Ebselen	0.2
2-(Methylseleno)benzanilide	0.8
Ebselen selenoxide	100
GSH	12
Selenomethionine	0.3
Methionine	20
Methionine selenoxide	>100
Sodium selenite	$>10^4$

[a] Data from Briviba *et al.* (1996) and Sies *et al.* (1997b).
[b] Concentration of the compound obtaining half-maximal inhibition.
[c] Phospholipid hydroperoxide GPx. Unpublished work, jointly with F. Ursini and M. Maiorino.

circles), GPx completely suppressed benzoate hydroxylation until 5 μM peroxynitrite had been infused, i.e., within the first minute of infusion in Fig. 3. Likewise, in the presence of 20 μM GSH, the effect of 10 μM peroxynitrite infused during 2 min was abolished by the same amount of GPx (solid triangles), and again the GSH/peroxynitrite ratio necessary for the inactivation of peroxynitrite in the presence of GPx was 2/1. These data (Sies *et al.*, 1997b) establish that GPx inactivates peroxynitrite in a catalytic reaction at the stoichiometry known for that of hydroperoxide reduction, i.e., the "classical" GPx reaction.

Table II shows that 330 nM GPx treated with peroxynitrite under steady-state conditions used in the experiments of Fig. 3 (cumulative concentration 13 μM) maintained the capability to reduce *tert*-butyl hydroperoxide, i.e., that the "classical" GPx activity was retained on exposure to a steady-state concentration of peroxynitrite. Further, exposure of 150 nM GPx in phosphate buffer at pH 7.3 to a bolus addition of peroxynitrite up to 30 μM did not detectably change the capability of GPx to reduce *tert*-butyl hydroperoxide.

Nitrite Formation from Peroxynitrite

As the spontaneous decay of peroxynitrite generates nitrate, the increase in the yield of nitrite rather than nitrate in the presence of selenocompounds is a measure of peroxynitrite reduction. The increase in the formation of ni-

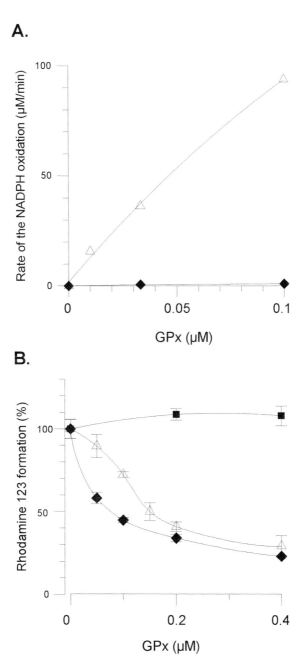

Figure 2 GSH peroxidase activity (A) and peroxynitrite-mediated rhodamine 123 formation (B) of GPx without (△) or with (◆) carboxymethylation of the selenocysteine in GPx by iodoacetate. Oxidation of dihydrorhodamine 123 was as in Fig. 1. In B, the effect of dialysis buffer after GPx dialysis (■) is presented as a contol. Modified from Sies *et al.* (1997b).

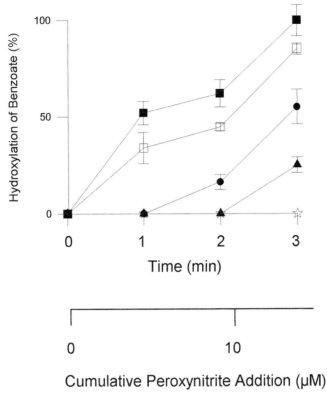

Figure 3 Protection by GSH peroxidase against hydroxylation of benzoate caused by a steady-state infusion of peroxynitrite. Peroxynitrite (cumulative concentration 13 μM) was infused over 3 min from a stock solution to yield a 0.2 μM steady-state concentration. The reaction mixture contained 10 mM benzoate and 0.1 mM DTPA in 0.5 M potassium phosphate buffer (pH 7.4) (■). Before peroxynitrite infusion, 10 μM GSH (□), 20 μM GSH (not shown), 330 nM GPx alone (not shown), 330 nM GPx in the presence of 10 μM GSH (●), or 330 nM GPx in the presence of 20 μM GSH (▲) was added. Infusion of peroxynitrite after decomposition at pH 7.4 did not cause hydroxylation of benzoate (☆). Modified from Sies *et al.* (1997b).

trite from peroxynitrite by GPx and GSH in the steady-state experiments shown in Fig. 3 is presented in the right-hand column of Table II; when the nitrite measurement was carried out without benzoate, results were similar. Correspondingly, the levels of nitrate were lowered (data not shown).

Ebselen

Ebselen, an organoselenium compound with GPx-like activity, inhibits the peroxynitrite-mediated oxidation of dihydrorhodamine 123 with a half-

Table II   GSH Peroxidase (GPx) Activity and Formation of Nitrite during Steady-State Exposure to Peroxynitrite[a]

Addition	GPx activity (U/l)		Formation of nitrite (μM) after exposure to ONOO$^-$
	Before exposure to ONOO$^-$	After exposure to ONOO$^-$	
GSH (20 μM)	0	0	0.3 ± 0.3
GPx (330 nM)	393 ± 12	386 ± 10	N.D.[b]
GPx plus GSH	390 ± 17	389 ± 16	4.0 ± 0.9

[a] The activity of GSH peroxidase was assayed as the rate of oxidation of NADPH (μM/min) before and after infusion of peroxynitrite to give a steady-state concentration of approximately 200 nM; at 3 min the cumulative concentration of peroxynitrite was 13 μM. The reaction mixture contained benzoate (10 mM) and DTPA (0.1 mM) in phosphate buffer (0.5 M) at pH 7.4 (see Fig. 3). GSH and GPx were present as indicated. The rates of spontaneous NADPH oxidation and spontaneous formation of nitrite were subtracted. Data are expressed as means ±SD (n = 3–6). Modified from Sies *et al.* (1997b).
[b] Not determined.

maximal inhibitory concentration of 0.2 μM, whereas the oxidation product, ebselen selenoxide, is practically ineffective (Table I).

Selenomethionine

Similarly, selenomethionine exhibits efficient protection against the peroxynitrite-mediated oxidation of dihydrorhodamine, whereas methionine is less effective (Table I). The oxidation of selenomethionine by peroxynitrite leads to the formation of methionine selenoxide, which does not protect against dihydrorhodamine 123 oxidation. Selenomethionine oxide can be reduced back to selenomethionine by thiols, i.e., GSH (Assmann *et al.*, 1998).

Selenomethionine generates a pronounced increase (up to 70% at 0.5 mM) in nitrite formation when 100 μM peroxynitrite was employed (Sies *et al.*, 1997b). This indicates successful competition with the spontaneous decay to nitrate.

Organotellurium Compounds

The activities of selenoorganic compounds in inhibiting dihydrorhodamine 123 oxidation, benzoate hydroxylation, and 4-hydroxyphenylacetate nitration are also found with a variety of organotelluric compounds (Briviba *et al.*, 1998). Regarding the compounds tested, bis(4-aminophenyl)-

telluride offered the most pronounced protection against dihydrorhodamine 123 oxidation, being 11 times more effective than selenomethionine.

When peroxynitrite was infused to maintain a steady-state concentration, bis(4-aminophenyl)telluride, in the presence of GSH, but neither bis(4-aminophenyl)telluride nor GSH alone, effectively inhibited the peroxynitrite-mediated hydroxylation of benzoate. The capabilities of protecting against peroxynitrite-induced oxidation and nitration reactions of a series of organotellurium compounds correlate with their glutathione peroxidase activity (Briviba *et al.*, 1998).

Regarding nitration reactions, bis(4-hydroxyphenyl)telluride was most effective in inhibiting 4-hydroxyphenylacetate nitration with a half-maximal inhibitory concentration about three to four times lower than that of selenomethionine or ebselen.

Protein Nitration in Cells

Western blots from human fibroblast lysates exposed to peroxynitrite using a monoclonal anti-3-nitrotyrosine antibody showed several bands of nitrated protein, e.g., 25 and 41 kDa, assigned to Mn-superoxide dismutase and actin, respectively (Sies *et al.*, 1997b). Reduced GPx, but not oxidized (untreated) GPx, and selenomethionine as well as bis(4-aminophenyl)telluride were protective against tyrosine nitration by peroxynitrite. Ebselen also protected, yet less efficiently. From such data (Fig. 4), the relative efficiencies in blocking protein nitration caused by peroxynitrite are found to be bis(4-aminophenyl)telluride > selenomethionine > ebselen. Protein nitration was abolished completely by 30 μM of reduced GPx (Fig. 4). This

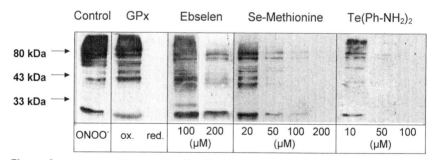

Figure 4 Nitration of human skin fibroblast lysates (1 mg/ml) by 200 μM peroxynitrite (control) as shown in Western blots with anti-3-nitrotyrosine monoclonal primary antibodies employed. Addition of the compounds as indicated before reacting with peroxynitrite led to different extents of inhibition of protein tyrosine nitration. Modified from Sies *et al.* (1997b) and Briviba *et al.* (1998).

is more effective than any of the low molecular weight compounds tested. As GPx is a tetramer, however, 30 μM of GPx corresponds to 120 μM of selenol, which still makes GPx more effective than ebselen and approximately as effective as selenomethionine.

DISCUSSION

Peroxynitrite Reductase

Selenoproteins carry out a variety of catalytic functions, many of which are redox reactions. A novel function for selenoproteins has been reported, the reduction of peroxynitrite (Sies *et al.*, 1997a,b). Studies were prompted by the observation of a very efficient reduction of peroxynitrite by ebselen (Masumoto and Sies, 1996a), exhibiting the highest second-order rate constant for a low molecular weight compound with peroxynitrite known so far, 2.0×10^6 $M^{-1}sec^{-1}$ (Masumoto *et al.*, 1996). In analogy to the reaction cycle for ebselen, Scheme 1A presents the proposed sequence: in the first step, the selenocysteine, probably as the selenolate, reacts with peroxynitrite to be oxidized to the corresponding selenenic acid, yielding nitrite (Table I; Fig. 3). However, peroxynitrous acid may also react to yield nitrous acid. The subsequent two steps in the reaction cycle are facile regeneration reactions at the expense of reducing equivalents provided by GSH in cells, known from the extensive work on GPx (see Flohé, 1989). Regarding the chemical mechanism, it might be concluded that the selenolate form of the selenocysteine residue is required. However, a selenol moiety is not strictly necessary for peroxynitrite reductase activity, in contrast to the GSH peroxidase action, because the carboxymethylated selenium derivative maintained activity (Fig. 2). This is in accord with the high rate constant obtained for 2-(methylseleno)benzanilide (Masumoto and Sies, 1996b) and for selenomethionine (Padmaja *et al.*, 1996) (Scheme 1B).

Physiological Significance

There is protection by selenoorganic compounds against peroxynitrite-induced single-strand breaks in plasmid DNA or base modifications sensitive to Fpg protein in bacteriophage DNA (Roussyn *et al.*, 1996; Epe *et al.*, 1996). It is possible that selenomethionine and selenocysteine residues in proteins in general may carry out similar functions, i.e., that selenoproteins or selenopeptides might have a biological function as a defense line against peroxynitrite (Briviba *et al.*, 1996; Sies and Masumoto, 1997; Sies *et al.*,

A.

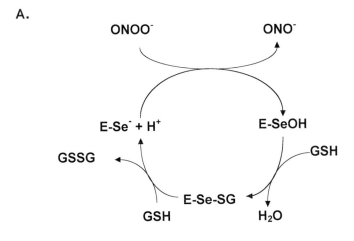

B.

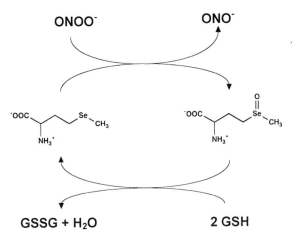

Scheme 1 Proposed catalytic mechanism of selenoperoxidases (A), selenomethionine (B), and diaryl tellurides, as exemplified by bis(4-aminophenyl)telluride (C), in the reduction of peroxynitrite to nitrite (or peroxynitrous acid to nitrous acid). The mechanism for selenoperoxidases is based on that established for GSH peroxidases and the mimic, ebselen (Flohé, 1989; Sies, 1993; Maiorino *et al.* 1995), which use ROOH and ROH as substrate and product, respectively. See Sies *et al.* (1997b) and Briviba *et al.* (1998).

1997b). A number of different selenopeptides and selenoproteins, many of them with still unknown function, have been described *in vivo*.

Although the 100- to 1000-fold higher second-order reaction rate constants of the selenium-containing compounds as compared to sulfur analogs

C.

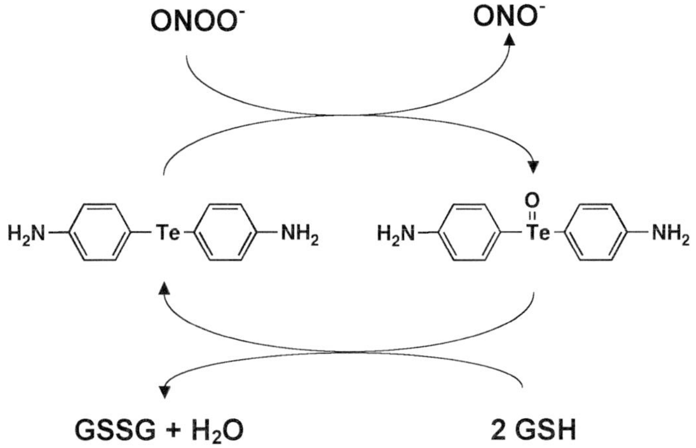

Scheme 1 Continued

make for a kinetic advantage, it should be considered that there are multiple other defense mechanisms against peroxynitrite in the organism. For example, there is prevention of the formation of peroxynitrite by control of nitric oxide synthase and by control of the level of nitric oxide by oxyhemoglobin and other binding sites, as well as control of superoxide levels by superoxide dismutase. Second, there are reactions of peroxynitrite, once formed, with other compounds such as ascorbate (Whiteman and Halliwell, 1996), GSH (Quijano *et al.*, 1997), or CO_2 (Zhu *et al.*, 1992; Gow *et al.*, 1996), all of which share in the modulation of potentially deleterious reactions caused by peroxynitrite.

Organoselenium Compounds

A special feature of the peroxynitrite reductase activity of selenoproteins resides in the catalytic nature and in the high efficiency of the reaction. The capacity of protecting against peroxynitrite-dependent reactions can be maintained in the presence of thiol equivalents at micromolar concentrations. While the reduction of methionine sulfoxide requires the enzymatic activity of methionine sulfoxide reductases (Levine *et al.*, 1996), glutathione is effective in reducing selenomethionine oxide (Assmann *et al.*, 1997), suggesting that the nonenzymatic regeneration of organoselenium compounds is sufficient. Thus, there is a low molecular weight defense system against peroxynitrite maintained by selenosubstituted methionine and cysteine residues in proteins using glutathione (Assmann *et al.*, 1998).

Organotellurium Compounds

As one proceeds down group 16 of the periodic table, organotellurium compounds with glutathione peroxidase-like activity as well as organosulfur or organoselenium compounds protect against oxidation and nitration reactions caused by peroxynitrite (Briviba *et al.*, 1998). The previously observed GPx-like activity of diaryl tellurides (Andersson *et al.*, 1993) has been ascribed to the ready oxidation of the heteroatom from the divalent to the tetravalent telluroxide state by hydrogen peroxide or organic hydroperoxides. As with GPx, regeneration of the active species then occurs via thiol reduction with disulfide formation, thus suggesting that diaryl tellurides act as scavengers of peroxynitrite by an oxygen transfer mechanism similar to that observed with hydroperoxides (Scheme 1C).

ACKNOWLEDGMENTS

We are grateful to Dr. Hiroshi Masumoto for helpful comments. This study was supported by the Deutsche Forschungsgemeinschaft, SFB 503, Project B1, and by the National Foundation for Cancer Research, Bethesda. V. S. Sharov was a Research Fellow of the Alexander von Humboldt Foundation, Bonn, Germany.

REFERENCES

Andersson, C. M., Hallberg, A., Brattsand, R., Cotgreave, I. A., Engman, L., and Persson, J. (1993). Glutathione peroxidase-like activity of diaryl tellurides. *Bioorg. Med. Chem. Lett.* **3**, 2553–2558.

Assmann, A., Briviba, K., and Sies, H. (1998). Reduction of methionine selenoxide to selenomethionine by glutathione. *Arch. Biochem. Biophys.* **349**, 201–203.

Beckman, J. S. (1996). The physiological and pathological chemistry of nitric oxide. *In* "Nitric Oxide: Principles and Actions" (J. Lancaster, ed.), pp. 1–82. Academic Press, San Diego.

Beckman, J. S., Ye, Y. Z., Anderson, P. G., Chen, J., Accavitti, M. A., Tarpey, M. M., and White, C. R. (1994). Extensive nitration of protein tyrosines in human atherosclerosis detected by immunohistochemistry. *Biol. Chem. Hoppe-Seyler* **375**, 81–88.

Briviba, K., Roussyn, I., Sharov, V. S., and Sies, H. (1996). Attenuation of oxidation and nitration reactions of peroxynitrite by selenomethionine, selenocystine and ebselen. *Biochem. J.* **319**, 13–15.

Briviba, K., Tamler, R., Klotz, L. O., Engman, L., Cotgreave, I. A., and Sies, H. (1998). Protection by organotellurium compounds against peroxynitrite-mediated oxidation and nitration reactions. *Biochem. Pharmacol.* **55**, 817–823.

Epe, B., Ballmaier, D., Roussyn, I., Briviba, K., and Sies, H. (1996). DNA damage by peroxynitrite characterized with DNA repair enzymes. *Nucleic Acid Res.* **24**, 4105–4110.

Flohé, L. (1989). The selenoprotein glutathione peroxidase. *In* "Glutathione: Chemical, Biochemical and Medical Aspects" (D. Dolphin, R. Poulson, and O. Avramovic, eds.), Part A, pp. 643–731. Wiley Interscience, New York.

Gow, A. J., Duran, D., Malcolm, S., and Ischiropoulos, H. (1996). Effects of peroxynitrite-induced protein modifications on tyrosine phosphorylation and degradation *FEBS Lett.* **385**, 63–66.

Kooy, N. W., Royall, J. A., Ischiropoulos, H., and Beckman, J. S. (1994). Peroxynitrite-mediated oxidation of dihydrorhodamine 123. *Free Radic. Biol. Med.* **16**, 149–156.

Koppenol, W. H., Kissner, R., and Beckman, J. S. (1996). Syntheses or peroxynitrite: To go with the flow or on solid grounds? *Methods Enzymol.* **269**, 296–302.

Levine, R. L., Mosoni, L., Berlett, B. S., and Stadtman, E. R. (1996). Methionine residues as endogenous antioxidants in proteins. *Proc. Natl. Acad. Aci. USA* **93**, 15036–15040.

MacMillan-Crow, L. A., Crow, J. P., Kerby, J. D., Beckman, J. S., and Thompson, J. A. (1996). Nitration and inactivation of manganese superoxide dismutase in chronic rejection of human renal allografts. *Proc. Natl. Acad. Sci. USA* **93**, 11853–11858.

Maiorino, M., Aumann, R., Brigelius-Flohé, R., Doria, D., Heuvel, V. D., J., McCarthy, J., Roveri, A., Ursini, F., and Flohé, L. (1995). Probing the presumed catalytic triad of selenium-containing peroxidases by mutational analysis of phospholipid hydroperoxide glutathione peroxidase (PHGPx). *Biol. Chem. Hoppe-Seyler* **376**, 651–660.

Masumoto, H., Kissner, R., Koppenol, W. H., and Sies, H. (1996). Kinetic study of the reaction of ebselen with peroxynitrite. *FEBS Lett.* **398**, 179–182.

Masumoto, H., and Sies, H. (1996a). The reaction of ebselen with peroxynitrite. *Chem. Res. Toxicol.* **9**, 262–267.

Masumoto, H., and Sies, H. (1996b). The reaction of 2-(methylseleno)benzanilide with peroxynitrite. *Chem. Res. Toxicol.* **9**, 1057–1062.

Padmaja, S., Squadrito, G. L., Lemercier, J. N., Cueto, R., and Pryor, W. A. (1996). Rapid oxidation of DL-selenomethionine by peroxynitrite. *Free Radic. Biol. Med.* **21**, 317–322.

Quijano, C., Alvarez, B., Gatti, R. M., Augusto, O., and Radi, R. (1997). Pathways of peroxynitrite oxidation of thiol groups. *Biochem. J.* **322**, 167–173.

Roussyn, I., Briviba, K., Masumoto, H., and Sies, H. (1996). Selenium-containing compounds protect DNA from single-strand breaks caused by peroxynitrite. *Arch. Biochem. Biophys.* **330**, 216–218.

Sies, H. (1993). Ebselen, a selenoorganic compound as glutathione peroxidase mimic. *Free Radic. Biol. Med.* **14**, 313–323.

Sies, H., and Masumoto, H. (1997). Ebselen as a glutathione peroxidase mimic and as a scavenger of peroxynitrite. *Adv. Pharmacol.* **38**, 229–246.

Sies, H., Masumoto, H., Sharov, V. S., and Briviba, K. (1997a). Defenses against peroxynitrite. *In* "Oxygen Homeostasis and Its Dynamics" (Y. Ishimura, ed.). Springer-Verlag, Tokyo.

Sies, H., Sharov, V. S., Klotz, L. O., and Briviba, K. (1997b). Glutathione peroxidase protects against peroxynitrite-mediated oxidations: A new function for selenoproteins as peroxynitrite reductase. *J. Biol. Chem.* **272**, 27812–27817.

Szabo, C., Ferrer Sueta, G., Zingarelli, B., Southan, G. J., Salzman, A. L., and Radi, R. (1997). Mercaptoethylguanidine and guanidine inhibitors of nitric-oxide synthase react with peroxynitrite and protect against peroxynitrite-induced oxidative damage. *J. Biol. Chem.* **272**, 9030–9036.

Whiteman, M., and Halliwell, B. (1996). Protection against peroxynitrite-dependent tyrosine nitration and alpha 1-antiproteinase inactivation by ascorbic acid: A comparison with other biological antioxidants. *Free Radic. Res.* **25**, 275–283.

Zhu, L., Gunn, C., and Beckman, J. S. (1992). Bactericidal activity of peroxynitrite. *Arch. Biochem. Biophys.* **298**, 452–457.

7 Selenium Peroxidases in Mammalian Testis

Matilde Maiorino

Dipartimento di Chimica Biologica
Viale G. Colombo
3. I- 35121-Padova, Italy

Too much selenium (Se) in the diet is detrimental to animal health. The most toxic compounds are Se inorganic salts and Se methionine, which are found in water and plants. However, low soil content results in deficiency syndromes. In humans, clinical manifestations of deficiency are Keshan's disease, a severe cardiomyopathy, Kashin–Beck disease, an osteoarthropathy (1), and cretinism, when associated with an iodine deficiency (2). The eradication of Keshan disease by dietary Se supplementation (3, 4) further strengthened the correlation between low soil content and the disease. Interestingly, however, other complicating factors, such as viruses, have been implicated to explain the seasonal recurrence of this disease (5, 6). To this end, the induction of virulence in certain viruses by selenodeficiency, as documented for the human Coxsackievirus B3, which becomes virulent after the infection of Se-deficient mice and maintains its virulence in normal animals (7–9), is one of the most important aspects of recent selenium research. This observation might provide a rationale for the evidence that different diseases are due to the deficiency of the same element and it may also be relevant for cancer research. Furthermore, *in vitro* experiments suggest a role for Se in atherosclerosis or aging (10), but convincing epidemiological studies of this aspect are still missing.

Historically, knowledge of the beneficial effects of Se has come from livestock. It is well known that, together with vitamin E, Se supplementation prevents liver necrosis, degeneration of skeletal and cardiac muscles, reduced growth rate, and infertility in cattle (1). This last aspect has been clarified in recent years by well-documented nutritional studies. In selenodeficiency, selenium levels decrease in other organs, but not in testis (11), suggesting that this element may have a peculiar function. Furthermore, with the progression of the deficiency, various degrees of degeneration appear in the seminiferous epithelium, first involving only the mitochondria of spermatids and spermatozoa (12), resulting in a complete disappearance of mature germinal cells (13).

Clearly, *in vivo,* the physiological functions of Se are accounted for by selenocysteine-containing proteins. Only three selenoproteins are characterized in mammalian testis, although at least twice as many have been detected after *in vivo* Se labeling in spermatozoa (14). These are the two glutathione peroxidases—GPx (EC 1.11.1.9) and PHGPx (EC 1.11.1.12)—and selenoprotein P, a protein containing 10–12 selenocysteine residues of unknown function, which has been found to be expressed in the interstitial cells of testis (i.e., Leydig cells) (15).

Regarding the glutathione peroxidase content in testis, immunological studies show that PHGPx is located in germinal cells, not in Leydig or Sertoli cells (16), but that the localization of cGPx in this tissue remains unknown. In a study evaluating the relative content of the activity of these two GPxs in mammalian testis, the response of these enzymes following hormonal manipulation was studied. A gel permeation column of the whole organ distinguished between cGPx and PHGPx activity on the basis of both the specificity of PHGPx for the substrate phosphatidylcholine hydroperoxide (PCOOH) and the different molecular weight. When testes from a prepuberal and adult animal were compared using this approach, a dramatic increase was observed only for the PHGPx peak in the adults, whereas cGPx, which was much lower than PHGPx in the adult, was not changed in respect to prepuberal animals (Fig. 1). Furthermore, the

---

Figure 1   Pattern of elution of the glutathione peroxidases from testes of an early puberal (A) and an adult (B) rat. Decapsulated testes were weighed, and 2 g of tissue was homogenized in 6 ml of 50 mM Tris–HCl, pH 7.5, 0.3 M KCl, 1 mM EDTA, 10% glycerol, 1% Triton X-100, containing 5 mM mercaptoethanol. After freeze-thawing and a brief centrifugation to eliminate unbroken cells, 5 ml of the supernatant, containing 25 mg/ml of proteins, was applied onto a gel permeation column (Superdex 75 prep. grade, Pharmacia) and equilibrated in the homogenization buffer. Fractions of 5 ml were collected and assayed spectrophotometrically in the coupled spectrophotometric test with NADPH, glutathione, and glutathione reductase, as described in Ursini *et al.* (24). In the activity assay, the glutathione concentration was 3 mM, and the final Triton X-100 concentration was adjusted to 0.1%. The reaction started with 16 μM phosphatidylcholine hydroperoxide (PCOOH) or 150 μM H$_2$O$_2$. Note that the analysis of the peroxidase activity with H$_2$O$_2$ in the isolated fraction reveals two peaks. The different MW together with the specific reactivity of the first and second peak with only H$_2$O$_2$ and with both H$_2$O$_2$ and PCOOH, allowed the identification of cGPx and PHGPx activity, respectively. Commercially available red blood cell cGPx (Sigma) and purified pig heart PHGPx (24) were used as MW markers in two independent runs and coeluted, with peaks I and II respectively,

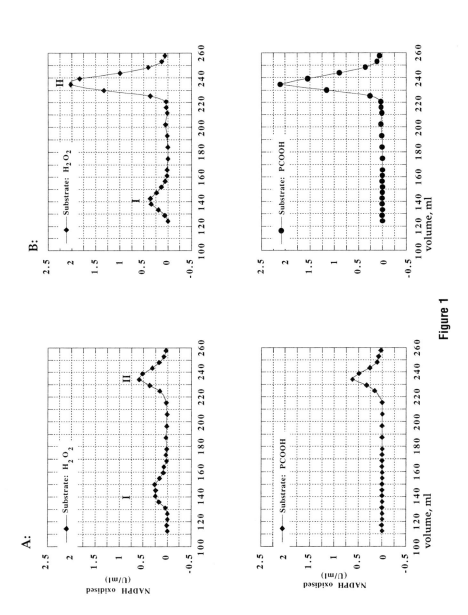

Figure 1

measurement of PHGPx activity in tissues other than testis demonstrated that only in this tissue does the dramatic burst of PHGPx activity occur following puberty, as the specific activity in other tissues is low in prepuberal animal and also remains low in adults (Table I), thus confirming that the selenoprotein PHGPx might have a peculiar role in testis.

The increased PHGPx activity following puberty, as observed in Fig. 1, was due to an increased protein synthesis, as judged by the higher amounts of mRNA and protein detectable, respectively, by Northern and Western blotting, in the adult with respect to prepuberal testis (not shown). This observation prompted us to explore the mechanism of induction of PHGPx in testis, with special regard to the possibility that PHGPx is under hormonal control. The PHGPx gene contains several consensus sequences for steroids (17), and a hormonal control of PHGPx in testis was also suggested by previous experiments, where it was shown that hCG treatment partially restores the drop of activity induced by hypophysectomy (16). For this purpose, rats were treated with ethane dimethane sulfonate (EDS), a compound that selectively destroys Leydig cells, the major androgen-producing cell type in testis (18), and the temporal relationship among Leydig cell destruction and testicular PHGPx activity/mRNA content was studied. The destruction of Leydig cells was monitored as disappearance of Leydig cell marker activity, the nonspecific phenylesterase. As reported in Table II,

Table I   Specific Activity of PHGPx in Prepuberal and Adult Rats[a]

| Organ | nmol NADPH oxidized/min/mg protein | |
	Prepuberal	Adult
Heart	4.8 ± 0.5	5.6 ± 0.7
Brain	5.5 ± 0.4	4.7 ± 0.7
Liver	6.5 ± 1.0	7.0 ± 0.9
Testis	60.0 ± 2.0	130.8 ± 6.0

[a] Organs were processed as indicated in the legend to Fig. 1. PHGPx activity was measured on the homogenate, following freezing–thawing and centrifugation, in the coupled spectrophotometric test with NADPH, glutathione (3 mM), glutathione reductase, and the specific substrate phosphatidylcholine hydroperoxide (16 μM). Protein determination was performed according to Besandoun and Weinstein (26).

Table II　Effect of Treatment with Ethane Dimethane Sulfonate (EDS) on Testicular Total Nonspecific Phenyl Esterase, cGPx, PHGPx Content, and PHGPx mRNA[a]

Days after EDS	Phenyl esterase	Total activity (%)		PHGPx/actin (cpm, %)
		cGPx	PHGPx	
0	100	100	100	100
1	87	152	99	100
3	7	148.6	102	99
7	7	122	92	102
14	7	109	73	85
28	19	110	60	71

[a] Testes from adult rats, receiving a single dose of EDS at day 0 [75 mg/kg, prepared according to Jackson and Jackson (18)], were processed for column chromatography at various times following the treatment, as indicated in the first column. The isolated fractions were tested spectrophotometrically for nonspecific phenyl esterase activity, with p-nitrophenyl acetate as the substrate, as reported by Meyer *et al.* (27), and glutathione peroxidases activity, with H_2O_2 as the substrate (all conditions as in Fig. 1). The area of the peaks of phenyl esterase activity, eluting in the void volume of the column, and of cGPx and PHGPx, eluting as in Fig. 1, were calculated, and values were expressed as the percentage of the corresponding peak in untreated rats (i.e., day 0 following the treatment) (columns 2, 3, and 4). The actual values of the areas, expressed in units per milliliter, of the peaks of nonspecific phenyl esterase, cGPx, and PHGPx activity were 16.75, 1.44, and 7.18, respectively. One unit is the amount of enzyme catalyzing the transformation of 1 μmol of substrate per minute at room temperature. Results obtained by Northern blotting are shown in column 5. This was performed on 20 μg of total mRNA extracted from testes under standard conditions. The blotting was probed separately with both the radiolabeled *Eco*RI fragment of pMM3, containing nucleotides 46–733 of pig heart PHGPx mRNA, and human β actin cDNA (Clontech). The ratio of the cpm in the corresponding spots, counted by an instant imager (Packard), was calculated, and values were expressed as the percentage of the ratio measured for untreated rats. Note that nonspecific phenyl esterase activity, the marker activity for Leydig cells, is practically negligible 3 days following EDS treatment, whereas PHGPx activity/mRNA starts to decrease only 7/14 days after this treatment. However, only a transient increase of cGPx activity was observed in this model.

Leydig cell marker activity was almost completely depleted 3 days after EDS administration, but this was not correlated with a variation in PHGPx activity and mRNA, which started to decrease significantly in only 1 or 2 weeks, respectively, following this treatment. In summary, data indicate that Leydig cell destruction depletes PHGPx activity/mRNA, but with temporal discrepancy in respect to the supposed androgen withdrawal (Table II). Clearly, all this does not argue for a *direct* modulation of PHGPx by androgens, and the explanation for the variation of PHGPx activity as observed Table II must be found elsewhere.

Interestingly, data also show that cGPx activity is generally not affected in the two models studied (Fig. 1 and Table II), except for a transient increase observed following EDS administration (Table II). Clearly, this is not dependent on hormone withdrawal, the increase of cGPx being only transient in the EDS model not being otherwise explained.

As already known from morphological studies, profound modifications, leading to an increased germ cell number, occur in the tubular epithelium of testis following puberty (19), and the role of androgens is to sustain germinal epithelium (20 and references therein). Our data actually indicate a strict parallelism between the condition of the germinal epithelium (better viewed as germ cell number) and PHGPx activity. Accordingly, PHGPx is low in young animals, when germinal epithelium is still not complete, but it rises in adults, when the germinal epithelium becomes complete. Furthermore, the enzyme is depleted several days following androgen withdrawal, when the germinal epithelium also starts to degenerate. Further strengthening this view, the treatment with EDS is reported to bring animals to sterility, due to low spermatozoa production, from 2 to 4 weeks from the beginning of the treatment (18), just in parallel with the decrease of PHGPx observed in our experiments. Thus, PHGPx seems to mark for the germinal cell content in testis. Also, our data argue for a peculiar capability of germ cells to produce more PHGPx in respect to other somatic cells, independently from the action of androgen, a view also strengthened by the observation that steroids failed to induce PHGPx activity/mRNA in proper receptor-bearing cells or in isolated testis incubated *in vitro* (not shown).

Nevertheless, our experiments indicate a very low content of cGPx, if any, both in Leydig cells, since the testicular cGPx content is not affected by Leydig cell destruction (Table II), and in germ cells, since cGPx enzymatic activity is not affected by disturbances of the latter. Therefore, the transient increase observed in Table II could rationally be explained only by the massive macrophages infiltration reported to occur in the tissue after the administration of EDS (21), which serves to remove the apoptotic Leydig cells thus generated. Fallout of this observation is that resident testicular macrophages, which can be numbered as $300/mm^2$ of interstitium in a normal adult rat testis (22), may contribute significantly to cGPx content in testis, as observed in Fig. 1.

Although an involvement in the cellular protection against the production of harmful free radicals, as classically proposed, cannot be excluded, GPxs unusual distribution in testis germ cells in respect to other somatic cells seems to point out a functional differentiation of these two proteins. The endogenous high level of PHGPx in male germ cells clearly suggests that the reaction catalyzed by PHGPx is more important here than in other somatic cells. At present, to bridge between classic and emerging concepts, it perhaps is not far from the truth to speculate that this reaction is related to a physiological function, which is, in turn, somehow linked to oxidative stress. This concept may be explained with the different specificity of PHGPx and cGPx for both the reducing and the oxidizing substrate.

PHGPx was classically described as a monomeric selenoprotein that reduces peculiarly complex hydroperoxides such as cholesterol and phospholipids, which are not substrates for cGPx. This property, which was later proposed for the monomeric nature of the determinant (23), suggested the name phospholipid hydroperoxide glutathione peroxidase. Nevertheless, from the time of its discovery, PHGPx exhibited a certain lack of specificity for the nature of the reducing agent, which, indeed, was not apparent for cGPx (24). This concept was further expanded in studies in which the molecular modeling of PHGx on the crystallized structure of cGPx revealed that the basic residues involved in the complex with glutathione in cGPx are deleted or mutated in PHGPx (23). Also, some synthetic dithiols were found to be better substrates than glutathione for isolated testis PHGPx (25).

In conclusion, a wide variety of evidence indicates that cGPx is a true glutathione peroxidase, using GSH exclusively as a reductant and small molecules, such as H_2O_2, as an oxidant(s). Since GSH is usually a very abundant compound in cells, cGPx seems to better accomplish the function of a *real* antioxidant protein. However, PHGPx is namely a peroxidase, but the reductant is at present undefined, and oxidants, such as phospholipid hydroperoxides and/or other hydroperoxides, are more complex molecules. In other words, the actual reaction of PHGPx may be better described as a *thiol* peroxidase. Interestingly, this specifically links PHGPx to still undefined thiol disulfide transitions. At present, the relation of all this to the physiology of germ cells is unknown, but certainly will be the matter for future exciting research.

REFERENCES

1. Combs, G. F., and Combs, S. B. (1986). "The Role of Selenium in Nutrition," pp. 532–543. Academic Press, New York/London.
2. Dumont, J. E., Corvilain, B., and Contempre, B. (1994). *Mol. Cell. Endocrinol.* **100**, 163–166.
3. Yang, G. Q., Ge, K. Y., Chen, J. S., and Chen, X. S. (1988). *World Rev. Nutr. Diet* **55**, 98–152.
4. Xia, Y., Hill, K. E., and Burk, R. F. (1989). J. *Nutr.* **119**, 1318–1326.
5. Bai, J., Wu, S., Ge, K., Deng, X., and Su, C. (1984). *Acta Acad. Med. Sin.* **2**, 29–31.
6. Tracy, S., Chapman, N. M., and Beck, M. A. (1991). *Rev. Med. Virol.* **1**, 145–154.

7. Beck, M. A., Kolbeck, P. C., Rohr, L. H., Shi, Q., and Morris, V. C. (1994). *J. Med. Virol.* **43**, 166–170.

8. Beck, M. A., Kolbeck, P. C., Shi, Q., Rohr, L. H., Morris, V. C., and Lavender, O. A. (1994). *J. Infect. Dis.* **170**, 351–357.

9. Beck, M. A., Shi, Q., Morris, V. C., and Lavender, O. A. (1995). *Nature Med.* **1**, 433–436.

10. Maiorino, M., Coassin, M., Roveri, A., and Ursini, F. (1989). *Lipids* **24**, 721–726.

11. Behne, D., Hofer, T., von Berswordt Wallrabe, R., and Egler, W. (1982). *J. Nutr.* **112**(9), 1682–1687.

12. Wallace, E., Calvin, H. I., Ploetz, K., and Cooper, G.-W. (1987). *In* "Selenium in Biology and Medicine" (G. F. Combs, O. A. Levander, J. E. Spallholz, and J. E. Oldfield, eds.) pp.181–196. AVI Publishing Co., Westport, CT.

13. Behne, D., Weiler, H., and Kyriakopoulos, A. (1996). *J. Reprod. Fertil.* **106**(2), 291–297.

14. Behne, D., Kyriakopulos, A., Kalcklösch, M., Weiss-Nowak, C., Pfeifer, H., Gessner, H., and Hammel, C. (1997). *Biomed. Environ. Sci.* **10**, 340–345.

15. Steinert, P., Bächner, D., and Flohé, L. (1997). *Biol. Chem.* **378**, S73.

16. Roveri, A., Casasco, A., Maiorino, M., Dalan, P., Calligaro, A., and Ursini, F. (1992). *J. Biol. Chem.* **267**, 6142–6146.

17. Brigelius-Flohé, R., Aumann, K.-D., Blöcker, H., Gross, G., Kieß, M., Klöppel, K.-D., Maiorino, M., Roveri, A., Schucklet, R., Ursini, F., Wingender, E., and Flohé, L. (1994). *J. Biol. Chem.* **269**(10), 7342–7348.

18. Jackson, C.-M., Jackson, H. (1984). *J. Reprod. Fertil.* **71**, 393–401.

19. Clermont, Y., and Perey, B. (1957). *Am. J. Anat.* **100**, 241–267.

20. Sharpe, R. M., Donakie, K., and Cooper, I. (1988). *J. Endocrinol.* **117**, 19–26.

21. Kerr, J. B., Donakie, K., and Rommerts, F. F. G. (1985). *Cell Tissue Res.* **242**, 145–156.

22. Hutson, J. C. (1990). *Biol. Reprod.* **43**, 885–890.

23. Ursini, F., Maiorino, M., Brigelius-Flohé, R., Aumann, K.-D., Roveri, A., Schomburg, D., and Flohé, L. (1995). *Methods Enzymol.* **252**, 38–53.

24. Ursini, F., Maiorino, M., and Gregolin, C. (1985). *Biochim. Biophys. Acta* **839**, 62–70.

25. Roveri, A., Maiorino, M., Nisii, C., and Ursini, F. (1994). *Biochim. Biophys. Acta* **1208**, 211–221.

26. Besnadoun, A., Weinstein, D. (1976). *Anal. Biochem.* **70**, 241–250.

27. Meyer, E. H. H., Forsgren, K., von Deimling O., and Engel, W. (1974). *Endocrinology* **95**, 1737–1739.

8 α-Lipoic Acid: Cell Regulatory Function and Potential Therapeutic Implications

Chandan K. Sen, Sashwati Roy, and Lester Packer

Membrane Bioenergetics Group
Department of Molecular and Cell Biology
University of California
Berkeley, California 94720

INTRODUCTION

α-Lipoic acid, also known as thioctic acid, 1,2-dithiolane-3-pentanoic acid, 1,2-dithiolane-3-valeric, or 6,8-thioctic acid, is a naturally occurring potent antioxidant (Packer *et al.*, 1995: Sen *et al.*, 1997). It is present as lipoyllysine in various natural sources. In the plant material studied, the lipoyllysine content was the highest in spinach (3.15 μg/g dry weight; 92.51 μg/mg protein). When expressed as weight per dry weight of lyophilized vegetables, the abundance of naturally existing lipoate in spinach is over three- and fivefold higher than that in broccoli and tomato, respectively. A lower concentration of lipoyllysine is also detected in garden pea, brussel sprouts, and rice bran. Lipoyllysine concentration has been found to be below detection limits in acetone powders of banana, orange peel, soybean, and horseradish, however. In animal tissues, the abundance of lipoyllysine in bovine acetone powders can be represented in the following order: kidney > heart > liver > spleen > brain > pancreas > lung. The concentration of lipoyllysine in bovine kidney and heart were 2.64 ± 1.23 and 1.51 ± 0.75 μg/g dry weight, respectively (Lodge *et al.*, 1997). Lipoic acid is also an integral component of the mammalian cell. It is present in trace amounts as lipoamide in at least five proteins where it is covalently linked to a lysyl residue. Four of these proteins are found in α-keto acid dehydrogenase complexes, the pyruvate dehydrogenase complex, the branched chain keto acid dehydrogenase complex, and the α-ketoglutarate dehydrogenase complex. Three lipoamide-containing proteins are present in the E2 enzyme dihydrolipoyl acyltransferase, which is different in each of the complexes and specific for the substrate of the complex (Packer *et al.*, 1997). One lipoyl residue is found in protein X, which is the same in each complex. The fifth lipoamide residue is present in the glycine cleavage system (Fujiwara *et al.*, 1991).

THE LIPOATE–DIHYDROLIPOATE REDOX CYCLE

The mitochondrial E3 enzyme, dihydrolipoyl dehydrogenase, reduces lipoate to dihydrolipoate (DHLA) in the presence of NADH. This enzyme shows a marked preference for the naturally occurring *R*-enantiomer of lipoate (Haramaki *et al.*, 1997). Lipoate is also a substrate for the NADPH-dependent enzyme glutathione reductase (Pick *et al.*, 1995). Glutathione reductase shares a high degree of structural homology with lipoamide dehydrogenase. Both are homodimeric enzymes with 50-kDa subunits conserved between all species. In contrast to dihydrolipoyl dehydrogenase, however, glutathione reductase exhibits a preference for the *S*-enantiomer of lipoate. Although lipoate is recognized by glutathione reductase as a substrate for reduction, the rate of reduction to DHLA is much slower than that of the natural substrate glutathione disulfide. Whether lipoate would be reduced in a NADH or NADPH-dependent mechanism is largely tissue specific. Thioredoxin reductase catalyzes the NADPH-dependent reduction of oxidized thioredoxin. Thioredoxin reductase from calf thymus and liver, human placenta, and rat liver has been observed to efficiently reduce both lipoate and lipoamide in NADPH-dependent reactions (Arner *et al.*, 1996). Under similar conditions at 20°C and pH 8.0, mammalian thioredoxin reductase reduced lipoic acid 15 times more efficiently than lipoamide dehydrogenase. The relative contribution of the three different enzymes known to reduce lipoate in mammalian cells is tissue and cell specific depending on the presence or absence of mitochondrial activity and of oxidized thioredoxin and GSSG (Fig 1).

PROGLUTATHIONE EFFECT

A major property of lipoic acid is that it can serve as a proglutathione agent and enhance the cellular glutathione (GSH) level (Han *et al.*, 1997; Sen *et al.*, 1997). Studies with human Jurkat T cells have shown that when

added to the culture medium, lipoate readily enters the cell where it is reduced to its dithiol form, DHLA. DHLA, a potent reducing agent, accumulates in the cell pellet and, when monitored over a 2-hr interval, the
dithiol is released to the culture medium (Handelman *et al.*, 1994). Following lipoate supplementation, extracellular DHLA reduces cystine outside
the cell to cysteine. The cellular uptake mechanism for cysteine by the ASC
system is approximately 10 times faster than that for cystine by the x_c^- system (Watanabe and Bannai, 1987). Thus, DHLA markedly improves cysteine availability within the cell, resulting in accelerated GSH synthesis (Fig
1) (Han *et al.*, 1997; Sen, 1998).

RADICAL SCAVENGING

Both lipoate and its reduced form DHLA have remarkable reactive
oxygen-detoxifying properties (Packer *et al.*, 1995). Lipoate and DHLA
scavenge several reactive species, including hydroxyl radicals, hydrogen
peroxide, hypochlorous acid, and singlet oxygen (Packer *et al.*, 1995; Matsugo *et al.*, 1996). In addition, lipoate/DHLA have transition metal chelation properties by virtue of which it may avert the transformation of relatively weak oxidants such as the superoxide anions and hydrogen peroxide
to the deleterious hydroxyl radical (Packer *et al.*, 1995).

ANTIOXIDANT DEFENSE NETWORK

Antioxidant reactions are essentially oxidation–reduction reactions in
which reactive forms of oxygen are reduced and thus scavenged by the antioxidants; in the process the antioxidant is oxidized to its functionally in-
ert form. Effective functioning of redox antioxidants requires the recycling
of the oxidized form of antioxidant to its potent reduced form. DHLA is a
strong reductant and is thus capable of recycling some of such oxidized antioxidants (Packer *et al.*, 1995). DHLA can directly regenerate ascorbate
and indirectly regenerate vitamin E from their respective oxidized radical
forms (Constantinescu *et al.*, 1993).

CELL CALCIUM AND NF-κB

Flow cytometric determination of intracellular Ca^{2+} concentration
($[Ca^{2+}]_i$) revealed that 0.25 mM hydrogen peroxide treatment results in a
marked calcium flux within the cell (Sen and Packer, 1996; Sen *et al.*, 1996).

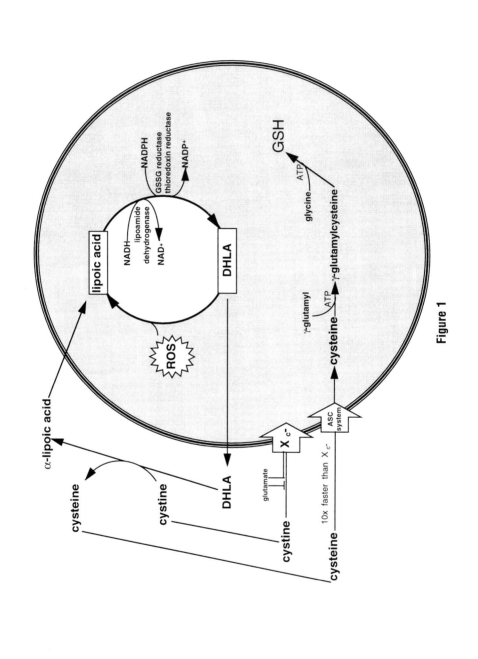

Figure 1

Using extracellular calcium chelators, it was observed that this flux is mainly contributed by calcium released from intracellular stores. Pretreatment of these cells with α-lipoic acid markedly protects against oxidant-induced disturbance in intracellular calcium homeostasis (Sen and Packer, 1996; Sen *et al.*, 1996).

Redox changes in cells, as during oxidative stress, triggers molecular response (Sen and Packer, 1996; Sen, 1998). NF-κB is a well-characterized redox-sensitive transcription factor, the function of which has been related to a number of clinical disorders. The activity of NF-κB is inducible in response to a wide range of stimuli, including peroxide, cytokines, phosphatase inhibitors, and viral products. It is suggested that reactive oxygen species may serve as a common intracellular messenger for NF-κB activation in response to the diverse range of stimuli (Sen and Packer, 1996). Treatment of Jurkat T cells with lipoate suppresses phorbol ester or tumor necrosis factor (TNF) α-induced activation of NF-κB in a dose-dependent manner (Suzuki *et al.*, 1992; Suzuki and Packer, 1994). This NF-κB inhibitory effect was also seen with DHLA. Direct addition of DHLA to the cell culture medium suppressed TNFα-induced NF-κB activation (Suzuki *et al.*, 1995). Both *R*- and *S*- enantiomers of lipoate were effective with respect to the NF-κB inhibitory function (Suzuki *et al.*, 1992; Suzuki and Packer, 1994). These molecular effects of lipoate are not simply mediated by enhanced cellular glutathione. Lipoate appears to act on certain specific molecular loci to produce these NF-κB regulatory effects. Observations suggest that the effect of lipoate on the oxidant-induced cell calcium response may contribute to the inhibition of NF-κB activated by hydrogen peroxide (Sen *et al.*, 1996).

GLUTAMATE CYTOTOXICITY

Because of a high plasma glutamate level, impaired cellular cystine uptake in human immunodeficiency virus-infected (HIV$^+$) patients has been suggested to be a causative factor of a low leukocyte-reduced GSH level in these patients. Elevated levels of extracellular glutamate also causes neu-

---

Figure 1 Schematic outline of the proposed mechanism by which lipoic acid increases cellular glutathione (GSH) levels. Lipoic acid enters cell, is reduced to dihydrolipoate (DHLA), and is released into the medium. DHLA subsequently reduces cystine to cysteine. The newly formed cysteine is taken up by the ASC transporter, thereby decreasing the cell's dependence on cystine uptake by the glutamate-sensitive x_c^- transport system. The efficient cysteine uptake by the ASC transport system leads to an increase of the influx of cysteine, the rate-limiting substrate in GSH synthesis, resulting in an elevation of cellular GSH levels. ROS, reactive oxygen species.

ronal damage and brain disorders, including stroke, epilepsy, and Parkinson's disease. Thus, strategies to protect against glutamate-induced cytotoxicity are expected to have a vast therapeutic potential (Sen *et al.*, 1998). Lipoate protects C6 glial cells from glutamate-induced cytotoxicity (Han *et al.*, 1997). In certain diseases, such as AIDS and cancer, elevated plasma glutamate lowers cellular GSH by inhibiting cystine uptake. Low (10–100 μM) concentrations of lipoate and lipoamide have been shown to bypass the adverse effects of an elevated extracellular glutamate (Han *et al.*, 1997; Sen *et al.*, 1997).

APOPTOSIS

Apoptosis, a form of active cell death, plays an important role in the development and regulation of the immune system. Oxidative stress is one of several stimuli that induces apoptosis in cells. DHLA and lipoamide have been reported to prevent methylprednisolone and etoposide-induced apoptosis in rat thymocytes (Bustamante *et al.*, 1995).

REDUCTIVE STRESS

In pathologies such as diabetes and ischemia–reperfusion, reductive [high NAD(P)H/NAD(P)$^+$ ratio] stress has been considered one of the major factors contributing to these metabolic disorders. Lipoate may lower reductive stress in pathologies such as diabetes and ischemic injury by utilizing cellular NAD(P)H for its reduction to DHLA, thus lowering the cellular NAD(P)H/NAD(P)$^+$ ratio (Fig. 2) (Roy *et al.*, 1997). Consistently, beneficial effects of lipoate supplementation have been observed in diabetes and ischemia–reperfusion injury (Jacob *et al.*, 1995; Panigrahi *et al.*, 1996).

Figure 2   Schematic representation of the influence of α-lipoic acid (LA) reduction to dihydrolipoate (DHLA) on the reducing equivalent homeostasis in Wurzburg T cells and suggested beneficial effects in hyperglycemia and hypoxia. An increase in the NADH to NAD$^+$ ratio has been observed in hyperglycemia because of an increased rate of oxidation of sorbitol to fructose and in hypoxic tissue due to impaired mitochondrial electron transport. An elevated NADH/NAD$^+$ ratio or reductive stress in cells inhibits several major metabolic pathways, such as glycolysis and fatty acid oxidation, and promotes the intracellular formation of reactive oxygen species (ROS). α-Lipoic acid decreases the NADH/NAD$^+$ ratio or reductive stress by rapidly consuming NADH during its reduction to DHLA. In addition, both α-lipoic acid and DHLA are potent antioxidants in that they directly scavenge ROS and recycle thioredoxin, glutathione, vitamin E, and vitamin C. Reprinted with permission of Roy S. *et al.* (1997). *Biochem. Pharmacol.* 53, 393–399.

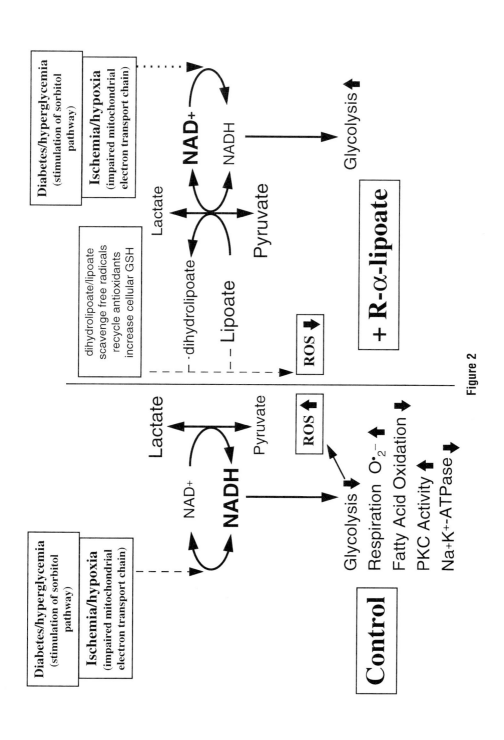

Figure 2

SUMMARY

α-Lipoic acid is a clinically safe supplement that has marked antioxidant and cell redox regulatory properties. A property that makes lipoate unique is that it is accepted as a substrate by at least three enzymes inside the human cell. In this way, lipoate can be continuously recycled to its reduced dyhydrolipoate form at the expense of cellular-reducing equivalents. This provides lipoate with a clear advantage when compared to other supplements of its class, such as N-acetylcysteine (Merin *et al.*, 1996; Sen, 1997; Sen *et al.*, 1997).

REFERENCES

Arner, E. S. J., Nordberg, J., and Holmgren, A. (1996). Efficient reduction of lipoamide and lipoic acid by mammalian thioredoxin reductase. *Biochem. Biophys. Res. Commun.* **225,** 268–274

Bustamante, J., Andrew, F., Slater, G., and Orrenius, S. (1995). Antioxidant inhibition of thymocyte apoptosis by dihydrolipoic acid. *Free Radic. Biol. Med.* **19,** 339–347.

Constantinescu, A., Han, D., and Packer, L. (1993). Vitamin E recycling in human erythrocyte membranes. *J. Biol. Chem.* **268,** 10906–10913.

Fujiwara, K., Okamura-Ikeda, K., and Motokawa, Y. (1991). Lipoylation of H-protein of the glycine cleavage system. *FEBS Lett.* **293,** 115–118.

Han, D., Handelman, G., Marcocci, L., Sen, C. K., Roy, S., Kobuchi, H., Flohe, L., and Packer, L. (1997). Lipoic acid increases de novo synthesis of cellular glutathione by improving cysteine utilization. *Biofactors* **6,** 321–338.

Han, C., Sen, C. K., Roy, S., Kobayashi, M., and Packer, L (1997). Protection against glutamate induced cytotoxicity in C6 glial cells by thiol antioxidants. *Am. J. Physiol.* **273,** R1771–1778.

Handelman, G. J., Han, D., Tritschler, H., and Packer, L (1994). Alpha-lipoic acid reduction by mammalian cells to the dithiol form, and release into the culture medium. *Biochem. Pharmacol.* **47,** 1725–1730.

Haramaki, N., Han, D., Handelman, G. J., Tritschler, H. J., and Packer, L. (1997). Cytosolic and mitochondrial systems for NADH- and NADPH- dependent reduction of α-lipoic acid. *Free Radic. Biol. Med.* **22,** 535–542.

Jacob, S., Henriksen, E. J., Schiemann, A. L., Simon, I., Clancy, D. E., Tritschler, H. J., Jung, W. I., Augustine, H. I., and Dietze, G. J. (1995). Enhancement of glucose disposal in patients with type 2 diabetes by alpha-lipoic acid. *Arzneim.-Forsch.* **45,** 872–874.

Lodge, L., Handelman, G. J., Konishi, T., Matsugo, S., Mathur, V. V., and Packer, L. (1997). Natural sources of lipoic acid: Determination of lipoyllysine released from protease-digested tissues by high performance liquid chromatography incorporating electrochemical detection. *J. Appl. Nutr.* **49,** 3–11.

Matsugo, S., Konishi, T., Masuo, D., Tritschler, H. J., and Packer, L. (1996). Re-evaluation of superoxide scavenging activity of dihydrolipoate and its analogues by chemiluminiscent method using 2-methyl-6-[p-methoxyphenyl]-3,7-dihydroimidazo-[1,2-a]pyrazine-3-one (MCLA) as a superoxide probe. *Biochem. Biophys. Res. Commun.* **227,** 216–220.

Merin, J. P., Matsuyama, M., Kira, T., Baba, M., and Okamoto, T. (1996). Alpha-lipoic acid blocks HIV-1 LTR-dependent expression of hygromycin resistance in THP-1 stable transformants. *FEBS Lett.* **394,** 9–13.

Packer, L., Roy, S., and Sen, C. K. (1997). Alpha-lipoic acid: A metabolic antioxidant and re-dox modulator of transcription. *In* "α-Lipoic Acid: A Metabolic Antioxidant and Redox Modulator of Transcription" (H. Sies, ed.), pp. 79–101. Academic Press, San Diego.

Packer, L., Witt, E. H., and Tritschler, H. J. (1995). α-Lipoic acid as a biological antioxidant. *Free Radic. Biol. Med.* **19**, 227–250.

Panigrahi, M., Sadguna, Y., Shivakumar, B. R., Kolluri, S., Roy, S., and Packer, L. (1996). Alpha-lipoic acid protects against reperfusion injury following cerebral ischemia in rats. *Brain Res.* **717**, 184–187.

Pick, U., Haramaki, N., Constantinescu, A., Handelman, G. J., Tritschler, H. J., and Packer, L. (1995). Glutathione reductase and lipoamide dehydrogenase have opposite stereospe-cificities for alpha-lipoic acid enantiomers. *Biochem. Biophys. Res. Commun.* **206**, 724–730.

Roy, S., Sen, C. K., Tritschler, H., and Packer, L. (1997). Modulation of cellular reducing equivalent homeostasis by alpha-lipoic acid: Mechanisms and implications for diabetes and ischemic injury. *Biochem. Pharmacol.* **53**, 393–399.

Sen, C. K. (1997). Nutritional biochemistry of cellular glutathione. *J. Nutr. Biochem.* **8**, 660–672.

Sen, C. K. (1998). Cellular thiols and redox regulated signal transduction. *Curr. Opin. Cell. Regul.*, in press.

Sen, C. K., and Packer, L. (1996). Antioxidant and redox regulation of gene transcription. FASEB J. 10, 709–720.

Sen, C. K., Roy, S., Han, D., and Packer, L. (1997). Regulation of cellular thiols in human lym-phocytes by alpha-lipoic acid: A flow cytometric analysis. *Free Radic. Biol. Med.* **22**, 1241–1257.

Sen, C. K., Roy, S., and Packer, L. (1996). Involvement of intracellular $Ca2^+$ in oxidant-in-duced NF-kappa B activation. *FEBS Lett.* **385**, 58–62.

Sen, C. K., Roy, S., and Packer, L. (1997). Therapeutic potential of the antioxidant and redox properties of alpha-lipoic acid. *In* "Therapeutic Potential of the Antioxidant and Redox Properties of Alpha-Lipoic Acid" (L. Montagnier, R. Olivier, and C. Pasquier, eds.), pp. 251–267. Dekker, New York.

Sen, C. K., Roy, S., and Packer, L. (1998). Oxidants and antioxidants in glutamate induced cy-totoxicity. *In* "Oxidants and Antioxidants in Glutamate Induced Cytotoxicity" (L. Packer and A. S. H. Ong, eds.). AOCS Press, Champaign, IL, in press.

Suzuki, Y. J., Agarwal, B. B., and Packer, L. (1992). Alpha-lipoic acid is a potent inhibitor of NF-kB activation in human T cells. *Biochem. Biophys. Res. Commun.* **189**, 1709–1715.

Suzuki, Y. J., Mizuno, M., Tritschler, H. J., and Packer, L. (1995). Redox regulation of NF-kappa B DNA binding activity by dihydrolipoate. *Biochem. Mol. Biol. Int.* **36**, 241–246.

Suzuki, Y. J., and Packer, L. (1994). "Alpha-Lipoic Acid is a Potent Inhibitor of NF-kappa B Activation in Human T-Cells: Does the Mechanism Involve Antioxidant Activities" (L. Packer and E. Cadenas, eds.), pp. 87–96, Hippocrates Verlag, Stuttgart.

Watanabe, H., and Bannai, S. (1987). Induction of cystine transport activity in mouse peri-toneal macrophages. *J. Exp. Med.* **165**, 628–640.

9 Natural Sources of Lipoic Acid in Plant and Animal Tissues

John K. Lodge and Lester Packer

Department of Molecular and Cell Biology
University of California
Berkeley, California 94720

INTRODUCTION

α-Lipoic Acid as a Cofactor in Oxidative Metabolism

α-Lipoic has been known to be an essential cofactor in oxidative metabolism for many years. It was first purified in 1951 by Reed and co-workers (1), after being recognized that this compound was responsible for a number of compounds displaying the same activity (2). The earliest of these was in 1937, when the term potato growth factor was coined to describe a substance necessary for the growth of bacteria (3). From these early studies it was noticeable that lipoic acid was a common constituent of normal animal and plant tissues and was in fact tentatively described as a vitamin after isolation. It is now known that both plants and animals can synthesize lipoic acid from octanoic acid (4), but the complete biosynthetic pathway is still unknown.

In eukaryotes, lipoic acid is bound covalently to a lysine residue of five distinct mitochondrial proteins (5). Three of these are E_2 subunits of the pyruvate dehydrogenase (PDC), the branched chain keto acid dehydrogenase, and the α-ketoglutarate dehydrogenase complexes (6–8). These complexes catalyze the oxidative decarboxylation of pyruvate into acetyl-CoA and α-ketoglutarate into succinyl-CoA, respectively. The PDC complex com-

Antioxidant Food Supplements in Human Health

prises a structural core of approximately 60 E_2 subunits to which multiple E_1 and E_3 components are attached (8). The E_1 component decarboxylates the substrate and transfers the acetyl group to E_2. The lipoate moiety of E_2 becomes acetylated and later transfers this group to coenzyme A, forming acetyl coenzyme A and dihydrolipoate. The E_3 component regenerates the oxidized form to continue the catalytic cycle. E_2 and E_3 are anchored together by protein X, which also contains the lipoyllysine moiety (8).

The final lipoamide moiety is found as the H-protein, part of the glycine cleavage system (9). In plants this is known as the glycine decarboxylase complex, where it catalyzes the conversion of glycine into CO_2, ammonium, and methylene tetrahydrofolate. Plants are also unique in that they contain chloroplastic PDC as well as mitochondrial PDC (10). Various amounts of each have been found to be species specific (10). Hence lipoate occupies a central position in metabolism as a regulator of carbon flow into the Krebb's cycle, resulting in the production of ATP (8).

α-Lipoic Acid as an Antioxidant

In the last decade lipoic acid and its reduced form, dihydrolipoic acid (DHLA), have emerged as powerful antioxidants *in vitro* (11). The lipoate couple, having a potential of -0.32 V, is a strong reductant and as such has the ability of interacting with many cellular thiols, antioxidants, and reactive oxygen species. The antioxidant network is an important line of defense against potential free radical attack *in vivo*. It has been shown in various *in vitro* systems that DHLA can regenerate ascorbate, which in turn recycles vitamin E (12, 13). Glutathione can also recycle ascorbate (14), but cannot recycle lipoic acid. DHLA, however, is able to reduce GSSG, and so lipoate can work at even higher levels of the antioxidant network, accepting electrons directly from NAD(P)H. Recycling, or protection, of vitamin E has also been shown *in vivo* in vitamin E-deficient mice (15). Lipoic acid supplementation prevents deficiency symptoms and sustains tissue vitamin E levels (15). Both LA and/or DHLA can scavenge a variety of reactive oxygen species, including hydroxyl, peroxyl, and superoxide radicals, hypochlorous acid, and singlet oxygen, as well as having the ability to chelate metal ions (11, 13, 16). Thus the lipoate/dihydrolipoate couple represents a powerful weapon in the fight against free radical attack.

Aims of This Work

Attention has been paid to exogenous lipoic acid as a therapeutic agent (11) and as a health supplement. However, it has not been clarified whether protein-bound lipoic acid itself may act as an antioxidant or as a source of free lipoic acid. Therefore, it is important first to quantify the amount of endogenous lipoic acid (in the form of lipoyllysine) in tissues for understanding its role in protection against oxidative stress and in energy metabolism. Also, because of the potential use of lipoic acid as a health supplement, it is of importance to ascertain the lipoic acid content of common ingredients in the diet.

METHODS OF ANALYSIS

Previous Methods for the Determination of α-Lipoic Acid

So far high-performance liquid chromatography (HPLC), gas chromatography (GC), and GC mass spectroscopy (GC-MS) methods have been developed to measure the contents of endogenous lipoic acid (17–21). Among these, GC and GC-MS methods are sensitive and specific for the measurement of endogenous lipoic acid, but they require the preparation of derivatives, which may decrease recovery, and GC-MS requires expensive equipment. According to previous reports, protein-bound lipoic acid has to be liberated by acidic or alkaline hydrolysis of protein-bound lipoic acid at high temperature. This, however, leads to decomposition and eventually to the decreased recovery of free lipoic acid (18, 21). To avoid the decomposition of lipoic acid in strong acidic or basic conditions, an enzyme immunoassay was developed for the analysis of protein-bound lipoic acid without the release of lipoic acid. However, this method is not as sensitive as previously thought, and preparation of the antilipoic acid antibody is a laborious process (22). It was also reported that endogenous lipoic acid in isolated lipoate-containing proteins was measured in form of lipoyllysine by enzymatic hydrolysis using HPLC equipped with an ultraviolet (UV) detector (330 nm) (17). This method has the advantage of increasing the recovery of endogenous lipoic acid. However, the UV detection system is not sensitive enough to detect the small amount of lipoyllysine from tissue hydrolysate containing many UV-absorbing compounds at 330 nm (17).

Detailed Procedure

The method involves (1) disrupting appropriate tissue samples and (2) treatment with proteases to digest proteins to their amino acids. Lipoyllysine is then extracted into ethanol, which is dried down, reconstituted into a mobile phase, and analyzed by HPLC-ECD. This is summarized in a flow diagram (Fig 1).

Samples were either obtained as lyophilized powders or lyophilized.

NOTES

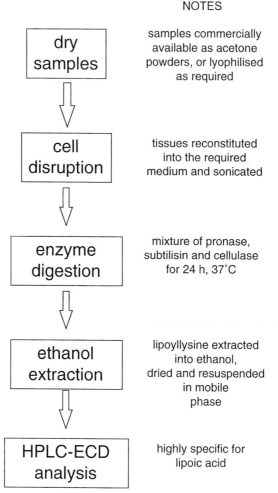

dry samples	samples commercially available as acetone powders, or lyophilised as required
cell disruption	tissues reconstituted into the required medium and sonicated
enzyme digestion	mixture of pronase, subtilisin and cellulase for 24 h, 37°C
ethanol extraction	lipoyllysine extracted into ethanol, dried and resuspended in mobile phase
HPLC-ECD analysis	highly specific for lipoic acid

Figure 1   Flow diagram summarizing the various steps in the measurement of protein-bound lipoic acid (lipoyllysine).

Three hundred milligrams of tissue sample was suspended in 3 ml of 0.1 M Tris–HCl (pH 7.6) and disrupted with an ultrasonicator (Branton) for approximately 5 min. Ten units of protease VIII (subtilisin Carlsberg from *Bacillus licheniformis*) and 10 U of protease XIV (pronase from *Streptomyces griseus*) were added to the disrupted samples. The samples were incubated for 24 hr at 37°C. Protease treatment for 48 hr did not increase the detected amounts of lipoyllysine (data not shown). With plant sources, cellulase (10 U) was also added to break down the cellulose cell wall. As this was found to double the amount of released lipoyllysine from the plant sources tested (data not shown), cellulose was therefore used routinely in all plant-based samples. Ethanol extraction of the digested samples was performed three times, and the supernatant was dried under nitrogen. The dried samples were then dissolved in 0.4 ml of mobile phase, and the precipitate was removed by centrifugation (15,000 rpm, 15 min). The final supernatants were used directly as the sources of lipoyllysine.

To prepare standards, the desired amounts of lipoyllysine were added to solutions of bovine serum albumin (BSA) (2.5 mg/ml), incubated, protease digested, and ethanol extracted, as described earlier. Standards in water alone were also used, in which case no pretreatment was used.

High-Performance Liquid Chromatography

The determination of lipoyllysine was carried out with a Alltima C_{18} column [5 μm, 4.6 mm (i.d.) $\times$ 150 mm] (Alltech Associates, Deerfield, IL). The mobile phase was 65% 0.15 M monochloroacetic acid (pH 2.8), 21% methanol, and 14% acetonitrile with a flow rate of 1.0 ml/min. The detection of lipoyllysine was performed using an amperometric electrochemical detection system (Bioanalytical System, West Lafayette, IN) equipped with a dual gold electrode coated with triple-distilled mercury (Merck), operating at a 0.5-V potential.

Quantitation was carried out by comparison of peak areas to the area of standard curves obtained from lipoyllysine standards in the concentration range of 0–2.5 μg/400 μl. This analysis was carried out using Turbochrome software (P. E. Nelson, Cupertino, CA).

Protein Determinations

Protein concentrations of the tissue samples were determined by the Lowry assay (Biorad DC Protein Assay), using BSA as the standard. Samples (300 mg/3 ml) were diluted 60-fold in order to be in the sensitivity range of the assay (0.05–1 mg/ml protein).

RESULTS

HPLC Analysis and Confirmation

A typical HPLC chromatogram used for the analysis of lipoyllysine is shown in Fig. 2 for a standard (A) and a typical sample (B). The peak for

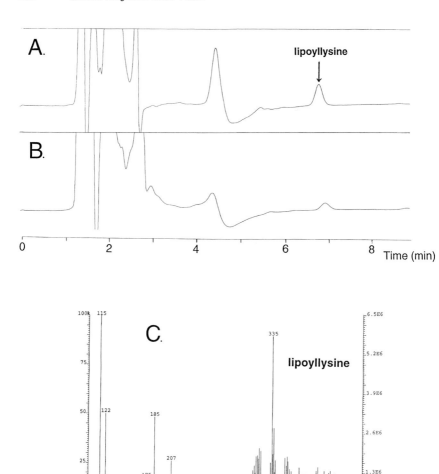

Figure 2 HPLC analysis and confirmation of lipoyllysine. A typical HPLC-ECD chromatogram for a lipoyllysine standard is shown (A), whereas a typical sample is shown (B). Collecting this peak fraction and analyzing it by FAB mass spectroscopy (C) confirm that the peak of interest is indeed lipoyllysine.

lipoyllysine is shown to be clean and free from contaminating signals that could interfere with analysis. The increased selectivity the detection system allows for the lipoyl group is clearly shown.

In order to verify that the peak in question is in fact lipoyllysine, the elutent from this peak was collected and analyzed by FAB mass spectroscopy. The analysis for a particular sample is shown in Fig. 2C. The lipoyllysine ion $[M+H]^+$ is shown as 335. Peaks at 207, 185, 132, and 115 correspond to the matrix. This confirms that the peak of interest is lipoyllysine.

Standards of lipoyllysine liberated from BSA served as a standard curve from which to measure the lipoyllysine content in samples. A new standard curve was constructed on each day of analysis. Each curve was found to be linear in the range covered (0.1 to 2.5 μg), with r values always at least 0.97.

Lipoyllysine Content of Plant and Animal Tissues

The lipoyllysine content of various animal tissues is shown in Table I. Values were found in the region of 0.1–2.6 μg/g dry weight in the lowest and highest amounts in the lung and kidney, respectively. Lipoyllysine was also present in high amounts in muscular tissues such as the heart, skeletal muscle, and intestine.

The lipoyllysine content of various plant tissues is shown in Table II. Values were found to be tissue specific, with by far the highest value in spinach. Other "green" tissues, such as broccoli, pea, and sprout, also contained appreciable amounts of lipoyllysine.

Because plants contain two isoforms of the keto acid dehydrogenase complex, in chloroplasts as well as in mitochondria, the goal was to analyze individual plant fractions. Spinach was chosen as the tissue of choice as it contained the highest lipoyllysine content. The results are shown in

Table I   Lipoyllysine Content of Various Animal Tissues

Source	μg/g dry weight	ng/mg protein
Kidney[a]	2.64 ± 1.23	50.57 ± 5.51
Heart[a]	1.51 ± 0.75	41.42 ± 2.76
Liver[a]	0.86 ± 0.33	15.49 ± 0.01
Spleen[a]	0.36 ± 0.08	5.69 ± 1.27
Brain[a]	0.27 ± 0.08	4.85 ± 1.69
Pancreas[a]	0.12 ± 0.05	1.97 ± 0.97
Lung[a]	0.12 ± 0.08	3.20 ± 0.04
Skeletal muscle[b]	0.97 ± 0.07	
Small intestine[b]	0.83 ± 0.16	
Large intestine[b]	1.18	
Stomach[b]	1.63	

[a] Bovine acetone powders.
[b] Lyophylized rat tissue.

Table II   Lipoyllysine Content of Various Plant Tissues

Source	μg/g dry weight	ng/mg protein
Spinach[a]	3.15 ± 1.11	92.51 ± 4.03
Broccoli[a]	0.94 ± 0.25	41.01 ± 1.02
Tomato[b]	0.56 ± 0.23	48.61 ± 1.69
Green pea[a]	0.39 ± 0.07	17.13 ± 1.23
Brussel sprouts[a]	0.39 ± 0.21	18.39 ± 2.42
Rice bran[a]	0.16 ± 0.02	4.44 ± 2.12
Banana[b]	nd[c]	
Orange peel[b]	nd	
Soybean[b]	nd	
Horseradish[b]	nd	
Egg yolk[b]	0.05 ± 0.07	1.17 ± 0.01
Yeast[b]	0.27 ± 0.05	4.49 ± 1.78
E. coli[b]	8.07	68.71 ± 11.24

[a] Lyophylized material.
[b] Acetone powders.
[c] Not determined.

Fig. 3A. It appears that spinach mitochondria contain almost twofold more lipoyllysine, by weight, than the chloroplasts. The total lipoyllysine content of spinach leaves can be accounted for by the chloroplast and mitochondrial fractions. A similar relationship is present if the results are presented in terms of protein content (Fig. 3B). In comparing mitochondria from two plant sources, a similar lipoyllysine content per protein was found; however, if this was compared to animal mitochondria (in this case rat liver), then a twofold higher lipoyllysine content per protein was found. Such differences are accountable in terms of the metabolic activity of plant and animals.

DISCUSSION

Naturally occurring lipoic acid is well recognized to play a fundamental role in metabolism, serving as a cofactor in enzyme complexes, which function at strategic points in carbohydrate metabolism, the citric acid cycle, and amino acid catabolism (8). On the last decade, experimental evidence has grown showing free lipoic acid as a powerful antioxidant and redox regulator (11) and, as such, to be a potentially useful therapeutic tool. It is important that the distribution and content of lipoic acid in plants and animals be ascertained.

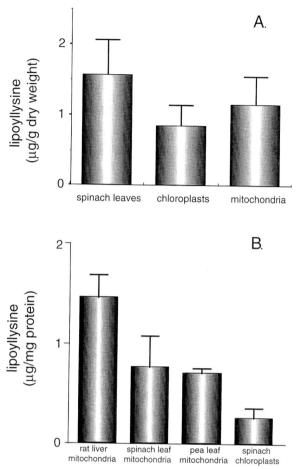

Figure 3   Fractionation of spinach leaves into mitochondrial and chloroplast fractions. (A) The lipoyllysine content is displayed as a function of initial wet weight. (B) The lipoyllysine content is displayed as per protein and is compared with other mitochondrial fractions.

Therefore, a method for measuring the naturally occurring protein-bound form of lipoic acid (lipoyllysine) (23) has been developed, and this method has been used to measure the lipoyllysine content of various plant and animal tissues (23).

In lipoate-containing enzymes (α-keto acid dehydrogenases), lipoic acid is bound covalently to a lysine residue. This is an important consideration for determination as the release of lipoic acid is crucial. Previous methods have tried to overcome this problem via hydrolysis with strong acid/base;

however, a larger amount of lipoic acid is lost by such methods, and indeed recoveries of only 50% have been reported (18), and no more than 70% have been found (17). To overcome this problem, proteolytic hydrolysis was introduced to liberate lipoic acid. This also has the advantage of liberating the actual protein-bound form (lipoyllysine). Lipoic acid from animal tissues has been determined previously, but with methods of detection such as refractive index, ultraviolet, and GC mass spectroscopy. The former methods are inadequate as tissue hydrolysates contain a large amount of contaminants, which absorb around 330 nm, and GC methods require prior derivatization. An electrochemical detection system was introduced to overcome these problems. The lipoate/dihydrolipoate couple has a high redox potential and, as such, is easily detectable electrochemicaly. Indeed, this system is very sensitive for the lipoate moiety, as only substances that are oxidized at the applied voltage will be detected. Hence the clean chromatograms. (Fig 2).

The lipoyllysine content of animal tissues is tissue specific and appears to correlate with metabolic activity. The highest amounts were detected in kidney, heart, and muscular tissues, and only one-tenth of this amount was found in the brain and pancreas. It is conceivable that highly metabolic tissues, such as muscular tissues, contain high amounts of lipoic acid, as they also contain high concentrations of mitochondria in order to provide ATP. Lipoamidase (lipoyl-x-hydrolase) is an enzyme that cleaves lipoic acid from lipoylated proteins. This enzyme is found in many tissues, including serum, liver, and brain, but the activity is tissue dependent (24). In the rat, the highest activities were found to be in kidney and liver (24), which correlates with lipoyllysine content. However, one might expect that a negative correlation would be observed, as this enzyme cleaves lipoyllysine.

Our results have a similar trend to those found previously for lipoic acid determination. In one such study, the variation between highest and lowest amounts was only 2-fold, whereas a 22-fold difference was reported, again highlighting the better sensitivity of our method.

The number of lipoyl domains of the E_2 component is species specific. For example *Escherichia coli* has three lipoylated domains, mammals have two, plants have two and yeast has but one (25). It is not known, however, to what extent repeating lipoyl domains are lipoylated *in vivo* and it is possible that there is additional lipoylation of these domains on exogenous lipoic acid supplementation, which may aid in the effectiveness of the compound.

Plant tissues are unique in that they contain an extra isoform of the pyruvate dehydrogenase complex. This is present in the chloroplasts. Therefore, both mitochondrial and chloroplastic PDC exist and hence lipoyllysine. The two isoforms are immunologically different and have different molecular masses that have been found to be related to the number of lipoyl domains in each unti (10). Hence the lipoyllysine content of plant tissues will be dependent on both the mitochondrial and the chloroplast content. Indeed, we have seen that the total amount of lipoyllysine in spinach leaves can be accounted for by the content in mitochondria (mt) and chloroplasts (cp). Interestingly, the relative amounts of mtPDC and cpPDC are different for each plant species (10). For example, spinach leaves contain 15% mtPDC and 85% cpPDC, whereas in pea the opposite is the case, with 87% mtPDC and 13% cpPDC (10) However, these ratios are only in terms of enzyme activity, as more lipoyllysine (and hence PDC) is found in mitochondria than in chloroplasts (of spinach). It is clear though that "green" tissues, i.e., those with mitochondria and chloroplasts, will be relatively rich in lipoic acid when compared to parts of the plant that are absent in chloroplasts, e.g., roots. Indeed, this was found to be the case (Table II). Even though spinach leaves were found to be the richest source of lipoic acid in plant tissues tested so far, the amount is still very small. To put this into perspective, it is estimated that in order to obtain 1 mg of lipoic acid, a human would have to consume 7 lb. (3.5 kg) wet weight of spinach. Such quantities are, of course, not exactly feasible, and so if lipoic acid is required as a health supplement, then it needs to be taken exogenously in concentrated amounts.

Bioavailability

Lipoic acid (as lipoyllysine) is present in both plant and animal tissues to varying degrees; however, does this form become bioavailable on ingestion and indeed how bioavailable is exogenously supplied free lipoic acid?

Lipoic acid is now known to be synthesized in plant and animal cells. The pathways are still not completely understood; however, it is known that the direct precursor is octanoic acid, and the sulfur source is probably cysteine or methionine. It has been shown that mitochondria can synthesize fatty acids (26) and that this may be involved in lipoic acid production. It was also shown that specific acyl carrier proteins in bacteria (27, 28) and mitochondria (28) attach lipoate to the dehydrogenase enzymes, thus providing a direct link between fatty acid synthesis in mitochondria and lipoate attachment.

Experimental procedures to release protein-bound lipoic acid are similar to what occurs once food is ingested, since it is subjected to low pH in the stomach and then digested with proteases in the intestine before absorption. From such early studies using acid hydrolysis, recovery of lipoic acid was at best no more than 70%. Therefore, it is likely that a similar low recovery of lipoic acid from the diet occurs. Given that dietary amounts of lipoic acid are low initially, it is clear that plasma levels of lipoic acid will also be low. A value of 15.8 μg/liter in plasma has been reported (29). Another consideration is that the turnover of protein-bound lipoic acid is very low, so that once the enzymes are lipoylated, they do not readily become a source of free lipoic acid.

Relatively few reports answer the question as to whether exogenously supplied lipoic acid reaches the tissues. One such report in vitamin E-deficient mice shows that lipoic acid, after 5 weeks supplementation, was found in the skin, liver, brain, and highest in the heart, in both oxidized and reduced forms (15). Another report has shown that lipoate, after either intraperitoneal or oral administration, is located first in the liver, but is then located to other tissues, namely the skeletal muscle (3). Studies on metabolism have shown that only a small percentage of administered lipoate is metabolized (31). This suggests that the fate of the majority of supplemented lipoic acid is in the unaltered form Therefore, it does seem likely that lipoic acid can potentially reach tissues and so exert its effects.

Future Considerations

There is no doubt that *in vitro* lipoic acid is a powerful antioxidant and redox regulator. Naturally occurring lipoic acid is bound to proteins as a cofactor and, as such, plays a fundamental role in energy metabolism. However, it is not known whether this protein-bound lipoic acid can participate in such an antioxidant role or whether it is possible to increase the content of protein-bound lipoic acid with supplementation. If so, then this lipoate could act as an antioxidant directly or as a pool of free lipoic acid inside cells. There is also the possibility that increasing the lipoate content of the mitochondrial proteins may also potentiate metabolism. Such questions highlight the importance of these lines of research.

REFERENCES

1. Reed, L. J., DeBusk, B. G., Gunsalus, I. C., and Hornberger, J. C. S. (1951). Crystalline α-lipoic acid: A catalytic agent associated with pyruvate dehydrogenase. *Science* **114**, 93–94.
2. Snell, E. E., and Broquist, H. P. (1949). On the probability of several growth factors. *Arch. Biochem. Biophys.* **23**, 326–328.
3. Snell, E. E., Strong, F. M., and Peterson, W. H. (1937). Growth factors for bacteria.

VI. Fractionation and properties of an accessory factor for lactic acid bacteria. *Bioch. J.* **31**, 1789–1799.

4. Carreau, J. P. (1979). Biosynthesis of lipoic acid via unsaturated fatty acids. *Methods Enzymol.* **62**, 152–158.

5. Reed, L. J. (1974). Multienzyme complexes. *Acc. Chem. Res.* **7**, 40–46.

6. Patel, M. S., and Roche, T. E. (1990). Molecular biology and biochemistry of pyruvate dehydrogenase complexes. *FASEB J.* **4**, 3224–3233.

7. Patel, M. S., and Smith, R. L. (1994). *In* "The Evolution of Antioxidants in Modern Medicine" (K. Schmidt, A. T. Diplock, and H. Ulrich, eds.), pp. 65–77. Hippocrates Verlag, Stuttgart.

8. Patel, M. S., Roche, T. E., and Harris, R. A. (1996). *In* "Molecular and Cell Biology Updates" (A. Azzi and L. Packer, eds.), pp. 321. Birkhauser Verlag, Boston.

9. Fujiwara, K., Okamura, K., and Motokawa, Y. (1979). Hydrogen carrier protein from chicken liver: Purification characterization, and role of its prosthetic group, lipolic acid, in the glycine cleavage reaction. *Arch. Biochem. Biophys.* **197**, 454–462.

10. Lernmark, U., and Gardestrom, P. (1994). Distribution of pyruvate dehydrogenase complex activities between chloroplasts and mitochondria from leaves of different species. *Plant Physiol.* **106**, 1633–1638.

11. Packer, L., Witt, E. H., and Tritschler, H. J. (1995). Alpha-lipoic acid as a biological antioxidant. *Free Radic. Biol. Med.* **19**, 227–250.

12. Kagan, V. E., Serbinova, E. A., Forte, T., Scita, G., and Packer, L. (1992). Recycling of vitamin E in human low density lipoproteins. *J. Lipid Res.* **33**, 385–397.

13. Kagan, V. E., Shvedova, A., Serbinova, E., Khan, S., Swanson, C., Powell, R., and Packer, L. (1992). Dihydrolipoic acid: A universal antioxidant both in the membrane and in the aqueous phase. *Biochem. Pharmacol.* **44**, 1637–1649.

14. Bast, A., and Haenen, G.R.M.M. (1988). Interplay betweeen lipoic acid and glutathione in the protection against microsomal lipid peroxidation. *Biochim. Biophys. Acta* **963**, 558–561.

15. Podda, M., Tritschler, H. J., Ulrich, H., and Packer, L. (1994). α-Lipoic acid supplementation prevents symptoms of vitamin E deficiency. *Biochem. Biophys. Res. Commun.* **204**, 98–104.

16. Suzuki, Y. J., Tsuchiya, M., and Packer, L. (1991). Thioctic Acid and Dihydrolipoic acid are novel antioxidants which interact with reactive oxygen species. *Free Radic. Res. Commun.* **15**, 255–263.

17. Hayakawa, K., and Oizumi, J. (1989). Determination of lipoyllysine derived from enzymes by liquid chromatography. *J. Chromatogr.* **490**, 33–41.

18. Kataoka, H., Hirabayashi, N., and Makita, M. (1993). Analysis of lipoic acid in biological samples by gas chromatography with flame photometric detection. *J. Chromatogr.* **615**, 197–202.

19. Matarese, R. M., Spoto, G., and Dupre, S. (1981). Hydrolysis of lipoylysine and recovery of lipoic acid by gas chromatography. *J. Appl. Biochem.* **3**, 372–377.

20. Mattulat, A., and Baltes, W. (1992). Determination of lipoic acid in meat of commercial quality. *Zeitsch. Lebensmitt. Untersuch. Forsch.* **194**, 326–329.

21. White, R. H. (1981). A gas chromatographic method for the analysis of lipoic acid in biological samples. *Anal. Biochem.* **110**, 89–92.

22. MacLean, A. I., and Bachas, L. G. (1991). Homogeneous enzyme immunoassay for lipoic acid based on the pyruvate dehydrogenase complex: A model for an assay using a conjugate with one ligand per subunit. *Anal. Biochem* **195**, 303–307.

23. Lodge, J. K., Youn, H.-D., Handelman, G. J., Konishi, T., Matsugo, S., Mathur, V. V., and Packer, L. (1997). Natural sources of lipoic acid: Determination of lipoyllysine released

from protease-digested tissues by high performance liquid chromatography incorporating electrochemical detection. *J. Appl. Nutr* **49**, 3–11.

24. Nilsson, L., and Kagedal, B. (1994). Lipoamidase and biotinidase activities in the rat: Tissue distribution and intracellular localization. *Eur. J. Clin. Chem. Clin. Biochem.* **32**, 501–509.

25. Perham, R. N. (1996). *In* "Alpha-Keto Acid Dehydrogenase Complexes" (M. S. Patel, T. E. Roche, and R. A. Harris, eds.), pp. 1–15., Birkhauser Verlag, Basel/Switzerland.

26. Wada, H., Shintani, D., and Ohlrogge, J. (1997). Why do mitochondria synthesize fatty acids? Evidence for involvement in lipoic acid production. *Proc. Natl. Acad. Sci. USA* **94**, 1591–1596.

27. Brody, S., Oh, C., Hoja, U., and Schweizer, E. (1997). Mitochondrial acyl carrier protein is involved in lipoic acid synthesis in *Saccharomyces cerevisiae*. *FEBS Lett.* **408**, 217–220.

28. Jordan, S. W., and Cronan, J. E., Jr. (1997). A new metabolic link: The acyl carrier protein of lipid synthesis donates lipoic acid to the pyruvate dehydrogenase complex in *Escherichia coli* and mitochondria. *J. Biol. Chem.* **272**, 17903–17906.

29. Shigeta, K., Hirauzumi, G., Wada, M., Oji, K., and Yoshida, T. (1961). Study of the serum level of thioctic acid in patients with various diseases. *J. Vitaminol.* **7**, 48–52.

30. Harrison, E. H., and McCormick, D. B. (1974). The metabolism of dl-[1,6-^{14}C]lipoic acid in the rat. *Arch. Biochem. Biophys.* **160**, 514–522.

31. Spence, J. T., and McCormick, D. B. (1976). Lipoic acid metabolism in the rat. *Arch. Biochem. Biophys.* **174**, 13–19.

10 Determination of Protein-Bound Lipoic Acid in Tissues by the Enzyme Recycling Method

Tetsuya Konishi* and Lester Packer†
*Department of Radiochemistry-Biophysics
Niigata College of Pharmacy
Niigata, 950-21 Japan
†Department of Molecular and Cell Biology
University of California at Berkeley
Berkeley, California 94720

The tissue level of lipoic acid was determined by a new enzymatic method. Bound lipoyl groups were liberated in the form of lipoyllysine by protease digestion and were assayed by lipoamide dehydrogenase (NADH : lipoamide oxidreductase; EC 1.8.1.4)-mediated NADH (nicotinamide dinucleotide) oxidation. NADH oxidation was coupled to reduction of the lipoyl disulfide group. The fluorescence kinetics of NADH oxidation was enhanced markedly by the addition of glutathione disulfide, recycling the enzyme-mediated lipoyl/dihydrolipoyl conversion. In the presence of a large excess of glutathione disulfide, NADH oxidation follows pseudo-first-order kinetics in terms of the lipoyllysine concentration. A good linear correlation is obtained between the oxidation rate and the lipoyl lysine concentration, up to 5 μM. This method was applied to the protease lysates of bovine, rat, and rabbit tissues to determine lipoyllysine levels. Kidney and liver were found to have the highest content of lipoic acid in the range of 3.9–4.6 nmol/g rat or rabbit wet tissue or 11.6–13.1 nmol/g bovine acetone powder.

INTRODUCTION

α-lipoic acid (lipoate) is a covalently bound disulfide-containing cofactor required for functions of the mitochondrial dehydrogenases such as pyruvate dehydrogenase, α-ketoglutarate dehydrogenase, branched chain α-keto acid dehydrogenase, and glycine cleavage enzyme complexes in eukaryotes and bacteria (1–3). The constrained unique dithiorane ring of lioic acid provides its characteristic redox property [redox potential of -0.32 V for DHLA/lipoate couple compared to -0.24 V for GSH/GSSG (glutathione disulfide) couple], establishing the base of its potential antioxidant activity (4–8).

Intracellular lipoate mainly exists in the protein-bound form, but it remains unclear whether the protein-bound lipoate functions as a physiological antioxidant, as does the free form lipoate, or whether any other protein-bound form exists other than dehydrogenases. Because lipoamidase, which cleaves the amide bond of protein bound lipoate to liberate free lipoate, is present in cells (9–11), it is reasonable to consider that the protein-bound form could be a source of free lipoate. To understand the physiological roles of the intracellular lipoate, it is important to know the precise amounts of protein-bound lipoate in biological tissues.

A new assay method for protein-bound lipoate has been developed that measures tissue lipoate levels of experimental animals (12, 13).

ASSAY PROCEDURE OF PROTEIN-BOUND LIPOATE

In the newly established method, tissue samples are digested by proteases and the tissue lysate containing total lipoate in the form of lipoyllysine or a lipoylated small peptide fragment is measured directly by NADH oxidation mediated by a GSSG-amplified lipoamide dehydrogenase reaction.

The principle of the lipoyllysine assay by the GSSG-amplified lipoamide dehydrogenase reaction is shown in Fig. 1.

Lipoamide dehydrogenase-mediated reduction of lipoyl (disulfide : lipoate) to the dihydrolipoyl group [dithiol : DHLA (dihydrolipoic acid)] accompanies oxidation of NADH. However, the NADH fluorescence changes were too small to determine lipoyllysine concentrations below 100 μM. The addition of GSSG is expected to reoxidize dihydrolipoyllysine back to lipoyllysine because of the redox potential difference, hence recycling the lipoamide dehydrogenase reaction, which leads to an amplification of the NADH fluorescence changes due to its oxidation.

Typically, the assay was carried out as follows. To an aliquot of buffer solution (100 mM sodium phosphate with 5 μM EDTA at pH 8.0) in a

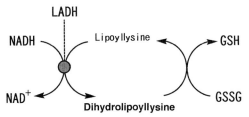

Figure 1   Principle of lipoate assay by the GSSG-amplified lipoamide dehydrogenase reaction. Lipoyllysine-dependent NADH oxidation mediated by lipoamide dehydrogenase was measured fluorometrically using exitation at 360 nm and emission at 465 nm. GSSG addition facilitates lipoamide dehydrogenase-mediated lipoyllysine/dihydrolipoyllysine recycling to amlify NADH oxidation.

quartz cuvette, 2 μl of NADH (40 nmol), 100 μl of GSSG (100 mmol), and successive aliquots of lipoyllysine standard were added to make the total reaction mixture 1.0 ml. After recording a 100% fluorescence level of NADH, 1 μl lipoamide dehydrogenase (2.8 units) was added to initiate NADH oxidation coupled to the lipoamide dehydrogenase-mediated lipoyl disulfide reduction. The NADH oxidation rate was determined from the slope of the fluorescence decay curve at 3 min after initiation of the reaction and was correlated with the lipoyllysine concentration in the reaction mixture.

As shown in Fig. 2, the NADH oxidation rate was correlated linearly with lipoyllysine concentrations up to 5 μM when more than 100 mM GSSG was present. The average CV (coefficient of variation; percent of SD per mean value) obtained for the determination of lipoyllysine levels in the range of 1–5 μM was 0.14. From the linearity of the standard curve, the detection limit of lipoyllysine was found to be about 100 nM under the reaction conditions used.

DETERMINATION OF LIPOYLLYSINE LEVELS IN TISSUE SAMPLES

To determine lipoyllysine levels in animal tissues, tissue samples from laboratory animals and commercially available bovine tissue acetone powders were treated with protease combinations (proteases Type VIII and XIV) for 12 hr at 37°C. The protein lysate was then subjected directly to the lipoamide dehydrogenase assay described earlier after the addition of an aliquot of lipoyllysine as an internal standard.

Because tissue lysates were too turbid to obtain a stable fluorescence baseline for NADH, lysates had to be solubilized with 0.3% (w/v) sodium dodecyl sulfate.

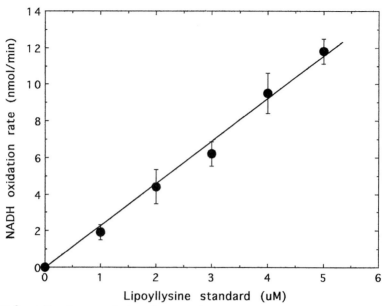

Figure 2   Calibration curve for lipoyllysine. In the presence of excess amounts of GSSG (100 or 200 mM), lipoamide dehydrogenase-mediated NADH oxidation was measured at different lipoyllysine concentrations. Each data point is the average of three to four determinations at 100 mM GSSG and one determination at 200 mM GSSG.

Further, it was found that the lipoyllysine dependence on NADH oxidation was found to be variable in different tissue lysates (Fig. 3). However, a good linearity was obtained between the lipoyllysine internal standard added and the rate of NADH oxidation in each different tissue lysate. Thus the standard addition method is necessary to adapt this assay for tissue samples.

Tissue levels of protein-bound lipoate determined as lipoyllysine are summarized in Table I. In all animals, kidney and liver were found to yield the highest lipoyl contents (3.9–4.6 nmol/g wet tissue for rat and rabbit, and 11.6–13.1 nmol/g bovine acetone powder), followed by brain and heart tissue (1.2–3.2 nmol/g wet tissue for rat and rabbit, and 6.4–6.6 nmol/g bovine acetone powder).

DISCUSSION

The present study revealed that levels of lipoate as lipoyllysine in bovine tissue acetone powder are in the range of 0.8 to 13.1 nmol/g dry tis-

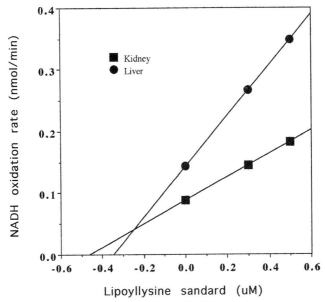

Figure 3 Lipoyllysine dependence of the lipoamide dehydrogenase reaction in different tissue samples. Each tissue sample was solubilized with 0.04% SDS.

sue sample. The highest level was found in kidney, followed by liver, spleen, brain, and heart, whereas the contents in lung and pancreas were low. In fresh tissues from rat and rabbit, lipoyllysine levels were determined in the range of 0.5 to 4.6 nmol/g wet tissues. Among the tissues examined, liver showed the highest content in both animals, followed by the kidney and brain. These values are comparable to those obtained by other methods reported elsewhere. For example, 0.5 to 2 μg lipoate/g wet tissue was determined in mouse tissues by gas chromatography (14). The tissue distribution showed the same tendency as in the present study, namely that kidney, heart, and liver showed the highest level among the tissues examined. Lipoyllysine contents in bovine tissue acetone powders were studied by the high-performance liquid chromatography (HPLC) electrochemical detection method (22); levels of 0.9–2.6 μg/g dry tissues were determined for kidney, heart, and liver tissues. The values obtained in the present study have the same tendency but are somewhat higher compared to these values. The differences are due to the principal differences between the two methods. In the present method, the complete digestion of tissues and the following extraction of freed lipoyllysine are not necessary, thus the recovery is high compared with HPLC analysis.

Table I Animal Tissue Levels of Protein-Bound Lipoate[a]

Tissue	Bovine		Rat		Rabbit	
	nmol/g tissue	μmol/mg protein	nmol/g tissue	μmol/mg protein	nmol/g tissue	μmol/mg protein
Kidney	13.14 ± 0.68	2.05 ± 0.11	3.87 ± 0.34	1.15 ± 0.10	4.00 ± 0.30	1.23 ± 0.09
Liver	11.62 ± 0.50	1.76 ± 0.08	4.57 ± 0.31	1.43 ± 0.10	4.18 ± 0.31	1.16 ± 0.09
Brain	6.55 ± 0.50	1.06 ± 0.08	2.22 ± 0.19	0.48 ± 0.04	3.23 ± 0.18	0.67 ± 0.04
Heart	6.38 ± 0.50	1.45 ± 0.11	1.23 ± 0.11	0.41 ± 0.04	1.82 ± 0.14	0.59 ± 0.05
Lung	2.13 ± 0.28	0.47 ± 0.06	0.51 ± 0.04	0.20 ± 0.02	1.12 ± 0.09	0.38 ± 0.03
Spleen	7.91 ± 0.64	1.02 ± 0.08	—	—	2.25 ± 0.25	0.82 ± 0.09
Pancreas	0.80 ± 0.07	0.54 ± 0.05	—	—	—	—

[a] Protein-bound lipoate was determined as lipoyllysine in the protease lysate of tissues by the method described in the text. Values are the mean and SD of three experiments.

In general, the tissue level of lipoyllysine was found to be higher in metabolically active tissues such as liver, kidney, brain, and heart. Therefore, it is possible that in addition to the energy metabolism of mitochondria through dehydrogenase reactions, certain roles in oxidative stress protection could be implicated for the membrane-bound lipoyllysine in these tissues.

REFERENCES

1. Snoep, J. L., Westphal, A. H., Benen, J. A., Teixeira de Mattos, M. J., Neijssel, O. M., and DeKok, A. (1992). *Eur. J. Biochem.* **203**, 245–250.
2. Pettit, F. H., Yeaman, S. J., and Reed, L. J. (1990). *J. Biol. Chem.* **265**, 8971–8974.
3. Pares, S., Cohen-Addad, C., Sieker, L., Neuburger, M., and Douce, R. (1994). *Proc. Natl. Acad. Sci. USA* **91**, 4850–4853.
4. Packer, L., Witt, E. H., and Tritschler, H. J. (1995). *Free Radic. Biol. Med.* **19**, 227–250.
5. Suzuki, Y. J., Tsuchiya, M., and Packer, L. (1991). *Free Radic. Res. Comm.* **15**, 255–263.
6. Scott, B. L., Aruoma, O. I., Evans, P. J., O'Neill, C., Vander Vliet, A., Tritschler, H., and Halliwell, B. (1994). *Free Radic. Res.* **20**, 119–133.
7. Matsugo, S., Konishi, T., Matuo, D., Tritschler, H. J., and Packer, L. (1996). *Biochem. Biophys. Res. Comm.* **227**, 216–220.
8. Kagan, V. E., Shvedova, A., Serbinova, E., Khan, S., Swanson, C., Powell, R., and Packer, L. (1992). *Biochem. Pharmacol.* **44**, 1637–1649.
9. Nilsson, L., and Kagedal, B. (1994). *Eur. J. Clin. Chem. Clin. Biochem.* **32**, 501–509.
10. Oizumi, J., and Hayakawa, K. (1997). *Biochem. J.* **166**, 427–434.
11. Yoshikawa, K., Hayakawa, K., Katsumata, N., Tanaka, T., Kimura, T., and Yamauchi, K. (1996). *J. Chromatogr.* **679**, 41–47.
12. Konishi, T., Handelman, G., Matsugo, S., Mathur, V. V., Tritschler, H. J., and Packer, L. (1996). *Biochem. Mol. Biol. Int.* **38**, 1155–1161.
13. Akiba, S., Matsugo, S., Packer, L., and Konishi, T. (1998). *Anal. Biochem.* **258**, 299–304.
14. Kataoka, H., Hirabayashi, N., and Makita, M. (1993). *J. Chromatogr.* **615**, 197–202.
15. Matarese, R. M., Spoto, G., and Dupre, S. (1981). *J. Appl. Biochem.* **3**, 372–377.
16. White, R. H. (1981). *Anal. Biochem.* **110**, 89–92.
17. White, R. H. (1980). *Biochemistry* **19**, 9–15.
18. Mattulat, A., and Baltes, W. (1992). *Z. Lebemsar Unters Forsch.* **194**, 326–329.
19. Shih, J. C., and Steinsberger, S. (1981). *Anal. Biochem.* **116**, 65–68.
20. White, K. H., Bleile, D. M., and Reed, L. S. (1980). *Biochem. Biophys. Res. Comm.* **94**, 78–84.
21. Teichert, J., and Preis, R. (1995). *J. Chromatogr.* **672**, 277–281.
22. Lodge, J. K., Han, D., Handelman, D. J., Konishi, T., Matsugo, S., Mathur, V. V., and Packer, L. (1997). *J. Appl. Nutr.* **49**, 3–11.

11 Reaction of Ubiquinols with Nitric Oxide

Enrique Cadenas,* Juan José Poderoso,†
Fernando Antunes,* and Alberto Boveris‡

* Department of Molecular Pharmacology and Toxicology
School of Pharmacy
University of Southern California
Los Angeles, California 90033

† Laboratory of Oxygen Metabolism
University Hospital
University of Buenos Aires
Buenos Aires, Argentina

‡ Department of Biophysics
School of Pharmacy and Biochemistry
University of Buenos Aires
Buenos Aires, Argentina

INTRODUCTION

Coenzyme Q or ubiquinone-10 is a redox component present in all mammalian cell membranes. In the inner mitochondrial membrane, ubiquinone plays a key role in shuttling electrons from complexes I and II to complex III (cytochrome b-c$_1$ segment) of the respiratory chain. In extramitochondrial membranes, ubiquinone may function in its reduced form (ubiquinol) as an antioxidant protecting unsaturated fatty acids from peroxidative damage. A comprehensive description of the antioxidant functions of ubiquinol requires consideration of the participation of its one-electron oxidation product, ubiseminquinone, in oxygen radical production, resulting in possible prooxidant effects (Kagan *et al.,* 1996). This chapter surveys some aspects of the general redox properties of ubiquinols and free radicals, with emphasis on nitric oxide (•NO), their interaction, and the fates of the ubisemiquinone radical thereby derived, as well as the biological significance of these processes in all ubiquinone-containing membranes.

REDOX PROPERTIES OF UBIQUINOLS

Like all quinones, ubiquinones are endowed with two main chemical properties that underlie their biological functions: on the one hand, their ability to undergo reversible one-electron transfers with intermediate formation of a semiquinone species (Fig. 1); on the other hand, their electrophilic character, the basis of which is the cross-conjugated system represented by the alternating single and double bonds, including exocyclic moieties (Brunmark and Cadenas, 1986). Whereas the former property applies to all ubiquinones, regardless of their R substituent, the latter—reactivity of ubiquinone with nucleophiles—is possible only with UQ_0 (R = H), for the presence of isoprenoids substituents (R) in other ubiquinones hinders nucleophilic attack across the double bond. Even in this case, the electron-donating properties of the adjacent —CH_3 substituent are expected to decrease substantially the electrophilic character.

The antioxidant and prooxidant properties of ubiquinone may be analyzed within the context of the redox transitions illustrated in Fig. 1. The $E(Q/Q^{\bullet-})$ and $E(Q^{\bullet-},2H^+/QH_2)$ values of -220 and 190 mV, respectively (Table I) (Rich and Bendall, 1980; Swallow, 1982), permit one to view the reactivity of ubiquinols with free radicals on thermodynamic grounds. Overall, it appears that the ubiquinol $\leftrightarrow$ ubisemiquinone transition may be associated with antioxidant functions, whereas the ubisemiquinone $\leftrightarrow$ ubiquinone redox reaction may be endowed with prooxidant properties.

Ubiquinol $\leftrightarrow$ Ubisemiquinone Redox Transition

The two major antioxidant properties of ubiquinol are linked to the reduction of peroxyl radicals and α-tocopheroxyl radicals within the

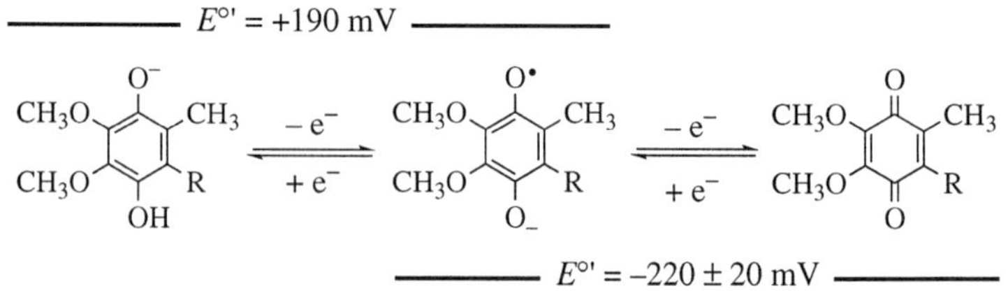

Figure 1   Redox transitions of ubiquinone.

Table I   Relevant Reduction Potentials for Ubiquinones and Nitrogen- and Oxygen-Containing Compounds

Redox couple	$E^{o'}$ (V)[a]
UQ/UQ$^{\bullet-}$	− 0.22
UQ$\bullet-$,H+/UQH−	+ 0.19
$O_2/O_2\bullet-$	− 0.16
$O_2\bullet-$,2H+/H_2O_2	+ 0.94
$HO_2\bullet$,H+/H_2O_2	+ 1.06
ROO$\bullet$,H+/ROOH	+ 1.00
NO$\bullet$/NO− (triplet)	+ 0.39
$NO_2\bullet$/NO_2−	+ 0.99
ONOO−/$NO_2\bullet$	+ 1.40

[a] $E^{o'}$ values at pH 7 taken Rich and Bendall (1980), Swallow (1982), Koppenol and Butler (1985), Brandt (1996), and Koppenol (1996)

ubiquinol ↔ ubisemiquinone transition. Reaction (1) implies that ubiquinol functions as chain-breaking antioxidant reducing peroxyl radicals to hydroperoxides (Frei *et al.,* 1990; Yamamoto *et al.,* 1990; Kagan *et al.,* 1990a); reaction (2) suggests the participation of ubiquinol in a concerted mechanism encompassing the reduction of the α-tocopheroxyl radical to tocopherol (Mukai *et al.,* 1990, 1993; Maguire *et al.,* 1992; Kagan *et al.,* 1990b). In the following reactions, ubiquinols are depicted in the monoanion form (UQH$^-$) because deprotonation of hydroquinones is a requisite condition for electron transfer (Stenken, 1979). In fact, the reaction sequence involving deprotonation → electron transfer → deprotonation (QH_2 → QH$^-$ + H$^+$; QH$^-$ − e$^-$ → QH•; QH• → Q$^{\bullet-}$ + H$^+$) is implicit to all mechanistic models for the oxidation of ubihydroquinone in the respiratory chain (Brandt, 1996) and, likely, in extramitochondrial membranes.

$$UQH^- + ROO\bullet \rightarrow UQ^{\bullet-} + ROOH \tag{1}$$

$$UQH^- + \alpha\text{-TO}\bullet \rightarrow UQ^{\bullet-} + \alpha\text{-TOH} \tag{2}$$

The reaction of ubiquinol-6 and ubiquinol-9 with peroxyl radicals proceeds with second-order rate constants of 3.2 and 3.4×10^5 $M^{-1}sec^{-1}$, respectively (Naumov and Khrapova, 1983). Second-order rate constants for the reaction of several ubihydroquinones with substituted phenoxyl (alkoxyl) radicals in different solvents range from 2.6×10^3 to 2.1×10^4 $M^{-1}\text{sec}^{-1}$ (Mukai *et al.*, 1993). These values indicate that ubiquinols may act as peroxyl radical scavengers, albeit more modestly than α-tocopherol does. Accordingly, the ratio of inhibition of autoxidation of egg phosphatidylcholine in liposomes by ubiquinol-3 and α-tocopherol was 0.7 [k_{inh} (Q_3H_2)/k_{inh} (α-TOH); Landi *et al.*, 1992].

Stopped-flow kinetic studies of vitamin E regeneration with ubiquinol-10 in solution involving α-tocopheroxyl radicals with different substitution patterns revealed that rate constant values for this interaction decreased as the total electron-donating capacity of the alkyl substituents on the aromatic ring of tocopheroxyls increased (Mukai *et al.*, 1990). An approximate k_2 value of 2.1×10^5 $M^{-1}\text{sec}^{-1}$ was calculated (Mukai *et al.*, 1990, 1993); this value is slightly lower than that for the recovery of vitamin E by ascorbate (1.5×10^6 $M^{-1}\text{sec}^{-1}$) (Packer *et al.*, 1979).

Although actual rate constants are not available, the ubiquinol $\leftrightarrow$ ubisemiquinone transition may also be coupled to reduction of the oxoferryl moiety in ferrylmyoglobin (Mordente *et al.*, 1994), ADP-Fe^{3+}-$O_2^{\bullet-}$ complexes used to initiate lipid peroxidation, and lipid alkyl radicals (Forsmark *et al.*, 1991). Likewise, reactions of ubiquinol with $O_2^{\bullet-}$ and its protonated form, $HO_2\bullet$, are certainly possible on thermodynamic grounds ($\Delta G^{\circ\prime}$ values of -15 and -20 kcal/mol, respectively, may be calculated from the reduction potential of the couples involved) (Table I). A rate constant of $\sim 10^4$ $M^{-1}\text{sec}^{-1}$ was reported for the oxidation of *p*-benzoquinol by $HO_2\bullet$ (Nadezhdin and Dunford, 1974). The reaction involving $O_2^{\bullet-}$ is of importance when considering the role of $O_2^{\bullet-}$ as a propagating species in ubiquinol oxidation (Öllinger *et al.*, 1990). The reaction involving $HO_2\bullet$, however, is potentially important when considering the role of this species in initiating lipid peroxidation (Antunes *et al.*, 1996).

Ubisemiquinone $\leftrightarrow$ Ubiquinone Redox Transition

k_3 values for the transfer of electrons from the ubisemiquinone radical anions to O_2 (autoxidation) are influenced profoundly by the quinone substitution pattern and, consequently, the one-electron reduction potential. For example, the k_3 value for the reaction involving durosemiquinone is $\sim 10^8$ $M^{-1}\text{sec}^{-1}$, whereas benzosemiquinone is unreactive toward O_2; in the latter instance, the backward reaction ($Q + O_2^{\bullet-}$) is far more efficient. A k_3 value of $\sim 10^3$ $M^{-1}\text{sec}^{-1}$ for the reaction involving electron transfer from the ubisemiquinone-6 anion to

$$UQ^{\bullet-} + O_2 \leftrightarrow UQ + O_2^{\bullet-} \tag{3}$$

O_2 has been reported (Bensasson and Land, 1973). The autoxidation of ubisemiquinone-10 proceeds with a free energy change of approximately −1.8 kcal/mol.

•NO ↔ NO⁻ Redox Transition

Of interest within the scope of this chapter is the reduction of nitric oxide to nitroxyl anion [or oxonitrate (1−); NO⁻]. •NO is a modest oxidant with a reduction potential of 0.39 V (Koppenol, 1996), a value that places its oxidant strength between those of the α-tocopheroxyl radical [$E(\alpha$-TO•,H⁺/α-TOH = 0.48 V] and the ascorbyl radical [$E(A^{\bullet-}$,H⁺/AH⁻ = 0.28 V]. At variance with the stronger oxidant peroxynitrite (ONOO⁻), there is not much evidence in the literature for the reduction of •NO to NO⁻. This may be due in part to the fast reaction of •NO with $O_2^{\bullet-}$ (reaction 4; $k_4 = 6.9 \times 10^9 \ M^{-1}sec^{-1}$) (Huie and Padmaja, 1993), which has received considerable attention because its potentially deleterious character is based on the chemical reactivity of the product, ONOO⁻.

$$O_2^{\bullet-} + \bullet NO \rightarrow ONOO^- \tag{4}$$

Reduction of •NO to NO⁻ takes place during the •NO-mediated H abstraction of the O-deprotonated hydroxylamine anion (NH$_2$O⁻) (Reaction 5) (Bonner, 1996). Perhaps the most significant biological process encompasses the •NO → NO⁻ transition is the interaction of •NO with Cu,Zn-superoxide dismutase (Reaction 6) (Murphy and Sies, 1991).

$$H_2NO^- + \bullet NO \rightarrow H\bullet NO^- + NOH \tag{5}$$

$$NO\bullet + SOD-Cu^+ \rightarrow NO^- + SOD-Cu^{2+} \tag{6}$$

The reduction of •NO by short-chain analogs of coenzyme Q (ubiquinol-0 and ubiquinol-2) in anaerobic conditions yields ubisemiquinone and NO⁻ (Reaction 7; $k_7 = 0.5–1.6 \times \sim 10^4 \ M^{-1}sec^{-1}$) with a stoichiometry of •NO consumed *per* ubiquinol of ~2 (Poderoso *et al.*, 1998a). In these conditions, the

$$UQH^- + NO\bullet \rightarrow UQ^{\bullet-} + H^+ + NO^- \tag{7}$$

ubisemiquinone species decays by disproportionation (Reaction 8), thus accounting for the

$$UQ^{\bullet-} + UQH\bullet \rightarrow UQ + UQH^- \tag{8}$$

stoichiometry. The rate of disproportionation is dependent on pH ranging 8.4×10^4 to $4.8 \times 10^7 \, M^{-1}sec^{-1}$ for the reaction involving the anionic or the protonated forms of the semiquinone, respectively (Land and Swallow, 1970). At neutral pH, the disproportionation involving the protonated and anionic forms contributes to the fast decay of the semiquinone species. Decay of ubisemiquinone via autoxidation (Reaction 3) appears to prevail in aerobic conditions.

EXTRAMITOCHONDRIAL UBIQUINOL AND FREE RADICALS

Following the pioneering studies by Lea and Kwietzy (1962) and Mellors and Tappel (1966), several research groups have provided a wealth of information and circumstantial evidence (see Lenaz *et al.*, 1990) regarding the antioxidant properties of ubiquinol. This evidence is derived largely from experiments with reconstituted membrane systems, mitochondrial membranes, and low-density lipoproteins (LDL) to observations on intact animals as well as in the clinical setting. Excellent reviews by Ernster and Dallner (1995) and Kagan *et al.* (1996) address in detail the multiple aspects inherent in the antioxidant functions of ubiquinol. Surprisingly, there are very little kinetic data on the mechanistic aspects by which this compound exerts its antioxidant activity, and the precise molecular and cellular mechanism(s) underlying the antioxidant function of ubiquinol in various biological membranes remains to be elucidated. Although the protective effects of ubiquinol against lipid peroxidation (encompassed by Reactions 1 and/or 2) are well documented, the fate of the ubisemiquinone radical form is not addressed.

As has been implied or suggested before (Cadenas *et al.*, 1992; Kagan *et al.*, 1996), the antioxidant functions of ubiquinol are encompassed by the $UQH^- \rightarrow UQ^{\cdot-}$ redox transition, whereas their prooxidant character arises from O_2 reduction coupled to the $UQ^{\cdot} \rightarrow UQ$ reaction (Fig. 2). These concepts are simply an expression of the physicochemical properties of ubiquinones. Whether ubiquinols elicit their antioxidant action per se (Reaction 1) (Ernster *et al.*, 1992) or in a concerted manner with α-tocopherol (Reaction 2) (Kagan and Packer, 1993) is a matter of debate; it is plausible, however, that both mechanisms, a chain-breaking antioxidant activity and recovery of vitamin E, are operative in a cellular setting.

As mentioned earlier, ubiquinone is not only present in the inner mitochondrial membrane, but in LDL, plasma membranes, and all intracellular membranes (its content in Golgi and lysosomal membranes being the highest) (Åberg *et al.*, 1992). Ubiquinone in membranes other than the inner mitochondrial membrane is found largely in the reduced state (Åberg *et al.*,

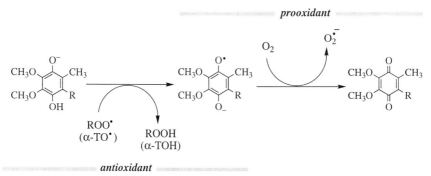

Figure 2   Antioxidant and prooxidant functions of ubiquinone.

1992), an observation suggesting that redox cycles of ubiquinone may play significant roles in cellular organelles and raising the question concerning its recovery. In this regard, two enzymic activities may be plausible candidates in maintaining ubiquinone in its reduced state in extramitochondrial membranes. First, a coenzyme Q reductase from liver plasma membrane has been purified and its role in transplasma membrane electron transport discussed. This is a 34-kDa protein sensitive to quinone site inhibitors, which requires coenzyme Q as an intermediate electron carrier between NADH and extracellular electron acceptors (Villalba *et al.*, 1995). Second, a cytosolic NADPH-dependent ubiquinone reductase appears to account for 68% of the total ubiquinone reductase activity of rat liver homogenates (Takahashi *et al.*, 1995, 1996). This enzyme reduces ubiquinone by a two-electron transfer mechanism and is distinguishable from other known microsomal, mitochondrial, and cytosolic quinone reductases (Takahashi *et al.*, 1995, 1996). Both enzymes might be involved in maintaining ubiquinone in a reduced state in extramitochondrial membranes and, consequently, may be critical for the antioxidant properties of this compound.

   The role of ubiquinol in low-density lipoproteins is important when addressing the antioxidant status of LDLs, their oxidative modification, and their potential involvement in the early stages of atherosclerotic lesions. Despite the lack of kinetic information on Reaction 1, the high effectivity of ubiquinol as an antioxidant in LDL particles may be due in part to the reductive elimination of lipid peroxyl radicals (i.e., ubiquinol as a chain-breaking antioxidant) (Ingold *et al.*, 1993). The autoxidation of ubisemiquinone referred to earlier is not necessarily associated with a prooxidant function within the LDL particle, as it was suggested that autoxidant of ubisemiquinone serves to export the radical character ($O_2^{\cdot-}$) into plasma, which subsequently reduces the α-tocopheroxyl radical

(Bowry and Stocker, 1993). The latter reaction occurs in a manner similar to that shown in a pulse radiolysis study for the water-soluble analog of vitamin E, Trolox C (Reaction 9; $k_9 = 4.5 \pm 0.5 \times 10^8 \ M^{-1}sec^{-1}$),

$$O_2^{\cdot -} + \alpha\text{-TO}\cdot + H^+ \rightarrow O_2 + \alpha\text{-TOH} \tag{9}$$

(Cadenas *et al.*, 1989). Viewed under this perspective, the antioxidant efficacy of ubiquinol in LDL is apparently higher than that of the α-tocopherol/ascorbate couple; however, the latter seems to be the major antioxidant system in LDL under physiological conditions, for only 50–60% of the LDL particles contain a molecule of ubiquinol (Mohr *et al.*, 1992).

From these observations it may be surmised that ubiquinone is the product ensuing from the antioxidant activity of ubiquinol in LDL. This, along with the facts that ubiquinone is present in LDL predominantly in its reduced form (Okamoto *et al.*, 1989; Stocker *et al.*, 1991) and that dietary supplementation with ubiquinone results in increased levels of ubiquinol in circulating lipoproteins (Mohr *et al.*, 1992), purports the requirement of systems in charge of the recovery and/or maintenance of this antioxidant. The transplasma membrane reductase activities present in different cells exposed to circulating LDL *in vivo* may participate in the reduction of lipoprotein ubiquinone; thus, hepatocytes and erythrocytes reduce ubiquinone-10 in LDL, albeit at slower rates, than they reduce ubiquinone-1 (Stocker and Suarna, 1993). Of note, this work considers that the enzymic reduction of ubiquinone is a direct two-electron transfer to ubiquinol. Whether the flavins of transplasma membrane reductases actually catalyze two-electron reduction of extracellular electron acceptors needs to be assessed; usually, these activities are examined by the univalent reduction of ferricyanide (which cannot enter cells); moreover, some quinones are reduced to their semiquinones by transplasma membrane redox systems in the erythrocyte (Pedersen *et al.*, 1990).

Regardless of whether the maintenance of ubiquinol involves one- or two-electron transfer reductases present in cytosol (Takahashi *et al.*, 1995, 1996) or in membranes (Villalba *et al.*, 1995), the antioxidant properties of this compound in extramitochondrial membranes may be widened by considering its reactivity toward •NO. The latter is a free radical formed in a variety of cell types by NO-synthase and is involved in an wide array of important physiological and pathophysiological phenomema (for a review see Gross and Wolin, 1995). The diffusibility of •NO and its concentration gradients within the cell (Beckman, 1996) provide a situation in which interception of this radical by physiological electron donors (antioxidants) may be of interest. In this connection, the reaction of •NO with ubiquinols (Reaction 7) takes on a new meaning, not only because of the wide distribution of ubiquinol in cell membranes and the diffusibility of •NO, but also because these redox systems may encompass the generation and removal of oxidants stronger than •NO and responsible for cytotoxicity.

Whatever the nature of the free radicals (peroxyl or •NO), their scavenging by UQH_2 yields a semiquinone (Figs. 2 and 3), which usually decays by electron transfer to O_2 to form $O_2^{\bullet-}$ (Reaction 3). Although •NO may initiate ubiquinol oxidation (Reaction 7), it is feasible that its further oxidation occurs via reaction 10 and that $ONOO^-$ is involved in the propagation steps. The reaction of •NO with $O_2^{\bullet-}$ appears to be relevant because (1) it accelerates semiquinone autoxidation (Reaction 3) on the effective removal of $O_2^{\bullet-}$ (Reaction 4) and (2) the product of semiquinone autoxidation, $O_2^{\bullet-}$, is a "reducing" radical $[E(O_2/O_2^{\bullet-}) = -0.156 \text{ V}]$ (Koppenol and Butler, 1985). After its reaction with •NO, a strong "oxidizing" species ($ONOO^-$) is formed $[E(ONOO^-/NO_2\bullet, HO^-) = 1.4 \text{ V}]$ (Koppenol, 1996), which propagates hydroquinone autoxidation (Fig. 3).

The oxidation of UQH_2 by $ONOO^-$ (Reaction 10) is thermodynamically favorable considering the

$$UQH^- + ONOO^- \rightarrow UQ^{\bullet-} + NO_2\bullet + HO^- \qquad (10)$$

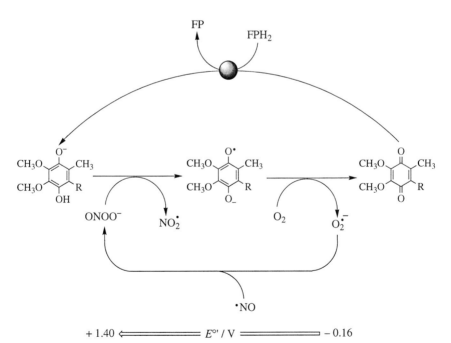

Figure 3 Antioxidant functions of extramitochondrial ubiquinol and nitric oxide.

reduction potentials of the couples involved $[E(UQ^{\bullet-}/UQH^-)] = 0.19$ V (Rich and Bendall, 1980; Brandt, 1996) and $E(ONOO^-/NO_2\bullet, HO^-) = 1.4$ V (Koppenol, 1996)]. Additionally, the oxidation of UQH_2 by the product of Reaction 10, nitrogen dioxide ($NO_2\bullet$) (Reaction 11), has no thermodynamic constrains $[E(NO_2\bullet/NO_2^-) = 0.99$ V] (Koppenol, 1996). For example, $NO_2\bullet$ was reported to react with thiols [having a reduction potential of 0.9 V (Surdhar and Armstrong, 1987), more positive than that of ubihydroquinone] with second-order rate constants of $\sim 10^8$ $M^{-1}sec^{-1}$ (Prütz *et al.*, 1985).

$$UQH^- + NO_2\bullet \rightarrow UQ^{\bullet-} + NO_2^- \tag{11}$$

This discussion suggests that at least three oxidants may be coupled to the $UQH^- \rightarrow UQ^{\bullet-}$ transition with different efficiency: $ONOO^- > NO_2\bullet > \bullet NO$ [based on their respective reduction potentials of 1.4, 0.99, and 0.39 V, respectively (Koppenol, 1996)]. On thermodynamic grounds, the antioxidant activity of ubiquinols—coupled to the regeneration via NADPH-ubiquinone reductase present in cytosol (Takahashi *et al.*, 1995, 1996) or coenzyme Q reductase in membranes (Villalba *et al.*, 1995)—may be an effective cellular device to prevent the oxidative damage elicited by a broad range of nitrogen radicals (Fig. 3). At variance with the scavenging of peroxyl radicals by ubiquinol, in this reaction scheme $O_2^{\bullet-}$ is not involved in hydroquinone propagation but—through its reaction with $\bullet NO$—in $ONOO^-$ detoxification. Another point of interest is that the reaction of $\bullet NO$ with alkyl, alkoxyl, and peroxyl radicals is very rapid ($k_{12} = 2 \times 10^9$ $M^{-1}sec^{-1}$) (Padmaja and Huie, 1993) and, certainly, four orders of magnitude higher than that of ubiquinol with peroxyl radicals. Reaction 12,

$$ROO\bullet + \bullet NO \rightarrow ROONO \tag{12}$$

together with the cell steady-state level of $\bullet NO$ ($\sim 10^{-7}$ M) and diffusibility of this species, may aid in the understanding of the effects of $\bullet NO$ on lipid peroxidation (Rubbo *et al.*, 1994).

INNER MITOCHONDRIAL MEMBRANE UBIQUINOL AND FREE RADICALS

The ubiquinol content in rat heart mitochondria under normoxic conditions may be calculated as $\sim 1.08 \times 10^{-4}$ M, assuming a mitochondrial content of UQH_2 of 2 nmol/mg protein in four resting state conditions (Boveris and Stoppani, 1970) and 54 mg mitochondrial protein/g rat heart (Costa *et al.*, 1988). In addition to its well-established role as an electron carrier in the respiratory chain, the ubiquinone pool may participate in antioxidant actions or be a source of oxygen radicals. This discrepancy is only apparent when considering that the former activity is the domain of the ubiquinol → ubisemiquinone transition and the latter is that of the ubisemiquinone → ubiquinone redox reaction.

Mitochondrial ubiquinone has been shown to protect mitochondrial membranes against lipid peroxidation: studies with ubiquinol homologs with different isoprenoid chain lengths showed that long-chain homologs were more effective than short-chain homologs (Kagan *et al.*, 1990a). This antioxidant function may be ascribed to a synergism with vitamin E (Kagan *et al.*, 1990b) or a direct radical scavenging activity (Ernster et al., 1992). Whatever the actual molecular mechanism, ubiquinol also provides protection against the oxidative damage to proteins and DNA ensuing from lipid peroxidation (Forsmark-Andrée *et al.*, 1995). Expectedly, the mitochondrial ubiquinone pool undergoes an oxidative modification or destruction during lipid peroxidation, an event that was suggested to be the primary cause of inactivation of the respiratory chain (Forsmark-Andrée *et al.*, 1997).

At variance with the recovery of ubiquinol in extramitochondrial membranes, which presumably involves a single enzymic activity, this process is complex in the respiratory chain: ubiquinol transfers one of its two electrons to the iron sulfur clusters in cytochrome reductase, shuttled further to cytochrome c_1 and cytochrome c. The second electron, within the ubisemiquinone → ubiquinone transition, is transferred to cytochromes b, which in turn reduce $UQ^{\cdot-}$ to UQH^{-}. Hence, at least in mitochondrial membranes, the ubisemiquinone formed during peroxyl radical (Reaction 1) or α-tocopheroxyl radical (Reaction 2) reduction may be "recovered" via reaction with the cytochrome b_{562} component of the mitochondrial "Q" cycle (Reaction 13). It may be hypothesized that the cycle encompassed by Reactions 1 or 2 and Reaction 13 (redox cycling in the classical sense) will provide an efficient antioxidant system.

$$UQ^{\cdot-} + b^{2+}{}_{562} + H^+ \rightarrow UQH- + b^{3+}{}_{562} \qquad (13)$$

Alternatively, the ubisemiquinone species may autoxidize (Reaction 3), a well-documented process, which appears to be an effective cellular source of $O_2^{\cdot-}$ and, on disproportionation, H_2O_2 (Boveris and Cadenas, 1975; Boveris *et al.*, 1976; Cadenas *et al.*, 1977; Forman and Boveris, 1982; Turrens *et al.*, 1985; Giulivi *et al.*, 1995; Lass *et al.*, 1997). Although it has been argued that changes in membrane fluidity, as it occurs in different toxicological and pathological situations, are required to observe ubisemiquinone autoxidation (Konstantinov *et al.*, 1987; Turrens *et al.*, 1991; Kagan *et al.*, 1996), there is consensus that this may be a major site for $O_2^{\bullet-}$ production and, consequently, H_2O_2 formation in the mitochondrial respiratory chain.

A potential interaction of •NO with inner membrane ubiquinol is relevant in view that •NO itself elicits important reversible changes in the mitochondrial electron-transfer chain (Cleeter *et al.*, 1994; Poderoso *et al.*, 1996), such as inhibition of cytochrome oxidase, binding to the cytochrome bc_1 segment, and, probably, inhibition of succinate dehydrogenase. These effects, which may lead to an increased level of ubisemiquinone and, consequently, an enhanced production of $O_2^{\bullet-}$ and H_2O_2, need to be analyzed within a sequence encompassing (a) competition between •NO and O_2 for cytochrome oxidase, (b) changes in the mitochondria electron carriers redox status, and (c) oxyradical production.

Hypoxic Conditions and Inhibition of Cytochrome Oxidase by •NO

Competition between O_2 and •NO for cytochrome oxidase should be expected to occur at $\sim 1 \times 10^{-6} \, M \, O_2$ (the anoxic edge, corresponding to the O_2 concentratin that decreases electron transfer by cytochrome oxidase by 50%, depending on mitochondrial metabolism) and at about $10^{-7} \, M$ •NO (the concentration that inhibits cytochrome oxidase activity to one-half). This notion provides some basis to speculate about a regulatory role of •NO in tissue O_2 uptake. A decrease in blood vessels pO_2 stimulates the release of •NO by the endothelium (Pohl and Busse, 1989), reaching vascular tissue concentrations of $1.3–2.7 \times 10^{-6} \, M$ (Horton *et al.*, 1994). These relatively high •NO levels ensure an endothelial–mitochondrial gradient for •NO diffusion into mitochondria, where it binds to and inhibits cytochrome oxidase and electron transfer at the ubiquinone–cytochrome bc_1 rgion of the respiratory chain. The inhibitory effect of •NO on mitochondrial electron transport is expected to be more profound at low O_2 tensions (Poderoso *et al.*, 1996; Cassina and Radi, 1996). Accordingly, inhibition of ATP synthesis, although transient, continues for longer periods at low O_2 concentrations, such as those in the cytosol (Takehara *et al.*, 1995; Okada *et al.*, 1996). The advantage of reducing tissue O_2 consumption becomes evident after considering that an •NO-inhibited cytochrome oxidase will allow O_2 to diffuse further along its gradient, reaching other mitochondria or cells, and lowering the steepness of the pO_2 gradient in the normoxic–anoxic transition (Poderoso *et al.*, 1996) and making the respiration rate sensitive to pO_2 (Brown, 1997).

Cytochrome b Redox Status and •NO

Cytochrome spectra are consistent with a multiple inhibition of the mitochondrial respiratory chain in the presence of •NO, predominantly involving the binding of these species to the cytochrome bc_1 segment and cytochrome oxidase. The reaction of •NO with the latter probably involves two binding sites (Torres *et al.*, 1995): ferrocytochrome a_3 and Cu_B^+, the latter having a lower affinity for •NO. The inhibition of NADH- and succinate-cytochrome *c* reductase activities by relatively high concentration of •NO (1.2×10^{-6} *M*) (Poderoso *et al.*, 1996) suggests that, in addition to the reported effect on cytochrome oxidase (Cleeter *et al.*, 1994), there is a second •NO-sensitive site in the common pathway for both reductases of the electron transfer chain. •NO elicits cytochrome *b* reduction in the presence of succinate while cytochromes aa_3 and *c* remain oxidized; this effect indicates inhibition of electron transfer at the O_2 side of cytochrome *b* (Poderoso *et al.*, 1996).

Formation of Oxyradicals by •NO-Inhibited Respiratory Chain

The previously described sequence builds a situation in which $O_2^{\bullet-}$ and H_2O_2 may be generated, probably involving autoxidizable compounds on the electron donor side of cytochrome *b*. Consistent with this notion, •NO-mediated inhibition of mitochondrial electron transfer resulted in an enhancement of $O_2^{\bullet-}$ production (Poderoso *et al.*, 1996). This effect is transient because the removal of •NO on its reaction with $O_2^{\bullet-}$, producing stoichiometric amounts of $ONOO^-$ (Reaction 4), would "release" cytochromes from the inhibitory effects. Considering the rates of •NO and $O_2^{\bullet-}$ production under physiological conditions and the short half-life of $ONOO^-$ (less than 1 sec), the intramitochondrial production of this species should remain at relatively low rates. This notion seems important given the report that at $ONOO^-$ 0.5–1 m*M* was able to inhibit succinate dehydrogenase, promoting H_2O_2 release (Radi *et al.*, 1994).

The potential reaction of •NO with ubiquinol or $O_2^{\bullet-}$ in the last step of the sequence may be analyzed as follows (Fig. 4). The rate of •NO utilization via Reaction 7 may be calculated, assuming the aforementioned concentration of ubiquinol in mitochondria in state 4 (Boveris and Stoppani, 1970) and a steady-state level of •NO of $\sim 1 \times 10^{-7}$ *M* (Beckman, 1996; Poderoso *et al.*, 1998b). Under these conditions, a rate of $\sim 1.7 \times 10^{-7}$ *M* sec^{-1} may be estimated. An elevated steady-state level of reduced ubiquinone may expectedly increase the utilization of •NO via Reaction 7 (Fig. 4*a*); such a situation may be achieved by •NO itself on inhibition of cytochrome oxidase (Cleeter *et al.*, 1994; Torres *et al.*, 1995; Poderoso *et al.*, 1996) and binding to the cytochrome bc_1 region, thereby resulting in a higher level of UQH_2 (Poderoso *et al.*, 1996). Alternatively, in hypoxia and anoxia, the mitochondrial ubiquinol level may be increased up to 1.85×10^{-4} *M* (Boveris and Stoppani, 1970) and the steady-state level of $O_2^{\bullet-}$ will

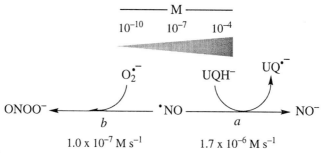

Figure 4   The reaction of •NO with ubiquinol and $O_2^{•-}$. Upper values correspond to increasing steady-state levels of $O_2^{•-}$, •NO, and ubiquinol.

be decreased to nearly zero. Under these conditions, •NO produced by endothelial cNOS may decay largely on its reaction with ubiquinol (Fig. 4a).

By comparison, the rate of utilization of •NO via Reaction 4 (Fig. 4b) may be calculated as 1.04×10^{-7} M sec^{-1}, assuming a steady-state level of $O_2^{•-}$ ~1.5×10^{-10} M (Boveris and Cadenas, 1997) and a k_4 value of 6.7×10^{-9} M^{-1}sec^{-1} (Hui and Padmaja, 1993). This rate may actually be an overestimation because it considers a high steady-state level of $O_2^{•-}$, calculated on the assumption that $O_2^{•-}$ decays only by a superoxide dismutase-catalyzed pathway.

The reversible inhibition of mitochondrial respiration rate by •NO apparently does not damage mitochondria (Borutaité and Brown, 1996). It has been proposed that mitochondria catalyzed •NO breakdown by two separate mechanisms and that both mechanisms were stimulated at very low O_2 tensions and presumably involved reductive reactions. One of these reductive pathways is sensitive to azide and cyanide and apparently involves the reduction of •NO (to N_2O?) by cytochrome oxidase (Borutaité and Brown, 1996). The other reductive pathway may be represented by the interaction of •NO with ubiquinol illustrated in Fig. 4a (Poderoso et al., 1998a). A different picture emerges in situations such as cardiac posthypoxic reoxygenation: ONOO$^-$, formed on prevalence of the pathway in Fig. 4b, was suggested to be the species responsible for a slow reversal of the •NO effect and the ensuing oxidative damage (Xie and Wolin, 1996). The irreversible inhibition of mitochondrial respiration by ONOO$^-$ apparently occurs at complexes I–III; it was suggested that a requirement for this inhibition would be the formation of ONOO$^-$ in the immediate vicinity of a redox site within the respiratory chain (Lizasoain et al., 1996). ONOO$^-$ in the mitochondrial matrix may be effectively reduced

$$E-Se^- + H^+ + ONOO^- \rightarrow E-SeOH + ONO^- \qquad (14)$$

to nitrite by a novel peroxynitrite reductase activity of glutathione peroxidase; the catalytic mechanism appears to involve the steps of the classical glutathione peroxidase using hydroperoxides as substrates, with the step responsible for $ONOO^-$ reduction being the oxidation of selenoate to seleninic acid in the enzyme (Reaction 14) (Sies *et al.*, 1997).

CONCLUDING REMARKS

Evaluation of the biological significance of the reaction of •NO with ubiquinol requires consideration of the cellular O_2 gradients, the steady-state levels of ubiquinols and •NO, and the distribution of ubiquinone in membranes, distinguishing all cell membranes from the inner mitochondrial membrane. This distinction is important because of the following: in the former cases, following its reaction with free radicals, ubiquinol may be recovered by a direct two-electron transfer catalyzed by cytosolic NADPH-ubiquinone reductase (Takahashi *et al.*, 1995, 1996) or membrane NADH-coenzyme Q reductase (Villalba *et al.*, 1995) (Fig. 3). In the latter case, if operative, ubiquinol may be recovered from ubisemiquinone by cytochrome b^{2+}_{562} (Reaction 13).

The wide range of effects elicited by •NO is achieved at concentrations of this species, ranging from 10^{-9} to 10^{-5} M: the activation of cytosolic guanylate cyclase (Ignarro, 1996), inhibition of cytochrome oxidase (Cleeter *et al.*, 1994; Torres *et al.*, 1995; Poderoso *et al.*, 1996), and effects on the cytochrome bc_1 region in mitochondria (Poderoso *et al.*, 1996) require progressively increasing concentrations of •NO. The reaction of •NO with ubiquinol in the respiratory chain may take place at concentrations of •NO over those needed for effective binding to cytochromes and favored by the resulting increase in ubiquinol. Hence, the reported antimycin-like effect of •NO on isolated mitochondria leading to an enhancement of H_2O_2 production (Poderoso *et al.*, 1996) may be reached at a critical concentration of •NO, enough to oxidize QH- via Reaction 7, and lower than that of Mn-SOD [0.3–1.1 $\times$ 10^{-5} M (Boveris and Cadenas, 1997)]. Above these •NO concentrations, inhibition of H_2O_2 production may be expected and $ONOO^-$ formation favored via Reaction 4. Whether $ONOO^-$, formed under these conditions, would elicit oxidative damage in mitochondria would depend on its site of generation (Lizasoain *et al.*, 1996), its interception by matrix GSH (Augusto *et al.*, 1994), and its reduction by the peroxynitrite reductase activity of mitochondrial glutathione peroxidase (Sies *et al.*, 1997).

It is worth noting that the effects encompassed by the sequence mentioned earlier are transient and reversible and that •NO-induced $O_2^{•-}$ production in mitochondria may establish a sensitive regulatory control of mitochondrial O_2 uptake and functions (Poderoso *et al.*, 1996), enabling mitochondria to act as sensors of O_2 over the physiological range (Brown and Cooper, 1994; Brown, 1995). Furthermore, the reductive pathway for •NO consumption in Fig. 4*a* may contribute to the reversibility of •NO effects on mitochondria.

ACKNOWLEDGMENT

Supported by Grant HL53467 from NIH. F.A. holds a fellowship from Praxis XXI/BPD/11778/97.

REFERENCES

Åberg, F., Appelkvist, E.-L., Dallner, G., and Ernster, L. (1992). Distribution and redox state of ubiquinones in rat and human tissues. *Arch. Biochem. Biophys.* **295**, 230–234.

Antunes, F., Salvador, A., Marinho, H. S., Alves, R., and Pinto, R. E. (1996). Lipid peroxidation in mitochondria inner membranes. I. An integrative model. *Free Radic. Biol. Med.* **21**, 917–943.

Augusto, O., Gatti, R. M., and Radi, R. (1994). Spin trapping studies of peroxynitrite decomposition and of 3-morpholinosydnonimime N-ethylcarbamide autoxidation. *Arch. Biochem. Biophys.* **310**, 118–125.

Beckman, J. S. (1996). The physiological and pathological chemistry of nitric oxide. *In* "Nitric Oxide: Principles and Actions (J. Lancaster, Jr., ed.), pp. 1–82. Academic Press, San Diego.

Bensasson, R., and Land, E. J. (1973). Optical and kinetic properties of semireduced plastoquinone and ubiquinone: Electron acceptors in photosynthesis. *Biochim. Biophys. Acta* **325**, 175–181.

Bonner, F. T. (1996). Nitric oxide gas. *Methods Enzymol.* **268**, 50–57.

Borutaité, V., and Brown, G. C. (1996). Rapid reduction of nitric oxide by mitochondria, and reversible inhibition of mitochondrial respiration by nitric oxide. *Biochem. J.* **315**, 1295–299.

Boveris, A., and Cadenas, E. (1975). Mitochondrial production of superoxide anions and its relationship to the antimycin-insensitive respiration. *FEBS Lett.* **54**, 311–314.

Boveris, A., and Cadenas E. (1997). Cellular sources and steady-state levels of reactive oxygen species. *In* "Oxygen, Gene Expression, and Cellular Function" (L. B. Clerch, and D. J. Massaro, eds.), pp 1–25. Dekker, New York.

Boveris, A., Cadenas, E., and Stoppani, A. O. M. (1976). Role of ubiquinone in the mitochondrial generation of hydrogen peroxide. *Biochem. J.* **156**, 435–444.

Boveris, A., and Stoppani, A. O. M. (1970). Inhibition of electron transport and energy transfer by 19-norethylnyltestosterone acetate. *Arch. Biochem. Biophys.* **141**, 641–655.

Bowry, V. W., and Stocker, R. (1993). Tocopherol-mediated peroxidation: The prooxidant effect of vitamin E on the radical-initiated oxidation of human low density lipoprotein. *J. Am. Chem. Soc.* **115**, 6029–6044.

Brandt, U. (1996). Bifurcated ubihydroquinone oxidation in the cytochrome bc_1 complex by protongated charge transfer. *FEBS Lett.* **387**, 1–6.

Brown, G. C. (1995). Hypothesis: Nitric oxide regulates mitochondrial respiration and cell functions by inhibiting cytochrome oxidase. *FEBS Lett.* **369**, 136–139.

Brown, G. C. (1997). Nitric oxide inhibition of cytochrome oxidase and mitochondrial respiration: Implications for inflammatory, neurodegenerative and ischaemic pathologies. *Mol. Cell Biochem.* **174**, 189–192.

Brown, G. C., and Cooper, C. E. (1994). Nanomolar concentrations of nitric oxide reversibly inhibit synaptosomal respiration by competing with oxygen at cytochrome oxidase. *FEBS Lett.* **356**, 295–298.

Brunmark, A., and Cadenas, E. (1986). Redox and addition chemistry of quinoid compounds and its biological implications. *Free Radic. Biol. Med.* **7**, 435–477.

Cadenas, E., Boveris, A., Ragan, C. I., and Stoppani, A. O. M. (1977). Production of superoxide radicals and hydrogen peroxide by NADH-ubiquinone reductase and ubiquinol-cytochrome c reductase from beef-heart mitochondria. *Arch. Biochem. Biophys.* **180**, 248–257.

Cadenas, E., Hochstein, P., and Ernster, L. (1992). Pro- and antioxidant functions of quinones and quinone reductases in mammalian cells. *Adv. Enzymol.* **65**, 97–146.

Cadenas, E., Merényi, G., and Lind, J. (1989). Pulse radiolysis study on the reactivity of trolox C phenoxyl radical with superoxide anion. *FEBS Lett.* **253**, 235–238.

Cassina, A., and Radi, R. (1996). Differential inhibitory action of nitric oxide and peroxynitrite on mitochondrial electron transport. *Arch. Biochem. Biophys.* **328**, 309–316.

Cleeter, M. W. J., Cooper, J. M., Darley-Usmar, V. M., Moncada, S., and Schapira, A. H. V. (1994). Reversible inhibition of cytochrome c oxidase, the terminal enzyme of the mitochondrial respiratory chain, by nitric oxide: Implications for neurodegenerative diseases. *FEBS Lett.* **345**, 50–54.

Costa, L. RE., Boveris, A., Koch, O. R., and Taquini, A. C. (1988). Liver and heart mitochondria in rats submitted to chronic hypobaric hypoxia. *Am. J. Physiol.* **255**, C123–C129.

Ernster, L., and Dallner, G. (1995). Biochemical, physiological, and medical aspects of ubiquinone function. *Biochim. Biophys. Acta* **1271**, 195–204.

Ernster, L., Forsmark, P., and Nordenbrand, K. (1992). The mode of action of lipid-soluble antioxidants in biological membranes: Relationship between the effects of ubiquinol and vitamin E as inhibitor of lipid peroxidation in submitochondrial particles. *J. Nutr. Sci. Vitaminol.* **548**, 41–46.

Forman, H. J., and Boveris, A. (1982). Superoxide radical and hydrogen peroxide in mitochondria. *In* "Free Radicals in Biology" (W. A. Pryor, ed.), Vol 5, pp. 65–90. Academic Press, San Diego.

Forsmark, P., Åberg, F., Norling, B., Nordenbrand, K., Dallner, G., and Ernster, L. (1991) Vitamin E and ubiquinol as inhibitors of lipid peroxidation in biological membranes. *FEBS Lett.* **285**, 39–43.

Forsmark-Andrée, P., Dallner, G., and Ernster, L. (1995). Endogenous ubiquinol prevents protein modification accompanying lipid peroxidation in beef heart submitochondrial particles. *Free Radic. Biol. Med.* **19**, 749–757.

Forsmark-Andrée, P., Lee, C.-P., Dallner, G., and Ernster, L. (1997). Lipid peroxidation and changes in the ubiquinone content and the respiratory chain enzymes of submitochondrial particles. *Free Radic. Biol. Med.* **22**, 391–400.

Frei, B., Kim, M. C., and Ames, B. N. (1990). Ubiquinol-10 is an effective lipid-soluble antioxidant at physiological concentrations. *Proc. Natl. Acad. Sci. USA* **87**, 4879–4883.

Giulivi, C., Boveris, A., and Cadenas, E. (1995). Hydroxyl radical generation during mitochondrial electron transfer and the formation of 8-hydroxy-desoxyguanosine in mitochondrial DNA. *Arch. Biochem. Biophys.* **316,** 909–916.

Gross, S., S., and Wolin, M. S. (1995). Nitric oxide: Pathophysiological mechanisms. *Annu. Rev. Physiol.* **57,** 737–769.

Horton, R. A., Ceppi, E. D., Knowles, R. G., and Titheradge, M. A. (1994). Inhibition of hepatic gluconeogenesis by nitric oxide: A comparison with endotoxic shock. *Biochem. J.* **299,** 735–739.

Huie, R. E., and Padmaja, S. (1993). The reaction of NO with superoxide. *Free Radic. Res. Commun.* **18,** 195–199.

Ignarro, L., J. (1996). Nitric oxide as a communication signal in vascular and neuronal cells. *In* "Nitric Oxide: Principles and Actions" (J. Lancaster, Jr., ed.), pp. 111–137. Academic Press, San Diego.

Ingold, K. U., Bowry, V. W., Stocker, R., and Walling, C. (1993). Autoxidation of lipids and antioxidation by α-tocopherol and ubiquinol in homogeneous solution and in aqueous dispersions of lipids: Unrecognized consequences of lipid particle size as exemplified by oxidation of human low density lipoprotein. *Proc. Natl. Acad. Sci. USA* **90,** 45–49.

Kagan, V. E., Nohl, H., and Quinn, P. J. (1996). Coenzyme Q: Its role in scavenging and generation of radicals in membranes. *In* "Handbook of Antioxidants" (E. Cadenas, and L. Packer, eds.) pp. 157–202. Dekker, New York.

Kagan, V. E., and Packer L. (1993). Electron transport regenerates vitamin E in mitochondria and microsomes via ubiquinone: An antioxidant duet. *In* "Free Radicals and Antioxidants in Nutrition" (F. Corongiu, S. Banni, M. A. Dessi, and C. Rice-Evans, eds.), pp. 27–36. Richelieu Press, London.

Kagan, V. E., Serbinova, E. A., Koynova, E. A., Kitanova, S. A., Tyurin, V. A., Stoytchev, T. S., Quinn, P. J., and Packer, L. (1990*a*). Antioxidant action of ubiquinol homologues with different isoprenoid chain length in biomembranes. *Free Radic. Biol. Med.* **9,** 117–126.

Kagan, V. E., Sebrinova, E. A., and Packer L. (1990*b*). Antioxidant effects of ubiquinones in microsomes and mitochondria are mediated by tocopherol recycling. *Biochem. Biophys. Res. Commun.* **169,** 851–857.

Konstantinov, A. A., Peski, A. V., Popova, E. Yu., Khomutov, G. B., and Ruuge, E. K. (1987). Superoxide generation by the respiratory chain of tumor mitochondria. *Biochim. Biophys. Acta* **894,** 1–10.

Koppenol, W. H. (1996). Thermodynamic reactions involving nitrogen-oxygen compounds. *Methods Enzymol.* **268,** 7–12.

Koppenol, W. H. (1997). The basic chemistry of nitrogen monoxide and peroxynitrite. *Free Radic. Biol. Med.* **25,** 385–391.

Koppenol, W. H., and Butler, J. (1985). Energetics of interconversion reactions of oxyradicals. *Adv. Free Radic. Biol. Med.* **1,** 91–131.

Land, E. J., and Swallow, A. J. (1970). One-electron reactions in biochemical systems as studied by pulse radiolysis. *J. Biol. Chem.* **245,** 1890–1894.

Landi, L, Cabrini, L., Fiorentini, D., Stefanelli, C., and Pedulli, G. F. (1992). The antioxidant activity of ubiquinol-3 in homogeneous solutions and in liposomes. *Chem. Phys. Lipids.* **61,** 121–130.

Lass, A., Agarwal, S., and Sohal, R. S. (1997). Mitochondrial ubiquinone homologues, superoxide radical generation, and longevity in different mammalian species. *J. Biol. Chem.* **272,** 19199–19204.

Lea, C. H., and Kwietny, A. (1962). The antioxidant properties of ubiquinone and related compounds. *Chem. Ind. (Lond.)* **24,** 1245–1246.

Lizasoain, I., Moro, M. A., Knowles, R. G., Darley-Usmar, V., and Moncada, S. (1996). Nitric oxide and peroxynitrite exert distinct effects on mitochondrial respiration which are differentially blocked by glutathione or glucose. *Biochem. J.* **314,** 877–880.

Maguire, J. J., Kagan, V. E., Serbinova, E. A., Ackrell, B. A., and Packer, L. (1992). Succinate-ubiquinone reductase-linked recycling of alpha-tocopherol in reconstituted systems and mitochondria: Requirement of reduced ubiquinone. *Arch. Biochem. Biophys.* **292**, 47–53.

Mellors, A., and Tappel, A. L. (1966). Quinones and quinols as inhibitors of lipid peroxidation. *Lipids.* **1**, 282–284.

Mohr, D., Bowry, V. W., and Stocker, R. (1992). Dietary supplementation with coenzyme Q10 results in increased levels of ubiquinol-10 within circulating lipoproteins and increased resistance of human low-density lipoprotein to the initiation of lipid peroxidation. *Biochim. Biophys. Acta* **1126**, 247–254.

Mordente, A., Santini, S. A., Miggiano, G. A. D., Martorana, G. E., Petitti, T., Minotti, G., and Giardina, B. (1994). The interaction of short-chain coenzyme Q analogs with different redox states of myoglobin. *J. Biol. Chem.* **269**, 27394–27400.

Mukai, K., Kikuchi, S., and Urano, S. (1990). Stopped-flow kinetic study of the regeneration reaction of tocopheroxyl radical by reduced ubiquinone-10 in solution. *Biochin. Biophys. Acta* **1035**, 77–83.

Mukai, K., Morimoto, H., Kikuchi, S., and Nagaoka, S. (1993). Kinetic study of free radical scavenging action of biological hydroquinones (reduced forms of ubiquinone, vitamin K, and tocopherol quinone) in solution. *Biochim Biophys. Acta* **1157**, 313–317.

Murphy, M. E., and Sies, H. (1991). Reversible conversion of nitroxyl anion or nitric oxide by superoxide dismutase. *Proc. Natl. Acad. Sci. USA* **88**, 10860–10864.

Nadezhdin, A. D., and Dunford, H. B. (1974). The oxidation of ascorbic acid by perhydroxyl radicals: A flash photolysis study. *Can. J. Chem.* **57**, 3017–3021.

Naumov, V. V., and Khrapova, N. G. (1983). Study of the interaction of ubiquinone and ubiquinol with peroxide radicals by the chemiluminescent method. *Biophysics.* **28**, 774–780.

Okada, S., Takehara, Y., Yabuki, M., Yoshioka, T., Yasuda, T., Inoue, M., and Utsumi, K. (1996). Nitric oxide, a physiological modulator of mitochondrial function. *Physiol. Chem. Phys. Med. NMR* **28**, 69–82.

Okamoto, T., Matsuya, T., Fukunaga, Y., Kishi, T., and Yamagami, T. (1989). Human serum ubiquinol-10 levels and relationship to serum lipids Intern. *J. Vit. Nutr. Res.* **59**, 288–292.

Öllinger, K., Buffinton, G., Ernster, L., and Cadenas, E. (1990). Effect of superoxide dismutase on the autoxidation of hydro- and semi-naphthoquinones. *Chem.-Biol. Interact.* **73**, 53–76.

Packer, J. E., Slater, T. F., and Wilson, R. L. (1979). Direct observation of a free radical interaction between vitamin E and vitamin E. *Nature* **278**, 737–738.

Padmaja, S., and Huie, R. E. (1993). The reaction of nitric oxide with organic peroxyl radicals. *Biochem. Biophys. Res. Commun.* **195**, 539–544.

Pedersen, J. Z., Marcocci, L., Rossi, L., Mavelli, I., and Rotilio, G. (1990). Generation of daunomycin radicals on the outer side of the erythrocyte membrane. *Biochem. Biophys. Res. Commun.* **168**, 240–247.

Poderoso, J. J., Carreras, M. C., Lisdero, C., Riobó, N., Schöpfer, F., and Boveris, A. (1996). Nitric oxide inhibits electron transfer and increases superoxide radical production in rat heart mitochondria and submitochondrial particles. *Arch. Biochem. Biophys.* **328**, 85–92.

Poderoso, J. J., Carreras, M. C., Schöpfer, F., Lisdero, C. D., Riobó, N. A., Giulivi, C., Boveris, A. D., Boveris, A., and Cadenas, E. (1998a). The reaction of nitric oxide with ubiquinol: Kinetic properties and biological significance. *Free Radical Biol. Med.*, in press.

Poderoso, J. J., Peralta, J. G., Lisdero, C. L., Carreras, M. C., Radisic, M., Schöpfer, F., Cadenas, E., and Boveris, A. (1998*b*). Nitric oxide regulates oxygen uptake and promotes hydrogen peroxide release by the isolated beating rat heart. *Am. J. Physiol.,* in press.

Prütz, W. A., Mönig, H., Butler, J., and Land, E. J. (1985). Reactions of nitrogen dioxide in aqueous model systems: Oxidation of tyrosine units in peptides and proteins. *Arch. Biochem. Biophys.* **243**, 125–134.

Pohl, U., and Busse, R. (1989). Hypoxia stimulates release of endothelium-derived relaxant factor. *Am. J. Physiol.* **256**, H1595–H1600.

Radi, R., Rodriguez, M., Castro, L., and Telleri, R. (1994). Inhibition of mitochondrial electron transport by peroxynitrite. *Arch. Biochem. Biophys.* **308**, 89–95.

Rich, P. R., and Bendall, D. S. (1980). The kinetics and thermodynamics of the reduction of cytochrome c by substituted p-benzoquinols in solution. *Biochim. Biophys. Acta* **592**, 506–518.

Rubbo, H., Radi, R., Trujillo, M., Telleri, R., Kalyanaraman, B., Barnes, S., Kirk, M., and Freeman, B. A. (1994). Nitric oxide regulation of superoxide and peroxynitrite-dependent lipid peroxidation: Formation of novel nitrogen-containing oxidized lipid derivatives. *J. Biol. Chem.* **269**, 26066–26075.

Sies, H., Sharov, V. S., Klotz, L.-O., and Briviba, K. (1997). Glutathione peroxidase protects against peroxynitrite-mediated oxidations: A new function for selenoproteins as peroxynitrite reductase. *J. Biol. Chem.* **272**, 27812–27817.

Stenken, S. (1979). Oxidation of phenolates and phenylenediamines by 2-alkanoxyl radicals produced from 1,2-dihydroxy- and 1-hydroxy-2-alkoxyalkyl radicals. *J. Phys. Chem.* **83**, 595–599.

Stocker, R., Bowry, V. W., and Frei, B. (1991). Ubiquinol-10 protects human low density lipoprotein more efficiently against lipid peroxidation than does α-tocopherol. *Proc. Natl. Acad. Sci. USA* **88**, 1646–1650.

Stocker, R., and Suarna, C. (1993). Extracellular reduction of ubiquinone-1 and -10 by human HepG2 and blood cells. *Biochim. Biophys. Acta* **1158**, 15–22.

Surdhar, P. S., and Armstrong, D. A. (1987). Reduction potentials and exchange reactions of thiyl radicals and disulfide anion radicals. *J. Phys. Chem.* **91**, 6532–6537.

Swallow, A. J. (1982). Physical chemistry of semiquinones. *In* "Function of Quinones in Energy Conserving Systems" (B. L. Trumpower, ed.), pp. 59–72. Academic Press, San Diego.

Takahashi, T., Okamoto, T., and Kishi, T. (1996). Characterization of NADPH-dependent ubiquinone reductase activity in rat liver cytosol: Effect of various factors on ubiquinone-reducing activity and discrimination from other quinone reductases. *J. Biochem.* **119**, 256–263.

Takahashi, T., Yamaguchi, T., Shitashige, M., Okamoto, T., and Kishi, T. (1995). Reduction of ubiquinone in membrane lipids by rat liver cytosol and its involvement in the cellular defense system against lipid peroxidation. *Biochem. J.* **309**, 883–890.

Takehara, Y., Kanno, T., Yoshioka, T., Inoue, M., and Utsumi, K. (1995). Oxygen-dependent regulation of mitochondrial energy metabolism by nitric oxide. *Arch. Biochem. Biophys.* **323**, 27–32.

Torres, J., Darley-Usmar, B., and Wilson, M. T. (1995). Inhibition of cytochrome *c* oxidase in turnover by nitric oxide: Mechanism and implications for control of respiration. *Biochem. J.* **312**, 169–173.

Turrens, J. F., Alexandre, A., and Lehninger, A. L. (1985). Ubisemiquinone is the electron donor for superoxide formation by complex III of heart mitochondria. *Arch. Biochem. Biophys.* **237**, 408–414.

Turrens, J. F., Beconi, M., Barilla, J., Chavez, U. B., and McCord, J. M. (1991). Mitochondrial generation of oxygen radicals during reoxygenation of ischemic tissues. *Free Radic. Res. Commun.* **12**, 681–689.

Villalba, J. M., Navarro, F., Córdoba, F., Serrano, A., Arroyo, A., Crane, F. L., and Navas, P. (1995). Coenzyme Q reductase from liver plasma membrane: purification and role in trans-plasma membrane electron transport. *Proc. Natl. Acad. Sci. USA* **92,** 4487–4891.

Xie, Y.-W., and Wolin, M. S. (1996). Role of nitric oxide and its interaction with superoxide in the suppression of cardiac muscle mitochondrial respiration. *Circulation* **94,** 2580–2586.

Yamamoto, Y., Komuro, E., and Niki, E. (1990). Antioxidant activity of ubiquinol in solution and phosphatidylcholine liposomes. *J. Nutr. Sci. Vitaminol.* **36,** 505–511.

12 Coenzyme-Q Redox Cycle as an Endogenous Antioxidant

Takeo Kishi, Takayuki Takahashi, and Tadashi Okamoto
Department of Biochemistry
Faculty of Pharmaceutical Sciences
Kobe Gakuin University
Kobe 651-2180, Japan

The coenzyme-Q (CoQ) reductase activity responsible for maintaining the redox state of cellular CoQ was investigated in rats to demonstrate the physiological role of intracellular CoQ as an endogenous antioxidant. As a result, a novel NADPH–CoQ reductase was found in cytosol that is more specific to longer chain CoQ homologs than DT-diaphorase and can reduce CoQ in lipid membranes. However, CoQ-10 given to rats increased cytosolic NADPH–CoQ reductase activity, as well as the cellular CoQ-10 pool. Furthermore, it subsequently depressed not only the formation of thiobarbituric acid-reacting substances (TBARS), but also the consumption of α-tocopherol and $CoQH_2$ in the rat livers caused by CCl_4. These observations led to the conclusion that the redox cycle of cellular CoQ by cytosolic NADPH-CoQ reductase acts as an endogenous antioxidant, at least in rat liver cells; consequently CoQ-10 supplementation may be beneficial in protecting cellular membranes from lipid peroxidation.

INTRODUCTION

It is well known that coenzyme-Q (=ubiquinone, CoQ) is biosynthesized *de novo* from *p*-hydroxybenzoate and isopentenylpyrophosphate (Momose and Rudney, 1972) in the inner mitochondrial membrane and that it serves as an essential electron and proton carrier in the respiratory chain there (Szarkowska, 1966). However, CoQ has been detected not only in mitochondria, but also in other subcellular fractions such as microsomes, Golgi apparatus, and plasma membranes (Crane and Morré, 1977; Kalén et al., 1987; Takahashi et al., 1993). In addition, some part of CoQ in these subcellular fractions is always present as the reduced, hydroquinone form ($CoQH_2$) (Takahashi et al., 1993). Furthermore, it has been demonstrated that CoQ in nonmitochondrial membranes is sequentially synthesized through the enzymes in the endoplasmic reticulum–Golgi membrane system (Teclebrhan et al., 1995) and is then translocated from this compartment to other cellular membranes and plasma lipoproteins (Kalén et al., 1987) by transfer vesicles. However, the physiological role of CoQ in subcellular fractions other than mitochondria is still not well understood, although it has been suggested that CoQ also serves as an electron carrier in the Golgi apparatus (Crane and Morré, 1977) and plasma membrane (Villalba et al., 1995). In the mitochondrial respiratory chain, CoQ transfers electrons from flavoprotein dehydrogenases to the cytochrome bc_1 complex by redox cycling between its quinone and hydroquinone forms via the semiubiquinone radical; chemically, the hydroquinone and semiquinone forms can function as hydrogen donors. Therefore, it has been hypothesized that another physiological role of CoQ in the form of $CoQH_2$ may be an endogenous antioxidant against lipid peroxidation (Ernster and Dallner, (1995; Ernster and Forsmark-Andrée, 1993). In fact, much evidence for the antioxidant role of CoQ has accumulated since the mid-1980s through observations in various *in vitro* and *in vivo* systems.

For example, Frei et al. (1990) and Yamamoto et al. (1990) showed that $CoQH_2$-10 is about as effective in scavenging free radicals in liposomal membranes as α-tocopherol and spares α-tocopherol by reducing the α-tocopheroxyl radical to α-tocopherol. Further, Stocker et al. (1991) and Suarna et al. (1993) showed that $CoQH_2$-10 protects human low-density lipoprotein (LDL) more efficiently against lipid peroxidation than α-tocopherol. The antioxidant action of $CoQH_2$ has also been observed in subcellular membranes such as mitochondria (Takeshige et al., 1980) and microsomes (Jakobsson et al., 1994). These *in vitro* observations suggest that $CoQH_2$, but not oxidized CoQ, serves as the first line of antioxidant defense against lipid peroxidation in biological systems.

However, CoQ of the oxidized form is also as active as $CoQH_2$ as an antioxidant in animals. Leibovitz et al. (1990) showed that *tert*-butyl hydroperoxide-induced lipid peroxidation in rat tissue slices was depressed by feeding the rats diets fortified with CoQ-10 before the experiment. In hepatic ischemia and subsequent reperfusion, Marubayashi et al. (1984) reported that a decrease in the reduced form of glutathione (GSH) was not observed in rats injected with CoQ-10 prior to the experiment. Beyer et al.

(1988) also showed that carbon tetrachloride (CCl_4)-induced lipid peroxidation in rats was depressed by prior supplementation of the rats with CoQ-10. A similar antioxidant effect of CoQ-10 supplementation has also been observed in acetoaminophen-treated mice (Amimoto *et al.*, 1995), exercise-loaded rats (Faff and Frankiewics, 1997), and male bicycle racers (Braun *et al.*, 1991). These observations indicate that CoQ-10 given to animals is easily reduced to $CoQH_2$-10 in the tissues. In fact, Okamoto *et al.* (1989) have reported that the oral administration of CoQ-10 increased the human serum level of $CoQH_2$-10. Sugino *et al.* (1989) also reported that CoQ-10 acts *in vivo* as an antioxidant after it has been reduced to $CoQH_2$-10 and suppresses hepatic damage by lipid peroxidation in endotoxicin-administered mice. Furthermore, Mohr *et al.* (1992) also showed that dietary supplementation with CoQ-10 results in increased levels of $CoQH_2$-10 within circulating human LDL and in increased resistance of the LDL to lipid peroxidation. Accordingly, the capability of $CoQH_2$ as an endogenous antioxidant is dependent on the regeneration rate of $CoQH_2$ from CoQ, which results from the antioxidant action. However, the enzyme responsible for the reduction of CoQ to $CoQH_2$ in animal tissues has not yet been identified, although some enzymes such as DT-diaphorase and mitochondrial respiratory enzymes are able to reduce CoQ to $CoQH_2$ (Ernster and Forsmark-Andrée, 1993).

Takahashi *et al.* (1992, 1995, 1996a,b) and Kishi *et al.* (1997) found that cytosol fractions from rat tissues have a novel NADPH-dependent CoQ reductase (NADPH-CoQ reductase) that is resistant to dicoumarol, an inhibitor of DT-diaphorase [=NAD(P)H: quinone acceptor oxidoreductase, EC 1.6.99.2], and to rotenone and antimycin A, inhibitors of respiratory enzymes. Furthermore, it was observed that a prolonged supplementation of rats with CoQ-10 caused significant increases in the NADPH-CoQ reductase activity in their livers and partially protected the rats from CCl_4-induced hepatoxicity. These results led to the speculation that redox cycling $CoQH_2$ by cytosolic NADPH-CoQ reductase may be a fundamental defense system against cellular oxidative stress.

This chapter describes the results that support such a NADPH-CoQ reductase-dependent CoQ redox cycle acting as an endogenous antioxidant.

MATERIALS AND METHODS

Chemicals and Animals

CoQ-9 and CoQ-10 were donated by Eisai Co. (Tokyo, Japan). $CoQH_2$-9 and $CoQH_2$-10 were prepared by reducing the corresponding CoQ homologs with sodium borohydride (Takahashi *et al.*, 1993). All other chemicals were of the highest grade commercially available. Specific pathogen-free, male Wistar rats (8 weeks old, 170–180 g body weight) were purchased from SLC Co., Shizuoka, Japan, and fed on Lavo MR Stock, a commercial feed (SLC Co., Shizuoka, Japan) until use.

Preparation of Subcellular Fractions and Liposomes

Subcellular fractions and liposomes containing $CoQH_2$-10 and 2,2'-azobis(2,4-dimethylvaleronitrile) (AMVN) were prepared as described previously (Takahashi *et al.*, 1995).

Determination of Low Molecular Weight Antioxidants

The amounts of $CoQH_2$-9, $CoQH_2$-10, CoQ-9, and CoQ-10 were determined by HPLC with electron capture detection according to the method of Okamoto *et al.* (1988). Levels of reduced glutathione, ascorbic acid (AsA), and α-tocopherol were determined as described elsewhere (Takahashi *et al.*, 1996b).

Enzyme Assays

NADPH-CoQ reductase activity was determined by measuring $CoQH_2$-10 formed in the presence of NADPH as described previously (Takahashi *et al.*, 1995). DT-diaphorase activity was determined by the reduction of 2,6-dichlorophenol-indophenol (DCPIP) in the presence of NAD(P)H (Ernster, 1967). The activities of catalase (EC 1.11.1.6), superoxide dismutase (SOD, EC 1.15.1.1), glutathione peroxidase (EC 1.11.1.9), glutathione *S*-transferase (EC 2.5.1.18), glutathione reductase (EC 1.6.4.2), and asparate and alanine aminotransferases (GOT, EC 2.6.1.1., and GPT, EC 2.6.1.2) were determined as described previously (Takahashi *et al.*, 1996b).

Lipid Peroxidation Assay

TBARS and conjugated dienes were measured as described elsewhere (Takahashi *et al.*, 1995).

Protein Content

Protein content was determined by the method of Lowry *et al.* (1951).

RESULTS

Distribution of Redox State of CoQ-10 in Rat Tissues and Subcellular Fractions

The reduced form of CoQ is known to be distributed broadly in tissues. Reportedly the tissue concentration and redox state of CoQ in human and rat vary from highest in heart to lowest in lung (Åberg *et al.*, 1992). However, it is not clear how the redox state of CoQ in the cell as a whole is related to the redox states of CoQ and CoQ reductase activities in individual subcellular compartments of the cell. Therefore, the reduction rates of $CoQH_2$/total CoQ ($tCoQ = CoQ + CoQH_2$) were measured in the subcellular compartments of rats. Results from the liver and kidney are shown in Table I. Rates were about 70% in liver and about 25% in kidney, irrespective of the subcellular compartments (Takahashi *et al.*, 1993). Similar results were obtained in other tissues, suggesting that CoQ in every subcellular fraction may be reduced by the same enzyme system.

Distribution of CoQ Reductase Activity in Rat Tissues and Subcellular Fractions

Some enzymes are capable of reducing CoQ to $CoQH_2$ in cells, such as DT-diaphorase and mitochondrial NADH/succinate-CoQ reductases (Cadenas, 1992). However, the reduction rates of CoQ in subcellular compartments from a tissue showed almost the same value (Table I). Therefore it is questionable as to whether the main function of these enzymes is for maintaining a certain redox level of cellular CoQ. Thus, CoQ reductase activity in rats was measured (Takahashi *et al.*, 1992, 1995), and the results of their

Table I   Reduction Rates of CoQ-9 and -10 in Subcellular Compartments of Rat Liver and Kidney[a]

Intracellular fractions	Liver		Kidney	
	CoQ-9	CoQ-10	CoQ-9	CoQ-10
Homogenate	80 ± 3	80 ± 7	25 ± 1	25 ± 1
Nuclei	67 ± 6	69 ± 5	23 ± 2	24 ± 4
Mitochondria	73 ± 7	74 ± 7	26 ± 1	27 ± 2
Lysosomes	67 ± 10	63 ± 9	25 ± 3	24 ± 3
Microsomes	69 ± 7	66 ± 9	24 ± 1	25 ± 3
Cytosol	74 ± 7	72 ± 5	27 ± 3	27 ± 3

[a] Values are means ± SEM (n = 3–11 rats) and are expressed as a percentage of the total amount of the CoQ homolog.

Table II CoQ Reductase Activity of Rat Liver Homogenate in the Presence of NAD(P)H as a Hydrogen Donor[a]

Hydrogen donor	None	0.2 mM NADH	0.2 mM NADPH
None	36 ± 4	147 ± 1	312 ± 7
+10 μM rotenone	35 ± 7	$118 \pm 9^{*b}$	306 ± 16
+1 μM antimycin A	38 ± 4	$167 \pm 10^{*b}$	310 ± 15
+5 μM dicoumarol	22 ± 4	$121 \pm 4^{\dagger b}$	$263 \pm 23^{*b}$

[a] CoQ reductase activity was determined by the reduction of CoQ-10 similarly to the determination method of NADPH-CoQ reductase activity as described in the text. Values are means ± SD of four experiments.
[b] Significant difference at $^{*}p < 0.05$ and $^{\dagger}p < 0.001$ (Student's unpaired t test), respectively, as compared with the corresponding control (none) containing no inhibitor.

liver homogenates are shown in Table II. CoQ reductase activity in the presence of NADPH was twice as high as that in the presence of NADH. Moreover, the NADPH-CoQ reductase activity was not inhibited by rotenone and antimycin A, which are inhibitors of the mitochondrial NADH-CoQ reductase and cytochrome bc_1 complex (=CoQH$_2$-cytochrome c reductase), respectively, although it was inhibited slightly by dicoumarol, an inhibitor of DT-diaphorase. Interestingly, 68% of the total cellular activity of NADPH-CoQ-10 reductase was located in the fraction of cytosol that surrounds the subcellular organelles, as shown in Table III.

Relation between CoQ Redox States and CoQ Reductase Activity in Rat Tissues

Results suggested that the NADPH-CoQ reductase in cytosol may be the enzyme responsible for the maintenance of the cellular CoQ redox state.

Table III Distribution of NADPH-CoQ Reductase Activity in Rat Liver Cells[a]

	NADPH-CoQ reductase activity	
Subcellular fraction	pmol/min/mg protein	% of total cellular activity
Nuclei	16.3 ± 20.9	5.8 ± 6.1
Mitochondria	40.7 ± 11.6	8.2 ± 4.0
Lysosomes	16.3 ± 20.9	2.6 ± 0.5
Microsomes	105 ± 64	15.0 ± 4.3
Cytosol	312 ± 36	67.9 ± 1.9

[a] Values are means ± SD of measurements obtained from three rats.

The next step was to investigate whether the enzyme activity is related to the redox state of cellular CoQ. Results are shown in Fig. 1. A good correlation was observed between the enzyme activities in the logarithmic scale and the reduction rates of CoQ-9 in rat tissues. The correlation coefficient was calculated as $r = 0.899$ ($p < 0.01$).

Enzymatic Properties of Cytosolic NADPH-CoQ Reductase

Enzymatic properties of NADPH-CoQ reductase and DT-diaphorase from rat livers are compared in Table IV. NADPH-CoQ reductase had a molecular weight of about 49,000 and an optimum pH of 7.4. The enzyme was very unstable even at $-20°C$, although it was stabilized greatly in the presence of NAD(P)H. The enzyme preferred NADPH to NADH as electron donor with a 16 times lower K_m. The most important points are that the enzyme hardly reduced 2,6-dichlorophenol-indophenol, an artificial electron acceptor, and that it was not inhibited by dicoumarol in contrast to DT-diaphorase. From these results, the NADPH-CoQ reductase appeared to be a novel enzyme clearly distinguishable from DT-diaphorase, which is also located mainly in cytosol.

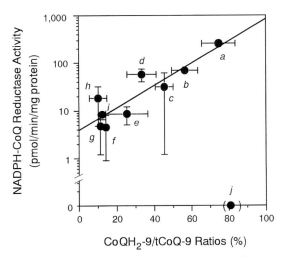

Figure 1   Relationship between CoQ-9 reduction rates and cytosolic NADPH-CoQ reductase activity in rat tissues. Alphabetical numbering represents the rat tissues: (a) liver, (b) small intestine, (c) testis, (d) kidney, (e) lung, (f) brain, (g) spleen, (h) heart, (i) skeletal muscle, and (j) plasma. The NADPH-CoQ reductase activity in cytosol was determined by the reduction rate of CoQ-10 as the substrate to CoQH$_2$-10. Values are means ± SD (bars) of three experiments. Reprinted with permission of Takahashi, T. et al. (1995). Biochem. J. **309**, 883–890.

Table IV   Enzymatic Properties of NADPH-CoQ Reductase and DT-Diaphorase from Rat Liver Cytosol

Property	NADPH-CoQ reductase	DT-diaphorase
Molecular weight	49,000	55,000
Optimum pH	7.4	5.7
Stability	Unstable	Stable
Electron donors (K_m)		
NADH	307 μM	38 μM^a
NADPH	19 μM	36 μM^a
Electron acceptors: DCPIP	No	Yes
Inhibitors		
Dicoumarol	Slightly	Yes
Rotenone	No	No

[a] Menadione (vitamin K_3), instead of DCPIP, was used as an electron acceptor.

Reduction of CoQ in Lipid Membranes by NADPH-CoQ Reductase and Its Antioxidant Action

In order to establish whether cytosolic NADPH-CoQ reductase reduces CoQ in lipid membranes and subsequently prevents lipid peroxidation, liposomes containing $CoQH_2$-10 and AMVN were incubated with rat liver cytosol and NADPH. As shown in Fig. 2, AMVN-induced lipid peroxidation in the liposomes was almost completely inhibited in the presence of both liver cytosol and NADPH, but not in the presence of either one. In addition, this antioxidant effect of cytosol with NADPH was highly related to the level of reduced CoQ-10 found in the liposomes. When the reaction mixture did not contain either cytosol or NADPH, the $CoQH_2$-10 in the liposomes had almost been completely used up before the AMVN-induced lipid peroxidation (formation of conjugated dienes) began to be observed. In the presence of both cytosol and NADPH, however, $CoQH_2$-10 was maintained at more than 40% the initial level throughout the incubation of up to 6 hr. These results indicate that cytosolic NADPH-CoQ reductase reduces CoQ-10 incorporated into the lipid membranes in the presence of NADPH and prevents lipid peroxidation.

A similar effect of NADPH-CoQ reductase on CoQ-10 in lipid membranes was also observed using rat liver microsomes (Takahashi *et al.*, 1995). Microsomesd with a low reduction rate of CoQ-9 were prepared by repeated freezing and thawing and were incubated with rat liver cytosol and NAD(P)H. Results are shown in Table V. NADPH alone (without cytosol) appeared to stimulate lipid peroxidation in the microsomes and, subsequently, the oxidation of $CoQH_2$-9. However, the simultaneous addition of cytosol and NADPH increased the levels of $CoQH_2$-9 to a great extent,

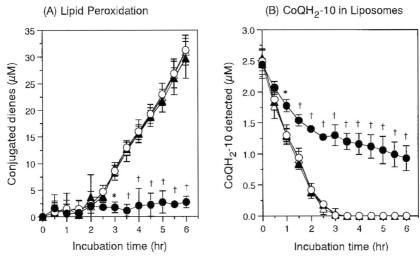

Figure 2 Effects of cytosolic NADPH-CoQ reductase and NADPH on AMVN-induced lipid peroxidation in lecithin liposomes containing $CoQH_2$-10. The complete reaction mixture (●) in 1 ml consisted of 50 mM Tris–HCl buffer (pH 7.4) containing 0.2 mM NADPH, 0.5 mM 2-mercaptoethanol, 1.45 mg of rat liver cytosol protein and lecithin liposomes, which consisted of 5 mM egg yolk lecithin, 1 mM AMVN, and 2.5 μM $CoQH_2$-10. Symbols: ●, the complete system; ○, omitted both cytosol and NADH; ▲ omitted cytosol; △, omitted NADPH. Reaction mixtures were incubated at 37°C for the indicated times. Values are means ± SD (bars) of four experiments. Superscripts represent a significant difference at $*p < 0.05$ and $^{\dagger}p < 0.01$ vs the reaction mixture with both cytosol and NADPH omitted. (A) Lipid peroxidation (conjugated dienes formed) and (B) $CoQH_2$-10 in liposomes were measured by HPLC methods as described in the text.

from 27% to more than 60% of tCoQ-9. In addition, this reduction of CoQ-9 was not inhibited by rotenone and dicoumarol. These results indicate that cytosol can also reduce CoQ-9 in cellular membranes, as well as lecithin liposomes; in other words, the NADPH-CoQ reductase in cytosol is the enzyme responsible for the reduction of cellular CoQ.

Supplementation with CoQ-10 Induces NADPH-CoQ Reductase in Rats and Prevents Radical-Induced Hepatitis

Reportedly, CoQ-10 supplements protect animals from lipid peroxidation damages caused by radical inducers such as CCl_4. Therefore, the effect of CoQ-10 supplements on CoQ redox levels and NADPH-CoQ reductase were investigated. Table VI shows the effect of CoQ-10 and CCl_4 injected into rats. Rats injected with CoQ-10 (CoQ rats) before CCl_4 treatment showed a significant decrease in the level of TBARS in liver and of GPT and GOT

Table V   Reduction of Endogenous CoQ-9 in Rat Liver Microsomes by Cytosol NADPH-CoQ Reductase with NAD(P)H as a Hydrogen Donor[a]

Reaction condition	Reduction rate (%)	
	NADPH	NADH
Before incubation	26.8 ± 3.7	26.8 ± 3.7
After incubation	62.9 ± 1.2	54.6 ± 3.6
$-$Cytosol	19.1 ± 2.1	26.9 ± 1.4
$+5\ \mu M$ dicoumarol	62.4 ± 3.1	51.1 ± 4.1
$+10\ \mu M$ rotenone	60.8 ± 2.4	53.2 ± 2.7

[a] The complete reaction mixture consisted of rat liver microsomes (0.25 mg protein), 1.45 mg rat liver cytosol (310 pmol/min/mg protein), and 0.2 mM NAD(P)H and contained 51.8 ± 1.2 ng of tCoQ-9. The reaction was carried out at 37°C for 15 min. Values are means $\pm$ SD of three experiments. The reduction rate was defined as $CoQH_2$-9/tCoQ-9.

in plasma and, to some extent, were resistant against the liver swelling in comparison with the rats not given CoQ-10 (control rats).

Effects of CoQ-10 and CCl_4 injection on rat liver antioxidant parameters are summarized in Table VII. Rats given CoQ-10 in advance of the CCl_4 injection showed marked increases in the levels of CoQ-10 and $CoQH_2$-10 in their livers, but all other antioxidant parameters, including

Table VI   Effect of CCl_4 Injection on Control and CoQ Rat Liver Indices[a]

Group	Plasma GOT (Karmen units)	Plasma GPT (Karmen units)	Liver weight gain (g)	TBARS formed (nmol/mg protein)
Control rats + saline	10.0 ± 2.0	4.8 ± 1.0	5.43 ± 0.35	1.00 ± 0.20
CoQ rats + saline	9.3 ± 2.4	4.5 ± 1.2	5.40 ± 0.19	1.15 ± 0.08
Control rats + CCl_4	7080 ± 1740	4410 ± 840	7.59 ± 0.37	1.82 ± 0.20
CoQ rats + CCl_4	3200 ± 530^{tb}	1540 ± 250^{tb}	$6.42 \pm 0.61^{*b}$	$1.06 \pm 0.32^{*b}$

[a] Rats ($n = 16$) were divided into two groups. One group (CoQ rats) was injected with CoQ-10 (2 mg/kg body weight) daily for 2 weeks. The other group (control rats) was sham operated with the vehicle of CoQ-10. Next, each group was subdivided into two. One group from each set was then injected with CCl_4 (2 ml/kg body weight) and the others were treated with saline. After 15 hr, the clinical parameters for CCl_4-induced hepatitis were determined as described in the text. Values are mean $\pm$ SD ($n = 4$).
[b] A significant difference at $^*p < 0.05$ and $^tp < 0.001$ vs the control rats + CCl_4.

Table VII   Effect of Supplementation with CoQ-10 on Levels of Antioxidant Factors in Rat Liver[a]

Antioxidant factor	Control rats CCl$_4$ treated	Control rats (None)	CoQ rats CCl$_4$ treated	CoQ rats (None)
CoQ-9 (total) (ng)	99 ± 3	(347 ± 18)	98 ± 14	(328 ± 17)
CoQH$_2$-9 (ng)	50 ± 5	(342 ± 16)	61 ± 4	(313 ± 18)
CoQ-10 (total) (ng)	100 ± 2	(35 ± 2)	125 ± 23	(155 ± 17[†b])
CoQH$_2$-10 (ng)	53 ± 6	(31 ± 2)	92 ± 4	(135 ± 16)
GSH (nmol)	58 ± 3	(6.76 ± 0.16)	62 ± 2	(6.50 ± 0.43)
Ascorbic acid (nmol)	31 ± 2	(1.60 ± 0.10)	33 ± 5	(1.53 ± 0.11)
α-Tocopherol (nmol)	59 ± 2	(0.22 ± 0.01)	73 ± 2[*b]	(0.22 ± 0.01)
Cu, Zn- and Mn-SOD (NU)	100 ± 13	(559 ± 109)	92 ± 15	(566 ± 81)
Catalase (μmol/min)	98 ± 5	(247 ± 32)	104 ± 7	(251 ± 5)
Glutathione peroxidase (nmol/min)	98 ± 2	(805 ± 46)	97 ± 3	(832 ± 17)
Glutathione S-transferase (nmol/min)	78 ± 2	(348 ± 19)	92 ± 5	(352 ± 16)
Glutathione reductase (nmol/min)	93 ± 2	(31 ± 2)	98 ± 4	(31 ± 2)

[a] Rats were divided into four groups treated with CoQ-10 and CCl$_4$ as described in Table VI. Values are means ± SEM ($n = 4$) and are expressed as a percentage of the corresponding level from the nontreated control group. Values in parentheses are the levels of antioxidant factors per milligram of protein in the group: nontreated control rats.

[b] A significant difference at $*p < 0.001$ and $†p < 0.01$ as compared with corresponding values in the control rats.

CoQ-9, were unchanged. Interestingly, the redox state of CoQ-10 in the livers was kept at the same level as those before the treatment.

The CCl$_4$ treatment dramatically depressed the liver cells of all low molecular antioxidants measured. However, supplementation with CoQ-10 in advance of the CCl$_4$ treatment saved α-tocopherol significantly and seemed to spare CoQH$_2$-9 and -10, which suggests that the reduction of CoQH$_2$ may be stimulated by the supplementation. However, it did not save GSH and ascorbic acid, which were located exclusively in the aqueous phase. Results suggest that CoQH$_2$ is not a very active antioxidant in the soluble fraction of cells.

Effects of CoQ-10 and CCl$_4$ treatments on liver quinone reductases are shown in Table VIII. CoQ-10 given to rats stimulated NADPH-CoQ reductase activity in cytosol significantly, but not the one in the homogenate that expresses the total cellular activity, including microsomal and mitochondrial enzyme activities. However, it did not stimulate or inhibit DT-diaphorase activity, which is induced by a number of xenobiotics, such as

Table VIII   Effect of CCl$_4$ on the Activity of DT-Diaphorase and NADPH-CoQ Reductase in Livers of Control and CoQ Rats[a]

UQ reductase	Control rats		CoQ rats	
	Before	After CCl$_4$	Before	After CCl$_4$
DT-diaphorase (homogenate)	100 ± 18 (36 ± 4)	105 ± 17	96 ± 16	98 ± 7
NADPH-CoQ reductase (homogenate)	100 ± 14 (269 ± 37)	52 ± 2	120 ± 10	57 ± 4
NADPH-CoQ reductase (cytosol)	100 ± 7 (355 ± 24)	36 ± 3	122 ± 2*[b]	57 ± 5[tb]

[a] Values are means ± SD and are expressed as a percentage of the enzyme activities (values in parentheses are specific activities of DT-diaphorase, nmol/min/mg protein, and NADPH-CoQ reductase, pmol/min/mg protein) before CCl$_4$ treatment in control rats ($n = 4$).

[b] A significant difference at *$p < 0.05$ vs the corresponding values of control rats, respectively.

3-methylcholanthrene (Cadenas *et al.*, 1992). CCl$_4$ treatment inhibited NADPH-CoQ reductase activity drastically, but the activity that remained in CoQ rats was significantly higher than that in control rats. These results suggest that the CoQH$_2$ redox cycle in CoQ rats is more active than that in control rats.

The total amounts of CoQH$_2$-9 and -10 lost due to CCl$_4$ treatment were estimated by comparing the levels of CoQH$_2$-9 and -10 (on the molar basis) in rat livers before and after CCl$_4$ treatment. Results are shown in Fig. 3. In all the subcellular fractions measured, the comsumption of CoQH$_2$ in rats given CoQ-10 was lower than those in sham-operated control rats. In rats given CoQ-10, the liver levels of TBARS formed and α-tocopherol consumed caused by CCl$_4$ treatment were both significantly lower as compared with those in the control rats mentioned earlier. If we postulate that the amounts of CCl$_4$-induced radicals formed in the liver were equal in both CoQ and control rats, then it is possible to conclude that the CoQ redox cycle (Fig. 4) in CoQ rats has been stimulated by cytosolic NADPH-CoQ reductase activated by exogenous CoQ-10.

DISCUSSION

Some enzymes may serve as CoQ reductases in animal cells. Microsomal NADH-cytochrome b$_5$ and NADPH-cytochrome P450 reductases, DT-diaphorase, and NADH dehydrogenases in mitochondria, Golgi complex,

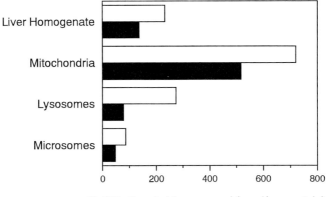

CoQH$_2$-9 and -10 consumed (pmol/mg protein)

Figure 3   Total amounts of CoQH$_2$-9 and -10 oxidized in rat livers by CCl$_4$ treatment. Symbols: open bar, control rats; closed bar, CoQ rats. Total amounts (on the molar basis) of CoQH$_2$-9 and -10 in control and CoQ rats were estimated as differences between the total amount of reduced CoQ in control (or CoQ) rats treated with CCl$_4$ and that in control (or CoQ) rats not treated with CCl$_4$.

and plasma membranes have been suggested as the enzymes responsible for the reduction of cellular CoQ (Ernster and Forsmark-Andrée, 1993). However, the enzyme incorporated into a subcellular compartment such as mitochondrial NADH-CoQ reductase is not likely to be the enzyme responsible for cellular CoQ reduction because there is no known means of transferring CoQH$_2$ formed in the mitochondria to other organelles. From this viewpoint, DT-diaphorase, which is located mainly in cytosol and to some degree in microsomes and mitochondria, appears to be the most appropriate candidate. Indeed, Beyer *et al.* (1996) and Landi *et al.* (1997) insisted that DT-diaphorase is the enzyme responsible for maintaining the cellular CoQ redox levels. They have shown that DT-diaphorase is able to generate and to maintain the reduced CoQ in uni- and multilamellar lipid vesicles containing a thermolabile radical inducer (Landi *et al.*, 1997) and that rat hepatocytes are more susceptible to doxorubicin-induced lipid peroxidation in the presence of 20 μ*M* dicoumarol, a strong inhibitor of DT-diaphorase (Beyer *et al.*, 1996), although the concentration is high enough to almost completely inhibit NADPH-CoQ reductase found in cytosol (Takahashi *et al.*, 1992). However, CoQ homologs with a long isoprenoid chain are not preferable substrates for DT-diaphorase (Ernster *et al.*, 1962). In this chapter, DT-diaphorase, i.e., dicoumarol-sensitive NAD(P)H-CoQ reductase activity, constituted only a small part of the total CoQ reductase activity in rat livers (Table II). Furthermore, this dicoumarol-

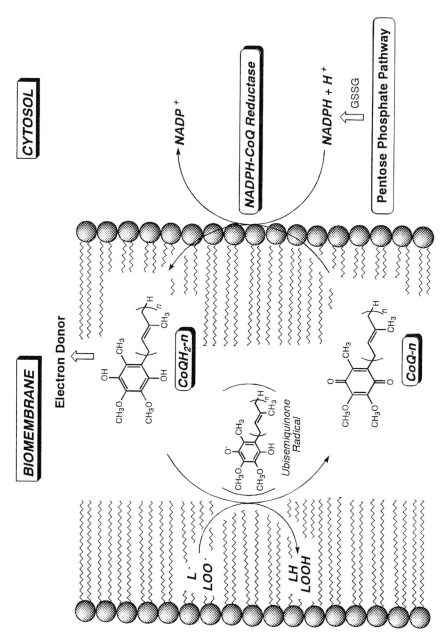

Figure 4 A postulated CoQ redox cycle by cytosolic NADPH-CoQ reductase as the cellular antioxidant system.

sensitive NAD(P)H-CoQ reductase activity in rat livers was not induced by giving the rats CoQ-10 (Table VIII), although DT-diaphorase is induced remarkably by giving rats polycyclic hydrocarbons such as 3-methylcholanthrene (Cadenas *et al.*, 1992). Considering these points, DT-diaphorase is unlikely to be the enzyme responsible for maintaining the redox state of cellular CoQ.

This chapter has demonstrated that the largest part of the total CoQ reductase activity in cytosol was attributable to a novel, dicoumarol-insensitive NADPH-CoQ reductase, which is also able to utilize NADH to some extent as a hydrogen donor. This enzyme regenerated $CoQH_2$ from CoQ incorporated in lecithin liposomes and rat liver microsomes, and subsequently inhibited lipid peroxidation of the lipid membranes (Fig. 2). Furthermore, it was observed that supplementation of rats with CoQ-10 significantly induced the NADPH-CoQ reductase in their lives (Table VIII). In addition, results of rats with CCl_4-induced hepatitis support the conclusion that this enzyme is likely the enzyme responsible for maintaining the redox state of cellular CoQ (Fig. 3).

It is very interesting that this enzyme prefers NADPH to NADH as a hydrogen donor, considering that glutathione reductase (EC 1.6.4.2) for the regeneration of GSH from GSSG also prefers NADPH as a hydrogen donor and that dehydroascorbic acid reductase for the reduction of dehydroascorbic acid to ascorbic acid requires GSH as a hydrogen donor. It is well known that intracellular NADPH is supplied mainly through the pentose phosphate pathway in cytosol, by which about 30% of glucose in livers is metabolized. Glucose-6-phosphate (G6P) dehydrogenase (EC 1.1.1.49), which catalyzes the oxidation of G6P to D-glucono-δ-lactone-6-phosphate in the presence of $NADP^+$, the first reaction of this pathway, is reported to be activated by GSSG (Eggleston and Krebs, 1975) formed during lipid peroxidation. Furthermore, increased activity of this enzyme also has been observed in livers of rats with CCl_4-induced hepatitis (Watanabe and Nagashima, 1983). However, phosphofructokinase (Gilbert, 1982) and pyruvate kinase (von Berkel *et al.*, 1973), key enzymes of glycolysis, which exclusively produces NADH, have been reported to be inactivated by GSSG.

In addition, it is known that thyroid hormones, which enhance oxygen consumption in various organs, stimulate hepatic G6P dehydrogenase (Miksicek *et al.*, 1982). These hormones also stimulate CoQ biosynthesis (Åberg *et al.*, 1996). Therefore, this biochemical background supports the notion that the redox state of cellular CoQ is maintained by NADPH-CoQ reductase and that it constitutes the first line of cellular antioxidant defense against lipid peroxidation, as shown in Fig. 4.

ACKNOWLEDGMENTS

This work was supported in part by a Grant-in-Aid for Science Research (08457617) from the Japanese Ministry of Education and a grant from the Science Research Fund from the Japan Private School Promotion Foundation.

REFERENCES

Åberg, F., Appelkvist, E. L., Dallner, G., and Ernster, L. (1992). Distribution and redox state of ubiquinones in rat and human tissues. *Arch. Biochem. Biophys.* **295**, 230–234.

Åberg, F., Zhang, Y., Teclebrhan, H., Appelkvist, E. L., and Dallner, G. (1996). Increases in tissue levels of ubiquinone in association with peroxisome proliferation. *Chem. Biol. Interact.* **99**, 205–218.

Amimoto, T., Matsura, T., Koyama, S. Y., Nakanishi, T., Yamada, K., and Kajiyama, G. (1995). Acetaminophen-induced hepatic injury in mice: The role of lipid peroxidation and effects of pretreatment with coenzyme Q_{10} and alpha-tocopherol. *Free Radic. Biol. Med.* **19**, 169–176.

Beyer, R. E. (1988). Inhibition by coenzyme Q of ethanol- and carbon tetrachloride-stimulated lipid peroxidation in vivo and catalyzed by microsomal and mitochondrial systems. *Free Radic. Biol. Med.* **5**, 297–303.

Beyer, R. E., Segura Aguilar, J., Di Bernardo, S., Cavazzoni, M., Feto, R., Fiorentini, D., Galli, M. C., Setti, M., Landi, L., and Lenaz, G. (1996). The role of DT-diaphorase in the maintenance of the reduced antioxidant form of coenzyme Q in membrane systems. *Proc. Natl. Acad. Sci. USA* **93**, 2528–2532.

Braun, B., Clarkson, P. M., Freedson, P. S., and Kohl, R. L. (1991). Effects of coenzyme Q10 supplementation on exercise performance, VO_2max, and lipid peroxidation in trained cyclists. *Int. J. Sport. Nutr.* **1**, 353–365.

Cadenas, E., Hochstein, P., and Ernster, L. (1992). Pro- and antioxidant functions of quinones and quinone reductases in mammalian cells. *Adv. Enzymol.* **65**, 97–146.

Crane, F., and Morré, D. J. (1977). Evidence for coenzyme Q function in Golgi membranes. In "Biomedical and Clinical Aspects of Coenzyme Q" (K. Folkers and Y. Yamamura, eds.), Vol. 1, pp. 3–14. Elsevier, Amsterdam.

Eggleston, L. V., and Krebs, H. A. (1974). Regulation of the pentose phosphate cycle. *Biochem. J.* **138**, 425–435.

Ernster, L. (1967). DT diaphorase. In "Methods in Enzymology" (R. W. Estabrook and M. E. Pullman, eds.), Vol. X, pp. 309–317. Academic Press, New York.

Ernster, L., and Dallner, G. (1995). Biochemical, physiological and medical aspects of ubiquinone function. *Biochim. Biophys. Acta* **1272**, 195–204.

Ernster, L., Danielson, L., and Ljunggren, M. (1962). DT diaphorase: 1. Purification from the soluble fraction of rat-liver cytoplasm, and properties. *Biochim. Biophys. Acta* **58**, 171–188.

Ernster, L., and Forsmark-Andrée, P. (1993). Ubiquinol: An endogenous antioxidant in aerobic organisms. *Clin. Invest.* **71**, S60–S65.

Faff, J., and Frankiewics Jozko, A. (1997). Effect of ubiquinone on exercise-induced lipid peroxidation in rat tissues. *Eur. J. Appl. Physiol.* **75**, 413–417.

Frei, B., Kim, M. C., and Ames, N. M. (1990). Ubiquinol-10 is an effective lipid-soluble antioxidant at physiological concentrations. *Proc. Natl. Acad. Sci. USA* **87**, 4879–4883.

Gilbert, H. F. (1982). Biological disulfides: The third messenger? *J. Biol. Chem.* **257**, 12086–12091.

Jakobsson, B. A., Åberg, F., and Dallner, G. (1994). Lipid peroxidation of microsomal and mitochondrial membranes extracted with n-pentane and reconstituted with ubiquinol, dolichol and cholesterol. *Biochim. Biophys. Acta* **1213**, 159–166.

Kalén, A., Norling, B., Appelkvist, E. L., and Dallner, G. (1987). Ubiquinone biosynthesis by the microsomal fraction from rat liver. *Biochim. Biophys. Acta* **926**, 70–78.

Kishi, T., Takahashi, T., and Okamoto, T. (1997). Cytosolic NADPH-UQ reductase-linked recycling of cellular ubiquinol: Its protective effect against carbon tetrachloride hepatoxicity in rat. *Mol. Aspects Med.* **18**, s71–s77.

Landi, L., Fiorentini, D., Galli, M. C., Segura-Aguilar, J., and Beyer, R. E. (1997). DT-diaphorase maintains the reduced state of ubiquinones in lipid vesicles thereby promoting their antioxidant function. *Free Radic. Biol. Med.* **22**, 329–335.

Leibovitz, B., Hu, M.-L., and Tappel, A. L. (1990). Dietary supplements of vitamin E, beta-carotene, coenzyme Q10 and selenium protect tissues against lipid peroxidation in rat tissue slices. *J. Nutr.* **120**, 97–104.

Lowry, O. H., Rosenbrough, N. J., Farr, A. L., and Randall, R. J. (1951). Protein measurement with the folin phenol reagent. *J. Biol. Chem.* **193**, 265–275.

Marubayashi, S., Dohi, K., Yamada, K., and Kawasaki, T. (1984). Changes in the levels of endogenous coenzyme Q homologs, alpha-tocopherol, and glutathione in rat liver after hepatic ischemia and reperfusion, and the effect of pretreatment with coenzyme Q$_{10}$. *Biochim. Biophys. Acta* **797**, 1–9.

Miksicek, R. J., and Towle, H. C. (1982). Changes in the rates of synthesis and messenger RNA levels of hepatic glucose-6-phosphate and 6-phosphogluconate dehydrogenases following induction by diet or thyroid hormone. *J. Biol. Chem.* **257**, 11829–11835.

Mohr, D., Bowry, V. W., and Stocker, R. (1992). Dietary supplementation with coenzyme Q$_{10}$ results in increased levels of ubiquinol-10 within circulating lipoproteins and increased resistance of human low-density lipoprotein to the initiation of lipid peroxidation. *Biochim. Biophys. Acta* **1126**, 247–254.

Momose, K., and Rudney, H. (1972). 3-Polyprenyl-4-hydroxybenzoate synthesis in the inner membrane of mitochondria from p-hydroxybenzoate and isopentenylpyro-phosphate. *J. Biol. Chem.* **247**, 3930–3940.

Okamoto, T., Fukunaga, Y., Ida, Y., and Kishi, T. (1988). Determination of reduced and total ubiquinones in biological materials by liquid chromatography with electro-chemical detection. *J. Chromatogr.* **430**, 11–19.

Okamoto, T., Matsuya, T., Fukunaga, Y., Kishi, T., and Yamagami, T. (1989). Human serum ubiquinol-10 levels and relationship to serum lipids. *Internat. J. Vit. Nutr. Res.* **59**, 288–293.

Stocker, R., Bowry, V. W., and Frei, B. (1991). Ubiquinol-10 protects human low density lipoprotein more efficiently against lipid peroxidation than does alpha-tocopherol. *Proc. Natl. Acad. Sci. USA* **88**, 1646–1650.

Suarna, C., Hood, R. L., Dean, R. T., and Stocker, R. (1993). Comparative antioxidant activity of tocotrienols and other natural lipid-soluble antioxidants in a homogeneous system, and in rat and human lipoproteins. *Biochim. Biophys. Acta* **1166**, 163–170.

Sugino, K., Dohi, K., Yamada, K., and Kawasaki, T. (1989). Changes in the levels of endogenous antioxidants in the liver of mice with experimental endotoxemia and the protective effects of the antioxidants. *Surgery* **105**, 200–206.

Szarkowska, L. (1966). The restoration of DPNH oxidase activity by coenzyme Q (ubiquinone). *Arch. Biochem. Biophys.* **113**, 519–525.

Takahashi, T., Okamoto, T., and Kishi, T. (1996a). Characterization of NADPH-dependent ubiquinone reductase activity in rat liver cytosol: Effect of various factors on ubiquinone-reducing activity and discrimination from other quinone reductases. *J. Biochem.* **119**, 256–263.

Takahashi, T., Okamoto, T., Mori, K., Sayo, H., and Kishi, T. (1993). Distribution of ubiquinone and ubiquinol homologues in rat tissues and subcellular fractions. *Lipids* **28**, 803–809.

Takahashi, T., Shitashige, M., Okamoto, T., Kishi, T., and Goshima, K. (1992). A novel ubiquinone reductase activity in rat cytosol. *FEBS Lett.* **314**, 331–334.

Takahashi, T., Sugimoto, N., Takahata, K., Okamoto, T., and Kishi, T. (1996b). Cellular antioxidant defense by a ubiquinol-regenerating system coupled with cytosolic NADPH-dependent ubiquinone reductase: Protective effect against carbon tetrachloride-induced hepatotoxicity in the rat. *Biol. Pharm. Bull.* **19**, 1005–1012.

Takahashi, T., Yamaguchi, T., Shitashige, M., Okamoto, T., and Kishi, T. (1995). Reduction of ubiquinone in membrane lipids by rat liver cytosol and its involvement in the cellular defense system against lipid peroxidation. *Biochem. J.* **309**, 883–890.

Takeshige, K., Takayanagi, R., and Minakami, S. (1980). Reduced coenzyme Q-10 as an antioxidant of peroxidation in bovine heart mitochondria. *In* "Biomedical and Clinical Aspects of Coenzyme Q" (Y. Yamamura, K. Folkers, and Y. Ito, eds.), Vol. 2, pp. 15–26. Elsevier, Amsterdam.

Teclebrhan, H., Jakobsson-Borin, Å., Brunk, U., and Dallner, G. (1995). Relationship between the endoplasmic reticulum-Golgi membrane system and ubiquinone biosynthesis. *Biochim. Biophys. Acta* **1256**, 157–165.

Villalba, J. M., Navarro, F., Córdoba, F., Serrano, A., Arroyo, A., Crane, F. L., and Navas, P. (1995). Coenzyme Q reductase from liver plasma membrane: Purification and role in trans-plasma-membrane electron transport. *Proc. Natl. Acad. Sci. USA* **92**, 4887–4891.

von Berkel, Th. J. C., Koster, J. F., and Hülsmann, W. C. (1973). Two interconvertible forms of L-type pyruvate kinase from rat liver. *Biochim. Biophys. Acta* **293**, 118–124.

Watanabe, A., and Nagashima, H. (1983). Glutathione metabolism and glucose 6-phosphate dehydrogenase activity in experimental liver injury. *Acta Med. Okayama* **37**, 463–470.

Yamamoto, Y., Komuro, E., and Niki, E. (1990). Antioxidant activity of ubiquinol in solution and phosphatidylcholine liposome. *J. Nutr. Sci. Vitaminol. Tokyo* **36**, 505–115.

13 Carotenoids: Occurrence, Biochemical Activities, and Bioavailability

Wilhelm Stahl and Helmut Sies

Institut für Physiologische Chemie I
Heinrich-Heine-Universität Düsseldorf
D-40001 Düsseldorf, Germany

INTRODUCTION

Carotenoids are natural colorants present in various fruits and vegetables, such as carrots, tomatoes, spinach, oranges, or peaches. Epidemiological studies have shown that the increased consumption of foods rich in carotenoids is correlated with a diminished risk for several diseases.

The antioxidant properties of carotenoids and their ability to induce gap junctional communication (GJC) have been discussed as possible biochemical mechanisms underlying their cancer-preventive effects. Carotenoids are efficient quenchers of singlet oxygen; this activity depends mainly on the number of conjugated double bonds present in the molecule. Natural and synthetic carotenoids are capable of inducing intercellular communication via gap junctions, a process involved in the growth control of transformed cells. Carotenoids induce GJC at concentrations of about 10^{-6} M, with six-membered ring compounds being more active than five-membered ring analogs. 4-Oxoretinoic acid, a decomposition product of canthaxanthin in cell culture, is active at even 10^{-8} M. Data suggest that oxidation products of carotenoids are the active compounds ultimately responsible for the effects of carotenoids on GJC.

The carotenoid pattern in fruits and human serum as well as the bioavailability of these compounds from natural sources are distinct, and little is known on what governs the specific patterns. Cryptoxanthin is the predominant carotenoid in citrus fruits, mainly present in esterified form. More than 30 different carotenoids and carotenol esters have been indentified in orange and tangerine juice applying matrix-assisted laser desorption time of flight (MALDI-TOF) mass spectrometry. Carotenoid esters contribute considerably to the supply with crypto-xanthin in the human. However, even after ingestion of a source high in cryptoxanthin esters, only the level of the free carotenoid increased in human serum and chylomicrons. β-Carotene, α-carotene, lycopene, and lutein are other major carotenoids in human serum and tissues. Several tissues such as liver, adrenal, or testes are rich in carotenoids, whereas lower amounts are detected in kidney, ovary, or brain.

Carotenoid bioavailability depends on several factors, including the food matrix. Ingestion of tomato paste was found to yield 2.5-fold higher lycopene peak concentrations than ingestion of fresh tomatoes. No difference was observed in the α- and β-carotene response. Thus, the bioavailability of lycopene is higher from tomato paste than from fresh tomatoes in humans.

OCCURRENCE OF CAROTENOIDS

Carotenoids are pigments widespread in nature and more than 600 different compounds have been identified in various organisms (Pfander, 1987; Olson and Krinsky, 1995). β-Carotene is the most prominent representative of this very lipophilic class of compounds. The basic structure of carotenoids consists of a tetraterpene skeleton, which may be cyclized at one end or both ends of the molecule; exemplary structures of these types of carotenoids are shown in Fig. 1. Cyclic end groups may be five- or six-membered ring systems. Carotenoids, which are composed only from carbon and hydrogen atoms, are collectively assigned as carotenes, e.g., β-carotene, γ-carotene, and lycopene. However, most natural carotenoids contain at least one oxygen function such as keto (violerythrin), hydroxy (lutein), or epoxy groups (violaxanthin) referred to as xanthophylls or oxocarotenoids. A common structural element of carotenoids is the extended system of conjugated double bonds, which is responsible for their color and some of their biological functions, such as antioxidant activity (Young and Britton, 1993).

Carotenoids are synthesized and stored in the photosynthetic apparatus of higher plants where they are involved in the light-harvesting system and in antioxidant defense against photooxidation (Demming-Adams, 1996). Animals and humans are not capable of synthesizing carotenoids but abß-Carotene

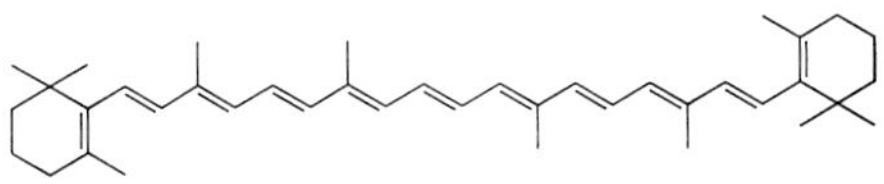

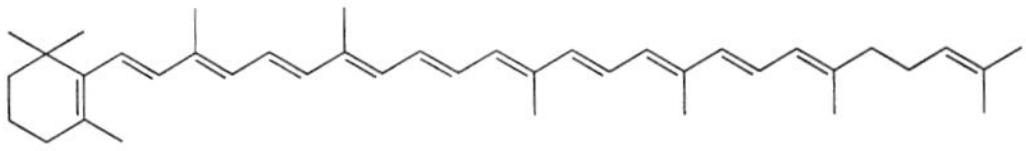

γ-Carotene

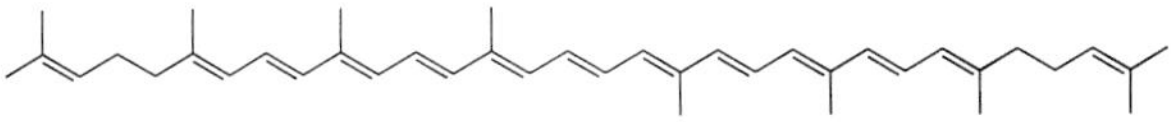

Lycopene

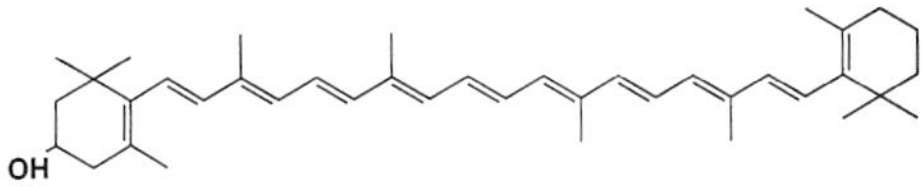

ß-Cryptoxanthin

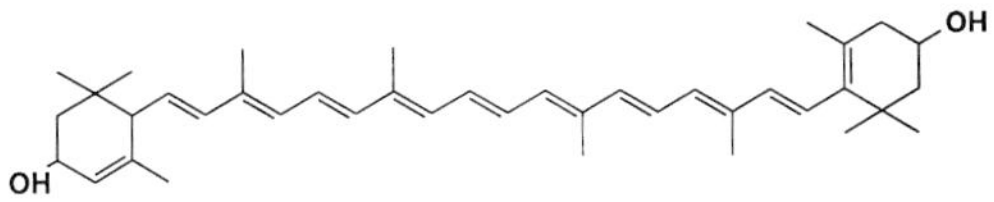

Lutein

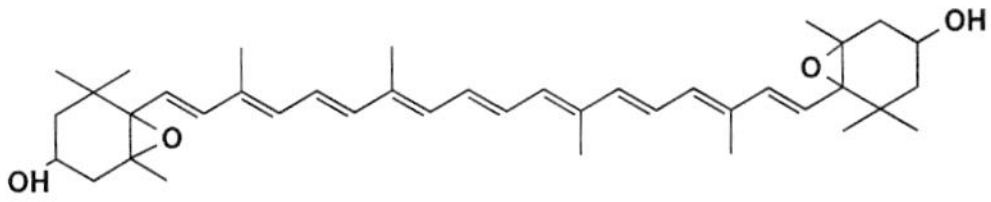

Violaxanthin

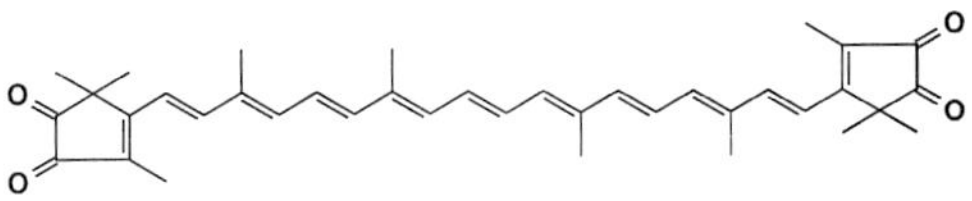

Violerythrin

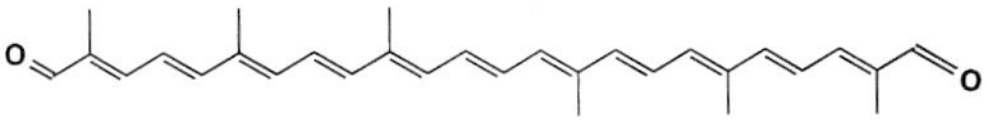

C-30-Dialdehyde

Figure 1 Structures of selected carotenoids.

sorb them from the diet and make use of provitamin A carotenoids for the vitamin A supply (Olson, 1994; De Pee and West, 1996).

Important dietary sources of carotenoids for the human are green leafy and orange to red vegetables as well as various fruits, including oranges, tangerines, or peaches (Mangels *et al.*, 1993). The provitamin A carotenoids β-carotene, α-carotene, and cryptoxanthin, as well as the nonprovitamin A compounds lutein, zeaxanthin, and lycopene are the major dietary carotenoids (Stahl and Sies, 1996), which are mainly provided by plant-based food. However, more than 35 different carotenoids have been identified in human plasma, indicating that additional carotenoids are supplied with the diet (Khachik *et al.*, 1992., 1997). Some of the more common dietary carotenoid sources are listed in Table I. Although β-carotene and lutein are found in many different kinds of fruits and vegetables, only a few products contain important other carotenoids such as lycopene or zeaxanthin. About 90% of dietary lycopene in the United States derives from tomatoes and tomato products (Chug-Ahuja *et al.*, 1993) with more than 50% from processed food. The major source for zeaxanthin is corn. Carrots are an important source for β-carotene along with spinach, broccoli, or green and red peppers.

Fruits also contain considerable amounts of the provitamin A carotenoids β-carotene and β-cryptoxanthin, and it has been suggested that they represent the most important source for vitamin A in developing countries (de Pee and West, 1996). Another major carotenoid found in fruits is lutein.

Hydroxylated carotenoids in fruits and vegetables may be present as either parent carotenols or esterified with various fatty acids (Khachik and Beecher, 1988). The carotenoid pattern of papaya and other fruits is dominated by carotenoid esters whereas only low amounts of free xanthophylls are found in some fruits. For instance, tangerines contain high amounts of β-cryptoxanthin present mainly in esterified form (Wingerath *et al.*, 1996). Major carotenoid esters in tangerines were identified as β-cryptoxanthin laurate, myristate, and palmitate. Aditionally, small amounts of lutein and zeaxanthin esters are detectable.

Carotenoids are also found in various kinds of seafood, including lobster and salmon (Liaaen-Jensen, 1990). Commercially, astaxanthin is used widely as a feed additive to obtain the typical color of salmon flesh (Bjerkeng, 1992). Various carotenoids are also detected in algae, some of which are capable of storing high amounts of these compounds. Halotolerant algae such as *Dunaliella salina* produce several carotenoids, mainly different geometrical isomers of β-carotene, which amount to more than 10% of the dry weight (Avron and Ben-Amotz, 1992).

Algae, fungi, bacteria, and other natural sources of carotenoids such as tomatoes or flowers are used to isolate carotenoids for the production of

Table I   Content of Carotenoids in Vegetables and Fruits (μg/100 g)[a,b]

Source	β-Carotene	α-Carotene	Lutein	Zeaxanthin	Cryptoxanthin	Lycopene
Vegetables						
Brussels sprouts	550		610			
Beans	380	70	490			
Broccoli	920		1610			
Cabbage	60		80			
Carrots	9700	3140	220			
Cucumber	300		670			
Greens	1950		3050			
Parsley	4520		5810			
Peas	440		1630			
Pepper, green	710		1120			
Pepper, orange	1130	640	2490	8480	780	
Sweet corn	60	60	520	437		
Spinach	4020		5870			
Tomato	440		80			2940
Fruits						
Apple	35		80	10	10	
Apricot	1770	40	100	30	230	
Grapefruit, pink	1310					3360
Mandarin	290	10	50	140	1770	
Orange	10		60	50	80	
Peach	100		80	40	90	
Watermelon	230		10			4100

[a] Modified from Hart and Scott (1995) and Mangels et al. (1993).
[b] Values presented are for orientation. Significant variances may exist within the same kind of fruits and vegetables.

supplements, food, and feed additives (Nelis and De Leenheer, 1991; Nir et al., 1993). A concentrated source of lutein and lutein esters is found in marigold (Rivas et al., 1989). Marigold petals contain 20 times higher concentrations of lutein than spinach. Further, large amounts of carotenoids are synthesized chemically for use in the food or feed industry.

BIOCHEMICAL ACTIVITIES OF CAROTENOIDS

There is increasing evidence that β-carotene and other carotenoids exhibit beneficial health effects in preventing the development of chronic diseases in humans (Blot *et al.*, 1993; Krinsky, 1994; van Poppel and Goldbohm, 1995; Flagg *et al.*, 1995: Gerster, 1995; Mayne, 1996). An increased consumption of a carotenoid-rich diet is associated with a diminished risk for some kinds of cancer and cardiovascular diseases. Adverse effects of long-term supplementation with β-carotene have been described in some high-risk groups (The Alpha-Tocopherol, Beta Carotene Cancer Prevention Study Group, 1994; Omenn *et al.*, 1996). It has been shown that heavy smokers have an increased risk for lung cancer when they are supplemented for about 5 years with 20 mg of β-carotene daily; asbestos workers were also identified as a risk group for β-carotene supplementation. Results of both studies, as well as possible reasons for their outcome, have been discussed controversially in the literature (Mayne *et al.*, 1996; Potter, 1997; Pastorino, 1997).

It has been suggested that carotenoids may be helpful in preventing photoaging of the skin and sunburn reactions (Ribaya-Mercado *et al.*, 1995; Biesalski *et al.*, 1996). β-Carotene supplements are widely used as oral sun protectants, and a daily intake over several weeks leads to increased carotenoid levels in human skin and plasma.

Lutein and zeaxanthin are the predominant carotenoids in the human macula lutea; none of the major human carotenes are detectable in this tissue (Bone *et al.*, 1997; Landrum *et al.*, 1997). Several lines of evidence indicate that carotenoids may protect from age-related macular degeneration (AMD) (Schalch, 1992; Seddon *et al.*, 1994; Mares-Perlman *et al.*, 1995). AMD is the leading cause of irreversible blindness among Americans 65 years and older. This disease is the cause of partial vision loss for 1 of every 20 people in the United States. A genetic component for an increased AMD risk has been identified (Allikmets *et al.*, 1997).

A number of biological effects have been attributed to carotenoids, including antioxidant activity (Sies and Stahl, 1995), influences on the immune system (Hughes *et al.*, 1997), control of cell growth and differentiation (Murakoshi *et al.*, 1989; Sharoni and Levy, 1996; Stivala *et al.*, 1996), and stimulatory effects on gap junctional communication (Bertram and Bortkiewicz, 1995). These effects are thought to be relevant with respect to their protective properties.

ANTIOXIDANT ACTIVITY OF CAROTENOIDS

Reactive oxygen species are formed in physiological processes and are capable of damaging biologically relevant molecules such as DNA, proteins, lipids, or carbohydrates (Sies, 1993; Halliwell,1996). These reactions are suggested to be involved in the pathobiochemistry of several diseases

(Gutteridge, 1993; Sies, 1997). A variety of antioxidant defense systems, including low molecular weight compounds, such as the carotenoids, are scavengers of reactive oxygen species; they are summarized under the term antioxidants.

Research on the antioxidant activity of carotenoids was sparked by the initial description of their property as singlet oxygen ($^1O_2$) quenchers (Foote and Denny, 1968) and their ability to trap peroxyl radicals (Burton and Ingold, 1984). The discovery that carotenoids inactivate $^1O_2$ was an important advance in understanding the biological effects of β-carotene and other carotenoids. The mechanism by which carotenoids protect biological systems against $^1O_2$-mediated damage appears to depend largely on physical quenching (Truscott, 1990; Stahl and Sies, 1993). In this process, the energy of the excited oxygen is transferred to the carotenoid molecule. The energy is dissipated through rotational and vibrational interactions between the excited carotenoid and the surrounding solvent to yield the ground state carotenoid and thermal energy. In the process of physical quenching the carotenoid molecule is not destroyed. It may undergo further cycles of singlet oxygen quenching, thus acting like a catalyst. Isomerization of the carotenoid may occur in the quenching process, via the lowest triplet state of the molecule. The quenching of singlet oxygen by β-carotene and other biological carotenoids occurs with rate constants approaching diffusion control; lycopene, the open-chain analog of β-carotene exhibited the highest rate constant (Di Mascio *et al.*, 1989). Thus, lycopene is the most efficient $^1O_2$ quencher among the biologically occurring carotenoids.

The quenching rate constants of several natural and synthetic carotenoids have been investigated to elucidate structure–activity relationships (Stahl *et al.*, 1997). All of the investigated carotenoids proved to be efficient quenchers of singlet molecular oxygen (Table II). The quenching rate constant depends on the number of conjugated double bonds present in the molecule. An empirical correlation exists between the π, π^* excitation energy and the carotenoid structure; the quenching ability of carotenoids depends on the excitation energy of their transition at longer wavelengths (Baltschun *et al.*, 1997). The C-20-dialdehyde polyene showed the lowest quenching rate constant. The C-40-dialdehyde, dinor-canthaxanthin, and violerythrin exhibited the highest quenching rate constants but also retrodehydro-β-carotene, echinenone, 3-hydroxy-β-carotene, 4-hydroxy-β-carotene, canthaxanthin, the C-30-dialdehyde, capsorubin, and β-carotene itself effectively quenched $^1O_2$. The C-40-dialdehyde is among the most potent quenchers of $^1O_2$.

An increased formation of $^1O_2$ has been discussed as a reason for skin lesions found in patients suffering from erythropoietic protoporphyria (EPP) (Truscott, 1990). EPP is a genetic defect affecting porphyrin biosyn-

Table II   Singlet Oxygen-Quenching Activity of Natural and Synthetic Carotenoids[a]

Compound	Number of fully conjugated C–C double bonds	$^1O_2$ quenching rate constant × 10^9 $(M^{-1} s^{-1})$	TEAC[b]
Control	—	—	—
β-Carotene	9	8	1.9
Echinenone	9	9	0.7
Canthaxanthin	9	9	0.02
4-OH-β-carotene	9	9	
Cryptoxanthin	9	9	2.0
Retrodehydro-β-carotene	10	10	
Capsorubin	9	8	
Dinor-canthaxanthin	9	13	
Violerythrin	9	12	
C-20-Dialdehyde	7	4	
C-30-Dialdehyde	11	10	
C-40-Dialdehyde	15	14	

[a] Modified from Stahl *et al.* (1997).
[b] Trolox equivalent antioxidant capacity; data taken from Miller *et al.* (1996).

thesis and leads to increased levels of protoporphyrin, which is thought to act as a photosensitizer (Mathews-Roth, 1978; Mathews-Roth *et al.*, 1974). Patients experience a burning and painful sensation of the skin on exposure to sunlight. The symptoms of EPP are treated efficiently with high doses of β-carotene (Mathews-Roth, 1986). Quenching of $^1O_2$ appears to be one of the biological mechanisms underlying skin protection.

Carotenoids are also able to inhibit free radical reactions (Burton and Ingold, 1984; Sies *et al.*, 1992; Palozza and Krinsky, 1992; Rice-Evans *et al.*, 1997). At low concentrations and at low partial pressures of oxygen, β-carotene, was found to inhibit the oxidation of model compounds initiated by peroxyl radicals (Kennedy and Liebler, 1992). This antioxidant activity of β-carotene, which is shared by other carotenoids as well (Lim *et al.*, 1992), may contribute to the protection of membranes from lipid peroxidation. It has been shown, however, that β-carotene is ineffective as an antioxidant when the compound is added to preformed lipid membranes (Liebler *et al.*, 1997)

Increased lipid peroxidation has been suggested to play a role in the pathophysiology of lung injury in patients with cystic fibrosis. High levels of malondialdehyde in the plasma of patients were associated with rather low levels of β-carotene (Lepage *et al.*, 1996). After application of β-

carotene for 2 months the malondialdehyde levels decreased to normal; β-carotene plasma levels increased about 50-fold. β-Carotene supplementation increases the antioxidant capacity of plasma (Meydani *et al.*, 1994). However, *in vitro* and *in vivo* data show that carotenoids have only limited effects in preventing low-density lipoproteins from lipid peroxidation (Gaziano *et al.*, 1995). A diminished susceptibility to oxidant stress was found in chick liver after application of β-carotene and zeaxanthin (Woodall *et al.*, 1997). Thus, carotenoids may exhibit important antioxidant functions in tissues.

Oxidative damage has been implicated in the pathogenesis of rheumatoid arthritis and systemic lupus erythematosus. In a prospective case-control study it was found that patients afflicted with both these diseases had lower levels of the lipophilic antioxidants β-carotene and α-tocopherol (Comstock *et al.*, 1997). In rheumatoid arthritis cases, β-carotene levels were about 30% lower than in controls.

Carotenoids trap radicals of different natures. An interaction between naturally occurring carotenoids and the trichloromethylperoxyl radical was described, producing a combination of a carotenoid radical cation and a radical adduct (Packer *et al.*, 1981). The radical adduct is less stable than the radical cation and decays to the cation with a rate constant of $1.8 \times 10^4 \text{sec}^{-1}$ (Hill *et al.*, 1997). A number of carotenoids were shown to react with the radical cation ABTS$^+$ [2,2′-azinobis(3-ethylbenzothiazoline-6-sulfonic acid] (Miller *et al.*, 1996). This scavenging ability becomes higher with increasing polarities and the number of conjugated double bonds of the carotenoid. An interaction of β-carotene with the lipid peroxyl radical leads to the formation of radical adducts, which have been identified *in vitro* (Liebler and McClure, 1996). The β-carotene radical reacts with oxygen to become a peroxyl radical, which is capable of autooxidizing β-carotene, and to react with lipids inducing the chain initiation. Several oxidation products of β-carotene, including apocarotenals and carotenoid epoxides, have been isolated and identified (Handelman *et al.*, 1991); Kennedy and Liebler, 1991). Other free radicals such as nitrogen dioxide, thiyl, and sulfonyl radicals are also scavenged rapidly by β-carotene (Everett *et al.*, 1997).

Most research has focused on the effects of β-carotene alone, but interactions between lipophilic antioxidant α-tocopherol have also been reported (Palozza *et al.*, 1992; Bohm *et al.*, 1997). In some model systems, a combination of α-tocopherol and β-carotene interacts synergistically to inhibit lipid peroxidation. However, little is known on antioxidant efficiency and cooperative interactions of other naturally occurring carotenoids with vitamin E.

CAROTENOID EFFECTS ON GAP JUNCTIONAL COMMUNICATION

Intercellular signaling is a prerequisite for coordinating biochemical functions in multicellular organisms. One pathway of communication comprises the direct coupling of cells organized in functional units. Such signaling is achieved by cell-to-cell channels, called gap junctions (for a review see Goodenough *et al.*, 1996). Gap junctions are water-filled pores, connecting the cytosol of two neighboring cells, which allow an exchange of low molecular weight compounds of < 1000 (Fig. 2). They are formed from two half-channels, each provided from one of the coupled cells. The half-channel consists of a hexamer of subunits, which belong to the gene family of connexins. Several physiological functions have been attributed to gap junctional communication: exchange of nutrients and ions between connected cells, transfer of electrical signals, and pathways for signaling compounds.

It has been speculated that GJC may play a role in carcinogenesis (Hotz-Wagenblatt and Shalloway, 1993; Holder *et al.*, 1993; Bertram, 1993). Tumor promotors such as phorbol esters are efficient inhibitors of this pathway of communication, whereas other compounds such as carotenoids and retinoids induce GJC (Wolf, 1992; Krutovskikh *et al.*, 1995). Interestingly, non-provitamin A carotenoids such as canthaxanthin or lycopene, as well as some synthetic carotenoid analogs, can also stimulate intercellular communication *in vitro* (Pung *et al.*, 1988; Hossain *et al.*, 1993; Pung *et al.*, 1993; Stahl *et al.*, 1997). In cell culture, carotenoids and

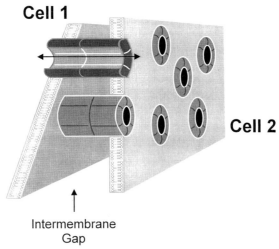

Figure 2   Gap junctional communication.

retinoids reversibly inhibit the progression of carcinogen-initiated fibroblasts to the transformed state (Hossain and Bertram, 1994). This inhibitory effect has been found to be related to an increased gap junctional communication induced by these compounds. Growing evidence shows that gap junctions play a role in the regulation of morphogenesis, cell differentiation, secretion of hormones, and growth (Yamasaki, 1995; Dahl *et al.*, 1995).

β-Carotene, as well as its natural and synthetic analogs retrodehydro-β-carotene, echinenone, 3-hydroxy-β-carotene, 4-hydroxy-β-carotene, and canthaxanthin, stimulates intercellular communication (Table III). The five-membered ring carotenoids exhibit only little effects; the analog of canthaxanthin, dinor-canthaxanthin, has less influence on GJC as compared to the parent compound. Straight-chain polyene dialdehydes do not significantly induce gap junctional communication. The influence of carotenoids on gap junctional communication does not correlate with their antioxidant activity (Zhang *et al.*, 1991; Stahl *et al.*, 1997), indicating that the two properties of carotenoids addressed in this chapter may operate independent of each other. The increase of intercellular communication by retinoids and carotenoids is accompanied by higher levels of connexin mRNA (Hanusch *et al.*, 1995; Clairmont *et al.*, 1996). This might be due to direct

Table III   Inductory Effects of Natural and Synthetic Carotenoids on Gap Junctional Communication (GJC)[a]

Compound	Number of fully conjugated C–C double bonds	GJC (% control)
Control	—	100
Six-membered ring		
β-carotene	9	530
Echinenone	9	409
Canthaxanthin	9	428
4-OH-β-carotene	9	388
Cryptoxanthin	9	307
Retrodehydro-β-carotene	10	522
Five-membered ring		
Capsorubin	9	107
Dinor-canthaxanthin	9	183
Violerythrin	9	148
Polyenes		
C-20-Dialdehyde	7	120
C-30-Dialdehyde	11	113
C-40-Dialdehyde	15	125

[a] Modified from Stahl *et al.* (1997).

or indirect influences of retinoids and carotenoids on the gene expression of connexins. Indirect effects have been observed in various cell systems, where retinoids modulate gap junctional communication by acting at the posttranscriptional level (Bex *et al.*, 1995; Clairmont *et al.*, 1996). Retinoids function by binding to nuclear receptors, which, in turn, bind to specific response elements within the regulatory region of their target genes (Mangelsdorf *et al.*, 1994; Chambon, 1996). However, it is still unclear whether retinoic acid receptors are involved directly in the regulation of the expression of gap junction proteins.

Biological activities of carotenoids apart from their antioxidant properties may be related to activities of intact compounds or to decomposition products of the parent carotenoid. Biological effects of the non-provitamin A carotenoid canthaxanthin and its decomposition products on GJC have been investigated to answer this question (Hanusch *et al.*, 1995). Decomposition fractions of canthaxanthin were isolated by preparative high-performance liquid chromatography and shown to be active in the cell communication assay. Two of the decomposition products were identified as *all-trans* and 13-*cis* 4-oxoretinoic acid. Both isomers of 4-oxoretinoic acid enhanced GJC in murine fibroblasts (C3H/10T1/2 cells) accompanied by an increase of the expression of connexin 43 mRNA. Therefore, it is concluded that the biological activity of canthaxanthin is at least in part due to the formation of active decomposition products such as 4-oxoretinoic acid. To obtain a significant induction of cell–cell communication in 10T1/2 cells, only 0.1% of the 10 μM incubation mixture with canthaxanthin needs to be converted to 4-oxoretinoic acid, which is active even at 10 nM.

BIOKINETICS OF CAROTENOIDS

The uptake of carotenoids from the diet is influenced by several factors, such as dietary fat, presence of fiber, or food processing (Erdman *et al.*, 1993; Olson, 1994; Wang, 1994; Parker, 1997). Similar to vitamin E, carotenoids are absorbed via the lymphatic pathway and therefore require the formation of micelles from fat and bile acids (Fig. 3). Thus, the intestinal uptake of these compounds is improved by the additional consumption of oil, margarine, or butter (Stahl and Sies, 1992). The particle size of uncooked food also influences carotenoid uptake. The bioavailabilty of carotenoids from pureed or finely chopped vegetables is considerably higher as compared to whole or sliced fresh vegetables. It has also been demonstrated that mild cooking increases the absorption of carotenoids, whereas additional heating can cause isomerization of the naturally occurring *all-trans* double bonds to *cis* configuration.

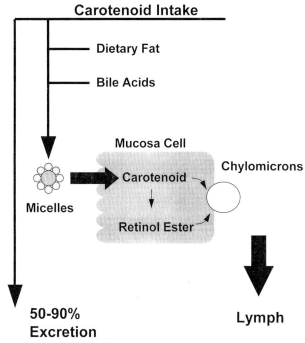

Figure 3 Absorption of carotenoids.

Little is known about the biological value of carotenoid *cis* isomers. It has been shown that the bioavailability of 9-*cis* β-carotene is far less as compared to its *all-trans* analog (Stahl *et al.*, 1995; You *et al.*, 1996). However, considerable amounts of this isomer are detectable in human tissues (Stahl *et al.*, 1992), and it was demonstrated that 9-*cis* β-carotene is a precursor for 9-*cis* retinoic acid (Hebuterne *et al.*, 1995), a high-affinity ligand for the RXR receptor (Mangelsdorf *et al.*, 1994). Heating of carotenoids in the presence of oxygen results in their oxidation. A variety of decomposition products (apocarotenals) are produced, some of which might still possess biological activity, in turn acting as precursors for biologically relevant retinoids.

As mentioned earlier, β-cryptoxanthin is a provitamin A carotenoid especially found in fruits such as oranges and tangerines. In the latter, carotenol fatty acid esters are predominant; major carotenoid esters are β-cryptoxanthin laurate, myristate, and palmitate (Wingerath *et al.*, 1996). Smaller amounts of lutein and zeaxanthin esters are also detectable. Additionally, β-carotene and low levels of free β-cryptoxanthin, lutein, and zeaxanthin are found in tangerine juice.

Upon ingestion of tangerine juice concentrate, rich in β-cryptoxanthin esters, increasing amounts of free β-cryptoxanthin were detected in chylomicrons and serum of the human (Wingerath *et al.*, 1995). Peak levels of the carotenoid in chylomicrons were reached at 6 hr; the concentration returned to basal levels 9 hr after ingestion. No β-cryptoxanthin esters were detected in chylomicrons or serum, indicating efficient cleavage in the intestine before the carotenoid is incorporated into lipoproteins by the liver.

Fatty acid esters of zeaxanthin and lutein were present in low amounts in tangerine juice. Similar to β-cryptoxanthin, no esters appeared in serum or chylomicrons, suggesting that the cleavage of carotenoid esters prior to release into the lymphatic circulation occurs generally in human oxocarotenoid biokinetics.

Lycopene bioavailability from a single dose of fresh tomatoes or tomato paste (23 mg lycopene), ingested together with 15 g corn oil, was compared by analyzing carotenoid concentrations in the human chylomicron fraction (Gärtner *et al.*, 1997). The lycopene isomer pattern was the same in both fresh tomatoes and tomato paste. The triacylglycerol response in chylomicrons was not statistically different after both treatments. Ingestion of tomato paste was found to yield 2.5-fold higher total and *all-trans* lycopene peak concentrations ($p < 0.05$ and $p < 0.005$, respectively) and 3.8-fold higher area under curve (AUC) responses ($p < 0.001$) than ingestion of fresh tomatoes. The same was calculated for lycopene *cis* isomers, but only the AUC response for the *cis* isomers was statistically significantly higher after the ingestion of tomato paste ($p < 0.005$). No difference was observed in the α- and β-carotene response. Thus, in humans, the bioavailability of lycopene is higher from tomato paste than from fresh tomatoes.

Differences in carotenoids bioavailability from various sources might in part explain results from epidemiological studies where protective effects toward prostate cancer were associated with specific food items (Giovannucci *et al.*, 1996). An increased intake of tomato sauce was correlated with a diminished risk, whereas no association was found with the consumption of tomato juice.

REFERENCES

Allikmets, R., Shroyer, N. F., Singh, N., Seddon, J. M., Lewis, R. A., Bernstein, P. S., Pfeiffer, A., Zabriskie, N. A., Li, Y., Hutchinson, A., Dean, M., Lupski, J. R., and Leppert, M. (1997). Mutation of the Stargardt disease gene (ABCR) in age-related macular degeneration. *Science* **277**, 1805–1807.

Avron, M., and Ben-Amotz, A. (1992). Dunaliella: Physiology, Biochemistry, and Biotechnology. CRC Press, London.

Baltschun, D., Beutner, S., Briviba, K., Martin, H.-D., Paust, J., Peters, M., Röver, S., Sies, H.,

Stahl, W., Steigel, A., and Stenhorst, F. (1997). Singlet oxygen quenching abilities of carotenoids. *Liebigs Ann. Recueil* 1887–1893.

Bertram, J. S., and Bortkiewicz, H. (1995). Dietary carotenoids inhibit neoplastic transformation and modulate gene expression in mouse and human cells. *Am. J. Clin. Nutr.* **62,** 1322S–1326S.

Bertram, J. S. (1993). Cancer prevention by carotenoids. *Ann. N.Y. Acad. Sci.* **691,** 177–191.

Bex, V., Mercier, T., Chaumontet, C., Gaillard-Sanchez, I., Flechon, B., Mazet, F., Traub, O., and Martel, P. (1995). Retinoic acid enhances connexin43 expression at the post-transcriptional level in rat liver epithelial cells. *Cell Biochem. Funct.* **13,** 69–77.

Biesalski, H. K., Hemmes, C., Hopfenmüller, W., Schmid, C., and Gollnick, H. P. M. (1996). Effects of controlled exposure of sunlight on plasma and skin levels of β-carotene. *Free Radic. Res.* **24,** 215–224.

Bjerkeng, B. (1992). Analysis of carotenoids in salmonids. *In* "Quality Assurance in the Fish Industry" (H. H. Huss, ed.), Elsevier, New York.

Blot, W. J., Li, J.-Y., Taylor, P. R., Guo, W., Dawsey, S., Wang, G.-Q., Yang, C. S., Zheng, S.-F., Gail, M., Li, G.-Y., Yu, Y., Liu, B., Tangrea, J., Sun, Y., Liu, F., Fraumeni, J. F., Zhang, Y.-H., and Li, B. (1993). Nutrition intervention trials in Linxian, China: Supplementation with specific vitamin/mineral combinations, cancer incidence, and disease-specific mortality in the general population. *J. Natl. Cancer Inst.* **85,** 1483–1492.

Bohm, F., Edge, R., Land, E. J., McGarvey, D. J., and Truscott, T. G. (1997). Carotenoids enhance vitamin E antioxidant efficiency. *J. Am. Chem. Soc.* **119,** 621–622.

Bone, R. A., Landrum, J. T., Friedes, L. M., Gomez, C. M., Kilburn, M. D., Menendez, E., Vidal, I., and Wang, W. (1997). Distribution of lutein and zeaxanthin stereoisomers in the human retina. *Exp. Eye Res.* **64,** 211–218.

Burton, G. W., and Ingold, K. U. (1984). β-Carotene: An unusual type of lipid antioxidant. *Science* **224,** 569–573.

Chambon, P. (1996). A decade of molecular biology of retinoic acid receptors. *FASEB J.* **10,** 940–954.

Chug-Ahuja, J. K., Holden, J. M., Forman, M. R., Mangels, A. R., Beecher, G. R., and Lanza, E. (1993). The development and application of a carotenoid database for fruits, vegetables, and selected multicomponent foods. *J. Am. Diet. Assoc.* **93,** 318–323.

Clairmont, A., Tessmann, D., and Sies, H. (1996). Analysis of connexin43 gene expression induced by retinoic acid in F9 teratocarcinoma cells. *FEBS Lett.* **397,** 22–24.

Comstock, G. W., Burke, A. E., Hoffman, S. C., Helzlsouer, K. J., Bendich, A., Masi, A. T., Norkus, E. P., Malamet, R. L., and Gershwin, M. E. (1997). Serum concentrations of α-tocopherol, β-carotene, and retinol preceding the diagnosis of rheumatoid arthritis and systemic lupus erythematosus. *Ann. Rheumat. Dis.* **56,** 323–325.

Dahl, E., Winterhager, E., Traub, O., and Willecke, K. (1995). Expression of gap junction genes, connexin40 and connexin43, during fetal mouse development. *Anat. Embryol.* **191,** 267–278.

Demming-Adams, B., Gilmore, A. M., and Adams, W. W. (1996). In vivo functions of carotenoids in higher plants. *FASEB J.* **10,** 403–412.

De Pee, S., and West, C. E. (1996). Dietary carotenoids and their role in combatting vitamin A deficiency: Review of the literature. *Eur. J. Clin. Nutr.* **50,** S38–S53.

Di Mascio, P., Kaiser, S., and Sies, H. (1989). Lycopene as the most efficient biological carotenoid singlet oxygen quencher. *Arch. Biochem. Biophys.* **274,** 532–538.

Erdman, J. W., Bierer, T. L., and Gugger, E. T. (1993). Absorption and transport of carotenoids. *Ann. N. Y. Acad. Sci.* **691,** 76–85.

Everett, S. A., Dennis, M. F., Patel. K. B., Maddix, S., Kundu, S. C., and Willson, R. L. (1996).

Scavenging of nitrogen dioxide, thiyl, and sulfonyl free radicals by the nutritional antioxidant β-carotene. *J. Biol. Chem.* **271**, 3988–3994.

Flagg, E. W., Coates, R. J., and Greenberg, R. S. (1995). Epidemiologic studies of antioxidants and cancer in humans. *J. Am. Coll. Nutr.* **14**, 419–426.

Foote, C. S., and Denny, R. W. (1968). Chemistry of singlet oxygen. VII. Quenching by β-carotene. *J. Am. Chem. Sci.* **90**, 6233–6235.

Gärtner, C., Stahl, W., and Sies, H. (1997). Increased lycopene bioavailability from tomato paste as compared to fresh tomatoes. *Am. J. Clin. Nutr.* **66**, 116–122.

Gaziano, J. M., Hatta, A, Flynn, M., Johnson, E. J., Krinsky, N. I., Ridker, P. M., Hennekens, C. H., and Frei, B. (1995). Supplementation with β-carotene in vivo and in vitro does not inhibit low density lipoprotein oxidation. *Atheroscler.* **122**, 187–195.

Gerster, H. (1994). β-Carotene, vitamin E, and vitamin C in different stages of experimental carcinogenesis. *Eur. J. Clin. Nutr.* **49**, 155–168.

Giovannucci, E., Ascherio, A., Rimm, E. B., Stampfer, M. J., Colditz, G. A., and Willett, W. C. (1995). Intake of carotenoids and retinol in relation to risk of prostate cancer. *J. Natl. Cancer Inst.* **87**, 1767–1776.

Goodenough, D. A., Goliger, J. A., and Paul, D. L. (1996). Connexins, connexons, and intercellular communication. *Annu. Rev. Biochem.* **65**, 475–502.

Gutteridge, J. M. C. (1993). Free radicals in disease processes: A compilation of cause and consequence. *Free Radic. Res. Commun.* **19**, 141–158.

Halliwell, B. (1996). Antioxidants in human health and disease. *Annu. Rev. Nutr.* **16**, 33–50.

Handelman, G. J., van Kuijk, F. J. G. M., Chatterjee, A., and Krinsky, N. I. (1991). Characterization of products formed during the autoxidation of β-carotene. *Free Radic. Biol. Med.* **10**, 427–437.

Hanusch, M., Stahl, W., Schulz, W. A., and Sies, H. (1995). Induction of gap junctional communication by 4-oxoretinoic acid generated from its precursor canthaxanthin. *Arch. Biochem. Biophys.* **317**, 423–428.

Hebuterne, X., Wang, X.-D., Johnson, E. J., Krinsky, N. I., and Russell, R. M. (1995). Intestinal absorption and metabolism of 9-cis-β-carotene in vivo: Biosynthesis of 9-cis-retinoic acid. *J. Lipid. Res.* **36**, 1264–1273.

Hill, T. J., Land, E. J., McGarvey, D. J., Schalch, W., Tinkler, J. H., and Truscott, T. G. (1995). Interactions between carotenoids and the $CCl_3O_2\cdot$ radical. *J. Am. Chem. Soc.* **117**, 8322–8326.

Holder, J. W., Elmore, E., and Barrett, J. C. (1993). Gap junction function and cancer. *Cancer Res.* **53**, 3475–3485.

Hossain, M. Z., and Bertram, J. S. (1994). Retinoids suppress proliferation, induce cell spreading, and up-regulate connexin43 expression only in postconfluent 10T1/2 cells: Implications for the role of gap junctional communication. *Cell Grow. Diff.* **5**, 1253–1261.

Hossain, M. Z., Zhang, L.-X., and Bertram, J. S. (1993). Retinoids and carotenoids upregulate gap-junctional communication: Correlation with enhanced growth control and cancer prevention. *Progr. Cell Res.* **3**, 301–309.

Hotz-Wagenblatt, A., and Shalloway, D. (1993). Gap junctional communication and neoplastic transformation. *Crit. Rev. Oncogen.* **4**, 541–558.

Hughes, D. A., Wright, A. J. A., Finglas, P. M., Peerless, A. C. J., Bailey, A. L., Astley, S. B., Pinder, A. C., and Southon, S. (1997). The effect of β-carotene supplementation on the immune function of blood monocytes from healthy male nonsmokers. *J. Lab. Clin. Med.* **129**, 309–317.

Kennedy, T. A., and Liebler, D. C. (1991). Peroxyl radical oxidation of β-carotene: Formation of β-carotene epoxides. *Chem. Res. Toxicol.* **4**, 290–295.

Khachik, F., and Beecher, G. R. (1988). Separation and identification of carotenoids and carotenol fatty acid esters in some squash products by liquid chromatography. 1. Quantification of carotenoids and related esters by HPLC. *J. Agric. Food Chem.* **36,** 929–937.

Khachik, F., Beecher, G. R., Goli, M. B., Lusby, W. R., and Smith, J. C. (1992). Separation and identification of carotenoids and their oxidation products in the extracts of human plasma. *Anal. Chem.* **64,** 2111–2122.

Khachik, F., Spangler, C. J., Smith, J. C., Canfield, L. M., Steck, A., and Pfander, H. (1997). Identification, quantification, and relative concentrations of carotenoids and their metabolites in human milk and serum. *Anal. Chem.* **69,** 1873–1881.

Krinsky, N. I. (1994). Carotenoids and cancer: Basic research studies. *In* "Natural Antioxidants in Human Health and Disease" (B. Frei, ed.), pp. 239–261. Academic Press, San Diego.

Krutovskikh, V. A., Mesnil, M., Mazzoleni, G. and Yamasaki, H. (1995). Inhibition of rat liver gap junction intercellular communication by tumor promoting agents in vivo. *Lab. Invest.* **72,** 571–577.

Landrum, J. T., Bone, R. A., and Kilburn, M. D. (1997). The macular pigment: A possible role in protection from age-related macular degeneration. *Adv. Pharmacol.* **38,** 537–556.

Lepage, G., Champagne, J., Ronco, N., Lamarre, A., Osberg, I., Sokol, R. J., and Roy, C. C. (1996). Supplementation with carotenoids corrects increased lipid peroxidation in children with cystic fibrosis. *Am. J. Clin. Nutr.* **64,** 87–93.

Liaaen-Jensen, S. (1990). Marine carotenoids-selected topics. *New J.Chem.* **14,** 747–759.

Liebler, D. C., and McClure, T. D. (1996). Antioxidant reactions of β-carotene: Identification of carotenoid-radical adducts. *Chem. Res. Toxicol.* **9,** 8–11.

Liebler, D. C., Stratton, S. P., and Kaysen, K. L. (1997). Antioxidant actions of β-carotene in liposomal and microsomal membranes: Role of carotenoid-membrane incorporation and α-tocopherol. *Arch. Biochem. Biophys.* **338,** 244–250.

Lim, B. P., Nagao, A., Terao, J., Tanaka, K., Suzuki, T., and Takama, K. (1992). Antioxidant activity of xanthophylls on peroxyl radical-mediated phospholipid peroxidation. *Biochem. Biophys. Acta* **1126,** 178–184.

Mangels, A. R., Holden, J. M., Beecher, G. R., Forman, M. R., and Lanza, E. (1993). Carotenoid content of fruits and vegetables: An evaluation of analytical data. *J. Am. Diet. Assoc.* **93,** 284–296.

Mangelsdorf, D. J., Umesono, K., and Evans, R. M. (1994). The retinoid receptors. *In* "The Retinoids, Biology, Chemistry, and Medicine" (M. B. Sporn, A. B. Roberts, and D. S. Goodman, eds.), pp. 319–350. Raven Press, New York.

Mares-Perlman, J. A., Brady, W. E., Klein, R., Klein, B. E. K., Bowen, P., Stacewicz-Sapuntzakis, M., and Palta, M. (1995). Serum antioxidants and age-related macular degeneration in a population-based case-control study. *Arch. Ophthalmol.* **113,** 1518–1523.

Mathews-Roth, M. M. (1978). Carotenoid pigments and the treatment of erythropoietic protoporphyria. *J. Infect. Diseas.* **138,** 924–927.

Mathews-Roth, M. M. (1986). Systemic photoprotection. *Dermatol Clin.* **4,** 335–339.

Mathews-Roth, M. M., Pathak, M. A., Fitzpatrick, T. B., Harber, L. C., and Kass, E. H. (1974). β-Carotene as an oral protective agent in erythropoietic protoporphyria. *J. Am. Med. Assoc.* **228,** 1004–1008.

Mayne, S. T. (1996). Beta-carotene, carotenoids, and disease prevention in humans. *FASEB J.* **10,** 690–701.

Mayne, S. T., Handelman, G. J., and Beecher, G. (1996). β-Carotene and lung cancer promotion in heavy smokers: A plausible relationship? *J. Natl. Cancer. Inst.* **88,** 1513–1515.

Meydani, M., Martin, A., Ribaya-Mercado, J. D., Gong, J., Blumberg, J. B., and Russell,

R. M. (1994). β-Carotene supplementation increases antioxidant capacity of plasma in older women. *J. Nutr.* **124,** 2397–2403.

Miller, N. J., Sampson, J., Candeias, L. P., Bramley, P. M., and Rice-Evans, C. A. (1996). Antioxidant activities of carotenes and xanthophylls. *FEBS Lett.* **384,** 240–242.

Murakoshi, M., Takayasu, J., Kimura, O., Kohmura, E., Nishino, H., Iwashima, A., Okuzumi, J., Sakai, T., Sugimoto, T., Imanashi, J., and Iwasaki, R. (1989). Inhibitory effects of α-carotene on proliferation of the human neuroblastoma cell line GOTO. *J. Natl. Cancer Inst.* **81,** 1649–1652.

Nelis, H. J., and De Leenheer, A. P. (1991). Microbial sources of carotenoid pigments used in foods and feeds. *J. Appl. Bacteriol.* **70,** 181–191.

Nir, Z., Hartal, D., and Rahev, Y. (1993). Lycopene from tomatoes. *Int. Food Ingr.* **6,** 45–51.

Olson, J. A. (1994). Absorption, transport, and metabolism of carotenoids in humans. *Pure Appl. Chem.* **66,** 1011–1016.

Olson, J. A., and Krinsky, N. I. (1995). The colorful fascinating world of carotenoids: Important biological modulators. *FASEB J.* **9,** 1547–1550.

Omenn, G. S., Goodman, G. E., Thornquist, M. D., Balmes, J., Cullen, M. R., Glass, A., Keogh, J. P., Meyskens, F. L., Valanis, B., Williams, J. H., Barnhart, S., Cherniack, M. G., Brodkin, C. A., and Hammar, S. (1996). Risk factors for lung cancer and for intervention effects in CARET, the beta-carotene and retinol efficacy trial. *J. Natl. Cancer Inst.* **88,** 1550–1559.

Packer, J. E., Mahood, J. S., Mora-Arellano, V. O., Slater, T. F., Willson, R. L., and Wolfenden, B. S. (1981). Free radicals and singlet oxygen scavengers: Reaction of a peroxyl radical with β-carotene, diphenyl furan and 1,4-diazobicyclo (2,2,2)-octane. *Biochem. Biophys. Res. Commun.* **98,** 901–906.

Palozza, P., and Krinsky, N. I. (1992). Antioxidant effects of carotenoids in vivo and in vitro: An overview. *Methods Enzymol.* **213,** 403–420.

Palozza, P., Moualla, S., and Krinsky, N. I. (1992). Effects of β-carotene and α-tocopherol on radical-initiated peroxidation of microsomes. *Free Radic. Biol. Med.* **13,** 127–136.

Parker, R. S. (1997). Bioavailability of carotenoids. *Eur. J. Clin. Nutr.* **51,** S86–S90.

Pastorino, U. (1997). β-Carotene and the risk of lung cancer. *J. Natl. Cancer Inst.* **89,** 456–457.

Pfander, H. (1997). "Key to Carotenoids." Birkhäuser Verlag, Basel.

Potter, J. D. (1997). β-Carotene and the role of intervention studies. *Cancer Lett.* **114,** 329–331.

Pung, A., Franke, A., Zhang, L.-X., Ippendorf, M., Martin, H.-D., Sies, H., and Bertram, J. S. (1993). A synthetic C22 carotenoid inhibits carcinogen-induced neoplastic transformation and enhances gap junctional communication. *Carcinogenesis* **14,** 1001–1005.

Pung, A., Rundhaug, J. E., Yoshizawa, C. N., and Bertram, J. S. (1988). β-Carotene and canthaxanthin inhibit chemically- and physically-induced neoplastic transformation in 10T1/2 cells. *Carcinogenesis* **9,** 1533–1539.

Ribaya-Mercado, J. D., Garmyn, M., Gilchrest, B. A., and Russell, R. M. (1995). Skin lycopene is destroyed preferentially over β-carotene during ultraviolet irradiation in humans. *J. Nutr.* **125,** 1854–1859.

Rivas, J. D. (1989). Reversed-phase high-performance liquid chromatographic separation of lutein and lutein fatty acid esters from marigold flower petal powder. *J. Chromatogr.* **464,** 442–447.

Schalch, W. (1992). Carotenoids in the retina: A review of their possible role in preventing or limiting change caused by light and oxygen. *In* "Free Radicals and Aging" (I. Emerit and B. Chance, eds.), pp. 280–298. Birkhäuser Verlag, Basel.

Seddon, J. M., Ajani, U. A., Sperduto, R. D., Hiller, R., Blair, N., Burton, T. C., Farber, M. D.,

Gragoudas, E. S., Haller, J., Miller, D. T., Yannuuzzi, L. A., and Willett, W. C. (1994). Dietary carotenoids, vitamins A, C, and E, and advanced age-related macular degeneration. *JAMA* **272**, 1413–1420.

Sharoni, Y. and Levi, J. (1996). Anticarcinogenic properties of lycopene. *In* "Natural Antioxidants and Food Quality in Atherosclerosis and Cancer Prevention" (J. T. Kumpulainen and J. Salonen, eds.), pp. 378–385. The Royal Society of Chemistry, Cambridge.

Sies, H., and Stahl, W. (1995). Vitamins E and C, β-carotene, and other carotenoids as antioxidants. *Am. J. Clin. Nutr.* **62**, 1315S-1321S.

Sies, H. (1993). Strategies of antioxidant defense. *Eur. J. Biochem.* **215**, 213–219.

Sies, H. (1997). "Antioxidants in Disease Mechanisms and Therapy." Academic Press, London.

Stahl, W., Nicolai, S., Briviba, K., Hanusch, M., Broszeit, G., Peters, M., Martin, H.-D., and Sies, H. (1997). Biological activities of natural and synthetic carotenoids: Induction of gap junctional communication and singlet oxygen quenching. *Carcinogenesis* **18**, 89–92.

Stahl, W., Schwarz, W., Sundquist, A. R., and Sies, H. (1992). *cis*-trans isomers of lycopene and β-carotene in human serum and tissues. *Arch. Biochem. Biophys.* **294**, 173–177.

Stahl, W., Schwarz, W., von Laar, J., and Sies, H. (1995). all-trans-β-Carotene preferentially accumulates in human chylomicrons and very low density lipoproteins compared with the 9-cis geometrical isomer. *J. Nutr.* **125**, 2128–2133.

Stahl, W., and Sies, H. (1992). Uptake of lycopene and its geometrical isomers is greater from heat-processed than from unprocessed tomato juice in humans. *J. Nutr.* **122**, 2161–2166.

Stahl, W., and Sies, H. (1993). Physical quenching of singlet oxygen and cis-trans isomerization of carotenoids. *Ann. N. Y. Acad. Sci.* **691**, 10–19.

Stahl, W., and Sies, H. (1993). Lycopene: A biologically important carotenoid for humans? *Arch. Biochem. Biophys.* **336**, 1–9.

Stivala, L. A., Savio, M., Cazzalini, O., Pizzala, R., Rehak, L., Bianchi, L., Vannini, V., and Prosperi, E. (1996). Effect of β-carotene on cell cycle progression of human fibroblasts. *Carcinogenesis* **11**, 2395–2401.

The Alpha-Tocopherol, Beta Carotene Cancer Prevention Study Group (1994). The effect of vitamin E and beta carotene on the incidence of lung cancer and other cancers in male smokers. *N. Engl. J. Med.* **330**, 1029–1035.

Truscott, T. G. (1990). The photophysics and photochemistry of the carotenoids. *J. Photochem. Photobiol. B Biol.* **6**, 359–371.

van Poppel, G., and Goldbohm, R. A. (1995). Epidemiologic evidence for β-carotene and cancer prevention. *Am. J. Clin. Nutr.* **62**, 1393S–1402S.

Wang, X.-D. (1994). Review: Absorption and metabolism of β-carotene. *J. Am. Coll. Nutr.* **13**, 314–325.

Wingerath, T., Stahl, W., Kirsch, D., Kaufmann, R., and Sies, H. (1996). Fruit juice carotenol fatty acid esters and carotenoids as identified by matrix-assisted laser desorption ionization (MALDI) mass spectrometry. *J. Agric. Food Chem.* **44**, 2006–2013.

Wingerath, T., Stahl, W., and Sies, H. (1995). β-Cryptoxanthin selectively increases in human chylomicrons upon ingestion of tangerine concentrate rich in β-cryptoxanthin esters. *Arch. Biochem. Biophys.* **324**, 385–390.

Wolf, G. (1992). Retinoids and carotenoids as inhibitors of carcinogenesis and inducers of cell-cell communication. *Nutr. Rev.* **50**, 270–274.

Woodall, A. A., Britton, G., and Jackson, M. J. (1996). Dietary supplementation with carotenoids: Effects on α-tocopherol levels and susceptibility of tissues to oxidative stress. *Br. J. Nutr.* **76**, 307–317.

Yamasaki, H. (1995). Non-genotoxic mechanisms of carcinogenesis: Studies of cell transformation and gap junctional intercellular communication. *Toxicol. Lett.* **77**, 55–61.

You, C.-S., Parker, R. S., Goodman, K. J., Swanson, J. E., and Corso, T. N. (1996). Evidence of cis-trans isomerization of 9-cis-β-carotene during absorption in humans. *Am. J. Clin. Nutr.* **64**, 177–183.

Young, A., and Britton, G. (1993). "Carotenoids in Photosynthesis." Chappman & Hall, London.

Zhang, L.-X., Cooney, R. V., and Vertram, J. S. (1991). Carotenoids enhance gap junctional communication and inhibit lipid peroxidation in C3H/10T1/2 cells: Relationship to their cancer chemopreventive action. *Carcinogenesis* **12**, 2109–2114.

14 Dietary Carotenoids and their Metabolites as Potentially Useful Chemoprotective Agents against Cancer

Frederick Khachik,* John S. Bertram,†
Mou-Tuan Huang,‡ Jed W. Fahey,¶ and Paul Talalay¶

*Department of Chemistry and Biochemistry
Joint Institute for Food Safety and Applied Nutrition
University of Maryland
College Park, Maryland 20742

†Cancer Research Center of Hawaii
University of Hawaii at Manoa
Honolulu, Hawaii 96813

‡Department of Chemical Biology
Laboratory for Cancer Research
College of Pharmacy
Rutgers, The State University of New Jersey
Piscataway, New Jersey 08854

¶Brassica Chemoprotection Laboratory
Department of Pharmacology and Molecular Science
Johns Hopkins University School of Medicine
Baltimore, Maryland 21205

INTRODUCTION

One class of food components that has been widely studied for their protective effect against cancer is carotenoids. Carotenoids are among the most widespread of the naturally occurring groups of pigments and are found in all families of the plant and animal kingdoms. To date, as many as

700 carotenoids have been isolated from various sources and their chemical structures have been characterized. During the past decade, the authors have isolated, identified, and quantified carotenoids from fruits and vegetables commonly consumed in the United States (Khachik *et al.,* 1986; Khachik and Beecher, 1988; Khachik *et al.,* 1989, 1992a,e). These studies have revealed that as many as 40 to 50 carotenoids may be available from the diet and absorbed, metabolized, or utilized by the human body (Khachik *et al.,* 1991). However, among these, only 13 *all-E-* and 12 *Z-* carotenoids are found routinely in human serum and milk (Khachik *et al.,* 1992b,c,d, 1995b, 1997c). In addition, there are 1 *Z-* and 8 *all-E-* carotenoid metabolites resulting from two major dietary carotenoids, lutein and lycopene, which have also been characterized by Khachik *et al.* (1992b,c,d, 1995b, 1997c). The correlation between dietary carotenoids and carotenoids found routinely in the extracts from human serum/plasma has revealed that only selected groups of carotenoids make their way into the human bloodstream. Some of these carotenoids are absorbed intact and others, such as lutein, zeaxanthin, and lycopene, are converted to several metabolites. Some dietary carotenoids are also present in human tissues, i.e., liver, lung, breast, and cervix (Khachik *et al.,* 1998a).

Evidence for the nutritional significance of carotenoids in the prevention of diseases such as cancer, cardiovascular, and macular degeneration (an age-related degenerative eye disease) has been obtained from various interdisciplinary studies, which may be classified as (a) epidemiological studies, (b) carotenoid distribution in fruits, vegetables, human serum, and milk, (c) carotenoids in human organs and tissues, (d) *in vitro* studies of chemopreventive properties, and (e) *in vivo* studies with rodents. Ever since the 1970s, it has been suggested that one mechanism by which carotenoids exert their biological activity in disease prevention was by functioning as an antioxidant. In 1992, for the first time, the authors reported on the isolation and characterization of several oxidation products of carotenoids in human plasma (Khachik *et al.,* 1992c). More recently the *in vivo* oxidation of specific carotenoids, has been demonstrated, e.g., lutein, zeaxanthin, and possibly lycopene in humans (Khachik *et al.,* 1995a, 1997d; Paetau *et al.,* 1998).

This chapter identifies several new optical isomers of carotenoid metabolities in human plasma and provides additional evidence for previously proposed metabolic oxidation–reduction reactions of dietary carotenoids in humans. This chapter also presents studies that indicate that carotenoids, in addition to their antioxidant mechanism of action, can exert their biological activity in disease prevention by other mechanisms. These are (a) gap junctional intercellular communications (GJC) (King *et al.,* 1997), (b) anti-inflammatory and antitumor promoting properties and (c) induction of detoxication (phase 2) enzymes.

EXPERIMENTAL PROCEDURES

Instrumentation

For gradient high-performance liquid chromatography (HPLC) analysis, a binary solvent delivery system interfaced into a rapid-scanning ultraviolet (UV)/visible photodiode array detector was employed.

Chromatographic Conditions

For separation of the benzoate derivatives of carotenoids and their metabolites, which were prepared either from extracts of human plasma or by synthesis, a Sumichiral OA-2000 (5 μm, 25 cm $\times$ 4 mm) column (Phenomenex, Torrance, CA) was employed. The column was protected with a Brownlee Cyano guard cartridge (5 μm, 3 cm $\times$ 4.6 mm). A combination of an isocratic and a two-step gradient at the flow rate of 0.70 ml/min affected the separations in about 100 min. Details of the isocratic and the gradient conditions are summarized in Table I.

Reference Samples of Carotenoids

Carotenoid standards were either synthesized or isolated from natural sources according to published procedures (Khachik *et al.,* 1986, 1992c, 1995b, 1997c, 1998b). (3*R*,6'*R*)-3-Hydroxy-β,ϵ-3'-one was prepared from

Table I   Chromatographic Conditions for Separation of a Mixture of Benzoate Derivatives of Carotenoids and Their Metabolites Prepared from an Extract of Human Plasma

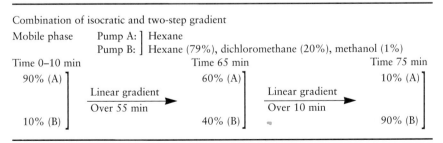

Combination of isocratic and two-step gradient

Mobile phase   Pump A: ⎤ Hexane
                Pump B: ⎦ Hexane (79%), dichloromethane (20%), methanol (1%)

Time 0–10 min		Time 65 min		Time 75 min
90% (A) ⎤	Linear gradient / Over 55 min →	60% (A) ⎤	Linear gradient / Over 10 min →	10% (A) ⎤
10% (B) ⎦		40% (B) ⎦		90% (B) ⎦

lutein by partial synthesis and (3*R*,6′*R*)-3-hydroxy-β,ε-caroten-3′-one was prepared by total synthesis (Khachik *et al.*, 1992c). Lactucaxanthin was isolated from an extract of Romaine lettuce (*Lactuca sativa*).

Preparation of Carotenoid Benzoates

Hydroxycarotenoids were dissolved in 2.0 ml of pyridine and cooled to 0°C under an atmosphere of nitrogen. A few drops of benzoyl chloride were added and the mixture was stirred for 30 min at 0°C and for 1 hr at room temperature. The product was partitioned between hexane and water. The hexane layer was washed with water several times, dried over sodium sulfate, and evaporated to dryness. The residue was dissolved in hexane (80%) and dichloromethane (20%) for HPLC analysis.

Preparation of Benzoate Derivatives of Carotenoids from an Extract of Human Plasma

Human plasma (American Red Cross, Baltimore, MD) was extracted according to published procedures (Khachik *et al.*, 1992c, 1997c). The extract was dried over sodium sulfate, dissolved in pyridine, and treated with benzoyl chloride; the product was worked up similar to the previous procedure.

Intercellular Communication Assays

Cell Culture Conditions

The human immortalized keratinocyte cell line HaCaT (Ryle *et al.*, 1989; Fitzgerald *et al.*, 1994) was used to examine carotenoid effects in human cells. All cultures were incubated at 37°C, in 5% carbon dioxide and 95% humidity. HaCaT cells were cultured in low calcium (0.1 mmol/liter) serum-free keratinocyte medium (GIBCO-BRL, Grand Island, NY) supplemented with epidermal growth factor (5 μg/ml), bovine pituitary extract (35 μg/ml), and insulin (5 μg/ml) in a monolayer culture until subconfluent. They were then harvested by trypsin/EDTA, resuspended in low calcium medium, and placed on Millicell-CM collagen-coated culture plate inserts (Millipore, Bedford, MA). Filters were incubated submerged in low calcium medium for approximately 7 days, at which point high calcium medium (1.15 mmol/liter), supplemented as described earlier, was placed below the filters while the surface of the filters were exposed to the atmosphere for 7 days. During this time a multilayered differentiating organotypic culture was produced. Cells were treated with carotenoids or retinoic acid dissolved in THF or acetone, respectively, using the precautions and procedures discussed previously (Cooney *et al.*, 1993) and added to culture medium concurrent with exposure to the air interface. Cells were refed and retreated every 2 days.

Protein Electrophoresis and Western Blotting

HaCaT cells were harvested and connexin 43 (Cx43) solubilized as described previously (Rogers *et al.*, 1990). Protein concentrations were determined by the Bio-Rad protein assay (Bio-Rad Laboratories, Hercules, CA), and equal amounts of protein were electrophoresed on 10% SDS polyacrylamide gels. Proteins were electroblotted onto an Immobilon PVDF membrane (Millipore) and subsequently incubated with a rabbit polyclonal antibody raised against a synthetic peptide corresponding to the C-terminal 15 residues of the predicted sequence of rat cardiac Cx43 (Beyer *et al.*, 1987). This sequence is 100% homologous in mouse, rat, and human. Bound antibody was visualized using a chemiluminescent detection system (Tropix, Bedford, MA) and recorded on X-ray film.

Phase 2 Enzyme Induction Assay

Purified carotenoids were dissolved in tetrahydrofuran (THF) and their quinone reductase (QR) inducer potency was determined by a coupled tetrazolium dye assay performed on digitonin extracts of Hepa 1c1c7 cells grown in microliter plates. The procedures have been described previously by Prochaska and Santamaria (1988) and Prochaska *et al.* (1992) and modified by Fahey *et al.* (1997). The protein content of digitonin extracts was determined using a bicinchoninic acid assay (Smith *et al.*, 1985). Carotenoids were tested over a concentration range of 0.78 to 100 μM and the final concentration of THF was 0.5% by volume. Activity is reported as follows. Concentration for doubling (CD) inducer activity is the amount of compound required to double the QR activity of a microtiter plate well, initially seeded with 10,000 Hepa 1c1c7 (murine hepatoma) cells and containing 0.15 ml of α-MEM culture medium amended with 10% fetal calf serum, streptomycin, and penicillin.

SEPARATION OF OPTICAL STEREOISOMERS OF CAROTENOIDS AND THEIR METABOLITES IN HUMAN PLASMA BY HPLC

Dietary hydroxycarotenoids and their metabolites, identified in human serum, contain two or three stereogenic centers in their molecules and would therefore be expected to exist as a number of optically active isomers. Although the authors have reported on the separation and identification of carotenoids and their metabolites from extracts of human serum and milk previously (Khachik *et al.*, 1992c, 1997c), the absolute configuration of these compounds has not been determined to date. Understanding the stereochemistry of carotenoids in serum can provide valuable information regarding the pathways leading to the formation of carotenoid metabolites in humans. Metabolites found in serum result from three major dietary carotenoids: lutein, zeaxanthin, and lycopene (Fig. 1). The chemical structures of carotenoid metabolites are shown in Fig. 2. Among these, only the absolute configurations of 3′-epilutein and the two dehydration products of dietary lutein, (3*R*,6′*R*)-3′,4′-didehydro-β,γ-carotene-3-ol and (3*R*,6′*R*)-2′3′-didehydro-β,ε-carotene-3-ol, are known (Khachik *et al.*, 1995b). In the case of the two lycopene metabolites, 2,6-cyclolycopene-1,5-diols A and B, only the relative but not the absolute configuration of these compounds at C-2, C-5, and C-6 is known at present (Khachik *et al.*, 1998b). The rest of the carotenoid metabolites shown in Fig. 2 are monoketo- and diketo-carotenoids that have also been isolated from hen's egg yolk by Matsuno *et al.* (1986). These investigators employed a Sumipax OA-2000 chiral HPLC column to separate the 3 stereoisomers of ε, ε-caroten-3,3′-dione and all 10 stereoisomers of lactucaxanthin [(3*S*,6*S*,3′*S*,6′*S*-ε, ε-carotene-3,3′-diol] (Fig. 1), a dihydroxycarotenoid with four stereogenic centers (Ikuno *et al.*, 1985). This was accomplished by derivatization of this dihydroxy-carotenoid with benzoyl chloride and separation of the resulting diben-zoates by chiral chromatography. A chiral HPLC column was employed to separate the benzoate derivatives of lutein, zeaxanthin, lactucaxanthin, and the metabolites of carotenoids typically found in human serum. The objective was to determine whether it was possible to simultaneously separate the benzoate derivatives of carotenoids as well as the optical isomers of their metabolites by chiral HPLC in such a complex mixture by derivatiz-ing all the mono- and dihydroxycarotenoids in an extract from human plasma with benzoyl chloride. Because of the lack of a hydroxyl group in ε, ε-caroten-3,3′-dione and the tertiary nature of the hydroxyl groups in 2,6-cyclolycopene-1,5-diols A and B (Fig. 2), these metabolites could not be derivatized with benzoyl chloride. Therefore, these compounds were mixed directly with a synthetic mixture of the benzoate derivatives of all the hy-droxycarotenoids in human serum for HPLC studies. Employing a combi-nation of isocratic and two-step gradient HPLC (Table I), the various opti-cal isomers of carotenoids and their metabolites in the just-described synthetic mixture were resolved adequately as shown in Fig. 3. The HPLC profile of an extract from human plasma derivatized with benzoyl chloride is shown in Fig. 4. The benzoate derivatives of monohydroxycarotenoids,

Dietary Carotenoids

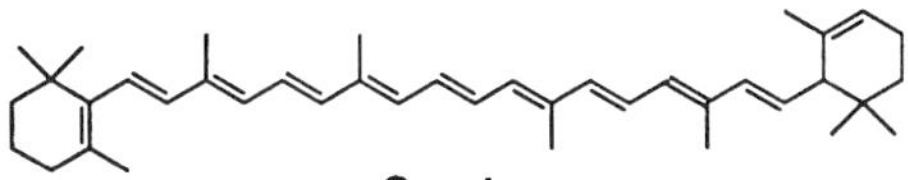

α-Carotene

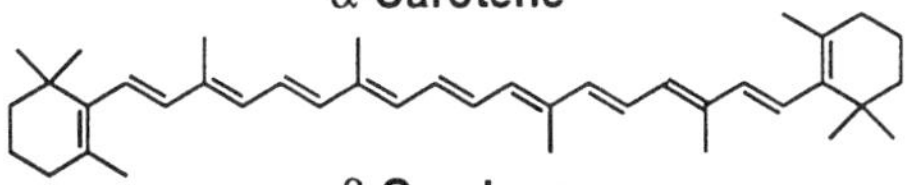

β-Carotene

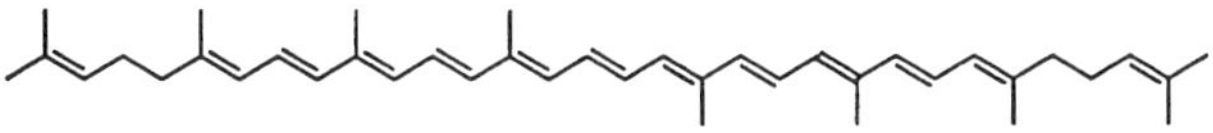

Lycopene

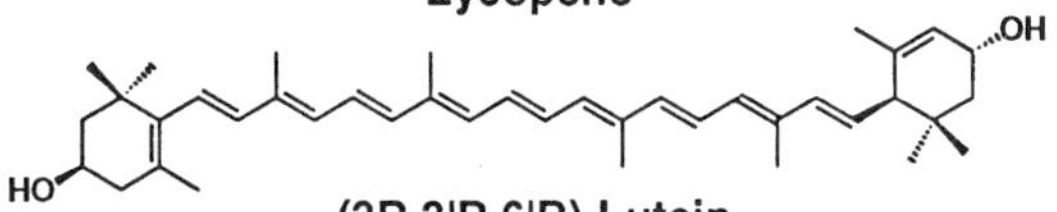

(3R,3'R,6'R)-Lutein

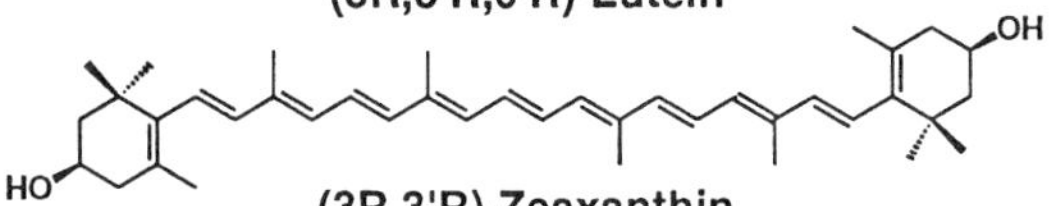

(3R,3'R)-Zeaxanthin

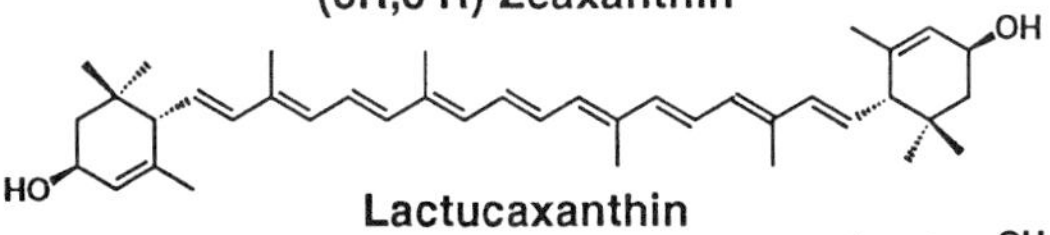

Lactucaxanthin

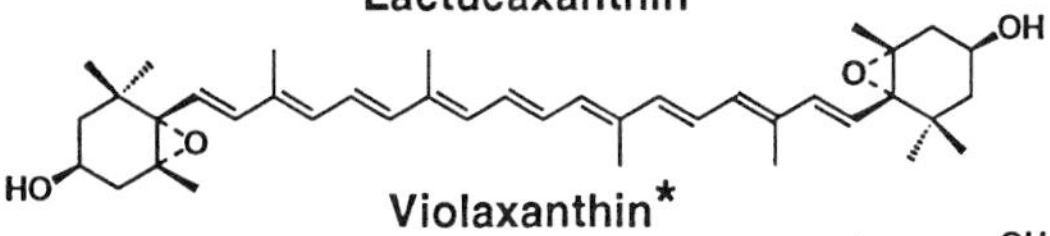

Violaxanthin*

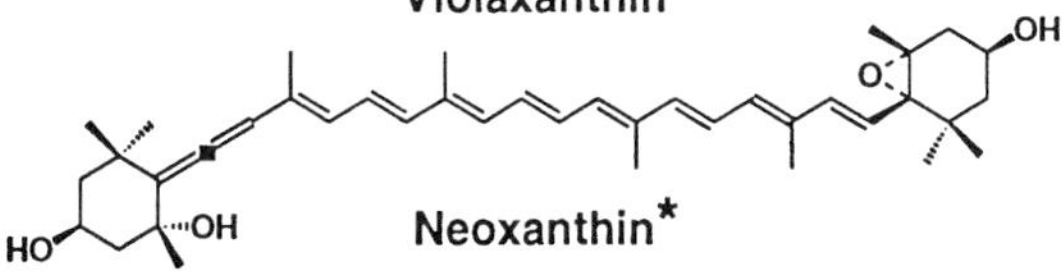

Neoxanthin*

Non-Carotenoids Tested

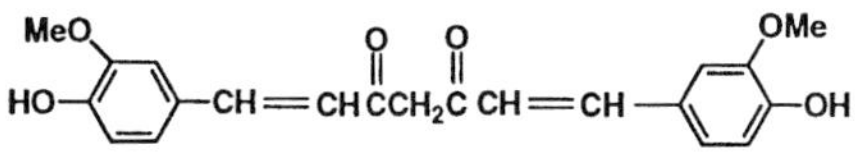

1,7-bis(4-Hydroxy-3-methoxyphenyl)-
1,6-heptadiene-3,5-dione (Curcumin)

$CH_3 - S(O) - (CH_2)_4 - N=C=S$

[(-)-1-Isothiocyanato-4(R)-(methylsulfinyl)
-butane] (Sulforaphane)

Figure 1 Chemical structures of some of the dietary carotenoids and two noncarotenoids, curcumin and sulforaphane, evaluated as inducers of phase 2 enzymes. Asterisks indicate carotenoids not absorbed into human serum/plasma.

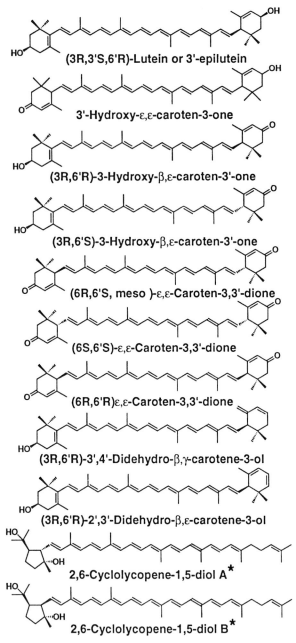

Figure 2 Chemical structures of carotenoid metabolites and their optical stereoisomers identified in an extract from human plasma. Absolute configurations of the various optical isomers of 3'-hydroxy-ε,ε-caroten-3-one are not known at present. Asterisks indicate that only the relative but not the absolute configuration of these compounds at C-2, C-5, and C-6 is known.

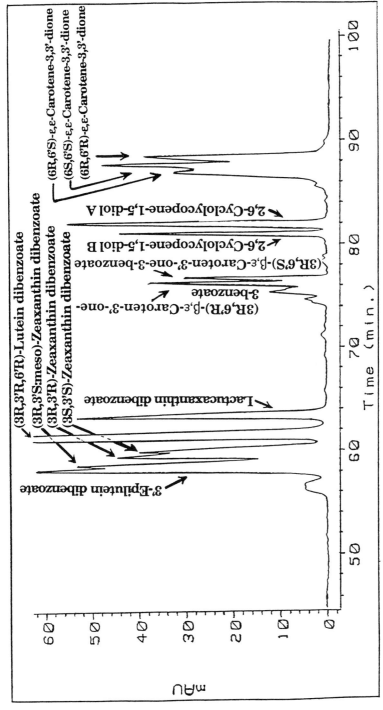

Figure 3 Chiral HPLC separation of optical isomers of a mixture of synthetic benzoate derivatives of certain carotenoids and their metabolites. For clarity of presentation, only the latter part of the HPLC profile is shown. Conditions are as described in the text.

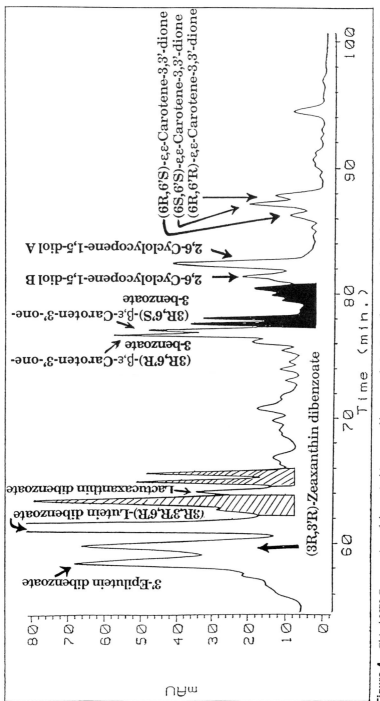

Figure 4 Chiral HPLC separation of the optical isomers of benzoate derivatives of carotenoids and their metabolites prepared from an extract of human plasma. For clarity of presentation, only the latter part of the HPLC profile is shown. Conditions are as described in the text. Shaded peaks are the Z-isomers of lutein and zeaxanthin dibenzoates and dark peaks are the optical isomers of ε,ε-caroten-3-one-3′-benzoate.

such as α- and β-cryptoxanthin, as well as the dehydration products of lutein, which are normally present in human plasma, were also prepared and examined by HPLC. However, due to early retention times (25–30 min), the separation of these compounds is not shown in the chromatograms in Figs. 3 and 4. Although 3′-hydroxy- ε, ε-caroten-3-one benzoate was not prepared by synthesis, the presence of the various optical isomers of this compound with three stereogenic centers was tentatively established by HPLC (dark peaks highlighted in Fig. 4). Individual carotenoid benzoates were separated by preparative HPLC and were identified by comparison of their circular dichroism spectra with those reported in the literature (Ikuno *et al.*, 1985; Matsuno *et al.*, 1986). Details of the structural elucidation will be described in a later publication. As shown in Fig. 4, there appears to be some HPLC peak overlap between 3′-epilutein dibenzoate and zeaxanthin dibenzoate. Therefore, lutein, zeaxanthin, and 3′-epilutein were isolated by preparative HPLC from an extract of human plasma (Khachik *et al.*, 1992c, 1992d) and then derivatized with benzoyl chloride separately. HPLC results indicated that in addition to dietary (3R,3′R,6′R)-lutein and its metabolite (3R,3′S,6′R)-lutein (3′-epilutein), none of the other optically active isomers of lutein were present in human plasma. However, in the case of dietary (3R,3′R)-zeaxanthin, approximately 2–3% of an optically inactive isomer of this compound, namely (3R,3′S,*meso*)-zeaxanthin, could be detected in plasma.

DIETARY CAROTENOIDS AND THEIR METABOLITES AS CANCER CHEMOPREVENTIVE AGENTS

There is very little disagreement within the scientific community regarding the protective effect of high consumption of fruits and vegetables and the lower risk for human cancers. However, the identity of the active components in such a diet and the mechanisms by which this protective effect may be provided are the subjects of intense research and much debate. Carotenoids are one such class of compounds, which are present at high concentrations in fruits and vegetables associated with risk reduction in many epidemiologic studies (Mayne, 1996). Although the number of dietary carotenoids is in excess of 40, it has been shown that 13 *all-E* and 12-Z dietary ones are actually absorbed into the bloodstream. These dietary carotenoids are metabolized to 8 *all-E* and 1-Z metabolites and, as a result, a total of 34 carotenoids can be detected in human serum and milk (Khachik *et al.*, 1997c). Despite the presence of all these dietary carotenoids and their metabolites in the serum, β-carotene has received much of the attention due to its provitamin A activity. The nutritional significance of

β-carotene and carotenoids in general in the prevention of cancer has been based solely on their provitamin A activity and, to some extent, on their antioxidant property. In 1991, Bertram and colleagues reported on the cancer preventive property of carotenoids by a different mechanism involving the upregulation of the expression of gap junctional communication proteins. However, to date, other mechanisms of chemoprotection by carotenoids, particularly their metabolites, have not been fully investigated. Possible mechanisms by which chemopreventive agents may work against events that are involved in the development of cancer have been reviewed by Kelloff *et al.* (1994).

The authors have examined the chemopreventive properties of major dietary carotenoids and their metabolites in light of some of these mechanisms and have established that these compounds may serve as potentially useful chemoprotective agents against cancer. The chemical structures of some of the dietary carotenoids investigated are shown in Fig. 1. Figure 1 also shows the structures of two noncarotenoids, curcumin and sulforaphane, which due to their unique and well-established role as chemoprotective agents were selected for comparative studies with carotenoids. α-Carotene, β-carotene, lycopene, lutein, and zeaxanthin are among the major dietary carotenoids found in human serum or plasma whereas only trace levels of lactucaxanthin can be detected. This is probably related to the dietary source of this compound, which is limited to Romaine lettuce (*L. sativa*). Although violaxanthin and neoxanthin are two major carotenoid epoxides found in fruits and green vegetables (Khachik *et al.*, 1986), these compounds have not been detected in human serum or plasma (Khachik *et al.*, 1992c, 1997c).

Studies with the carotenoids shown in Fig. 1 and some of their metabolites (Fig. 2) have revealed that these compounds may impart their biological activity by several mechanisms. These are (a) antioxidant function, (b) gap junctional intercellular communication, (c) anti-inflammatory and antitumor promoting properties, and (d) induction of detoxication (phase 2) enzymes.

Antioxidant Function

One hypothesis regarding the role of carotenoids as cancer chemopreventive agents is based on their antioxidant capability to quench singlet oxygen and other oxidizing species and to inhibit lipid peroxidation, thus preventing further promotion and replication in the neoplastic cell. If a free radical mechanism is involved in the initiation and promotion of carcinogenesis, carotenoids such as lutein, zeaxanthin, and lycopene may participate in quenching peroxides and protecting cells from oxidative damage.

This would be expected to result in the formation of a number of oxidative metabolites of these carotenoids. In 1992, for the first time, the authors reported on the isolation and characterization of oxidation products of carotenoids in human plasma (Khachik *et al.,* 1992c) and conducted several human supplementation studies to demonstrate the *in vivo* oxidation of certain dietary carotenoids such as lutein and zeaxanthin to their metabolites (Khachik *et al.,* 1995a, 1997d). In a more recent publication, the structures of two oxidative metabolites of lycopene, 2,6-cyclolycopene-1,5-diols A and B (Fig. 2) in human serum and milk, were established (Khachik *et al.,* 1997c, 1998b) and a mechanism for the formation of these metabolites was proposed (Khachik *et al.,* 1997d). Further evidence for possible *in vivo* oxidation of lycopene to these metabolites was obtained from a study with purified lycopene supplements involving healthy human volunteers (Paetau *et al.,* 1998). Based on these studies, the authors have proposed the most likely pathways leading to the formation of the oxidative metabolites of carotenoids in humans (Khachik *et al.,* 1995a, 1997d). These metabolic pathways are summarized in Fig. 5. With the exception of the metabolites of lycopene (2,6-cyclolycopene-1,5-diols A and B) (Khachik *et al.,* 1997d) and the two dehydration products of lutein (Khachik *et al.,* 1995b) reported previously, the carotenoid metabolites, shown in Fig. 5, can be formed by three types of reactions. These are (1) allylic oxidation of the ϵ-end group of lutein to give an α,β-unsaturated ketocarotenoid, (2) reduction of the resulting ketocarotenoid with epimerization at C-3, and (3) double bond isomerization in the β-end group of dietary zeaxanthin or other intermediate carotenoid metabolites to give an ϵ-end group. While the oxidation reactions remove the stereogenic centers in the carotenoid metabolites, the reduction and double bond isomerization reactions create new stereogenic centers. As described earlier, all the optical isomers of monoketo- and diketocarotenoids appear to be present in the extracts of human plasma. These results are in complete agreement with earlier proposed metabolic pathways of dietary lutein and zeaxanthin in humans. Another carotenoid that may serve as an antioxidant is lactucaxanthin (Fig. 1). Due to the presence of two allylic hydroxyl groups in its molecule, lactucaxanthin would be expected to oxidize readily to mono- and diketocarotenoids; metabolic pathways of this compound are not shown in Fig. 5. Because of its limited dietary source (Romaine lettuce), only a trace amount of lactucaxanthin has been detected in human serum (Khachik *et al.,* 1997c). It is imperative to point out that although only 2–3% of (3*R*,3'*S,meso*)-zeaxanthin has been found inhuman plasma, the presence of this compound completes the cycle of metabolic reactions and interconversion of dietary lutein and zeaxanthin in humans.

This chapter presents additional evidence for the *in vivo* oxidation of

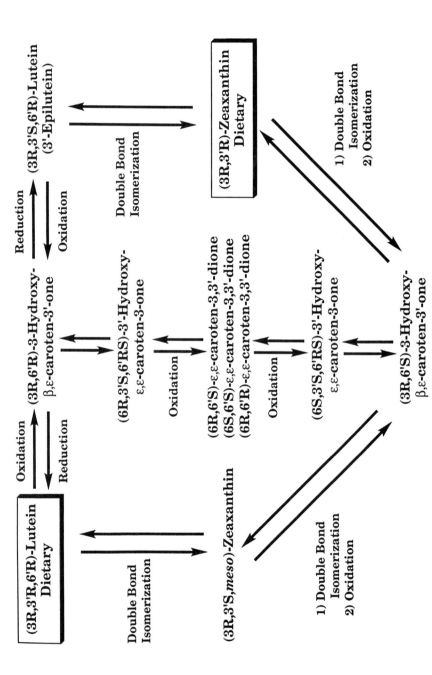

Figure 5 Metabolic pathways of dietary lutein and zeaxanthin in humans.

carotenoids by further elucidating the stereochemistry of some of their oxidative metabolites. However, the pharmacokinetics and pharmacodynamics of the oxidative metabolites of carotenoids, as well as the nature of the enzymes that may be involved in these metabolic reactions, are not known at present and they clearly need to be investigated.

Gap Junctional Intercellular Communications

Clincal studies have confirmed the experimental evidence that retinoids can act as cancer preventive agents (Hong *et al.*, 1990); however, the strong epidemiological association between dietary carotenoids and a decreased risk of cancer has not been substantiated in clinical trials of β-carotene (Mayne, 1996). The authors have been investigating the mechanistic basis for cancer prevention by these agents using a mouse assay for neoplastic transformation. In these studies, it has been demonstrated that certain retinoids (Merriman and Bertram, 1979) and carotenoids, including β-carotene (Bertram *et al.*, 1991), can inhibit the carcinogen-induced neoplastic transformation of 10T½ cells and that this inhibition correlates with increased gap junctional intercellular communication (Hossain *et al.*, 1989; Zhang *et al.*, 1991). In both cases the observed increase in GJC is mediated through an increase in the gap junctional protein connexin 43 (Cx43) at the mRNA and protein level (Rogers *et al.*, 1990; Zhang *et al.*, 1992). This increase in GJC is believed to be of functional significance to their actions as chemopreventive agents, as many separate studies have demonstrated an association between GJC and growth control (Neveu and Bertram, 1997).

An important question arises of how do carotenoids elevate expression of this gene? Because β-carotene is converted readily in mammals to retinoids, many of its effects have been considered to be mediated through its breakdown, either spontaneously or enzymatically to retinoids. However, the demonstration that several non-provitamin A carotenoids, including the acyclic lycopene, also mediate responses similar to retinoic acid (Zhang *et al.*, 1992) has led to a reevaluation of this concept. While conversion of many carotenoids to compounds known to have retinoid-like properties is certainly feasible on a chemical basis, it is more difficult to imagine such a conversion in the case of the straight-chain carotenoid lycopene (Fig. 1), which is also active in mouse cells (Bertram *et al.*, 1991). As described later, a possible explanation for this surprising activity of lycopene is its conversion, after ingestion to a novel five-membered ring cyclic metabolite known as 2,6-cyclolycopene-1,5-diol (Fig. 2) as a consequence of oxidation and subsequent rearrangement (Khachik *et al.*, 1997d).

As indicated earlier, lutein is another abundant dietary carotenoid that undergoes a series of oxidation–reduction and double bond isomerization reactions after ingestion. Found in green leaves, lutein was reported previously to possess only a moderate ability to inhibit transformation and increase Cx43 expression in 10T½ cells in comparison with β-carotene (Bertram *et al.*, 1991), indicating that its chemopreventive potential may be low. However, lutein has been shown to convert partially to zeaxanthin in the course of its metabolism (Khachik *et al.*, 1995a). The authors have shown that zeaxanthin has increased activity in elevating Cx43 expression in comparison with lutein. Most studies demonstrating biological effects of carotenoids relevant to carcinogenesis have been conducted in experimental animals or animal cell cultures. In order to more closely address the question of their effects in humans, studies have been conducted in human keratinocytes in organotypic culture. Under these conditions, cells differentiate to form a multilayered tissue with many of the characteristics of intact human skin. For reasons of reproducibility, availability, and ease of use, these new studies did not utilize primary keratinocytes derived from newborn foreskins, instead the authors used the immortalized human keratinocyte cell line HaCaT. Studies by others had shown this line to closely resemble normal human keratinocytes in its differentiation pattern in organotypic culture (Ryle *et al.*, 1989). Preliminary studies using immunofluorescent techniques had also demonstrated expression of connexin 43 in suprabasal cells (unpublished data); moreover, levels of Cx43 in HaCaT cells as detected by Western blotting were consistently higher than those detected in cell cultures of normal human keratinocytes (unpublished results). This may simply reflect a more homogeneous population of cells or that HaCaT cells are better adapted to cell culture.

The following section shows that products derived from the oxidation of dietary carotenoids possess enhanced biological activity in comparison to their parent dietary compounds.

Carotenoids and Their Effects on Human Keratinocytes

HaCaT cells respond to retinoic acid and β-carotene by increasing Cx43 expression. As shown in Fig 6A, retinoic acid caused a major increase in Cx43. HaCaT cultures were treated for 7 days, harvested, and analyzed by Western blotting using an antibody specific for Cx43. Digital image analysis showed this to be 2.2-fold at 10^{-7} M and 4.1-fold at 10^{-6} M (lanes 2 and 3, respectively). β-Carotene was less potent: no effect was seen at 10^{-7} M, whereas 10^{-6} and 10^{-5} M caused 1.6- and 3.1-fold increases, respectively (lanes 6 and 7). It was shown previously that retinoic acid treatment of intact human skin causes increased expression of Cx43 (Guo *et al.*, 1992), thus this response of HaCaT cells mirrors the *in vivo* situation. By extension, these data suggest that β-carotene can also produce this response.

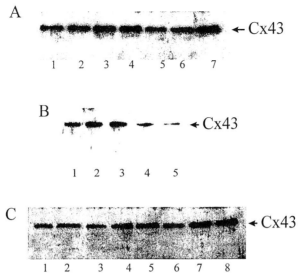

Figure 6 Induction of Cx43 by retinoic acid, carotenoids, and their metabolites in organotypic HaCaT cell cultures. Confluent HaCaT cells were grown in organotypic culture as described previously (King *et al.*, 1997). Cells were treated with retinoic acid or carotenoids for 7 days. Cells were then harvested and lysed, and the extracts were solubilized as described in the methods section. Gel electrophoresis, Western transfer, and detection of Cx43 were performed as described previously (Rogers *et al.*, 1990). Equal amounts of protein (25 μg) were loaded in each lane. (A) Lane 1, solvent control 0.5% THF; lanes 2–4, retinoic acid 10^{-7}, 10^{-6}, and 10^{-5} *M*, respectively; lanes 5–7, β-carotene 10^{-7}, 10^{-6}, and 10^{-5} *M*, respectively; (B) Lane 1, THF control; lanes 2 and 3, zeaxanthin 10^{-6} and 10^{-5} *M*, respectively; lanes 4 and 5, lutein 10^{-6} and 10^{-5} *M*, respectively. (C) Lane 1, control (no treatment; lane 2, THF 0.5% lanes 3–5, lycopene 10^{-7}, 10^{-6}, and 10^{-5} *M*, respectively; lanes 6–8, 2,6-cyclolycopene-1,5-diol 10^{-7}, 10^{-6}, and 10^{-5} *M*, respectively. Modified from King *et al.* (1997).

Lycopene, lutein, and their oxidation products induce Cx43 in human cells. To investigate if lycopene and lutein modulated Cx43 expression in human keratinocytes, cultures were treated as described earlier and were then subjected to Western blotting. As shown in Fig. 6B, although lutein did not elevate Cx43 expression above that seen in solvent controls (lanes 4 and 5), zeaxanthin 10^{-6} *M* (lane 2) caused a 5.3-fold increase in expression in comparison with lutein at the same concentration (lane 4). These results are in accord with previous studies in murine cells in which lutein was only marginally active (Bertram *et al.*, 1991; Zhang *et al.*, 1991) and suggest strongly that its conversion to zeaxanthin increases its biological effects greatly.

The actions of lycopene are shown in Fig. 6C. Treatment with lycopene itself resulted in increased expression (2-fold) of Cx43 at 10^{-6}–10^{-5} M (lanes 4 and 5) but not at the lowest concentration. However, the metabolite of lycopene, 2,6-cyclolycopene-1,5-diol, caused a marginal increase in expression at 10^{-7} M and a 2.7 and 3.1-fold increase at 10^{-6} and 10^{-5} M concentrations, respectively (lanes 6–8). These results demonstrate that the oxidation of lycopene to 2,6-cyclolycopene-1,5-diol results in a more active compound. It has not been determined what proportion, if any, of added lycopene becomes converted to 2,6-cyclolycopene-1,5-diol under conditions of cell culture. If substantial oxidation of lycopene occurs, the apparent activity of lycopene may be a consequence of this conversion.

These data strongly suggest that zeaxanthin and 2,6-cyclolycopene-1,5-diol are more potent than their dietary counterparts, lutein and lycopene, in their ability to upregulate connexin 43. However, the caveat must be added that the stability of these metabolites in cells and in culture medium has yet to be investigated. Should they be significantly more stable and/or attain higher cellular concentrations than lutein or lycopene, respetively, this factor must also be considered as contributing to the observed effects.

Anti-inflammatory and Antitumor Promoting Properties

Due to the detailed and time-consuming experiments involved, the authors' anti-inflammatory and antitumor promoting studies to date have only focused on two major dietary carotenoids: lycopene and lutein. However, in order to elucidate structure/activity relationships, the anti-inflammatory and antitumor properties of other carotenoids and their metabolites need to be investigated.

Anti-inflammatory Property: Inhibitory Effect of Lycopene and Lutein on TPA-Induced Inflammation

12-O-Tetradecanoyphorbol-13-acetate (TPA) induces many biochemical, molecular, and morphological changes in mouse skin. Many of these changes appear to be associated with skin tumor promotion. For example, topical application of TPA to the skin of mice induces skin inflammation rapidly and results in an increase in epidermal ornithine decarboxylase activity, epidermal DNA synthesis, the number of epidermal cell layers, production of hydrogen peroxide, and c-Fos and c-Jun oncogene expression. Chemicals that inhibit these TPA-induced biochemical, molecular, or morphological changes usually inhibit TPA-induced tumor promotion in mouse skin. These biochemical, molecular, and morphological changes may be useful as biomarkers for the early detection of inhibitors of skin tumor promotion (Huang, *et al.*, 1992, 1994,1997).

The authors investigated the effect of purified lycopene and lutein on topical application of TPA-induced inflammation and tumor promotion in mouse skin. The possibility that lycopene or lutein could inhibit TPA-dependent inflammation was evaluated by studying the effect of lycopene and lutein on TPA-induced edema of mouse ears. Topical application of 0.07 or 0.36 mg of lycopene, together with 0.5 nmol TPA to ears of mice, inhibited TPA-induced edema of mouse ears by 72 or 100%, respectively. Topical application of 0.07 or 0.36 mg of lutein with 0.5 nmol TPA to ears of mice inhibited TPA-induced edema of mouse ears by 20 or 47%, respectively. Application of 0.07 or 0.36 mg of curcumin, a known potent anti-inflammatory agent (Huang *et al.*, 1988), with 0.5 nmol TPA, inhibited TPA-induced edema of mouse ears by 81 or 100%, respectively. These results indicate that the anti-inflammatory activity of lycopene is equal to that of curcumin, followed by lutein, which exhibits somewhat lesser anti-inflammatory property.

Antitumor Promoting Property: Inhibitory Effect of Lycopene and Lutein on TPA-Induced Tumor Promotion in Mouse Skin

In another experiment, the protective effect of lycopene and lutein on a two-stage mouse skin tumorigenesis model was investigated. The number of skin tumors in mice was measured every 2 weeks for a duration of 14 weeks; results after 6 and 14 weeks are summarized in Table II. The skin tumors were initiated in CD-1 mice with a single dose of 200 nmol of 7,12-dimethylbenz[*a*]anthracene (DMBA) and were promoted with 5 nmol TPA twice a week for a duration of 14 weeks, which resulted in the development of an average of 11.3 skin tumors per mouse. Topical application of 1.2 or 2.5 μmol of lycopene with 5 nmol TPA twice a week for 14 weeks to DMBA-treated mice inhibited the number of skin tumors per mouse by 23 or 74%, respectively (Table III). This resulted in a reduction of the percentage of mice with skin tumors by 7 or 43%, respectively. Topical application of 1.2 or 2.5 μmol lutein with 5 nmol TPA to the backs of DMBA-treated mice twice a week for 14 weeks inhibited an average number of skin tumors per mouse by 46 or 51%, respectively. Consequently, the percentage of mice with skin tumors was reduced by 12 or 21%, respectively. Although curcumin is a very potent inhibitor of TPA-induced skin inflammation and tumor promotion, the bioavailability of curcumin is very low due to its poor absorption and very short half-life, therefore limiting its application as a chemoprotective agent against cancer. However, lycopene and lutein are two of the most abundant carotenoids in the diet and a steady and high concentration of these compounds can be established readily in

Table II Inhibitory Effects of Lycopene and Lutein on 12-*O*-Tetradecanoylphorbol-13-acetate (TPA)-Induced Skin Tumor Promotion in CD-1 Mice Initiated Previously with DMBA[a]

Treatment	6 weeks		14 weeks	
	Tumors/mouse	Percentage of mice with tumors	Tumors/mouse	Percentage of mice with tumors
DMBA + TPA	1.29 ± 0.5	33	11.33 ± 2.4	76
+ Lycopene (1.25 μmol)	0.62 ± 0.3 (52%)	29 (12%)	8.67 ± 2.1 (23%)	43 (17%)
+ Lycopene (2.50 μmol)	0.38 ± 0.2 (71%)	19 (42%)	3.00 ± 1.2 (74%)	24 (54%)
+ Lutein (1.25 μmol)	0.48 ± 0.3 (63%)	14 (58%)	6.14 ± 1.9 (46%)	29 (44%)
+ Lutein (2.5 μmol)	0.05 ± 0.1 (96%)	5 (85%)	5.52 ± 1.6 (51%)	14 (73%)

[a] Female CD-1 mice (8–9 weeks old; 30 per group) were initiated with 200 nmol of 7,12-dimethyl benz[*a*]anthracene (DMBA). One week later, mice were promoted with 2.5 nmol of TPA or 2.5 nmol of TPA together with inhibitor in 200 μl acetone twice a week for 14 weeks. Skin tumors greater than 1 mm (in diameter) were counted and recorded. Data are presented as the mean $\pm$ SE. Data in parentheses are percentage of inhibition.

the human serum or plasma by supplementation (Khachik *et al.*, 1995a, 1997d; Paetau *et al.*, 1998). As a result of these characteristics, lycopene and lutein can serve as potent cancer preventative agents in many and at a variety of endogenous target sites. Future studies are planed to investigate the effect of purified lycopene and lutein on lung, stomach, and colon tumorigenesis in mice.

Induction of Detoxication (Phase 2) Enzymes

Detoxication (phase 2) enzymes, including quinone reductase [QR; NAD(P)H: (quinone acceptor) oxidoreductase, EC 1.6.99.2], are induced transcriptionally in many mammalian cells by low concentrations of a wide variety of chemical agents and such an induction is associated with a reduced susceptibility to chemical carcinogenesis. Eight chemical classes are currently recognized as inducers by Prestera *et al.* (1993). This section reports that carotenoids and their metabolites represent a new class of inducers and that some of these compounds are very potent. Bioassay of the induction of quinone reductase activity demonstrates that there is high inducer activity from some of these carotenoids. The inducer activity of carotenoids and selected metabolites with curcumin and sulforaphane as positive controls is shown in Table III. The definition of a unit of inducer activity and the concentration required for doubling that activity (CD) is described in the experimental section. It is interesting to note that an acyclic carotenoid, lycopene, which lacks any functional group, appears to be a reasonably good inducer, suggesting that the conjugated polyene chain itself has inducer ability. If one considers α-carotene (CD = 100 μM) with 10 conjugated and one isolated double bond to be the "parent" carotenoid, then by simply altering the position of one double bond in the end groups, the phase 2 enzyme inducer potency in β-carotene (CD = 14 μM) with 11 double bonds is increased sevenfold. This, of course, should be considered in the absence of any other structural changes. However, because it is well established that the β-end groups of β-carotene are not quite in the same plane as the polyene chain in this compound, there is only a limited overlap between the π double bonds of the β-end groups with the conjugated system in this compound.

A similar increase in potency (sixfold) was observed by the introduction of hydroxyl groups in the end groups of lutein (CD = 16.5 μM) in comparison with α-carotene (CD = 100 μM) with an identical chromophoric system. An additional twofold increase in potency in comparison with lutein (CD = 16.5 μM) occurs if the allylic hydroxyl group in this compound is oxidized to give an α,β-unsaturated keton, (3R,6'R)-3-hydroxy-β,ϵ-caroten-3'-one (CD = 7.3μM). The inducer activity of (6RS,6'RS)-ϵ, ϵ-caroten-3,3'-dione (CD = 2.5 μM) with two α,β-unsaturated ketones in

Table III   Phase 2 Enzyme Inducer Activities of Some Dietary Carotenoids and Their Metabolites in Comparison to Curcumin and Sulforaphane

Entry	Compound	CD^a (μM)
1	α-Carotene	100
2	Lycopene	25
3	Violaxanthin	17
4	(3R,3$'R$,6$'R$)-Lutein	16.5
5	β-Carotene	14
6	2,6-Cyclolycopene-1,5-diol (90% A + 10% B)[b,c]	14
7	(3R,6$'R$)-3-Hydroxy-β,ϵ-caroten-3$'$-one[b]	7.3
8	(6RS,6$'RS$)-ϵ-,ϵ-Carotene-3,3$'$-dione[b,d]	2.5
9	(3R,3$'R$)-Zeaxanthin	2.2
10	Neoxanthin	2.0
11	Curcumin	5.0
12	Sulforaphane	0.2

[a] CD (inducer activity) is defined in the experimental procedures.
[b] Represents carotenoid metabolites.
[c] A synthetic mixture of lycopene metabolites in approximately the same ratio found in human serum.
[d] A synthetic mixture of (6R,6$'S$)-, (6S,6$'S$)-, and (6R,6$'R$)-ϵ,ϵ-caroten-3,3$'$-dione in approximately the same ratio found in human serum.

comparison with lutein (CD = 16.5 μM) is increased nearly sevenfold even if the chromophores are slightly different in these two compounds. Addition of the two hydoxyl groups and the epoxide moieties to the β-end group of β-carotene (CD = 14 μM) results in violaxanthin (CD = 17 μM) with roughly the same inducer activity. This is probably due to an increase in the inducer activity of β-carotene by introduction of the hydroxyl groups and to a decrease in activity by the removal of the double bonds in the β-end groups of this compound. However, in neoxanthin (CD = 2.0 μM), the extended conjugated polyene due to the presence of the allenic (two cumulative) double bonds, as well as the three hydroxyl groups, results in substantial inducer activity. If the hydroxylation of α-carotene (CD = 100 μM) is accompanied by the extension of conjugation to 11 double bonds, the potency is increased dramatically, as is the case in zeaxanthin (CD = 2.2 μM).

2,6-Cyclolycopene-1.5-diol (90% diol A and 10% diol B), a mixture of metabolites of lycopene with 10 conjugated double bonds, a five-membered ring system, and two hydroxyl groups has an inducer activity of CD = 14 μM and is roughly twice as potent as lycopene (CD = 25 μM). Thus, the extent of conjugation, the addition of oxygen functionality to the rings, and the presence of an α,β-unsaturated ketone are all important and apparently additive determinants of inducer potency. Although the observed potencies do not follow the ranking for "Trolox equivalent antioxidant capacities" (ABTS* radical cation quenching) (Miller *et al.*, 1996), they are in general agreement with observed singlet quenching activities (Stahl *et al.*, 1997). Activities apparently increase with extension of the chromophore and maximum overlap of the π molecular orbitals in C=C double bonds. However, lycopene (the only acyclic compound tested) does not have the inducer potency one would predict based on the number of conjugated double bonds or ABTS radical scavenging capacity.

In the authors' system, zeaxanthin is only 10-fold less potent than sulforaphane (CD = 0.2 μM), which is the most potent naturally occurring inducer of the phase 2 enzymes. Zeaxanthin has comparable activity to benzyl isothiocyanate (data not shown) and is considerably more active than inducers such as 1,2-dithiole-3-thione (not shown) or curcumin (CD = 5 μM), both of which are currently in clinical trials by the U.S. National Cancer Institute for evaluation as chemopreventive agents (Kelloff *et al.*, 1996).

Carotenoids, therefore, appear to have considerable potential to induce phase 2 detoxication enzymes and may have potential significance as means of detoxifying xenobiotics. Some of the major dietary carotenoids and their metabolites found in serum (Khachik *et al.*, 1997c), as well as those that are apparently not detected in the blood (Khachik *et al.*, 1997b), such as neoxanthin and violaxanthin, have now been evaluated in Hepa 1c1c7 cells as described earlier. The relevance of these findings is bolstered by the fact that serum and tissue concentrations of carotenoids are in the micromolar range and are therefore comparable to those required to produce phase 2 enzyme induction.

CONCLUSION

In the course of 5 years, the authors have repeatedly emphasized the importance of other dietary carotenoids in addition to β-carotene and presented evidence for the nutritional significance of this important class of nutrients in the prevention of cancer (Khachik *et al.*, 1992c, 1995a, 1997b,c,d) as well as macular degeneration, a degenerative eye disease (Khachik *et al.*, 1997a). This was initially based on our knowledge of the widespread distribution of carotenoids in fruits and vegetables as well as the consistent presence of these compounds and their metabolites in human serum, milk, and tissues at relatively high concentrations. In addition, in

---

* 2,2'-Axinobis(3-ethylbenzothiazoline-6-sulfonic acid) diammonium salt.

several human bioavailability and metabolic studies with selected carotenoids, such as lutein, zeaxanthin, and lycopene, the authors established the *in vivo* oxidation of these carotenoids and proposed metabolic pathways for the formation of their oxidative metabolites. Until recently, the presumed mechanisms of action for the protective role of carotenoids were their antioxidant function and their ability to upregulate connexin 43 gene expression responsible for gap junctional communications.

The authors have presented preliminary evidence for the presence of the optical stereoisomers of dietary lutein and zeaxanthin in human plasma and have provided additional evidence to substantiate earlier proposed oxidative–reductive pathways for the metabolism of these compounds in humans. They have also reviewed the most recent findings with regard to the mechanism of action of carotenoids and their metabolites by upregulation of the gene expression of gap junction proteins.

However, perhaps the most fascinating findings of the studies reported here are the discovery of three other important biological properties of carotenoids and their metabolites that may relate to their role in cancer chemoprotection: (1) anti-inflammatory property, (2) antitumor promoting property, and (3) induction of the detoxication (phase 2) enzymes. It is also remarkable that among the carotenoids found in human serum, $(3R,3'R)$-zeaxanthin (dietary) exhibits the highest activity by these two unrelated mechanisms of action. The mechanisms described here also revealed that various carotenoids and their metabolites exhibit different degrees of activity in a structure-dependent manner. Furthermore, in some cases, these activities are more pronounced with carotenoid metabolites than with dietary carotenoids.

Therefore, future studies with carotenoids should focus on developing an understanding of the bioavailability, metabolism, function, and mechanisms of action, efficacy, and interaction of all the serum carotenoids. These studies will allow us to carefully assess the potential protective effect of carotenoids and their metabolites against cancer and ultimately develop a mixture of purified carotenoids as chemopreventive agents for human supplementation. Such a mixture should closely resemble the relative distribution of carotenoids in the serum of healthy individuals consuming a diet rich in fruits and vegetables.

REFERENCES

Bertram, J. S., Pung, A., Churley, M., Kappock, T. J. I., Wilkins, L. R., and Cooney, R. V. (1991). Diverse carotenoids protect against chemically induced neoplastic transformation. *Carcinogenesis* 12, 671–678.

Beyer, E. C., Paul, D. L., and Goodenough, D. A. (1987). Connexin 43: A protein from rat heart homologous to a gap junction protein from liver. *J. Cell Biol.* **105**, 2621–2629.

Cooney, R. V., Kappock, T. J., Pung, A., and Bertram, J. S. (1993). Solubilization, cellular uptake, and activity of β-carotene and other carotenoids as inhibitors of neoplastic transformation in cultured cells. *Methods Enzymol.* **214**, 55–68.

Fahey, J. W., Zhang, Y., and Talalay, P. (1997). Broccoli sprouts: An exceptionally rich source of inducers of enzymes that protect against chemical carcinogens. *Proc. Natl. Acad. Sci. USA* **94**, 10367–10372.

Fitzgerald, D. J., Fusenig, N. E., Boukamp, P., Piccoli, C., Mesnil, M., and Yamasaki, H. (1994). Expression and function of connexin in normal and transformed human keratinocytes in culture. *Carcinogenesis* **15**, 1859–1866.

Guo, H., Acevedo, P., Parsa, D. F., and Bertram, J. S. (1992). The gap-junctional protein connexin 43 is expressed in dermis and epidermis of human skin: Differential modulation by retinoids. *J. Invest. Dermatol.* **99**, 460–467.

Hong, W. K., Lippman, S. M., Itri, L. M., Karp, D. D., Lee, J. S., Byers, R. M., Schantz, S. P., Kramer, A., Lotan, R., Peters, L. J., Dimery, I. W., Brown, B. W., and Goepfert, H. (1990). Prevention of second primary tumors with isotretinoin in squamous-cell carcinoma of the head and neck. *N. Engl. J. Med.* **323**, 795–800.

Hossain, M. Z., Wilkens, L. R., Mehta, P. P., Loewenstein, W. R., and Bertram, J. S. (1989). Enhancement of gap junctional communication by retinoids correlates with their ability to inhibit neoplastic transformation. *Carcinogenesis* **10**, 1743–1748.

Huang, M.-T., Ho, C.-T., Ferraro, T., Finnegan-Olive, T., Lou, Y.-R., Wang, Z. Y., Mitchell, J. M., Laskin, J. D., Newmark, H., Yang, C. S., and Conney, A. H. (1992). Inhibitory effect of topical application of a green tea polyphenol fraction on tumor initiation and promotion in mouse skin. *Carcinogenesis* **13**, 947–954.

Huang, M.-T., Ho, C.-T., Wang, Z. Y., Stauber, K., Ma, W., Georgadis, C., Laskin, J. D., and Conney, A. H. (1994). Inhibition of skin tumorigenesis by rosemary and its constituents carnosol and ursolic acid. *Cancer Res.* **54**, 701–708.

Huang, M.-T., Ma, W., Lou, Y.-R., Lu, Y.-P., Chang, R., Newmark, H., and Conney, A. H. (1997). Inhibitory effects of curcumin on tumorigenesis in mice. *In* "Recent Development in Curcumin Pharmacochemistry" (S. Pramono, U. A. Jenie, R. S. Sudibyo, and D. Gunawam, eds.), pp. 47–63. Proceedings of the International Symposium on Curcumin Pharmacochemistry, Aditya Media, Yogyakarta.

Huang, M.-T., Smart, R. C., Wong, C.-Q., and Conney, A. H. (1988). Inhibitory effect of curcumin, chlorogenic acid, caffeic acid, and ferulic acid on tumor promotion in mouse skin by 12-O-tetradecanoylphorbol-13-acetate. *Cancer Res.* **48**, 5941–5946.

Ikuno, Y., Maoka, T., Shimizu, M., Komori, T., and Matsuno, T. (1985). Direct diastereomeric resolution of carotenoids: All ten stereoisomers of tunaxanthin (ϵ,ϵ-carotene-3,3'-diol). *J. Chromatogr.* **328**, 387–391.

Kelloff, G., J., Boone, C. W., Steele, V. E., Crowell, J. A., Lubet, R., and Sigman, C. C. (1994). Progress in cancer chemoprevention: Prospectives on agent selection and short-term clinical intervention trials. *Cancer Res.* **54**(Suppl.), 2015s–2024s.

Kelloff, G. J., Crowell, J. A., Hawk, E. T., Steele, V. E., and Lubet, R. A. (1996). Clinical development plans for cancer chemopreventive agents: Curcumin. *J. Cell Biochem.* **26**(Suppl.), 72–85.

Khachik, F., Beecher, G. R., and Whittaker, N. F. (1986). Separation, identification, and quantification of the major carotenoid and chlorophyll constituents in the extracts of several green vegetables by liquid chromatography. *J. Agric. Food Chem.* **34**(4), 603–616.

Khachik, F., and Beecher, G. R. (1988). Separation and identification of carotenoids and carotenol fatty acid esters in some squash products by liquid chromatography. I. Quantification of carotenoids and related esters by HPLC. *J. Agric. Food Chem.* **36**(5), 929–937.

Khachik, F., Beecher, G. R., and Lusby, W. R. (1989). Separation, identification, and quantification of the major carotenoids in extracts of apricots, peaches, cantaloupe, and pink grapefruit by liquid chromatography. *J. Agric. Food Chem.* **37**(6), 1465–1473.

Khachik, F., Beecher, G. R., Goli, M. B., and Lusby, W. R. (1991). Separation, identification, and quantification of carotenoids in fruits, vegetables and human plasma by high performance liquid chromatography. *Pure Appl. Chem.* **63**(1), 71–80.

Khachik, F., Beecher, G. R., Goli, M. B., and Lusby, W. R. (1992a). Separation and quantification of carotenoids in foods. *In* "Methods in Enzymology" (L. Packer, ed.), Vol. 213A, pp. 347–359. Academic Press, New York.

Khachik, F., Beecher, G. R., Goli, M. B., Lusby, W. R., and Daitch, C. E. (1992b). Separation and quantification of carotenoids in human plasma. *In* "Methods in Enzymology" (L. Packer, ed.), Vol. 213A, pp. 205–219. Academic Press, New York.

Khachik, F., Beecher, G. R., Goli, M. B., Lusby, W. R., and Smith, J. C. (1992c). Separation and identification of carotenoids and their oxidation products in extracts of human plasma. *Anal Chem.* **64**(18), 2111–2122.

Khachik, F., Englert, G., Daitch, C. E., Beecher, G. R., Lusby, W. R., and Tonucci, L. H. (1992d). Isolation and structural elucidation of the geometrical isomers of lutein and zeaxanthin in extracts from human plasma. *J. Chromatogr. Biomed. Appl.* **582**, 153–166.

Khachik, F. Goli, M. B., Beecher, G. R., Holden, J., Lusby, W. R., Tenorio, M. D., and Barrera, M. R. (1992e). The effect of food preparation on qualitative and quantitative distribution of major carotenoid constituents of tomatoes and several green vegetables. *J. Agric. Food Chem.* **40**(3), 390–398.

Khachik, F., Beecher, G. R., and Smith, J. C., Jr. (1995a). Lutein, lycopene, and their oxidative metabolites in chemoprevention of cancer. *J. Cell. Biochem.* **22**, 236–246.

Khachik, F., Englert, G., Beecher, G. R., and Smith, J. C., Jr. (1995b). Isolation, structural elucidation, and partial synthesis of lutein dehydration products in extracts from human plasma. *J. Chromatogr. Biomed. Appl.* **670**, 219–233.

Khachik, F., Spangler, C. J., Smith, J. C., Jr., Canfield, L. M., Pfander, H., and Steck, A. (1997c). Identification, quantification, and relative concentrations of carotenoids, their metabolites in human milk and serum. *Anal. Chem.* **69**, 1873–1881.

Khachik, F., Steck, A., and Pfander, H. (1997d). Bioavailability, metabolism, and possible mechanism of chemoprevention by lutein and lycopene in humans. *In* "Food Factors for Cancer Prevention" (H. Ohigashi, T. Osawa, J. Terao, S. Watanabe, and T. Yoshikawa eds.), pp. 542–547. Springer-Verlag, Tokyo.

Khachik, F., Nir, Z., Ausich, R. L., Steck, A., and Pfander, H. (1997b). Distribution of carotenoids in fruits and vegetables as a criterion for the selection of appropriate chemopreventive agents. *In* "Food Factors for Cancer Prevention" (H. Ohigashi, T. Osawa, J. Terao, S. Watanabe, and T. Yoshikawa eds.), pp. 204–208. Springer-Verlag, Tokyo.

Khachik, F., Bernstein, P., and Garland, D. L. (1997a). Identification of lutein and zeaxanthin oxidation products in human and monkey retinas. *Invest. Ophthalmol. Vis. Sci.* **38**, 1802–1811.

Khachik, F., Askin, F. B., and Lai, K. (1998a). Distribution, bioavailability, and metabolism of carotenoids in humans. *In* "Phytochemicals: a New Paradigm" (W. R. Bidlack, S. T. Omaye, M. S. Meskin, D. Jahner eds.). Technomic Publishing, Lancaster, PA, pp. 77–96.

Khachik, F., Steck, A., Niggli, U., and Pfander, H. (1998b). Partial synthesis and the structural elucidation of the oxidative metabolites of lycopene identified in tomato paste, tomato juice, and human serum. *J. Agric. Food Chem.* **46**, in press.

King, T. J., Khachik, F., Bortkiewicz, H., Fukishima, L. H. Morioka, S., and Bertram, J. S. (1997). Metabolites of dietary carotenoids as potential cancer preventive agents. *Pure Applied Chem.* **69**, 2135–2140.

Matsuno, T., Hirono, T., Ikuno, Y., Maoka, T., Shimizu, M., and Komori, T. (1986). Isolation of three new carotenoids and proposed metabolic pathways of carotenoids in hen's egg yolk. *Comp. Biochem. Physiol. B* **84** 477–481.

Mayne, S. T. (1986). Beta-carotene, carotenoids, and disease prevention in humans. *FASEB J.* **10**, 690–701.

Merriman, R., and Bertram, J. S. (1979). Reversible inhibition by retinoids of 3-methycholan-threne-induced neoplastic transformation in C3H10T1/2 cells. *Cancer Res.* **39**, 1661–1666.

Miller, N. J., Sampson, J., Candeias, L. P., Bramley, P. M., and Rice-Evans, C. A. (1996). Antioxidant activities of carotenes and xanthophylls. *FEBS Lett.* **384**, 240–242.

Neveu, M., and Bertram, J. S. (1998). Gap junctions and neoplasia. *In* "Gap Junctions" (E. L. Hertzberg, ed.). JAI Press, Greenwich. in press.

Paetau, I., Khachik, F. Brown, E. D., Beecher, G. R., Kramer, T. R., Chittams, J., and Clevidence, B. A. (1998). Chronic ingestion of lycopene-rich tomato juice or lycopene supplements significantly increases plasma concentrations of lycopene and related tomato carotenoids in humans. *Am J. Clin. Nutr.*, **68**. In press.

Prestera, H. J., and Santamaria, A. B. (1993). The electrophile counter attack response: Protection against neoplasia and toxicity. *Adv. Enzyme Regul.* **33**, 281–296.

Prochaska, H. G., and Santamaria, A. B. (1988). Direct measurement of NAD(P)H:quinone reductase from cells cultured in microliter wells: A screening assay for anticarcinogenic enzyme inducers. *Anal. Biochem.* **169**, 328–336.

Prochaska, H. G., Santamaria, A. B., and Talalay, P. (1992). Rapid detection of inducers of enzymes that protect against carcinogens. *Proc. Natl. Acad. Sci. USA* **89**, 2394–2398.

Rogers, M., Berestecky, J. M., Hossain, M. Z., Guo, H. M., Kadle, R., Nicholson, B. J., and Bertram, J. S. (1990). Retinoid-enhanced gap junctional communication is achieved by increased levels of connexin 43 mRNA and protein. *Mol. Carcinogen.* **3**, 335–343.

Ryle, C. M., Breitkreutz, D., Stark, H. J., Leigh, I. M., Steinert, P. M., Roop, D., and Fusenig, N. E. (1989). Density-dependent modulation of synthesis of keratins 1 and 10 in the human keratinocyte line HACAT and in ras-transfected tumorigenic clones. *Differentiation* **40**, 42–54.

Smith, P. K., Krohn, R. I., Hermanson, G. T., Mallia, A. K., Gartner, F. H., Provenzano, M. D., Fujimoto, E. K. Goeke, N. M., Olson, B. J., and Klenk, D. C. (1985). Measurement of protein using bicinchoninic acid. *Anal. Biochem.* **150**, 76–85.

Stahl, W., Swantje, N., Briviba, K., Hanusch, M., Broszeit, G., Peters, M., Martin, H. D., and Sies, H. (1997). Biological activities of natural and synthetic carotenoids: Induction of gap junctional communication and singlet oxygen quenching. *Carcinogenesis* **18**(1), 89–92.

Zhang, L. X., Cooney, R. V., and Bertram, J. S. (1991). Carotenoids enhance gap junctional communication and inhibit lipid peroxidation in C3H/10T1/2 cells: Relationship to their cancer chemopreventive action. *Carcinogenesis* **12**, 2109–2114.

Zhang, L., Cooney, R. V., and Bertram, J. S. (1992). Carotenoids up-regulate connexin43 gene expression independent of their pro-vitamin A or antioxidant properties. *Cancer Res.* **52**, 5707–5712.

15 Cancer Prevention by Natural Carotenoids

H. Nishino

Department of Biochemistry
Kyoto Prefectural University of Medicine
Kamigyoku, Kyoto 602, Japan

Epidemiological investigations have shown that cancer risk is inversely related to the consumption of green and yellow vegetable in which various carotenoids are distributed. Among them, about 14 carotenoids have been identified in human plasma and tissues. Thus, the author evaluated the biological activities of these carotenoids and found that some showed more potent activity than β-carotene, the most extensively studied carotenoid to date to suppress the process of carcinogenesis.

Some of the natural carotenoids, such as phytoene, are unstable when purified and thus it is very difficult to examine their biological activities. In such cases, stable production of these carotenoids in target cells may be helpful in more accurately evaluating their biological properties. In this context, a new method for the *de novo* synthesis of phytoene in animal cells was developed.

α-CAROTENE

α-Carotene showed higher activity than β-carotene in suppressing tumorigenesis in skin, lung, liver, and colon.

In a skin tumorigenesis experiment, a two-stage mouse skin carcinogenesis model was used. Seven-week-old ICR mice had their backs shaved with an electric clipper. One week after initiation with 100 μg of 7,12-

dimethylbenz[*a*]anthracene (DMBA), 1.0 μg of 12-O-tetradecanoylphorbol-13-acetate (TPA) was applied twice a week. α- or β-carotene (200 nmol) was applied with each TPA application. Greater potency of α-carotene than β-carotene was observed. The percentage of tumor-bearing mice in the control group was 68.8%, whereas the percentages of tumor-bearing mice in the groups treated with α- and β-carotene were 25.0 and 31.3%, respectively. The average number of tumors per mouse in the control group was 3.73, whereas the α-carotene-treated group had 0.27 tumors per mouse ($p < 0.01$, Students' *t* test). β-Carotene treatment also decreased the average number of tumors per mouse (2.94 tumors per mouse), but the difference from the control group was not significant.

The greater potency of α-carotene than β-carotene in the suppression of tumor promotion was confirmed by another two-stage carcinogenesis experiment, i.e., the 4-nitroquinoline 1-oxide (4NQO)-initiated and glycerol-promoted ddY mouse lung carcinogenesis model. 4NQO (10 mg/kg body weight) was given by a single sc injection on the first experimental day. Glycerol (10% in drinking water) was given continuously from experimental week 5 to week 30. α- or β-carotene (at the concentration of 0.05%) or a vehicle as a control was mixed as an emulsion into drinking water during the promotion stage. The average number of tumors per mouse in the control group was 4.06, whereas the α-carotene-treated group had 1.33 tumors per mouse ($p < 0.001$). β-Carotene treatment did not show any suppressive effect on the average number of tumors per mouse, but rather induced a slight increase (4.93 tumors per mouse).

In a liver carcinogenesis experiment, a spontaneous liver carcinogenesis model was used. Male C3H/He mice, which have a high incidence of spontaneous liver tumor development, were treated for 40 weeks with α- or β-carotene (at the concentration of 0.05%, mixed as an emulsion into drinking water) or a vehicle as a control. The mean number of hepatomas was decreased significantly by α-carotene treatment as compared to that in the control group; the control group developed 6.31 tumors per mouse, whereas the α-carotene-treated group had 3.00 tumors per mouse ($p < 0.001$). However, the β-carotene-treated group did not show a significant difference from the control group, although a tendency toward a decrease was observed (4.71 tumors per mouse).

As a short-term experiment to evaluate the suppressive effect of α-carotene on colon carcinogenesis, the effect of N-methylnitrosourea (MNU, three intrarectal administration of 4 mg in week 1)-induced colonic aberrant crypt foci formation was examined in Sprague–Dawley (SD) rats. α- or β-carotene (6 mg, suspended in 0.2 ml of corn oil, intragastric gavage daily) or a vehicle as a control as administered during weeks 2 and 5. The mean number of colonic aberrant crypt foci in the control group was 62.7, whereas the α- or β-carotene-treated group had 42.4 (significantly different from the control group: $p < 0.05$) and 56.1, respectively. Thus, the greater potency of α-carotene than β-carotene was also observed in this experimental model.

LUTEIN

Lutein, the dihydroxy form of α-carotene, is found in a variety of vegetables, such as kale, spinach, and winter squash, and fruits, such as mango, papaya, peaches, prunes, and oranges.

Epidemiological studies in the Pacific islands indicate that people with a high intake of all three, β-carotene, α-carotene, and lutein, have the lowest risk of lung cancer (1).

Thus, the effect of lutein on lung carcinogenesis was examined. Lutein showed antitumor-promoting activity in a two-stage carcinogenesis experiment, initiated with 4NQO and promoted with glycerol, in the lungs of ddY mice. 4NQO (10 mg/kg body weight), dissolved in a mixture of olive oil and cholesterol (20:1), was given by a single sc injection on the first experimental day. Glycerol (10% in drinking water) was given continuously as the tumor promoter from experimental week 5 to week 30. Lutein, 0.2 mg in 0.2 ml of a mixture of olive oil and Tween 80 (49:1), was given by oral intubation three times a week during the tumor promotion stage (25 weeks). Treatment with lutein showed a tendency to decrease lung tumor formation; the control group developed 3.07 tumors per mouse, whereas the lutein-treated group had 2.23 tumors per mouse.

The antitumor-promoting activity of lutein was confirmed by another two-stage carcinogenesis experiment; i.e., it showed antitumor-promoting activity in a two-stage carcinogenesis experiment, initiated with DMBA and promoted with TPA and mezerein, in the skin of ICR mice. At 1 week after initiation with 100 μg of DMBA, TPA (10 nmol) was applied once, and then mezerein (3 nmol for 15 weeks, and 6 nmol for a subsequent 15 weeks) was applied twice a week. Lutein (1 μmol, molar ratio to TPA = 100) was applied twice (45 min before and 16 hr after TPA application). At experimental week 30, the average number of tumors per mouse in the control group was 5.50, whereas the lutein-treated group had 1.91 tumors per mouse ($p < 0.05$) (2).

Lutein also inhibited the development of aberrant crypt foci in SD rat colon induced by MNU (three intrarectal administrations of 4 mg in week 1). Lutein (0.24 mg, suspended in 0.2 ml of corn oil, intragastric gavage daily) or a vehicle as a control as administered during weeks 2 and 5. The mean number of colonic aberrant crypt foci in the control group at week 5 was 69.3, whereas the lutein-treated group had 40.2 (significantly different from control group: $p < 0.05$).

ZEAXANTHIN

Zeaxanthin, the dihydroxy form of β-carotene, is found in foods such as corn and various vegetables. Because awareness of zeaxanthin as a beneficial carotenoid was achieved only recently, available data for zeaxanthin are few.

Some antitumor promoter features of zeaxanthin have been elucidated. For example, zeaxanthin suppressed the TPA-induced expression of early antigen of the Epstein–Barr virus in Raji cells; ID_{50} was about 6 μM. TPA-enhanced ^{32}Pi incorporation into phospholipids of cultured cells was also inhibited by zeaxanthin; ID_{50} was about 40 μM.

The anticarcinogenic activity of zeaxanthin *in vivo* was also examined. For example, in a preliminary experiment, it was found that spontaneous liver carcinogenesis in C3H/He male mice was suppressed with zeaxanthin (at the concentration of 0.005%, mixed as an emulsion into drinking water); the control group developed 1.75 tumors per mouse, whereas the zeaxanthin-treated group had 0.08 tumors per mouse.

LYCOPENE

Lycopene is found predominantly in tomatoes and tomato products. Recently, an exceptionally high singlet oxygen quenching ability to lycopene was found (3, 4).

Epidemiological studies in the elderly Americans indicate that high tomato intake is associated with 50% reduction of mortality from cancers at all sites (5). A case-control study in Italy showed potential protection against cancers of the digestive tract (6) with a high consumption of lycopene in the form of tomatoes. An inverse association between a high intake of tomato products and prostate cancer risk was also reported (7).

Studies on the anticarcinogenic activity of lycopene in animal models were carried out with mammary gland, liver, lung, skin and colon. The study in mice with a high rate of spontaneous mammary tumors showed that an intake of lycopene delayed and reduced tumor growth (8). Spontaneous liver carcinogenesis in C3H/He male mice was also suppressed. Treatment for 40 weeks with lycopene (at the concentration of 0.005%, mixed as an emulsion into drinking water) resulted in the significant decrease of liver tumor formation; the control group developed 7.65 tumors per mouse, whereas the lycopene-treated group had 0.92 tumors per mouse ($p <$ 0.005).

Lycopene showed antitumor-promoting activity in a two-stage carcinogenesis experiment, initiated with 4NQO and promoted with glycerol, in the lungs of ddY mice. Lycopene [0.2 mg in 0.2 ml of mixture of olive oil and Tween 80 (49 : 1)] was given by oral intubation three times a week during the tumor promotion stage. Treatment with lycopene resulted in the significant decrease of lung tumor formation; the control group developed 3.07 tumors per mouse, whereas the lycopene-treated group had 1.38 tumors per mouse ($p < 0.05$).

The antitumor-promoting activity of lycopene was confirmed by another two-stage carcinogenesis experiment; i.e., it showed antitumor-promoting activity in a two-stage carcinogenesis experiment, initiated with DMBA and promoted with TPA, in the skin of ICR mice. From 1 week after initiation with 100 μg of DMBA, 1.0 μg ($=1.6$ nmol) of TPA was applied twice a week for 20 weeks. Lycopene (160 nmol, molar ratio to TPA $= 100$) was applied with each TPA application. At experimental week 20, the average number of tumors per mouse in the control group was 8.53, whereas the lycopene-treated group had 2.13 tumors per mouse ($p < 0.05$).

Lycopene also inhibited the development of aberrant crypt foci induced with MNU (three intrarectal administration of 4 mg in week 1) in SD rat colon. Lycopene (0.12 mg, suspended in 0.2 ml of corn oil, intragastric gavage daily) or a vehicle as control was administered during weeks 2 and 5. The mean number of colonic aberrant crypt foci in the control group at week 5 was 69.3, whereas the lycopene-treated group had 34.3 (significantly different from the control group: $p < 0.05$).

β-CRYPTOXANTHIN

β-Cryptoxanthin is found in food such as oranges and is one of the major carotenoids detectable in human blood. Preliminary experimental data show that β-cryptoxanthin suppresses skin tumor promotion in mice. Thus, it seems worthy to study its biological activities more precisely.

PHYTOENE

Phytoene, which is detectable in human blood, was proved to suppress tumorigenesis in the skin (9). It was suggested that the antioxidative activity of phytoene may play an important role in its action mechanism. In order to confirm the mechanism, studies should be carried out; however, phytoene becomes unstable when purified, making such studies very difficult. Therefore, stable production of these carotenoids in target cells, which may be helpful for the evaluation of their biological properties, was tried. As the phytoene synthase encoding gene, *crtB,* has already been cloned from *Erwinia uredovora* (10), it was used for the expression of the enzyme in animal cells. Mammalian expression plasmids, pCAcrtB, to transfer the *crtB* gene to mammalian cells, were constructed as follows. First, the sequence around the initiation codon of the crtB gene on the plasmid pCRT-B was modified by polymerase chain reaction using the primers to replace the original bacterial initiation codon TTG with CTCGAGCCACCATG, which is a composite of the typical mammalian initiation codon ATG preceded by the Kozak consensus sequence and a *Xho*1 recognition site. The *Xho*l linker, which harbors a cohesive end for the *Eco*R1 site, was ligated to the *Eco*R1 site at the 3′ end of the *crtB* gene, and the 969-bp *Xho*1 fragment was cloned into the *Xho*1 site of the expression vecter pCAGGS. The resulting plasmid, pCAcrtB, drives the *crtB* gene by the CAG promoter (modified chicken β-actin promoter coupled with a cytomegalovirus immediate early enhancer). In the pCAGGS vecter, a rabbit β-globin polyadenylation signal is provided just downstream of the *Xho*1 cloning site. Plasmids were transfected by either electroporation or lipofection. For the gene transfer to NIH/3T3 cells, which were cultured in Dulbecco's modified minimum essential medium (DMEM) supplemented with 4 mM L-glutamine, 80 U/ml penicillin, 80 mg/ml streptomycin, and 10% calf serum (CS), the parameter for electroporation using a Gene Pulser (Bio-Rad) was set at 1500 V/25 mF with a DNA concentration of 12.5–62.5 μg/ml. Lipofection was carried out using LIPOFECTAMINE (GIBCO BRL) according to the protocol supplied by the manufacturer. pCAcrtB or pCAGGS was cotransfected with the plasmid pKOneo, which harbors a neomycin resistance encoding gene (kindly provided by Dr. Douglas Hanahan, University of California, San Francisco).

NIH/3T3 cells transfected with pCAcrtB showed the expression of 1.5 kb mRNA from the *crtB* gene as a major transcript. Those transcripts were not present in the cells transfected with the vector alone. For analysis of phytoene by high-performance liquid chromatography (HPLC), the lipid fraction, including phytoene, was extracted from cells (10^7–10^8). The sample was subjected to HPLC (column: 3.9 by 300 mm, Nova-pakHR, 6m C18, Waters) at a flow rate of 1 ml/min. To detect phytoene, ultraviolet (UV) absorbance of the eluate at 286 nm was measured by a UV detector (JASCO875). Phytoene was detected as a major peak in the HPLC profile in NIH/3T3 cells transfected with pCcrtB, but not in control cells. Phytoene was identified by UV and field desorption mass spectra.

Because lipid peroxidation is considered to play a critical role in tumorigenesis and because it was suggested that the antioxidative activity of phytoene may play an important role in its mechanism of anticarcinogenic action, the level of phospholipid peroxidation induced by oxidative stress in cells transfected with pCAcrtB or with vector alone was compared. Oxidative stress was imposed by culturing the cells in a Fe^{3+}/adenosine 5'-diphosphate (ADP)-containing medium [374 mM iron(III) choloride, 10 mM ADP dissolved in DMEM] for 4 hr. Cells were then washed three times with a Ca^{2+}- and Mg^{2+}-free phosphate-buffered saline [PBS(−)], harvested by scraping, washed once with PBS(−), suspended in 1 ml of PBS(−), and freeze-thawed once. The lipid fraction was extracted from the cell suspention twice with 6 ml of chloroform/methanol (2:1). The chloroform layer was collected, and dried with sodium sulfate. The sample was evaporated, and its residue was dissolved in a small volume of HPLC solvent (2-propanol:n-hexane:methanol:H_2O = 7:5:1) and then subjected to chemiluminescence-HPLC (CL-HPLC). The lipid was separated with the column (Finepack SIL NH2-5,250 × 4.6 mm i.d. JASCO) by eluting with a HPLC solvent (see earlier discussion) at a flow rate of 1 ml/min at 35°C. A postcolumn chemiluminescent reaction was carried out in a mixture of 10 mg/ml cytochrome c and 2 mg/ml luminol in a borate buffer (pH 10.0) at a flow rate of 1.1 ml/min. To detect lipids, UV absorbance of the eluate at 210 nm was measured with a UV-8011 detector (TOSOH), and chemiluminescence was detected with a CLD-110 detector (Tohoku Electric Ind.). The phospholipid hydroperoxidation level in the cells transfected with pCAcrtB, and confirmed to produce phytoene by HPLC, was lower than that in the cells transfected with vector alone. Thus, the antioxidative activity of phytoene in animal cells was confirmed.

It is of interest to test the effect of the endogenous synthesis of phytoene on the malignant transformation process, which is newly triggered in noncancerous cells. Thus, a study was carried out on the NIH/3T3 cells producing phytoene for its possible resistance against oncogenic insult imposed by transfection of the activated *H-ras* oncogene. Plasmids with the activated *H-ras* gene were transfected to NIH/3T3 cells with or without phytoene production, and the rate of transformation focus formation in 100-mm-diameter dishes was compared. As a result, it was proved that the rate of transformation focus formation induced by the transfection of activated *H-ras* oncogene was lower in phytoene-producing cells than in control cells.

It may be possible to employ this type of experimental method for the evaluation of anticarcinogenic activity of other phytochemicals, as the cloning of genes for the synthesis of various kinds of substances in vegetables and fruits has already been accomplished. It is particularly useful in evaluating the biological activity of unstable phytochemicals, such as phytoene and other carotenoids.

Valuable chemopreventive substances, including carotenoids, may be produced in a wide variety of foodsy by means of biotechnology; this new concept may be named "biochemoprevention." As a prototype experiment, phytoene synthesis in animal cells is demonstrated. Because phytoene produced in animal cells was proved to prevent oxidative damage of cellular lipids, it may become a valuable factor in animal foods for the reduction of the formation of oxidized oils, which are hazardous for health, as well as to keep freshness, resulting in the maintenance of the good quality of foods. Furthermore, phytoene-containing foods are valuable in preventing cancer, as phytoene is known as an anticarcinogenic substance. It may become one of the fundamental methods for the development of novel animal foods that have the potential to prevent diseases, such as cancer, to promote health.

ACKNOWLEDGMENTS

This work was supported in part by grants from the Program for Promotion of Basic Research Activities for Innovative Biosciences, the Program of Fundamental Studies in Health Sciences of the Organization for Drug ADR Relief, R&D Promotion and Product Review, the Ministry of Health and Welfare (The second-term Comprehensive 10-year Strategy for Cancer Control), the Ministry of Education, Science and Culture, SRF, and the Plant Science Research Foundation, Japan. This study was carried out in collaboration with the research groups of Kyoto Prefectural University of Medicine, Akita University College of Allied Medical Science, Kyoto Pharmaceutical University, National Cancer Center Research Institute, Dainippon Ink & Chemicals, Inc., Lion Co., and Kirin Brewery Co., Japan, and LycoRed Natural Products Industries, Ltd., Israel, and Dr. Frederick Khachik, U.S. Department of Agriculture.

REFERENCES

1. Le Marchand, L., Hankin, J. H., Kolonel, L. N., Beecher, G. R., Wilkens, L. R., and Zhao, L. P. (1993). *Cancer Epidemiol. Biomark. Prev.* **2**, 183–187.
2. Nishino, H. (1998). *J. Cell. Biochem.,* in press.
3. Stahl, W., and Sies, H. (1993). *Ann. N.Y. Acad. Sci.* **691**, 10–19.
4. Ukai, N., Lu, Y., Etoh, H., Yagi, A., Ina, K., Oshima, S., Ojima, F., Sakamoto, H., and Ishiguro, Y. (1994). *Biosci. Biotech. Biochem.* **58**, 1718–1719.
5. Colditz, G. A., Branch, L. G., and Lipnick, R. J. (1985). *Am. J. Clin. Nutr.* **41**, 32–36.
6. Franceschi, S., Bidoli, E., La Veccia, C., Talamini, R., D'Avanzo, B., and Negri, E. (1994). *Int. J. Cancer* **59**, 181–184.
7. Giovannucci, E., Ascherio, A., Rimm, E. B., Stampfer, M. J., Colditz, G. A., and Willett, W. C. (1995). *J. Natl. Cancer Inst.* **87**, 1767–1776.
8. Nagasawa, K., Mitamura, T., Sakamoto, S., Yamamoto, K. (1995). *Anticancer Res.* **15**, 1173–1178.
9. Mathews-Roth, M. M. (1982). *Oncology* **39**, 33–37.
10. Misawa, N., Nakagawa, M., Kobayashi, K., Yamano, S., Izawa, Y., Nakamura, K., and Harashima, K. (1990). *J. Bacteriol.* **172**, 6704–6712.

16 Screening of Phenolics and Flavonoids for Antioxidant Activity

Catherine Rice-Evans

International Antioxidant Research Centre
Guy's, King's and St. Thomas' School of Biomedical Sciences
London SE1 9RT, United Kingdom

INTRODUCTION

The contribution of free radical-mediated processes to the pathogenesis of human disease is indicated by biomarkers of oxidative damage to lipids, protein, or DNA (Fig. 1) detected in patients with atherosclerosis, certain cancers, neurodegenerative diseases, and lung disorders, especially those with an inflammatory component to their etiology. A range of reactive oxygen species (ROS) and reactive nitrogen species (RNS) have been implicated in the mechanisms of damage associated with disease development, including superoxide radical, hydrogen peroxide, hypochlorite, hydroxyl radical, ferryl heme protein species, lipid alkoxyl and peroxyl radicals, peroxynitrite, nitric oxide, and nitrogen dioxide radical. Defense systems against damage-induced by ROS/RNS fall into three categories:

- preventive antioxidants that suppress free radical formation
- radical-scavenging antioxidants that inhibit initiation of chain reactions and intercept chain propagation
- antioxidants involved in repair processes.

In this chapter, the focus is on the former two categories of preventive and radical scavenging antioxidants.

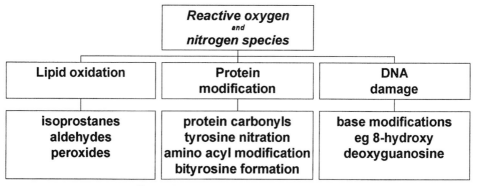

Figure 1   Biomarkers of oxidative damage.

REQUISITE CHARACTERISTICS FOR EFFECTIVE ANTIOXIDANTS

The requisite characteristics for effective antioxidant molecules include a number of structural features.

1. The presence of hydrogen-/electron-donating substituents with appropriate reduction potentials, in relation to those of the redox couples of the radicals to be scavenged (Bors *et al.*, 1990, 1995; Jovanovic *et al.*, 1994, 1995; Steenken and Neta, 1982)
2. The ability to delocalize the resulting radical (Bors *et al.*, 1990), whether a phenoxyl radical such as those derived from α-tocopherol or butylated hydroxytoluene, a aryloxyl radical such as those derived from flavonoids, a polyunsaturated hydrocarbon chain radical such as β-carotene, or a thiyl radical such as dihydrolipoic acid
3. The transition metal-chelating potential (Thompson and Williams, 1976, Morel *et al.*, 1993; Afanas'ev *et al.*, 1989; Paganga *et al.*, 1996) dependent on the nature of the functional groups and their arrangement within the molecule

Accessibility of the antioxidant to the site of action is another important consideration, which is defined by the lipophilicity or hydrophilicity of the antioxidant or the partition coefficient. For example, α-tocopherol is a much more effective chain-breaking antioxidant in scavenging lipid peroxyl radicals than vitamin C (reviewed in Niki, 1996). The interaction of antioxidant radicals with other antioxidant molecules, thus sparing the original antioxidant from depletion, is also an important factor defined by the relative position of the antioxidant radical in the pecking order of relative reducing abilities (Buettner, 1993), which is the most likely mode of action of ascorbate in enhancing the peroxyl radical scavenging potential of α-tocopherol through the reduction of the α-tocopheroxyl radical.

The screening of flavonoids and phenolics from natural sources for their bioactivities as antioxidants is assessed by determining their direct free radical scavenging activity as hydrogen- or electron-donating molecules, the ability to chelate transition metal ions, the efficacy as chain-breaking antioxidants in scavenging lipid peroxyl radicals, and the scavenging of reactive nitrogen species, which might involve nitration, hydroxylation, or oxidation of the antioxidant molecule. Flavonoids are hydrogen-donating antioxidants by virtue of the reducing properties of the multiple hydroxyl groups attached to aromatic ring structures, along with their abilities to delocalize the resulting antioxidant radical within the structure.

DIRECT REDUCING PROPERTIES THROUGH ELECTRON/H DONATION

The electron-/H-donating properties of flavonoids have been the focus of the basis of their antioxidant action. The understanding of their free radical scavenging properties is best approached through structure–antioxidant activity relationships. The ability of flavonoids to act as antioxidants by electron donation depends directly on the reduction potentials of their radicals and inversely on the reactivities of the flavonoid molecules with oxygen, as the generation of peroxyl radicals will propagate oxidative reactions. These concepts have been reviewed excellently (Jovanovic *et al.*, 1998). The range of reduction potentials of selected flavonoid radicals, relative to those of ascorbic acid and α-tocopherol, is shown in Table I, from

Table I   Reduction Potentials of Flavonoid Radicals

	$E_7{}^a$	TEAC^b
Quercetin	0.33	4.7
Epigallocatechin gallate	0.43	4.8
Epigallocatechin	0.42	3.8
Taxifolin	0.5	1.9
Catechin	0.57	2.4
Luteolin	(0.6)	2.1
Rutin	0.6	2.4
Kaempferol	0.75	1.3

^{a} From Jovanovic *et al.* (1998).
^{b} Rice-Evans *et al.* (1996).

the more oxidizable quercetin [E_7 0.33] to kaempferol [0.75] (Jovanovic *et al.*, 1998). Spectral studies have revealed that the radical site is on the B ring and that the A ring has no influence on the properties of the radicals from the B ring in flavonoids with a saturated C ring, such as catechin. In flavone radicals in which the B ring with a 3′,4′-dihydroxycatechol structure is conjugated through the C ring 2,3-double bond, evidence suggests that the radical is on the catechol B ring and that spectra resemble those of 3,4-dihydroxycinnamate radicals (Steenken and Neta, 1982). The antioxidant activity of flavonoids depends critically on the part of the molecule with more efficient electron-donating properties. In most flavonoids this is the B ring (Jovanovic *et al.*, 1998). Thus, the antioxidant activity of flavonoids as electron or hydrogen donors relates to the reduction potentials and reactivities of the substituent hydroxyl groups. A number of chemical studies on the reactivity of flavonoids with a range of radicals and the stability of the resultant antioxidant radicals have also emphasized specifically the role of the catechol structure in the B ring and unsaturation in the C ring.

A simple chemical assay broadly applicable in determining the hierarchy of antioxidant activities of polyphenols measures their ability to reduce the ABTS$^{·+}$ radical cation (Rice-Evans *et al.*, 1996; Rice-Evans and Miller, 1997; Miller and Rice-Evans, 1996, 1997). The scavenging of the ABTS$^{·+}$ radical cation by pure compounds is determined relative to that of Trolox as an antioxidant standard, with the Trolox equivalent antioxidant activity (TEAC) being defined as the concentration of Trolox with the same antioxidant activity as 1 mM concentration of the substance under investigation (Miller and Rice-Evans, 1996). Table I indicates the antioxidant activities of a range of flavonoids (Rice-Evans *et al.*, 1996) in comparison with the reduction potentials of the flavonoid radicals. Results show that the measured reducing properties of the phenolics, in terms of their antioxidant activities in this particular assay system, are consistent with the reduction potentials of their radicals. The highest activity for the flavonol quercetin relates to the activity of the *o*-dihydroxy structure in the B ring and the unsaturated C ring, with its 3-hydroxyl group, linking with the 5,7-dihydroxy structure in the A ring. Absence or glycosylation of the 3-hydroxyl group, as in luteolin and rutin, respectively, influences greatly their antioxidant potentials and increases the reduction potential of their radicals. The substitution of the catechol structure in the flavon-3-ol B ring by a phenolic ring with a single 4′-hydroxyl group, as in kaempferol, further decreases the antioxidant activity and increases the reduction potential. Hesperidin, with a 4′-methoxy group in the B ring and a glycosylated 7 position in the A ring, gives a lower TEAC than kaempferol, due to the latter substitution, and an approximately similar reduction potential. Catechin, with the five hydroxyl groups across the structure in the same positions as those of quercetin, has a TEAC value of the same order as that of luteolin (as expected since the 3-hydroxyl group on the saturated ring of catechin does not contribute) and a closely similar reduction potential. Taxifolin, predicted to have a similar antioxidant activity, has a slightly lower value with a marginally more favorable reduction potential. Epigallocatechin gallate contributes a similar TEAC value to the scavenging of the model $ABTS^{\cdot+}$ radical as quercetin, with the gallic acid moiety contributing to the increase over that of epicatechin; the reduction potential is low but not to the extent of that of quercetin. Epigallocatechin, with one additional hydroxyl group in the B ring compared with epicatechin, has a TEAC of 3.8 n*M*, much enhanced in relation to catechin, consistent with a lower reduction potential than the latter.

TRANSITION METAL CHELATION: EFFECTIVE CHELATION VERSUS REDUCING ACTIVITIES

Studies have suggested previously that the key functional groupings on flavonoid molecules (Thompson and Williams, 1976; Afanas'ev *et al.*, 1989) that can participate in metal chelation are the catechol 3',4'-dihydroxy grouping in the B ring, the 3-hydroxy, 4-keto structure in the C ring, and the possibility of chelation across the 5-hydroxy, 4-keto structure between the A and the C rings, as depicted schematically for quercetin (Fig. 2).

The screening of the activity of polyphenolics in chelating copper and the indication of their relative efficacies as preventive antioxidants are demonstrated by comparing four structurally related flavonoids: quercetin, rutin, luteolin, and kaempferol (Fig. 3). Quercetin and rutin are structurally similar in that rutin is the 3-rhamnoglucoside of quercetin; luteolin is 3-

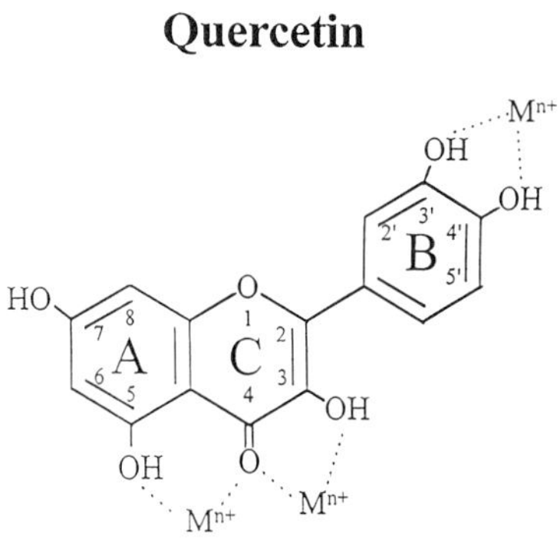

Quercetin

Figure 2 Proposed binding sites for transition metals to flavonoids.

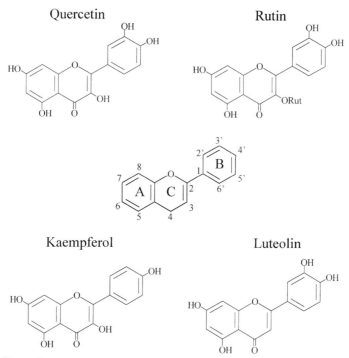

Figure 3 Chemical structures of quercetin, rutin, luteolin, and kaempferol.

desoxyquercetin. Kaempferol is structured similarly to quercetin except that it lacks the catechol structure in the B ring through the absence of the 3′-hydroxyl group. Interactions of copper ions with the flavones and 3-hydroxy flavones produced bathochromic shifts of the B ring component of the structure in the visible region. Flavones luteolin and rutin contain the *ortho*-3′,4′-dihydroxy structure in the B ring but lack the free 3-OH group (in the former case it is absent, in the latter it is glycosylated). On interaction with copper ions, new red-shifted peaks are assumed characteristic of a copper-flavone chelate, and the reversibility of the reaction and regeneration of the original flavonoid spectrum in the presence of EDTA confirms this (Brown et al., 1998). Ion chelation probably occurs through the catechol structure. However, the ability of putative antioxidants to chelate iron or copper should be balanced against the ability to reduce these transition metal ions and thus potentially drive prooxidant actions.

Spectroscopic studies indicate the chelation of copper by quercetin; titrating out the bound copper produces a spectrum indicative of oxidized quercetin. In view of the structural similarities of quercetin to its 3-

rutinoside (rutin), the involvement of the 3-hydroxyl group (the only structural difference between them) in the oxidation of the molecule is suggested. The likely structure of the resulting quinone is shown in Fig. 4. Major spectral changes also take place on the binding of copper to kaempferol, although the characteristic copper-chelate peak is much less pronounced. Thus the major contributor to copper chelation in this family of flavones is the B ring catechol structure, absent from kaempferol, rather than the 4-keto group in the C ring with either of its adjacent hydroxyl groups, which are present in the kaempferol structure (Brown et al., 1998). On removal of copper from the kaempferol, a dramatic change in the spectrum is observed, consistent with oxidation of the molecule at the 3- and 4′-hydroxyl groups as seen for quercetin. These observations conform with suggestions that the 3-OH group in flavonol structures should be blocked to render a compound that is unable to redox cycle transition metal ions (van Acker *et al.*, 1996).

SCAVENGING OF REACTIVE NITROGEN SPECIES

The ability to inhibit peroxynitrite-dependent nitration of tyrosine provides a useful assay to screen various compounds for their effectiveness in scavenging peroxynitrite and/or the nitrating species derived from it (Whiteman and Halliwell, 1996). Peroxynitrite is a well-known toxic oxidizing and nitrating species produced from rapid interaction between superoxide radical and nitric oxide (Huie and Padmaja, 1993; Beckman *et al.*, 1994). Peroxynitrite at physiological pH is protonated to form peroxynitrous acid, which is highly unstable and may react via a vibrationally excited intermediate or decay rapidly to form a mixture of products by homolytic dissociation to $NO_2\bullet$ and $\bullet OH$ or via heterolytic dissociation to form the nitron-

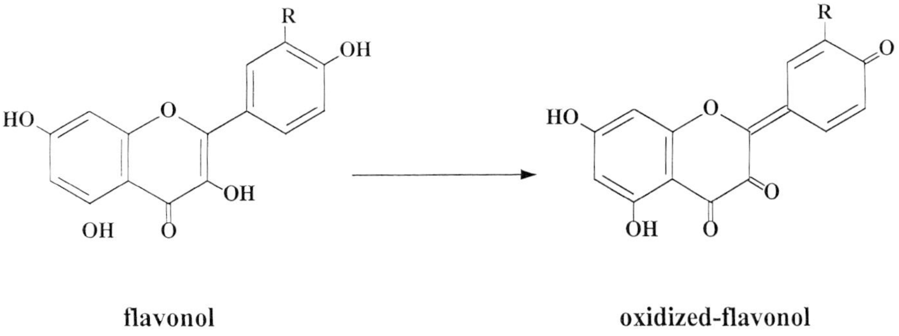

flavonol   oxidized-flavonol

Figure 4 Structure after oxidation of flavon-3-ols with 4′-hydroxyl groups.

ium ion (Squadrito and Pryor, 1995; Ischiropoulos *et al.*, 1992). The phenolic compound tyrosine is especially susceptible to peroxynitrite-dependent reactions forming 3-nitrotyrosine.

Two families of phenolics were screened for their abilities to scavenge peroxynitrite (Pannala *et al.*, 1997, 1998): the hydroxycinnamates, major constituents of fruit (Macheix and Fleuriet, 1990) and the precursors of flavonoids in the plant synthetic pathway, and the flavanols catechin (C), epicatechin (EC) (major polyphenolic constituents of red wine and green tea), epigallocatechin (EGC), and their gallate esters, epicatechin gallate (ECG) and epigallocatechin gallate (EGCG) (major monomeric constituents of teas). Competition studies between catechin polyphenols and tyrosine (100 μM) with peroxynitrite (500 μM), compared with gallic acid and with Trolox as the reference standard, were undertaken by measuring the inhibition of tyrosine nitration by high-performance liquid chromatography (HPLC). At higher concentrations of the polyphenols (50–100 μM), the reduction of tyrosine nitration was close to 100%. At lower concentrations (10 μM), the reduction of tyrosine nitration ranged from 20 to 40% (Table II).

Results show that ECG, EGCG, and gallic acid are the most effective peroxynitrite scavengers from the catechin polyphenolic series. The antioxidant protection imparted by these compounds is probably mediated by the direct competition with tyrosine for nitration. The possibility of nitration occurring on the chromanol ring is diminished due to a decreased delocalization resulting from the presence of two hydroxyl groups that are meta to each other. It is therefore likely that the majority of any nitration reactions involving these antioxidants will occur on the B ring, potentially at the 2′ and 5′ positions and to a lesser extent at the 6′ position. Of all the compounds tested, EGC was shown to be the least likely to undergo nitration due to the presence of an additional OH group in the 5′ position. The cor-

Table II   Extent of Inhibition of Peroxynitrite-Mediated Tyrosine Nitration by Catechin Polyphenols (10 μM)

	Inhibition (%)
Epicatechin gallate	38.1±3.6
Epigallocatechin gallate	32.1±7.5
Gallic acid	32.1±1.9
Epicatechin	22.9±3.3
Epigallocatechin	19.9±2.0
Trolox	13.6±2.9

responding gallate esters (EGCG and ECG), in addition to the potential for nitration in the basic catechin ring, can also undergo further nitration reactions on the gallic acid moiety and hence enhance their peroxynitrite scavenging properties. However, studies with catecholates suggest that a likely mechanism of scavenging might be through oxidation of the polyphenol, as seen with hydroxycinnamates.

Screening of hydroxycinnamates, caffeic acid (3,4-dihydroxycinnamic acid), chlorogenic acid (the quinic acid ester of caffeic acid), ferulic acid, (3-hydroxy, 4-methoxy cinnamic acid), and *p*-coumaric acid, (4-hydroxycinnamic acid), for the inhibition of tyrosine nitration by peroxynitrite demonstrates that caffeic and chlorogenic acids, the dihydroxy-substituted compounds, are more active than monohydroxycinnamates, such as ferulic and *p*-coumaric acids (Table III). In order to establish if the hydroxycinnamates were competing with and inhibiting tyrosine nitration by becoming nitrated themselves, they were exposed to peroxynitrite in the absence of tyrosine and the products were examined by ultraviolet/visible spectroscopy and HPLC analysis. Results show that the monohydroxycinnamates, ferulic acid and *p*-coumaric acid, are nitrated as evidenced by the appearance of absorbance increases at 430 nm indicative of the nitrophenolate species at pH 7 and detectable by HPLC. Ferulic acid, with its effective electron-donating group at the 3 position, has a greater ability to stabilize the phenoxyl radical than *p*-coumaric acid, which accounts for its higher activity. It can also undergo nitration rections due to partial activation of the 5-position by the 4-hydroxyl group. However, the catecholate derivatives, caffeic and chlorogenic acids, exhibited spectral changes but not with the associated characteristics indicative of a nitrated phenolic ring. Rather, the catecholates seem to suppress nitration reactions by reducing the reactive nitrogen species, themselves being oxidized through electron donation and possible quinone formation. Thus, on screening of phenolic phytochemicals for antioxidant activity against reactive nitrogen species,

Table III   Extent of Inhibition of Peroxynitrite-Mediated Tyrosine Nitration by Hydroxycinnamates (50 μM)

	Inhibition (%)
Caffeic acid	80.6±2.6
Chlorogenic acid	70.8±6.5
Ferulic acid	55.7±8.6
p-Coumaric acid	45.9±4.9
Trolox	51.7±4.9

the lack of appearance of nitrated antioxidants does not necessarily imply lack of activity.

α-Tocopherol is also known to be oxidized by peroxynitrite to give the appropriate quinone in aqueous solution; this suggests a sequential 2-electron oxidation (Hogg *et al.*, 1993, 1994). The differential reactivities of the tocopherol monophenolic family have been observed between α- and γ-tocopherol, in which γ-tocopherol is nitrated through interaction with peroxynitrite, as predicted earlier, whereas α-tocopherol reduces peroxynitrite through a nitration-independent process (Christen *et al.*, 1997). In the case of α-tocopherol, no positions are available on the phenolic ring to accommodate nitration as an appropriate mechanism of scavenging.

INTERCEPTION OF CHAIN PROPAGATION BY SCAVENGING LIPID PEROXYL RADICALS

Studies (too numerous to mention or to which to do justice) have been undertaken in lipophilic systems to screen polyphenolics as inhibitors of lipid peroxidation and to establish structural criteria for their activities in enhancing the stability of lipids, fatty acid dispersions, especially methyl linoleate, low-density lipoproteins (LDL), and oils toward autoxidation or oxidation (Dziedzic and Hudson, 1983; Hudson and Lewis, 1983; Torel *et al.*, 1986; Pokorny 1987; Chimi *et al.*, 1991; Shahidi and Wanasundara, 1992; Terao *et al.*, 1994; Salah *et al.*, 1995; Miura *et al.*, 1995; Brown *et al.*, 1998). Depending on the system used in the screening test to promote oxidation, the specific mode of inhibition of oxidation might relate to the direct scavenging of lipid alkoxyl and peroxyl radicals by hydrogen donation from the phenolic compound, the regeneration of α-tocopherol through the reduction of the α-tocopheroxyl radical, or the chelation of transition metal ions. However, one essential criterion, which is independent of preventative roles through metal chelation, is the partition coefficient of the polyphenolics and their accessability to lipid radicals. To study the antioxidant activity of polyphenols as scavengers of propagating lipid peroxyl radicals, no initiating radical species were added. Many researchers have applied transition metal ions or heme proteins as catalysts of the oxidative and reductive decomposition of lipid hydroperoxides. The use of azo initiators for inducing lipid oxidation is perhaps the most appropriate approach for screening chain-breaking antioxidant properties.

It has been suggested that for optimal inhibition of lipid peroxidation, a catechol group in the B ring, a 2-3 double bound conjugated with the 4-

oxo function and 3- and 5-hydroxyl groups are agreed structural criteria. Perhaps a more appropriate definition might include aspects of the former structural criteria that underlie effective hydrogen donation to lipid peroxyl radicals, e.g., with appropriate structural arrangements for delocalization of the resulting aryloxyl radical, but the lipophilicity and the accessibility of the antioxidant to the lipid radicals to be scavenged are also essential features.

Structure–activity relationships for the catechin/catechin gallates in suppressing LDL oxidation have revealed an association exemplifying the expected importance of the lipophilicity of the flavonoid (Salah *et al.*, 1995). The hierarchy for inhibition is

$$ECG \approx EGCG \approx EC \approx C > EGC > GA$$

Although epigallocatechin is a highly effective H donor in terms of its relative ability to scavenge the $ABTS^{\cdot+}$ radical cation in the aqueous phase (Salah *et al.*, 1995; Rice-Evans and Miller, 1997), its weaker influence in suppressing LDL oxidation is entirely consistent with its lower partition coefficient. For this family of phytochemicals, others have also shown EGC to be the least effective in protecting LDL from copper-mediated oxidation, with ECG being the most effective, with a threefold decrease in the IC_{50}, consistent with the author's findings (Terao *et al.*, 1991).

Brown *et al.* (1998) have also demonstrated that the relative reactivities of structurally related flavonoids in inhibiting lipid peroxidation within low-density lipoproteins are dependent on a combination of structurally dependent criteria. Comparing the inhibitory effects of quercetin, luteolin, rutin, and kaempferol, the overall results demonstrated that the sequence of effectiveness in the case of copper-mediated LDL oxidation is

$$quercetin \approx luteolin \approx rutin > kaempferol$$

and for metmyoglobin-induced LDL oxidation is

$$luteolin > quercetin > rutin > kaempferol$$

Although quercetin is apparently a more effective reducing agent and hydrogen donor than luteolin, rutin, or kaempferol, the superior protective effects of luteolin against lipid peroxidation in the heme protein-mediated oxidation system in which a metal chelate is not a consideration suggest that chelation plays a significant role in protection by quercetin against copper-induced lipid peroxidation. Indeed, the effects of polyphenols on the propagation rate of lipid peroxidation have implicated interactions with copper ions. In addition, partition coefficients (Table IV) suggest that the polar glycoside rutin would be least accessible to the peroxidizing lipid radicals; however, in contrast, it is a more efficient hydrogen donor than kaempferol, as mentioned previously.

In the case of chain-breaking antioxidants, antioxidant activity has been shown to decrease with increasing concentration. In the presence of low concentrations of antioxidants, the breaking of chain reactions is the predominant mechanism (Pokorny, 1987).

$$LOO\cdot + AH \rightarrow LOOH + A\cdot$$

$$LO\cdot + AH \rightarrow LOH + A\cdot$$

At high concentrations of antioxidant the rates of these reactions reach a constant value because of the limited content of LOO· and LO· radicals. Under these conditions, reactions such as

$$A\cdot + O_2 \rightarrow AOO\cdot$$

become more important, and a low level of free radicals is produced by reaction with lipid molecules

$$AOO\cdot + L\text{-}H \rightarrow LOOH + L\cdot$$

$$A\cdot + L\text{-}H \rightarrow LOOH + L\cdot$$

The antioxidant activity is reduced and losses of antioxidants by these side reactions increase. At very high antioxidant levels, these side reactions may predominate. Antioxidants with low reduction potentials are inclined to react in this manner (Pokorny, 1987). The prooxidant effect of α-

Table IV   Partition Coefficients of Structurally Related Flavones and Flavon-3-ols[a]

	K_{part}
Quercetin	1.2
Luteolin	22.2
Kaempferol	69.5
Rutin	0.37

[a] From Brown *et al.* (1998).

tocopherol at high levels has been demonstrated by Kochar (1993). The re-activity of α-tocopherol in the presence of copper ions has been shown to be that of a prooxidant in the presence of low-density lipoproteins at low radical fluxes rather than that of a lipid chain-breaking antioxidant. Similar stimulatory oxidative responses have been demonstrated with ferulic acid under similar conditions (Bourne and Rice-Evans, 1997).

Phenolic antioxidants have been suggested to be the most active antioxidants from natural sources. Looking at the chemistry, unsubstituted phenol is practically inactive. Substitution of the phenolic ring with alkyl groups increases the electron density and enhances the reactivity with electron acceptors. The presence of a second hydroxyl group in the aromatic ring improves the antioxidant activity in the case of 1,2- or 1,4-disubstituted compounds, but 1,3-dihydroxy derivatives are less active. Thus many synthetic antioxidants are derived from substituted phenols, including alkyl phenols, alkoxy phenols that contain one free and one alkylated hydroxyl group, usually methoxy (e.g., ferulic acid), polyphenols containing *o*- or *p*-dihydroxy groups, and phenols containing condensed rings, e.g., benzopyran ring, chroman ring, anthocyanins, flavones, and polymeric derivatives.

SUMMARY

Studies described herein indicate the multiple properties that flavonoids possess for scavenging reactive oxygen and nitrogen species. In considerations of the use and design of effective antioxidants, it should be noted that phenolics with the most extensive reducing properties, while potentially predicted to be the most effective antioxidants, can be involved in oxidation processes, especially on interaction with transition metals and at high antioxidant concentrations.

REFERENCES

Afanas'ev, I. B., Dorozhko, A. I., Brodskii, A. V., Kostyuk, V. A., and Potapovitch, A. I. (1989). Chelating and free radical scavenging mechanisms of inhibitory action of rutin and quercetin in lipid peroxidation. *Biochem. Pharmacol.* 38, 1763–1769.

Beckman, J. S., Chen, J., Ischiropoulos, H., and Crow, J. P. (1994). Oxidative chemistry of peroxynitrite. *Meth. Enzymol.* 233, 229–240.

Bors, W. and Saran, M. (1987). Radical scavenging by flavonoid antioxidants. *Free Rad. Res. Comm.* 2, 289–294.

Bors, W., Heller, W., Michel, C., and Saran, M. (1990). Flavonoids as antioxidants: Determination of radical scavenging efficiencies. *Methods Enzymol.* 186, 343–355.

Bors, W., Michel, C., Schikora, S. (1995). Interaction of flavonoids with ascorbate and determination of their univalent redox potentials: A pulse radiolysis study. *Free Radic. Biol. Med.* **19**, 45–52.

Bourne, L. C. and Rice-Evans, C. A. (1997). The effect of the phenolic antioxidant ferulic acid on the oxidation of low density lipoproteins depends on the pro-oxidant used. *Free Rad. Res.* **27**, 337–344.

Brown, J. E., Khodr, H., Hider, R. C., and Rice-Evans, C. (1998). Structural-dependence of flavonoid interactions with copper ions: Implications for their antioxidant properties. *Biochem. J.* **330**, 1173–1178.

Buettner, G. R. (1993). The pecking order of free radicals and antioxidants: Lipid peroxidation, α-tocopherol and ascorbate. *Arch. Biochem. Biophys.* **300**, 535–543.

Chimi, H., Cillard, J., Cillard, P., and Rahmani, M. (1991). Peroxyl and hydroxyl radical scavenging activity of some natural phenolic antioxidants. *J. Amer. Oil. Chem. Soc.* **68**, 307–312.

Christen, S., Woodall, A., Shigenaga, M., Sothwell-Keely, P., Duncan, M. W., and Ames, B. N. (1997). γ-Tocopherol traps mutagenic electrophiles such as NO_x and complements α-tocopherol: Physiological implications. *Proc. Natl. Acad. Sci.* **94**, 3217–3222.

DeWhalley, C., Rankin, S. M., Hoult, J. R., Jessup, W., Leake, D. (1990). Flavonoids inhibit the oxidative modification of low density lipoproteins by macrophages. *Biochem. Pharmacol.* **39**, 1743–1750.

Dziedzic, S. Z., Hudson, B. J. F. (1983). Polyhydroxychalcones and flavanones as antioxidants for edible foods. *Food Chem.* **12**, 205–212.

Eiserich, J. P., Cross, C. E., Jones, A. D., Halliwell, B., and Van der Vliet, A. (1996). Formation of nitrating and chlorinating species by reaction of nitrite with hypochlorous acid—a novel mechanism for nitric oxide-mediated protein modification. *J. Biol. Chem.* **271**, 19199–19208.

Graf, E. (1992). Antioxidant potential of ferulic acid. *Free Rad. Biol. Med.* **13**, 435–448.

Hogg, N., Darley-Usmar, V., Wilson, M. T., and Moncada, S. (1993). The oxidation of α-tocopherol in human LDL by the simultaneous generation of superoxide and nitric oxide. *FEBS Lett.* **326**, 199–203.

Hogg, N., Joseph, J., and Kalyanaraman, B. (1994). The oxidation of α-tocopherol and trolox by peroxynitrite. *Arch. Biochem. Biophys.* **314**, 153–158.

Hudson, B. J. F., and Lewis, L. I. (1983). Polyhydroxy flavonoid antioxidants for edible oils. Structural criteria for activity. *Food Chem.* **10**, 47–55.

Huie, R. E. and Padmaja, S. (1993). The reaction of NO with superoxide. *Free Rad. Res. Commun.* **18**, 195–199.

Ischiropoulos, H., Zhu, L., Chen, J., Tsai, M., Martin, J. C., Smith, C. D., and Beckman, J. S. (1992). Peroxynitrite mediated tyrosine nitration catalyzed by superoxide dismutase. *Arch. Biochem. Biophys.* **298**, 431–437.

Jovanovic, S. V., Steenken, S., Tosic, M., Marjanovic, B., and Simic, M. G. (1994). Flavonoids as antioxidants. *J. Am. Chem. Soc.* **116**, 4846–4851.

Jovanovic, S., Hara, Y., Steenken, S., and Simic, M. G. (1995). Antioxidant potential of gallocatechins. A pulse radiolysis study. *J. Am. Chem. Soc.* **117**, 9881–9888.

Jovanovic, S., Steenken, S., Simic, M., and Hara, Y. (1998). Antioxidant properties of flavonoids: Reduction potentials and electron transfer reactions of flavonoid radicals. In "Flavonoids in Health and Disease" (C. Rice-Evans and L. Packer, eds.) pp. 137–161. Marcel Dekker, New York.

Kochhar, S. P. (1993). Deterioration of edible oils, fats and foodstuffs. In: (G. Scott, ed.) "Atmospheric Oxidation and Antioxidants" pp. 71–131. Elsevier Science Publishers. Amsterdam, The Netherlands.

Macheix, J. J., Fleuriet, A., and Billot, J. (1990). *Fruit phenolics*. CRC Press, Boca Raton, FL.

Miller, N. J., and Rice-Evans, C. (1996). Spectrophotometric determination of antioxidant activity. *Redox Rep* **2**, 161–171.

Miller, N. J. and Rice-Evans, C. (1997). Factors influencing the antioxidant activity determined by the ABTS$^{.+}$ radical cation assay. *Free Rad. Res.* **26**, 195–197.

Miura, S., Watanabe, J., Sano, M., Tomita, T., Osawa, T., Hara, Y., Tomita, I. (1995). Effects of various natural antioxidants on the Cu^{2+}-mediated oxidative modification of low density lipoproteins. *Biol. Pharm. Bull.* **18**, 1–4.

Morel, I., Lescoat, G., Cogrel, P., Sergent, O., Pasdeloup, N., Brissot, P., Cillard, P., and Cillard, J. (1993). Antioxidant and iron-chelating activities of the flavonoids catechin, quercetin and diosmetin on iron-loaded rayt hepatocyte cultures. *Biochem. Pharmacol.* **45**, 13–19.

Niki, E. (1996). α-Tocopherol. In "Handbook of antioxidants." (L. Packer and E. Cadenas, eds.) pp. 3–25. Marcel Dekker, New York.

Paganga, G., Al-Hashim, H., Khodr, H., Scott, B. C., Aruoma, O. I., Hider, R. C., Halliwell, B., and Rice-Evans, C. (1996). Mechanisms of the antioxidant activities of quercetin and catechin. *Redox Report* **2**, 359–364.

Pannala, A. S., Razaq, R., Halliwell, B., Singh, S., Rice-Evans, C. (1998). Inhibition of peroxynitrite-dependent nitration by hydroxycinnamates: Nitration or electron donation? *Free Radic. Biol. Med.* (In the press).

Pannala, A. S., Rice-Evans, C. A., Halliwell, B., and Singh, S. (1997). Inhibition of peroxynitrite-mediated nitration by catechin polyphenols. *Biochem. Biophys. Res. Commun.* **232**, 164–168.

Pokorny, J. (1987). Major factors affecting the antioxidant of lipids. In "Autoxidation of Unsaturated Lipids," (H. Chan, ed.) pp. 141–206. Academic Press, London.

Rice-Evans, C., and Miller, N. (1997). Measurement of the antioxidant status of dietary constituents, low density lipoproteins and plasma. *Prosta. Leukotr. Ess. Fatty Acids* **57**, 499–505.

Rice-Evans, C., Miller, N. J., and Paganga, G. (1996). Structure-antioxidant activity relationships of flavonoids and phenolic acids. *Free Rad. Biol. Med.* **20**, 933–956.

Salah, N., Miller, J. J., Paganga, G., Tijburg, L. L., Bolwell, G. P., Rice-Evans, C. (1995). Polyphenolic flavonols as scavengers of aqueous phase radicals and as chain-breaking antioxidants. *Arch. Biochem. Biophys.* **322**, 339–346.

Shahidi, F., Wanasundara, P. K. J. (1992). Phenolic antioxidants. *Crit. Rev. Food Sci. Nutr.* **32**, 67–103.

Squadrito, G. L., and Pryor, W. A. (1995). The chemistry of peroxynitrite—a product from the reaction of nitric oxide with superoxide. *Am. J. Physiol.* **268**, 699–722.

Steenken, S., and Neta, P. (1982). One-electron redox potentials of phenols, hydroxphenols and aminophenols and related compounds of biological interest. *J. Phys. Chem.* **86**, 3661–3667.

Terao, J., Piskuli, M., Yao, Q. (1994). Protective effect of epicatechin, epicatechin gallate and quercetin on lipid peroxidation in phospholipid bilayers. *Arch. Biochem. Biophys.* **308**, 278–284.

Thompson, M., Williams, C. R., and Elliot, G. E. P. (1976). Stability of flavonoid complexes of Cu(II) and flavonoid antioxidant activity. *Anal. Chim. Acta.* **85**, 375–381.

Torel, J. O., Cillard, J., Cillard, P. (1986). Antioxidant activity of flavonoids and reaction with peroxyl radical. *Phytochemistry.* **25**, 383–387.

van Acker, S. A. B. E., van den Berg, D.-J., Tromp, M. N. J. L., Griffoen, D. H., van Bennekom, W. P., van der Vijgh, W. J. F., and Bast, A. (1996). Structural aspects of antioxidant activity of flavonoids. *Free Rad. Biol. Med.* **20**, 331–342.

Whiteman, M., and Halliwell, B. (1996). Protection against peroxynitrite-dependent tyrosine-nitration and α1-antiproteinase activation by ascorbic acid. A comparison with other biological antioxidants. *Free Rad. Res.* **25**, 275–283.

17 Dietary Flavonoids as Plasma Antioxidants on Lipid Peroxidation: Significance of Metabolic Conversion

Junji Terao

Department of Nutrition
School of Medicine
The University of Tokushima
Tokushima 770-8503, Japan

INTRODUCTION

Flavonoids are distributed widely in plant foods such as vegetables and fruits. They possess a unique C6–C3–C6 structure (diphenylpropane structure) with phenolic OH groups, and more than 4000 different varieties has been identified as natural flavonoids (Fig. 1). In plant foods, flavonoids are present mainly as glycosides in which phenolic hydrogen or hydrogens are substituted to the sugar moiety. Aglycone is released from glycosides in the digestive tract by the hydrolytic action of microflora. Flavonoids can be classified as calcones, flavone, flavanones, flavanols, and flavonols. Quercetin, a typical flavonol present in a wide variety of vegetables such as broccoli, onions and lettuce, contains phenolic OH groups at the 5 and 7 position in the A ring and at the 3′ and 4′ position in the B ring (Fig. 2). Although quercetin glycosides containing a sugar group at the 3 position are often found in plant foods, the site for sugar binding is not restricted to the 3 position in naturally occurring flavonol-type flavonoids. However, tea catechins are flavanol-type flavonoids in which phenolic OH groups are bound to the 5 and 7 position (Fig. 3). The hydroxyl group at the 3 posi-

Antioxidant Food Supplements in Human Health

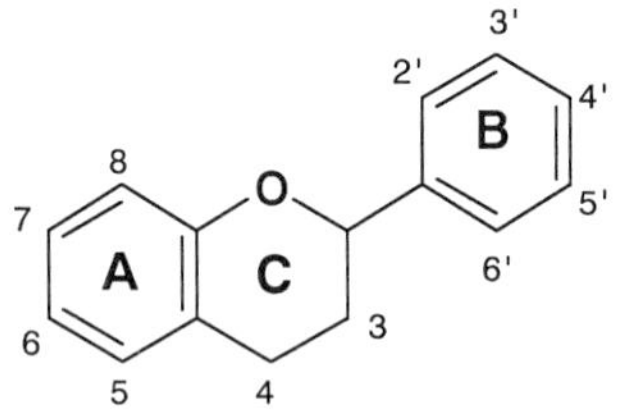

Figure 1   Basic structure of flavonoids.

tion is esterified frequently by gallic acid (epicatechin gallate and epigallo-catechin gallate).

The average intake of flavonoids by humans has been estimated as 25 mg/day (Hertog *et al.*, 1993b). This value covers only five aglycones involving quercetin. Thus, the antioxidant activity of flavonoids has attracted much attention in relation to their physiological functions. In particular, dietary flavonoids are expected to help in the prevention of coronary heart disease, as epidemiological studies have shown the inverse relationship between the intake of dietary flavonoids and coronary heart disease (Hertog *et al.*, 1993a; Knekt *et al.*, 1996). The so-called French paradox is at least

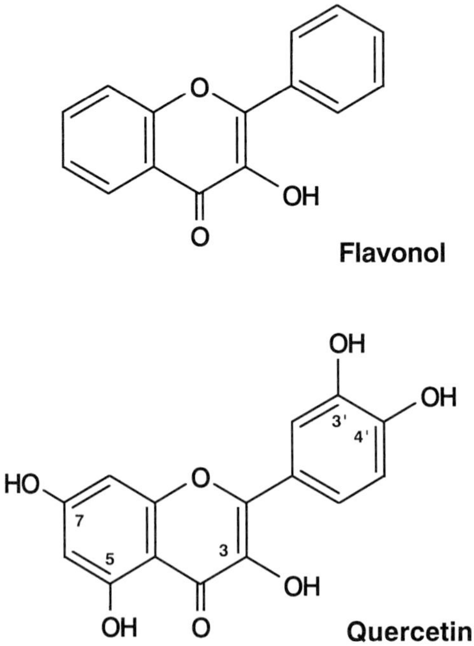

Figure 2   Structures of flavonol and quercetin.

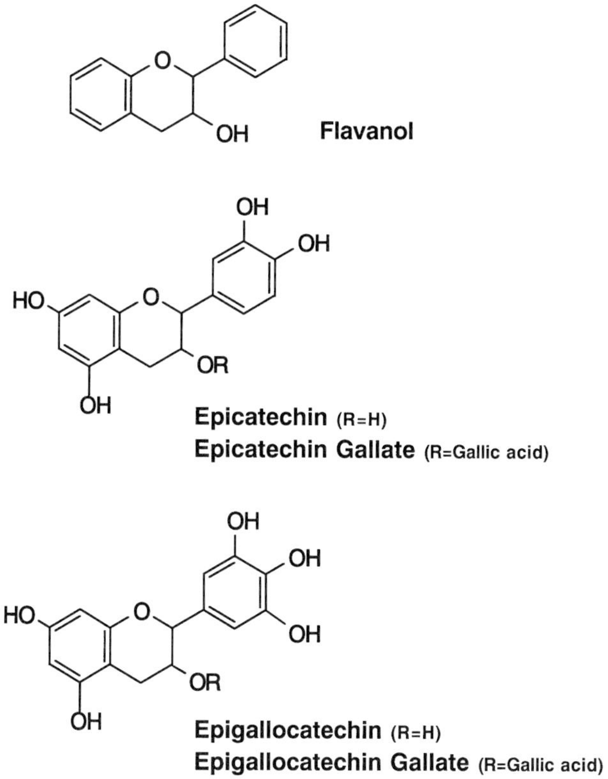

Figure 3 Structures of flavanol and tea catechins.

partly related to the consumption of red wine rich in flavonoids and other phenolic compounds. It is, however, necessary to know the biodynamics of flavonoids after intake for the estimation of *in vivo* antioxidant activity. This chapter describes the antioxidant activity of dietary flavonoids, especially quercetin and (-)-epicatechin, from the standpoint of antioxidant defense on lipid peroxidation in blood plasma and plasma lipoproteins.

STRUCTURE–ACTIVITY RELATIONSHIP OF FLAVONOIDS IN THE INHIBITION OF LIPID PEROXIDATION

Polyphenolic flavonoids can scavenge reactive oxygen radicals, such as the hydroxyl radical and superoxide anion radical, by donating a hydrogen atom or electron (Rice-Evans *et al.*, 1996). Their antioxidant activity seems to be substantially dependent on pH because phenolic OH groups are dissociated with the elevation of pH, and the electron-donating capacity of phenolic OH group in flavonoids increases with its dissociation (Mukai *et al.*, 1996). Furthermore, Bors *et al.* (1990) proposed that three structural groups are important determinants for radical scavenging: (1) the *o*-dihydroxyl structure (catechol structure) in the B ring, which is the obvious radical target site, (2) the 2,3 double bond in conjunction with 4-oxo function, which is responsible for electron delocalization, and (3) the additional presence of both the 3- and the 5-OH group for maximal radical scavenging potentials and the strongest radical absorption. In particular, the *o*-dihydroxyl structure is essential for radical scavenging in flavanols and flavanones that have no 2,3 double bonds. Quercetin and epicatechin both contain a *o*-dihydroxyl structure, indicating that these two flavonoids can act as effective inhibitors on oxygen radical-mediated lipid peroxidation by scavenging such radicals.

In general, lipid peroxidation proceeds via a radical chain reaction consisting of chain initiation and chain propagation.

Chain initiation
$$LH + X\bullet \rightarrow L\bullet + XH$$
$$L\bullet + O_2 \rightarrow LOO\bullet$$

Chain propagation
$$LOO\bullet + LH \xrightarrow{k_p} LOOH + L\bullet$$
$$L\bullet + O_2 \rightarrow LOO\bullet$$

Chain interruption
$$LOO\bullet + InH \xrightarrow{k_{inh}} LOOH + In\bullet$$

(LH; polyunsaturated lipids; X•; reactive oxygen radicals; LOO•, lipid peroxyl radicals; LOOH, lipid hydroperoxides; InH, chain-interrupting antioxidant).

Results using a radical generator-induced radical chain reaction of methyl linoleate hydroperoxidation in solution have demonstrated that both quercetin and (-)-epicatechin can interrupt chain propagation (Terao *et al.*, 1994). However, their rate of scavenging the chain-carrying lipid peroxyl radical (LOO•) is much lower than that of vitamin E (α-tocopherol) because the ratio of the rate constant for chain interruption (k_{inh}) to that of chain propagation (k_p) is five to six times lower than that of α-tocopherol. This indicates that these flavonoids interrupt radical chain oxidation much more slowly than vitamin E in homogeneous solutions.

Reactive oxygen radicals and/or transition metal ions such as copper and iron are likely to be responsible in the initial process of lipid peroxidation. In particular, the transition metal ion may play an important role in the chain initiation because the lipid alkoxyl radical (LO•) and hydroxyl radical (•OH) are created by the reaction of preformed lipid hydroperoxides and hydrogen peroxide with transition metal iron, respectively.

$$LOOH + M^{n+} \rightarrow LO\bullet + OH^- + M^{(n+1)+}$$
$$HOOH + M^{n+} \rightarrow \bullet OH + OH^- + M^{(n+1)+}$$

Considerable studies have shown that some flavonoids can inhibit lipid peroxidation occurring in membraneous phospholipids by scavenging chain-initiating oxygen radicals or chelating transition metal ion (Afanas'ev *et al.*, 1989; Mora *et al.*, 1990; Ratty and Das, 1988; van Acker *et al.*, 1996).

In the biphasic microenvironment constituting core lipids and water phase, such as biomembranes and plasma lipoproteins, localization of flavonoids should be taken into account for understanding their antioxidant activity. Vitamin E seems to be located within the membrane lipids or lipoprotein particles because of its high lipophilicity. However, it is demonstrated that flavonoid aglycone interacts in the polar surface region of the phospholipid bilayers in membranes (Ratty *et al.*, 1988; Saija *et al.*, 1995). It was also implied that quercetin and (-)-catechins served as powerful antioxidants against lipid peroxidation when phospholipid bilayers were exposed to aqueous oxygen radicals (Terao *et al.*, 1994). In this case, they scavenge chain-initiating aqueous radicals effectively, thus protecting phospholipid bilayers from chain propagation occurring in the membranes.

OXIDATIVE MODIFICATION OF PLASMA LIPOPROTEINS AND ITS INHIBITION BY FLAVONOIDS

The French paradox, a lack of positive correlation between a high intake of saturated fat and the occurrence of coronary heart disease, is at least partly related to the consumption of red wine rich in phenolic compounds (Kinsella *et al.*, 1993). Actually, (-)-catechin, (-)-epicatechin, and quercetin are present as flavonoids in red wine. In recent years, the oxidative damage of plasma lipoproteins is believed to participate in the development of atherosclerosis. The oxygen radical-induced modification of low-density lipoprotein (LDL) and its subsequent unregulated uptake by macrophase in the subendothelial space of the arteries via scavenger receptor-mediated pathways have been implicated in foam cell formation in the early stage of atherosclerosis (Steinberg *et al.*, 1989). The inhibition of oxidative modification of LDL is expected to be one strategy for the prevention of atherosclerosis. For this purpose, improvement of the resistance of LDL to oxidative attack should be sought. *In vitro* experiments have shown that morin (Jessup *et al.*, 1990), quercetin (Nege-Salvayre *et al.*, 1988), and catechins (Miura *et al.*, 1995) can act as effective inhibitors for transition metal ion-induced or macrophage-induced lipid peroxidation in human LDL. It was also demonstrated that the polyphenolic fraction obtained from red wine (Frankel *et al.*, 1993, 1995) and grape juice (Lanningham-Foster *et al.*, 1995) inhibited copper ion-induced oxidation of LDL. It is likely that flavonoids scavenge chain-initiating active oxygens or chelate metal ion in the surface of LDL particles by locating in the interface of water phase and lipoprotein particles.

An alternative pathway proposed as the initial step for LDL oxidation *in vivo* is the 15-lipoxygenase (15-LOX)-mediated enzymatic pathway (Kühn *et al.*, 1992). Evidence for this includes the finding that human and rabbit atherosclerotic lesions exhibit measurable levels of 15-LOX mRNA and protein, as well as 15-LOX final products (Ylä-Herttuala *et al.*, 1991; Folcik *et al.*, 1995). If 15-LOX plays an essential role in LDL lipid peroxidation *in vivo*, then 15-LOX inhibitors may be promising compounds for suppressing atherosclerosis progression. Several flavonoids, including quercetin, are known to be effective LOX inhibitors. An *in vitro* experiment using rabbit reticulocyte 15-LOX demonstrated that quercetin inhibits the formation of cholesteryl ester hydroperoxide (CE-OOH) much more effectively than does ascorbic acid or α-tocopherol, which was preloaded to LDL (Edson *et al.*, 1998). Thus, quercetin seems to inhibit 15-LOX-induced LDL oxidation more efficiently than ascorbic acid and α-tocopherol. In addition, quercetin, quercetin 3-*O*-β-glucoside (Q3G), and quercetin 7-*O*-β-glucoside (Q7G) exhibited higher inhibitory effects than did quercetin 4'-*O*-β-glucosides (Q4'G). This means that the o-dihydroxyl structure in the 3' and 4' position of the B ring takes part in elevating the inhibitory activity of quercetin on the 15-LOX reaction, although the OH group at the 3 and 7 position does not affect its inhibitory effects.

FLAVONOIDS IN THE DIGESTIVE TRACT AND THEIR INTESTINAL ABSORPTION

The behavior of flavonoids in the digestive tract still remains equivocal. However, it is suggested that their solubility in bile acid micelles is an important determinant for intestinal absorption. Flavonoid glycosides are likely to be hardly absorbed in the small intestine because their hydrophilicity lowers solubility in micelles. However, the β-glucosidase activity of microflora induces the hydrolysis of glycosides, resulting in aglycone in the large intestine (Tamura *et al.*, 1980). Enterobacteria such as bacteroides distasonis B, Uniform B, and Ovatus, seem to be responsible for β-glucosidase activity (Bokkenheuser *et al.*, 1987). Furthermore, a part of aglycone is subjected to ring scission by the enterobacteria. 3,4-Dihydroxylphenyl acetic acid, homovanilic acid, and hydrophenylacetic acid were identified as ring scission products from quercetin (Fig. 4) (Booth *et al.*, 1956). After such a conversion, flavonoids elevate lipophilicity, resulting in high solubility in bile acid micelles. It was reported that the rat experiments show absorption of rutin (quercetin-3-rutinoside) to be slower than that of aglycone and quercetin (Manach *et al.*, 1997). This phenomenon was explained by the idea that hydrolysis in the large intestine is required for the absorption of rutin in the digestive tract. However, Hollman *et al.* (1995) indicated from an experiment involving healthy ileostomy volunteers whose quercetin source resulted from the intake of quercetin glucoside-rich onion that quercetin glucosides are absorbed more easily than aglycone. His group also demonstrated that the quercetin (including metabolites and glycosides) level reached a maximum of 0.6 μM after the intake of fried onion (corre-

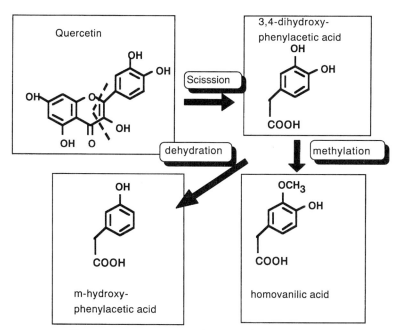

Figure 4 Decomposition of quercetin by enterobacteria.

sponding to 64 mg quercetin) in 2.9 hr (Hollman *et al.*, 1996). Paganga and Rice-Evans (1997) claimed that plasma from nonsupplemented humans contains quercetin and quercetin glycosides at the level of 0.5–1.6 μM. These contradictory results on the absorption of aglycone and glycosides call attention to the understanding of absorption of flavonol-type flavonoids from plant food (Fig. 5).

However, catechins have no glycosidic moiety, and several groups (Das, 1971; Unno *et al.*, 1996) have demonstrated that tea catechins are absorbed easily because of the lack of a sugar moiety. Efficiency of the absorption of flavonoids in the intestinal tract may depend on their individual structures, although there is little knowledge on the relationship between structure and intestinal absorption.

METABOLIC CONVERSION AND ENTEROHEPATIC CIRCULATION

Flavonoids absorbed from the intestinal tract are bound to albumin molecules and transported to the liver through the lymph duct or portal vein.

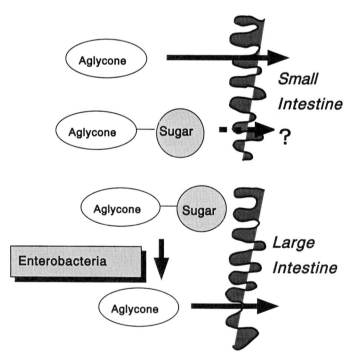

Figure 5   Absorption and metabolic pathway of aglycone and glycosides.

Flavonoids and their decomposition products are subjected to metabolic conversion in the liver, which includes methylatin of a hydroxyl group and reduction of a carbonyl group as well as conjugation with glucuronic acid and/or sulfuric acid (Fig. 6) (Hackett *et al.*, 1986). Manach *et al.* (1995) detected glucurono and sulfoconjugates in plasma of the rat administered quercetin and rutin. Furthermore, they found that conjugates of quercetin and isorhamnetin (3'-O-methylated quercetin) were present complexed with albumin in rat plasma and that their concentrations reached 115 μM when rats were fed a diet containing 0.25% (Manach, 1996), i.e., 20% of quercetin was absorbed from the digestive tract and the glucuronyl and sulfate conjugate of quercetin and isorhametin, 3'-O-methyl quercetin, was detected as glucurono and sulfoconjugates in bile and urine within 48 hr. These conjugates are excreted from urine as final metabolites and part of them is transferred to bile from the liver and finally excreted by the digestive tract (Ueno *et al.*, 1983). Such metabolites seem to be reabsorbed after hydrolysis and/or ring scission

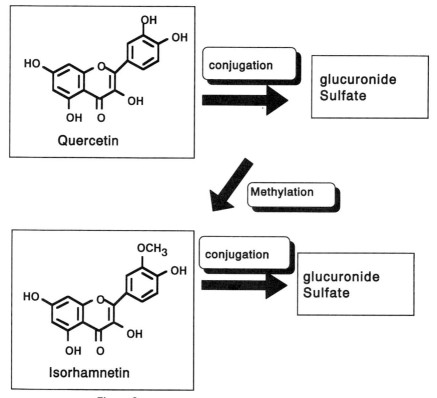

Figure 6 Proposed metabolic pathway of quercetin.

(enterohepatic circulation) (Hackett *et al.,* 1986). Thus, an appreciable amount of flavonoids from the diet can enter into the blood circulation as metabolites by this enterohepatic circulation mechanism.

It is generally believed that the liver is the main tissue for the metabolism of flavonoids. However, intestinal mucosa, kidney, and other tissues also possess enzymatic activity for metabolizing flavonoids, i.e., glucuronyl conjugation (UDP-glucuronyltransferase), O-methylation (O-methyltransferase), and hydroxylation (cytochrome P450) (Hackett and Griffiths, 1983). The authors measured UDP-glucuronyltransferase (UGT), phenosulfotransferase (PST), and catechol-O-methyltransferase (COMT) in the liver, kidney, lung, and intestinal mucosa (Piskula *et al.,* 1998). The highest UGT activity among all the tissues was found in the intestinal mucosa of the small and large intestines. In contrast, the highest activities for PST and COMT were found in the liver. It was therefore proposed that ingested flavonoids are subjected to glucuronidation in intestinal mucosa as the first step of metabolic conversion under normal conditions (Fig. 7).

Flavonoids enter the portal vein exclusively in the glucuronide-conjugated form. Next, following sulfation in the liver and methylation in the liver and kidney, metabolites are excreted from the body via bile or urine. This scheme is supported by the fact that the accumulation of (-)-epicatechin glucuronide preceded its sulfate or methylated products in the rat plasma after the administration of (-)-epicatechin. As described earlier because the antioxidant activity is partly associated with the number of OH groups and their structural relationship in molecules, it is likely that at least some of the antioxidant activity is lost during metabolic conjugation. The antioxidant activity of flavonoids has often been demonstrated in *in vitro* experimentation. However, because of their rapid and effective metabolic alteration, it is difficult to demonstrate such activity in *in vivo* systems.

METABOLITES AS ANTIOXIDANTS IN BLOOD PLASMA

The free radical scavenging activity of all flavonoids is essentially related to their phenolic hydrogens. However, following the oral intake of flavonoids such as quercetin and (-)-epicatechin, they occur in blood circulation mainly as conjugated derivatives, e.g., glucurono and sulfoconjugates with substituted phenolic hydrogen. Thus, it remains equivocal whether flavonoids retain any antioxidative properties after intestinal absorption and metabolic conversion. It has been found that orally administered (-)-epicatechin enhances the oxidative resistance of rat plasma (Edson *et al.,* 1998). In this experiment, the plasma pool was obtained 1 and 6 hr after intragastric (-)-epicatechin administration at 10 or 50 mg/rat and was

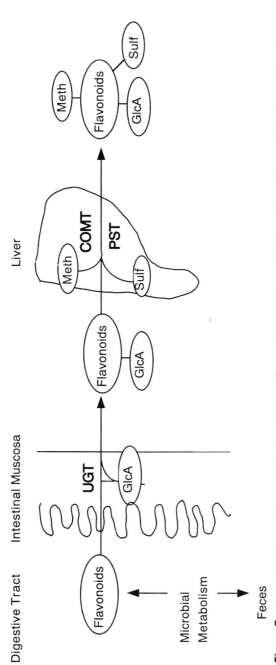

Figure 7 Proposed scheme of the metabolic fate of dietary flavonoids; COMT, catechol-O-methyltransferase; GlcA, glucuronide moiety; Meth, methyl moiety; PST, phenolsulfotransferase; Sul, sulfate moiety; UGT, uridine 5′-diphosphoglucuronosyltransferase.

exposed to copper sulfate or 2,2′-azobis(2-amidinopropane)dihydro-chloride (AAPH). Compared to the control group, plasma obtained from (-)-epicatechin administration was more resistant to lipid peroxidation because of cholesteryl ester hydroperoxide (CE-OOH) accumulation. The plasma level of nonmetabolized (-)-epicatechin is too low or not detectable to explain directly the effect of orally administered (-)-epicatechin. It is therefore likely that some plasma metabolites of (-)-epicatechin act as antioxidants when plasma is exposed to reactive oxygen radicals. As mentioned earlier, the *o*-dihydroxyl structure at the 3′ and 4′ positions in the B ring is the obvious target site for all flavonoids, including (-)-epicatechin. It is reported that O-methylation occurs at the 3′ position of the B ring in the process of metabolic conversion of (-)-epicatechin (Hackett, 1982). It is also suggested that the *o*-dihydroxyl structure in the B ring is involved in metal chelating by (-)-epicatechin (Afanas'ev *et al.*, 1989). Thus O-methylation seems to be responsible for the loss of antioxidant activity. The *in vivo* antioxidant activity of (-)-epicatechin after oral administration can be attributed to the conjugate or the conjugates containing the *o*-dihydroxyl structure.

In the case of quercetin, the 3′-O-methylated product (isorhamnetin) is found in blood plasma as the major metabolite. Similar to (-)-epicatechin, quercetin is essentially present as glucurono and sulfoconjugates of free quercetin and 3′-O-methylated quercetin. Thus, the conjugates of free quercetin seem to take a major part in the *in vivo* antioxidant activity of quercetin from plant foods.

Conjugation with glucuronic acid or sulfuric acid and O-methylation are common steps in the metabolic pathway of polyphenolic flavonoids (Cook *et al.*, 1996). However, some of the antioxidant activity of flavonoids still remains during the metabolic process and it may affect positively the antioxidative defense of blood plasma.

CONCLUSION

Polyphenolic flavonoids can exert inhibitory effects on lipid peroxidation by scavenging chain-initiating oxygen radicals as well as chain-propagating lipid peroxyl radicals. Because of their physical properties, flavonoids seem to be localized near the interface between the lipid phase and the water phase in heterogeneous biological systems. However, the efficiency of intestinal absorption and following metabolic conversion should be taken into account for the estimation of *in vivo* activity. It is suggested that flavonoids are already subjected to glucuronyl conjugation in the intestinal mucosa during absorption. Some conjugates containing a free phenolic group are likely to possess antioxidant activity to some extent in blood circulation.

REFERENCES

Afanas'ev, I. B., Dorozhko, A. I., Brodski, A. V., *et al.* (1989). Chelating and free radical scavenging mechanisms of inhibitory action of rutin and quercetin in lipid peroxidation. *Biochem. Pharmacol.* **38**, 1763–1769.

Bokkenheuser, V. D., Shacketon, C. H. L., and Winter, J. (1987). Hydrolysis of dietary flavonoide glycosides by strains of intestinal bacteroides from humans. *Biochem. J.* **248**, 953–956.

Booth, A. N., Murray, C. W., Junes, F. T., and DeEds, F. (1956). The metabolic fate of ratin and quercetin in the animal body. *J. Biol. Chem.* **233**, 251–257.

Bors, W., Heller, W., and Michel, C. (1990). Flavonoids as antioxidants: Determination of radical-scavenging efficiencies. *Methods Enzymol.* **186**, 343–355.

Cook, N. C., and Samman, S. (1996). Flavonoids: Chemistry, metabolism, cardioprotective effects, and dietary sources. *J. Nutr. Biochem.* **7**, 66–76.

Das, N. P. (1971). Studies on flavonoid metabolism; absorption and metabolism of (+)-catechin in man. *Biochem. Pharmacol.* **20**, 3435–3445.

Edson, L. S., Piskula, M., and Terao, J. (1998). Enhancement of antioxidative activity of rat plasma by oral administration of (-)-epicatechin. *Free Radic. Biol. Med.* **24**, 1209–1216.

Edson, L. S., Tsushida, T., and Terao, J. (1998). Inhibition of mammalian 15-lipoxygenase-dependent lipid peroxidation in low-density lipoprotein by quercetin and quercetin monoglucosides. *Arch. Biochem. Biophys.* **349**, 313–320.

Folcik, V. A., Nivar-Aristy, R. A., Krajewski, L. P., and Cathcart, M. K. (1995). Lipoxygenase contributes to the oxidation of lipids in human atherosclerotic plaques. *J. Clin. Invest.* **96**, 504–510.

Frankel, E. N., Kanner, J., German, J. B., *et al.* (1993). Inhibition of oxidation of human low-density lipoprotein by phenolic substances in red wine. *Lancet* **341**, 454–457.

Frankel, E. N., Waterhouse, A. L., and Pierre, L. (1995). Teissedre: Principal phenolic phytochemicals in selected california wines and their antioxidant activity in inhibiting oxidation of human low-density lipoproteins. *J. Agric. Food Chem.* **43**, 890–894.

Hackett, A. M. (1982). *In* "Plant Flavonoids in Biology and Medicine: Biochemical, Pharmacological and structure activity relationship" (V. Cody, E. Middleton, and J. B. Hardorne, eds.) Aran Press, New York.

Hackett, A. M., and Griffiths, L. A. (1983). The metabolism and excretion of (+)-[^{14}C]cyanidanol-3 in man following oral administration. *Xenobiotica* **13**, 279–286.

Hertog, M. G. L., Feskens, E. J. M., Hollman, P. C. H., *et al.* (1993a). Dietary antioxidant flavonoids and risk of coronary heart disease. *Lancet* **342**, 1007–1011.

Hertog, M. G. L., Holman, P. C. H., Katan, M. D., and Kromhout, D. (1993b). Intake of potentially anticarcinogenic flavonoids and their determinants in adults in the Netherland. *Nutr. Cancer* **20**, 19–21.

Hollman, P. C., Gaag, M., Mengelers, M. J. B., *et al.* (1996). Absorption and free disposition kinetics of the dietary antioxidant quercetin in man. *Free Radic. Biol. Med.* 21, 703–707.

Hollman, P. C. H., Vries, J. H. M., van Leeuwen, S. D., *et al.* (1995). Absorption of dietary quercetin glycosides and quercetin in healthy ileostomy volunteers. *Am. J. Clin. Nutr.* **62**, 1276–1282.

Jessup, W., Rankin, S. M., Whalley, C. D., *et al.* (1990). α-Tocopherol consumption during low-density-lipoprotein oxidation. *Biochem. J.* **265**, 399–405, 1990.

Kinsella, J. E., Frankel, E., German, B., *et al.* (1993). Possible mechanisms for the protective role of antioxidants in wine and plant foods. *Food Technol. April,* 85–89.

Knekt, P., Jarvinen, R., Reunanen, A., and Maatela (1996). Flavonoid intakes and coronary, mortality in Finland a cohort study *Br. Med. J.* **312**, 478–481.

Kühn, H., Belkner, J., Wiesner, R., Schewe, T., Lankin, V. Z., and Tikhaze, A. K. (1992). Structure elucidation of oxygenated lipids in human atherosclerotic lesions. *Eicosanoids* **5**, 17–22.

Lanningham-Foster, L., Chen, C., Chance, D. S. *et al.* (1995). Grape extract inhibits lipid per-oxidation of human low-density lipoprotein. *Biol. Phrm. Bull.* **18**, 1347–1351.

Manach, C., Morand, C., Taxier, O. Favier, M. C., Agullo, G., Deminge, C., Rederat, F., and Rèmèsy, C. (1995). Quercetin metabolites in plasma of rats fed diets containing ratin and quercetin. *J. Nutr.* **125**, 1911–1922.

Manach, C., Régérat, F., Taxier, P., Agullo, G., Demingne, C., and Rèmèsy, C., (1996). Bioavailability, metabolism and physiological impact of 4-oxa-flavonoids. *Nutr. Res.* **16**, 517–544.

Manach, C., Texier, O., Régérat, F., Agullo, G., Demingné, C., and Rèmèsy, C. (1996). Dietary quercetin is recovered in rat plasma as conjugated derivatives of isorhamnetin and quercetin. *J. Nutr. Biochem.* **7**, 375–380.

Miura, S., Watanabe, J., Sano, M., *et al.* (1995). Effects of various natural antioxidants on the Cu^{2+}-mediated oxidative modification of low-density lipoprotein. *Biol. Pharm. Bull.* **18**, 1–4.

Mora, A., Paya, M., Rios, J. L., *et al.* (1990). Structure-activity relationship of poly-methoxyflavons and other flavonoids as inhibitors of non-enzymatic lipid peroxidation. *Biochem. Pharmacol.* **40**, 793–797.

Mukai, K., Oka, W., Egawa, Y., *et al.* (1996). A kinetic study of the free-radical scavenging action of flavonoids in aqueous triton X-100 micellar solution. *In* "Proceedings of the International Symposium on Natural Antioxidants: Molecular Mechanisms and Health Effects" (L. Packer, M. G., Traber, and W. Xin, eds.), pp. 557–568. AOCS Press, Champaign, IL.

Nege-Salvayre, A., and Salvayre, R. (1988). Quercetin prevents the cytotoxixity of oxidized LDL on lymphoid cell lines. *Free Radic. Biol. Med.* **12**, 101–106.

Paganga, G., and Rice-Evans, C. A. (1997). The identification of flavonoids as glycosides in human plasma. *FEBS Lett.* **401**, 78–82.

Piskula, M., and Terao, J. (1998). Accumulation of (-)-epicatechin metabolites in rat plasma after oral administration and distribution of conjugated enzymes in rat tissues. **128**, 1172–1178. *J. Nutr.*, in press.

Ratty, A. K., and Das, N. P. (1988). Effects of flavonoids on nonenzymatic lipid peroxidation: Structure-activity relationship. *Biochem. Biomed. Metab. Biol.* **39**, 69–79.

Ratty, A. K., Sunamoto, J., and Das, N. P. (1988). Interaction of flavonoids with 1,1-dipheyl-2-picrylhydazyl free radical, liposomal membranes and soybean lipoxygenase-1. *Biochem. Pharmacol.* **37**, 989–995.

Rice-Evans, C. A., Miller, N. J., and Paganga, G. (1996). Structure-antioxidant activity relationships of flavonoids and phenolic acids. *Free Radic. Biol. Med.* **20**, 933–956.

Saija, A., Scalese, M., Lanza, M., *et al.* (1995). Flavonoids as antioxidant agents: Importance of their interaction with biomembranes. *Free Radic. Biol. Med.* **19**, 481–486.

Steinberg, D., Parthasarathy, S., Carew, T. E., Khoo, J. C., and Witztum, J. L. (1989). Beyond cholesterol: Modification of low-density lipoprotein that increase its atherogenicity. *N. Engl. J. Med.* **320**, 915–924.

Tamura, G., Gold, C., Ferr-Luzi, A., and Ames, B. N. (1981). *Proc. Natl. Acad. Sci. USA* **77**, 4961.

Terao, J., Piskula, M., and Yao, Q. (1994). Protective effect of epicatechin, epicatechin gallate, and quercetin on lipid peroxidation in phospholipid bilayers. *Arch. Biochem. Biophys.* **308**, 278–284.

Ueno, I., Nakawa, N., Nirono, I. J. (1983). Metabolic fate of [^{14}C]quercetin in the AC1 rat. *Japan J. Exp. Med.* **53**, 41–50.

Unno, T., and Takeo, T. (1995). Absorption of (-)-epigallocatechin gallate into the circulation system of rats. *Biosci. Biotech. Biochem.* **59**, 1558–1559.

van Acker, S. A. B. E., van den Berg, D.-J., and Tromp, M. N. J. L. (1996). Structural aspects of antioxidant activity of flavonoids. *Free Radic. Biol. Med.* **20**, 331–342.

Ylä-Herttuala, S., Rosenfeld, M.. E., Parthasarathy, S., Sigal, E., Sarkioja, T., Witztum, J. L., and Steinberg, D. (1990). Colocalization of 15-lipoxygenase mRNA and protein with epitopes of oxidized low density lipoprotein in macrophase-rich areas of atherosclerotic lesions. *Proc. Natl. Acad. Sci. USA* **87**, 6959–6963.

18 Flavonoids in Foods

Gary R. Beecher

Food Composition Laboratory
Beltsville Human Nutrition Research Center
Agricultural Research Service, USDA
Beltsville, Maryland 20705

Many commonly consumed fruits, vegetables, and beverages contain various classes of flavonoids and in varying quantities. Most fruits that are red or purple, such as red grapes, cherries, and blueberries, have substantial quantities of several anthocynanidins. Many of these same foods also contain flavonols, primarily quercetin. Catechins (flavan-3-ols) are present in red grapes and their processed product, red wine, as well as in other red fruits, such as apples. Citrus foods and juices have a unique class of flavonoids, flavanones (hesperetin and naringenin), as well as quercetin. Onions have the highest concentration of quercetin (about 300 mg/kg food) of any common food that has been analyzed. Other vegetables that contain quercetin or kaempferol, but at much lower quantities, include broccoli and kale. Teas provide high concentrations of catechins and small amounts of the flavonol quercetin. Most flavonoids in foods are conjugated to a carbohydrate moiety; the type of carbohydrate and its linkage to flavonoids is varied, which leads to many different molecular structures. A variety of dimers, trimers, oligomers, and polymers of flavonoids are present in many foods. There is a dearth of comprehensive data on the flavonoid content of foods. This is due primarily to the lack of a single, robust analytical system that measures all of the prominent food flavonoids, as aglycones, in a single run. Development of such a system, comprehensive analysis of foods, and subsequent summarization of food flavonoid data into a convenient database will provide the appropriate tools to ascertain the impact of dietary flavonoids on health maintenance and disease prevention.

INTRODUCTION

More than 4000 chemically unique flavonoids have been identified in plants. These low molecular weight compounds are found in fruits, vegetables, nuts, seeds, and flowers, as well as in several beverages, and are important constituents of the human diet. They have important effects in plant biochemistry, acting as antioxidants, enzyme regulators, precursors of toxic substances, pigments, and light screens, to name a few (Middleton and Kandaswami, 1994). Selected flavonoids have been shown in numerous *in vitro* and *in vivo* experiments to have antiallergic, anti-inflammatory, antiviral, and antioxidant activities (Middleton, 1996). In addition, some flavonoids have been shown to exert significant anticancer activity, including anticarcinogenic and prodifferentiative activities (Attaway, 1994). Flavonoid intake has been shown to be inversely related to cardiovascular disease (CVD) risk in epidemiologic studies conducted in the Netherlands and Finland. However, similar studies conducted in the United States and the United Kingdom have demonstrated either no association or a positive relationship (Katan, 1997). Nonetheless, a considerable body of evidence suggests that plant flavonoids may be health-promoting, disease-preventing dietary compounds (Middleton, 1996).

The prominent flavonoids in foods are characterized by several subclasses, including anthocyanidins, flavanols, flavonones, flavones, flavonols, and their metabolic precursors, chalcones (Robinson, 1963; Kuhnau, 1976). The general structure of flavonoids is two benzene groups connected by a three-carbon (propane) bridge. With the exception of chalcones, all flavonoids found in foods have a pyran ring (oxygen-containing heterocyclic ring), which is formed by the addition of oxygen to position 2 of chalcones and subsequent cyclization of the three-carbon chain with the "A" ring (Fig. 1). The various subclasses of flavonoids are derived from this basic structure by changing the oxidation state and substitution (primarily hydroxylation) of the propane portion of the molecule. Thus, flavanols are at the lowest oxidation state (saturated pyran ring) but are substituted with a hydroxyl group at position three (flavan-3-ols or catechins), whereas flavones and flavonols (3-OH flavones) are the most highly unsaturated with a 2,3 double bond and a keto group at carbon 4. All flavonoids have hydroxyl groups at positions 4′, 5, and 7 and the corresponding positions of chalcones. Within each subclass, specific flavonoid molecules are identified based on additional substitutions. For example, quercetin is a flavonol with an additional hydroxyl substitution at the 3′ position (3,5,7,3′,4′-pentahydroxy flavone) and catechin is also a flavan-3-ol with an additional hydroxyl substitution at the 3′ position (3,5,7,3′,4′-pentahydroxy flavan). There are a limited number of flavonoids within each class that are promi-

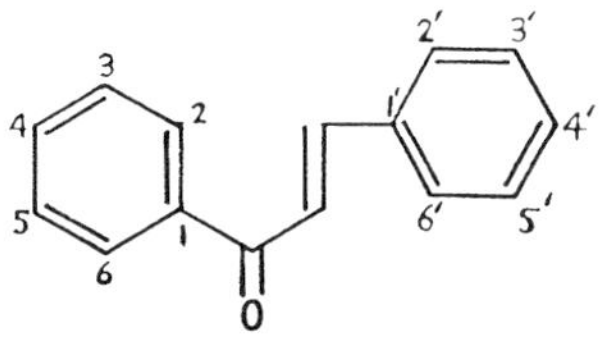

Chalcone

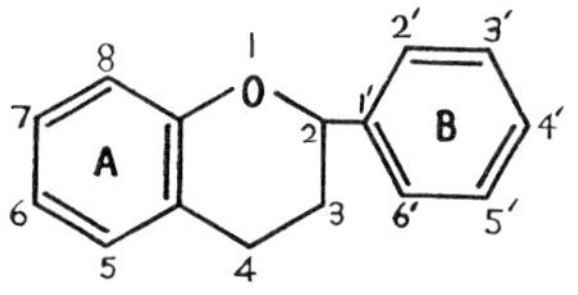

Flavonoid

Figure 1 Basic structures and numbering systems for chalcones and flavonoids.

nent in plant foods commonly consumed by human beings. These include about three anthocyanidins (cyanadin, delphinidin, malvidin), three flavan-3-ols (catechin, epicatechin, epigallocatechin), two flavanones (hesperetin, naringenin), two flavones (apigenin, luteolin), three flavonols (kaempferol, myricetin, quercetin), and three chalcones.

In nature, particularly in foods, most flavonoids are conjugated to mono- and disaccharides. Thus one of the most common forms of quercetin in foods is rutin, which is quercetin-3-O-β-rutinoside where rutinoside is a disaccharide of rhamnose and glucose (6-O-α-L-rhamnosyl-D-glucose). Other derivatives of flavonoids include glucose, galactose, methoxyl, and gallate. The 3 hydroxyl of flavonoids is the most commonly conjugated position; in its absence, as in flavones and flavanones, the 7 position is often conjugated, i.e., narirutin. In addition, double conjugates of selected flavonoids have been identified i.e., quercetin-3,4'-O-bis-β-D-glucoside in foods such as tea (Bailey *et al.*, 1990).

In several foods and beverages, flavonoids have been identified as dimers, oligomers, and polymers. The most prominent of these in commonly consumed foods are the procyanidins, a subclass of proanthocyanidins, characterized by catechin monomers linked through carbon–carbon bonds to form dimers, oligomers, and polymers (Porter, 1994). Because it is subjected to oxidizing conditions during production, black tea contains several unique dimers, i.e., theaflavins, theacitrins, and bisflavanols, as well as a large family of polymers, the thearubigens (Graham, 1992; Davis *et al.*, 1997). The biological importance of these complex flavonoids is unknown.

Nonetheless, they often constitute a large proportion of the flavonoid content of a food or beverage.

FLAVONOID CONTENT OF FOODS AND BEVERAGES

As outlined earlier, the flavonoid content of foods and beverages is very complex. Consequently, this section has been divided into three subsections in order to permit a more definitive discussion of their flavonoid content.

Flavonoid Aglycone Content

Flavonoids, as well as other compounds, devoid of their carbohydrate moieties are referred to as aglycones. The primary biological activities of flavonoids are resident in their aglycone forms; polymerization or conjugation with carbohydrates may attenuate the biological response. Flavonoids are widespread in fruits, vegetables, and beverages such as tea. The distribution of the prominent flavonoid subclasses of these foods is shown in Tables I–IV. The flavonoid content for only selected examples of each food category is shown. There is a dearth of data on the complete flavonoid profile for foods. In general, the anthocyanadin concentration in fruits is the highest of all the flavonoids (Table I). Anthocyanadins, at the pH of most

Table I Anthocyanadin Content of Selected Fruits, Vegetables, and Beverages[a]

Food/beverage	Total anthocyanidin [mg/100 g (ml)]
Fruits	
Blueberry	· 25–495
Cherry, sweet	350–450
Grapes, red	30–750
Vegetables	
Cabbage, red	25
Onion, red	9–21
Beverages	
Cranberry juice	2–9
Wine, red	10–100

[a] Values expressed on a fresh weight basis or as consumed [mg/100 g(ml)]. Data adapted from Wang *et al.* (1997).

fruits and vegetables, give those foods a red or purple color. Thus a good indication of the presence of anthocyanadins in a food is the intensity of its red to purple color, i.e., blueberries, red cabbage, and red wine (Table I). An exception is lycopene, which also contributes red color, but to a limited number of foods. These include tomatoes and tomato products, watermelon, and pink grapefruit. Because of their molecular structure, all other flavonoids in commonly consumed foods are colorless. Anthocyanadins are not prominent in those vegetable foods that conduct photosynthesis, i.e., kale, broccoli, and so on, or in citrus fruits, i.e., oranges, grapefruit, and so on.

Flavanols, a subclass of flavonoids found in foods in nearly as high a concentration as the anthocyanadins, are characterized by catechin, epicatechin, and similar compounds. These flavonoids are found in a limited number of foods and beverages (Table II). It is interesting to note that black tea contains about one-half the concentration of flavanols as green tea. This may be due to the oxidative formation of flavanol dimers, oligomers, and polymers during the processing of black tea (see Section II).

The flavonol content of selected fruits, vegetables, and beverages is tabulated in Table III. This subclass of flavonoids is characterized primarily by kaempferol, myricetin, and quercetin. Flavonols are distributed widely in fruits, vegetables, and beverages such as tea. The concentrations, however, of flavonols in foods are substantially lower than other flavonoid subclasses. As a result, several portions of vegetables would need to be consumed to equal the intake of flavonoids in a single portion of deeply colored fruit. Nonetheless, quercetin is one of the most comprehensively studied

Table II Flavanol Content of Selected Beverages[a]

Beverage	Total flavanol[b] (mg/100 ml)
Tea, black[c]	9–349
Tea, green[c]	16–719
Wine, red[d]	5–19

[a] Values are expressed as beverages are consumed (mg/100 ml).
[b] Total flavanol is the sum of the concentrations of epicatechin, epigallocatechin, epicatechin-3-gallate, and epigallocatechin-3-gallate.
[c] Data from summary in Bronner and Beecher (1998).
[d] Data from Goldberg *et al.* (1996).

Table III   Flavonol Content of Selected Fruits, Vegetables, and Beverages[a]

Food/beverage	Total flavonol[b] [mg/100 g(ml)]
Fruits	
Blueberry[c]	2–3
Cherry, sweet	2
Vegetables	
Broccoli	10
Kale	11
Onion	35
Beverages	
Tea, black	2–5
Tea, green	3–5
Wine, red[d]	1–10

[a] Values expressed as foods and beverages are consumed [mg/100 g(ml)]. Data from Hertog et al. (1992a, 1993b), except where noted.
[b] Total flavonol is the sum of the concentrations of kaempferol, myricetin, and quercetin expressed as their aglycones.
[c] Data from Bilyk and Sapers (1986).
[d] Additional data incorporated from Goldberg et al. (1996).

flavonoids relative to food chemistry and biological activity. Two closely related flavones, apigenin and luteolin, are often discussed with the common food flavonols. However, these flavonoids have been identified and quantified only in a limited number of foods, i.e., celery and red bell peppers (Hertog et al., 1992a,b).

Citrus fruits and juices contain a unique subclass of flavonoids, the flavanones (Table IV). The prominent flavanones in these foods are hesperidin, naringin, and narirutin as well as their corresponding aglycones, hesperetin and naringenin. The concentration of flavanones in citrus foods, in general, is about the same as flavonols in vegetables but is somewhat lower than the level of anthocyanidins in fruits and flavanols in teas. Citrus foods contain several other flavonoids in qualitative and quantitative patterns that are unique to each plant species. These characteristics have been suggested as a sensitive tool for use in citrus juice authentication and in detection of adulteration (Mouly et al., 1994; Robards et al., 1997). In general, these unique flavonoids are found at quite low concentrations and, as a result, are minor

Table IV   Flavanone Content of Selected Fruit Juices[a]

Fruit juice	Total flavanone[b] (mg/100 ml)]
Grapefruit juice	8–43
Orange juice	4–54

[a] Values expressed as juices are consumed (mg/100 ml). Data from summary in Bronner and Beecher (1995).
[b] Total flavanone is the sum or the concentrations of hesperidin, naringin, and narirutin expressed as their aglycones, hesperetin, and naringenin, respectively.

contributors to the total daily flavonoid intake from commonly consumed foods.

Data tabulated in Tables I–IV show large ranges in the flavonoid content of foods. Flavonoid metabolism in plants is sensitive to many factors, including species, cultivar, fertility, and amount of sunlight. Many flavonoid-containing foods are grown in the United States, but many are also imported from countries where unique cultivars may be grown and environmental conditions may vary. As a result, the flavonoid content may also be unique, which contributes to the variation of tabulated data.

Flavonoid Conjugates

Most flavonoids occur in living cells almost exclusively as glycosides and primarily as O-glycosides (Herrman, 1976). The preferred bonding site of the sugar moiety to flavonoids is the 3 position, if it is hydroxylated, or the 7 position in the absence of a 3-OH group (flavanones and flavones). A limited number of common foods contain bisglycosides as the prominent form of flavonoids. D-Glucose is the most common sugar residue conjugated to flavonoids, but galactose, rhamnose, arabinose, xylose, and the disaccharide, rutin, have also been identified. To complicate these structures even more, acylation of sugars occurs with acetic, benzoic, hydroxycinnamic, hydroxybenzoic, and malonic acids as well as other acids. The formation of glycosides generally depends on light. As a result, the highest concentrations of these compounds generally occur in leaves and other structures exposed to light while only traces are found in parts of the plant below the soil surface or shaded from light. Onions are an exception in that a majority of the flavonoids in this bulb occur as glycosides.

The qualitative flavonoid glycoside content of fruits and vegetables has been reviewed (Herrman, 1976; Robards and Antolovich, 1997). The structures of several typical quercetin glycosides are shown in Fig. 2. Quercetin-4'-O-β-D-glucoside and quercetin-3-4'-O-bis-β-D-glucoside are the primary glycosylated forms of quercetin in onions (Price and Rhodes, 1997). Quercetin-3-O-β-D-galactoside is one of several glycosylated forms of quercetin in apples (Lister *et al.*, 1994) whereas quercetin-3-O-β-rutinoside is a major constiuent of tea, apples, and wine (Hollman, 1997). Similar gly-

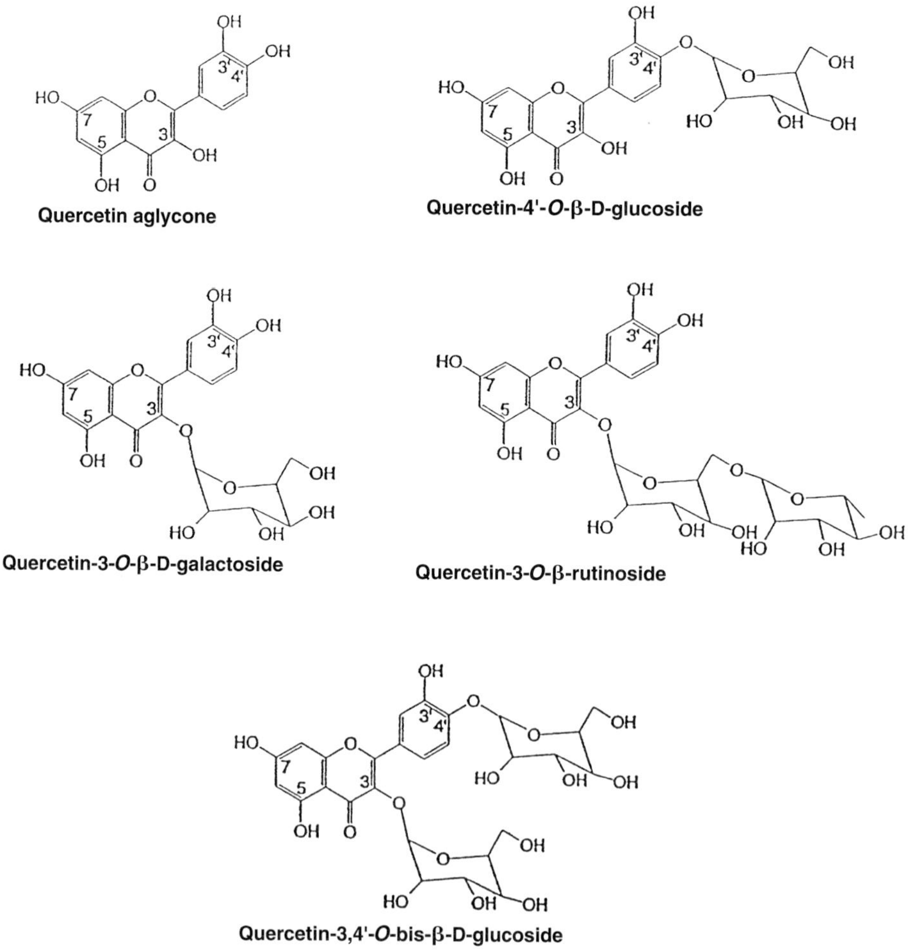

Quercetin aglycone

Quercetin-4'-*O*-β-D-glucoside

Quercetin-3-*O*-β-D-galactoside

Quercetin-3-*O*-β-rutinoside

Quercetin-3,4'-*O*-bis-β-D-glucoside

Figure 2   Typical quercetin glycosides found in foods.

cosides have been identified for other flavonoids prominent in commonly consumed foods. Research by Lister *et al.* (1994) as well as Price and Rhodes (1997) provides examples of the limited quantitative analyses that have been conducted in this complex area.

Flavonoid Dimers, Oligomers, and Polymers

A large number of dimers, trimers, and oligomers of flavonoids characterized by carbon–carbon linkages have been isolated from plant sources. Proanthocyanidins are a general class of these compounds which contain unique flavonoids and have characteristic linkages (Porter, 1994). Procyanidins are a subclass of proanthocyanidins and contain monomeric units of catechin linked from carbon four of one monomer to carbon six or carbon eight (usual linkage) of a second monomer. Procyanidins have been identified in such commonly consumed foods as apples, grapes, tea, peanuts, wine, strawberries, cinnamon, and cocoa (see Hammerstone *et al.*, 1998). The structure of a procyanidin dimer (B1) is shown in Fig. 3. The stereochemistry of the hydroxyl at carbon 3 of each catechin and the carbon–carbon linkage of the monomers as well as the number of monomers in the molecule determine the procyanidin number. Modern chromatographic techniques [high-performance liquid chromatography (HPLC)] have been used to separate and quantify procyanidin dimers and trimers in foods (Lunte *et al.*, 1988; Lister *et al.*, 1994). Hammerstone *et al.* (1998) have separated and identified dimers through decamers of procyanidins in extracts of cocoa and chocolate employing a normal-phase HPLC interfaced to an atmospheric pressure ionization–electrospray (API-ES) mass spectrometry system. Advances in instrumentation, such as these, will ultimately permit routine analysis of foods for these important flavonoids.

Black tea contains several classes of flavanol-derived dimers that result from the oxidative conditions present during processing. One of the most prominent subclasses of dimers are the theaflavins (Fig. 3). They constitute 3–6% of the solids of black tea, contribute to its quality, and are formed by the condensation and rearrangement of the oxidized "B" rings of epicatechin and epigallocatechin (Graham, 1992; Balentine *et al.*, 1997). The 3-O-gallate derivatives of these two catechins, epicatechin-3-gallate and epigallocatechin-3-gallate, undergo analogous reactions to form theaflavin gallates. A new subclass of dimers, the theacitrins, have been identified and characterized in black tea by British scientists (Davis *et al.*, 1997). These dimers are characterized by a triple ring system (Fig. 3) and are postulated to be derived from the condensation of two oxidized gallate-containing flavanols, epigallocatechin or epigallocatechin-3-gallate, followed by intramolecular cyclization, rearrangement, and hydration (Davis *et al.*, 1997). A series of bisflavanols, the theasinensins, characterized by C–C linkages of

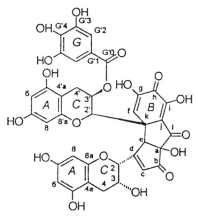

Procyanidin B1

Theaflavin

Theacitrin A

Figure 3 Examples of flavonoid dimers identified in foods and beverages.

the 6′ positions of epigallocatechins have also been identified in black tea (Graham, 1992). Their concentrations are relatively low and, as a result, probably contribute little to tea quality. Other bisflavonoids have also been identified in foods and medicinals (Packer *et al.*, 1998). Several other subclasses of flavanol dimers have been identified in oolong tea, a tea produced with short periods of oxidation (Graham, 1992; Balentine *et al.*, 1997).

Black tea is also a rich source of flavonoid polymers, the thearubigens. These complex, water-soluble, large molecules (1–40 kDa) constitute 12–18% of the solids of black tea and are also postulated to be formed during the oxidative conditions of black tea processing (Graham, 1992). They form insoluble complexes with solutions of caffeine and iron, a characteristic used as a qualitative test for these compounds. The definitive structures of thearubigens are unknown.

DIETARY INTAKE OF FLAVONOIDS

An often reported value of about 1 g per day average daily intake of flavonoids from foods in the United States was calculated by Kuhnau (1976). It is important to point out that this investigator included the consumption of biflavans (flavonoid dimers), estimated at 460 mg per day, in the overall intake of flavonoids. The intake of the remaining monoflavonoids was about equally distributed among anthocyanins, catechins, and 4-oxoflavonoids (flavanones, flavones, flavonols). The analysis of the flavonoid content of foods employing sensitive analytical techniques and the development of databases that summarize these values have resulted in more accurate estimates of flavonoid consumption. Hertog *et al.* (1993) estimated the intake of three flavonols (kaempferol, myrisiten, quercetin) and two flavones (apigenin, luteolin) at 23 mg per day for a group of elderly Dutch. Bronner and Beecher (1995) calculated the average intake of flavanones from citrus juices by Americans to be about 23 mg per day. Dragsted *et al.* (1997) carefully evaluated the consumption of all common classes of flavonoids from foods in Denmark. They estimated that the Danish diet provides about 100 mg per day of flavonoids. In general, these values suggest that the average daily consumption of flavonoids from foods is considerably lower than 1 g per day originally estimated by Kuhnau (1976).

The foregoing discussion has been limited to flavonoid consumption from foods. Another significant source of dietary flavonoids may be "naturally derived" supplements, such as *Ginkgo bilboa*, Pyncnogenol, and St. John's wort. At this time there is a dearth of data on the accurate consumption of these types of products and only limited information on their flavonoid content. These circumstances greatly complicate the estimation of flavonoid intake from a rapidly growing market of health-related products.

RESEARCH NEEDS

The association between flavonoid intake and human health effects ultimately can be tested only when detailed knowledge of the flavonoid content of foods, beverages, and dietary supplements is known and tabulated in convenient, easily accessible databases. Heretofore, measurement of the flavonoid content of foods has been part of food science and horticultural studies that investigated the role of production and food processing on the levels of these important plant constituents. As a result, analytical systems have been developed for each subclass of flavonoids, which requires several analyses of a food when it contains several subclasses of these important phytonutrients. A robust analytical system is required that will accurately and precisely measure all of the prominent flavonoids observed in commonly consumed plant foods and dietary supplements. Finally, data on the flavonoid content of foods and other dietary components must be tabulated and summarized in a centrally located and easily accessible database. Such a database must contain relevant statistical information about flavonoid data as well as sufficient details about the food or natural product from which data were derived so that it can be unambiguously identified and easily merged with other food composition databases.

REFERENCES

Attaway, J. A. (1994). Citrus juice flavonoids with anticarcinogenic and antitumor properties. *In* "Food Phytochemicals for Cancer Prevention. I. Fruits and Vegetables" (M. T. Huang, T. Osawa, C. T. Ho, and R. T. Rosen, eds.), ACS Symposium Series 546, pp. 240–248. American Chemical Society, Washington, DC.

Bailey, R. G., McDowell, I., and Nursten, H. E. (1990). Use of HPLC photodiode-array detector in a study on the nature of black tea liquor. *J. Sci. Food Agric.* **52**, 509–525.

Balentine, D. A., Wiseman, S. A., and Bouwens, S. C. M. (1997). The chemistry of tea flavonoids. *Crit. Rev. Food Sci. Nutr.* **37**, 693–704.

Bilyk, A., and Sapers, G. M. (1986). Varietal differences in the quercetin, kaempferol, and myricetin contents of highbush blueberry, cranberry, and thornless blackberry fruits. *J. Agric. Food Chem.* **34**, 585–588.

Bronner, W. E., and Beecher, G. R. (1995). Extraction and measurement of prominent flavonoids in orange and grapefruit juice concentrates. *J. Chromatogr.* **705**, 247–256.

Bronner, W. E., and Beecher, G. R. (1998). Method for determining the content of catechins in tea infusions by high-performance liquid chromatography. *J. Chromatogr. A* **805**, 137–142.

Davis, A. L., Lewis, J. R., Cai, Y., Powell, C., Davies, A. P., Wilkins, J. P. G., Pudney, P., and Clifford, M. N. (1997). A polyphenolic pigment from black tea. *Phytochemistry* **46**, 1397–1402.

Dragsted, L. O., Strube, M., and Leth, T. (1997). Dietary levels of plant phenols and other nonnutritive components: Could they prevent cancer. *Eur. J. Cancer Prev.* **6**, 522–528.

Goldberg, D. M., Tsang, E., Karumanchiri, A., Diamandis, E. P., Soleas, G., and Ng, E. (1996). Method to assay the concentrations of phenolic constituents of biological interest in wines. *Anal. Chem.* **68**, 1688–1694.

Graham, H. N. (1992). Green tea composition, consumption, and polyphenol chemistry. *Prev. Med.* **21**, 334–350.

Hammerstone, J. F., Lazarus, S. A., Mitchell, A. E., Rucker, R., and Schmitz, H. H. (1998).

Identification of procyanidins in cocoa and chocolate using high performance liquid chromatography/mass spectrometry. Submitted for publication.

Herrmann, K. (1976). Flavonols and flavones in food plants: A review. *J. Food Technol.* **11,** 433–448.

Hertog, M. G. L., Feskens, E. J. M., Hollman, P. C. H., Katan, M. B., and Kromhout, D. (1993a). Dietary antioxidant flavonoids and risk of coronary heart disease: The Zupten Elderly Study. *Lancet* **342,** 1007–1011.

Hertog, M. G. L., Hollman, P. C. H., and Katan, M. B. (1992a). Content of potentially anticarcinogenic flavonoids of 28 vegetables and 9 fruits commonly consumed in the Netherlands. *J. Agric. Food Chem.* **40,** 2379–2383.

Hertog, M. G. L., Hollman, P. C. H., and van de Putte, B. (1993b). Content of potentially anticarcinogenic flavonoids of tea infusions, wines, and fruit juices. *J. Agric. Food Chem.* **41,** 1242–1246.

Hertog, M. G. L., Hollman, P. C. H., and Venema, D. P. (1992b). Optimization of a quantitative HPLC determination of potentially anticarcinoghic flavonoids in vegetables and fruits. *J. Agric. Food Chem.* **40,** 1591–1598.

Katan, M. B. (1997). Flavonoids and heart disease. *Am. J. Clin. Nutr.* **65,** 1542–1543.

Kuhnau, J. (1976). The flavonoids: A class of semi-essential food components: Their role in human nutrition. *World Rev. Nutr. Diet* **24,** 117–191.

Lister, C. E., Lancaster, J. E., and Sutton, K. H. (1994). Developmental changes in the concentration and composition of flavonoids in skin of a red and green apple cultivar. *J. Sci. Food Agric.* **64,** 155–161.

Lunte, S. M., Blankenship, K. D., and Read, S. A. (1988). Detection and identification of procyanidins and flavanols in wine by dual-electrode liquid chromatography–electrochemistry. *Analyst* **113,** 99–102.

Middleton, E., Jr. (1996). Biological properties of plant flavonoids: An overview. *Intl. J. Pharmacognosy* **34,** 344–348.

Middleton, E., Jr., and Kandaswami, C. (1994). The impact of plant flavonoids on mammalian biology: Implications for immunity, inflammation and cancer. *In* "The Flavonoids: Advances in Research Since 1986" (J. B. Harborne, ed.), pp. 619–652. Chapman & Hall, New York.

Mouly, P. P., Arzouyan, C. R., Gaydou, E. M., and Estienne, J. M. (1994). Differentiation of citrus juices by factorial discriminant analysis using liquid chromatography of flavanone glycosides. *J. Agric. Food Chem.* **42,** 70–79.

Packer, L., Saliou, C., Droy-Lefaix, M.-T., and Christen, Y. (1998). Ginkgo biloba extract Egb 761: Biological actions, antioxidant activity, and regulation of nitric oxide synthase. *In* "Flavonoids in Health and Disease" (C. A. Rice-Evans and L. Packer, eds.), pp. 303–341. Dekker, New York.

Porter, L. J. (1994). Flavans and proanthocyanidins. *In* "The Flavonoids: Advances in Research Since 1986" (J. B. Harborne, ed.), pp. 23–55. Chapman & Hall, New York.

Price, K. R., and Rhodes, M. J. C. (1997). Analysis of the major flavonol glycosides present in four varieties of onion (*Allium cepa*) and changes in composition resulting from autolysis. *J. Sci. Food Agric.* **74,** 331–339.

Robards, K., and Antolovich, M. (1997). Analytical chemistry of fruit bioflavonoids; a review. *Analyst* **122,** 11R–34R.

Robards, K., Li, X., Antolovich, M., and Boyd, S. (1997). Characterization of citrus by chromatographic analysis of flavonoids. *J. Sci. Food Agric.* **75,** 87–101.

Robinson, T. (1963). "The Organic Constituents of Higher Plants." Burgess, Minneapolis.

Wang, H., Cao, G., and Prior, R. L. (1997). Oxygen radical absorbing capacity of anthocyanins. *J. Agric. Food Chem.* **45,** 304–309.

19 Dietary Flavonoids and Interaction with Physiologic Antioxidants

Piergiorgio Pietta* and Paolo Simonetti†

*ITBA-CNR
Milan, Italy

†diSTAM
University of Milan
Milan, Italy

INTRODUCTION

Evidence continues to emerge suggesting that fruits and vegetables are protective against oxidative diseases (Rice-Evans *et al.*, 1995). In addition to antioxidant vitamins and minerals, fruits and vegetables contain flavonoids and related phenolics, which are currently the subject of much interest (Rice-Evans and Packer, 1997; Blot *et al.*, 1996). Dietary flavonoids include various classes with specific structural characteristics (that are closely related to their activities) and different prevalence in our diet.

The relative importance of these potentially active components of our diet is not completely understood, and knowledge of their kinetics (absorption, metabolism, distribution, and excretion) and dynamics (effects on biological parameters) in humans is important in properly evaluating their potential.

This chapter attempts to provide evidence on absorption and metabolism of catechins and flavonols and on the relationship between kinetic data and the modifications of plasma parameters that are indicative of the antioxidant status.

Antioxidant Food Supplements in Human Health

FLAVONOID OCCURRENCE IN FOODS AND BEVERAGES

The most important polyphenols in foods and beverages are flavonoids, which consist mainly of catechins, proanthocyanidins, anthocyanidins, flavonols, flavones, flavanones, and their glycosides.

Catechins are distributed widely in plants; however, they are only rich in tea leaves, where catechins may constitute up to 25% of dry leaf weight. Catechins of green tea include the flavanols epicatechin, epigallocatechin, and their gallate esters (Fig. 1). During fermentation in the preparation of black tea, oxidative polymerization of flavanols occurs with the formation of theaflavin, theaflavingallates, thearubigins, and epitheaflavic acid (Fig. 2) (Graham, 1992).

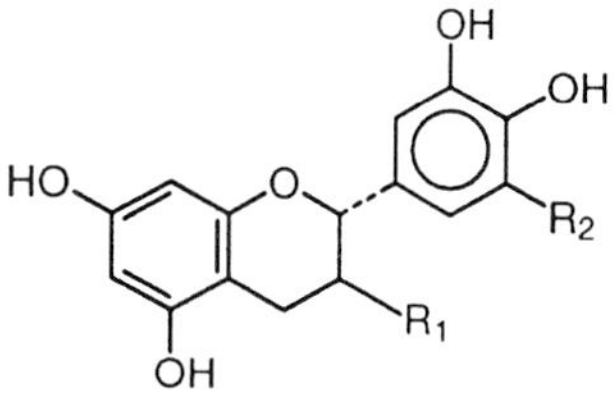

Catechins

Compound		R_1	R_2	M.W.
Epicatechin	EC	OH	H	290.3
Epigallocatechin	EGC	OH	OH	306.4
Epicatechin-gallate	ECg	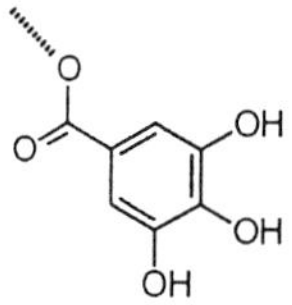 (gallate ester)	H	442.4
Epigallocatechin-gallate	EGCg	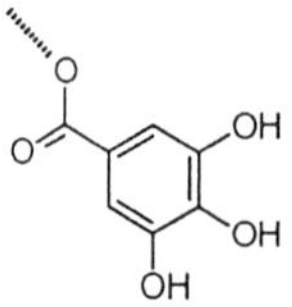 (gallate ester)	OH	458.4
Catechin	C	OH	H	290.3

Figure 1   Structure of catechins.

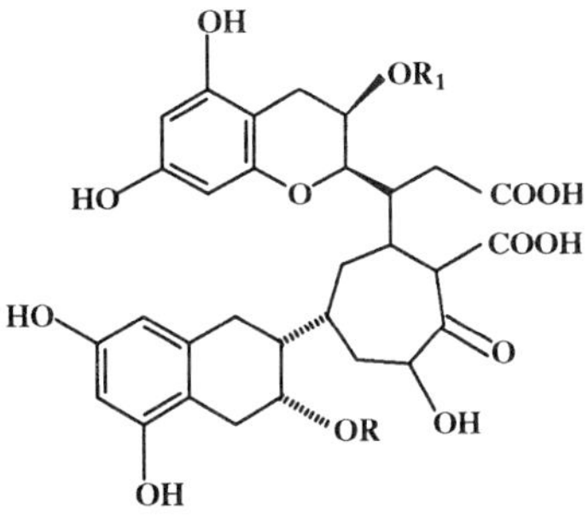

Thearubigin and related compounds

Compound	R	R₁
Thearubigin	H	H
Thearubigin-3-gallate	galloyl	H
Thearubigin-3'-gallate	H	Galloyl
Thearubigin-3,3'-digallate	galloyl	Galloyl

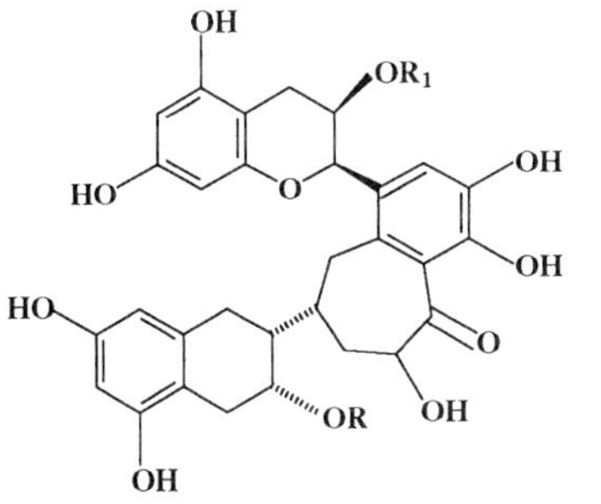

Theaflavin and related compounds

Compound	R	R₁
Theaflavin	H	H
Theaflavin-3-gallate	galloyl	H
Theaflavin-3'-gallate	H	galloyl
Theaflavin-3,3'-digallate	galloyl	galloyl

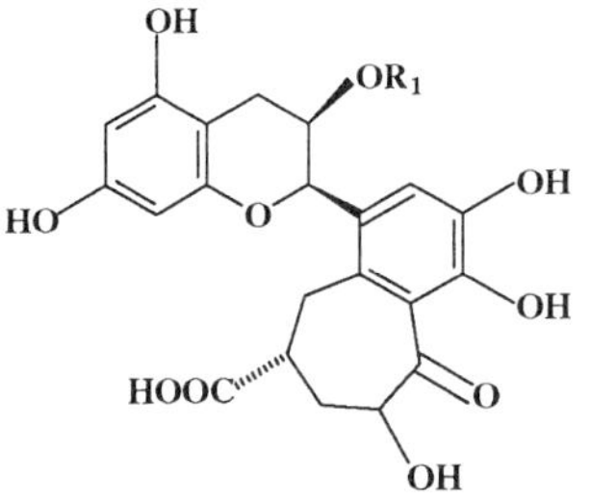

Epitheaflavic acid and derivatives

Compound	R₁
Epitheaflavic acid	H
Epitheaflavic gallate	gallate

Figure 2 Structures of thearubigin, theaflavin, epitheaflavic acid, and related compounds.

Proanthocyanidins (Fig. 3) are oligomers consisting of flavan-3-ol units. They are distributed largely in foods such as apples, grapes, strawberries, plums, and barley (Haslam, 1989). Proanthocyanidins are also present in grape seeds (Maffei Facino *et al.*, 1994) and in *Pinus maritima* bark (Masquelier, 1965).

Proanthocyanidins

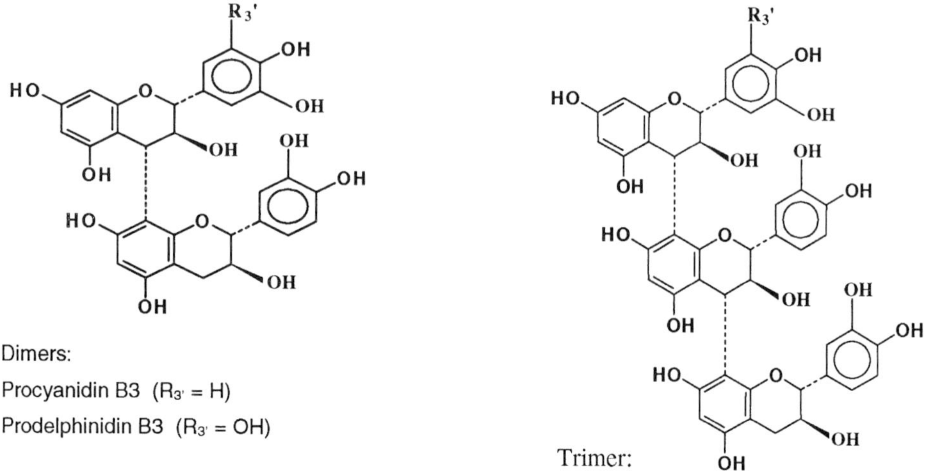

Dimers:

Procyanidin B3 ($R_{3'}$ = H)

Prodelphinidin B3 ($R_{3'}$ = OH)

Trimer:

Figure 3   Structures of proanthocyanidins.

Anthocyanidins (Fig. 4) are natural plant pigments, and their glycosides are known as the anthocyanins. These are largely responsible for the red, purple, and blue colors of flowers and fruits of higher plants, and are rich in berries and the red grape (Mazza *et al.*, 1993).

Flavonols, flavones, and their glycosides (Fig. 5) occur mainly in the leaves and in the outer parts of plants, with exception for onion tubers. Quercetin glycosides predominate in fruits and are also abundant in vegetables in which glycosides of kaempferol, myricetin, and luteolin are also present (Hermann, 1976; Pietta *et al.*, 1995b).

In addition to these classes, dietary flavonoids include isoflavones, which occur mainly in soy products (Reinlin and Block, 1996). Closely related to flavonoids are some derivatives of cinnamic acid, such as caffeic, *p*-couramic, and ferulic acids (Fig. 6); these acids may acylate the sugar moiety of the glycoside and are effective antioxidants (Hermann, 1989; Graf, 1992). Dietary flavonoids are generally considered nonnutrients; however, they are important components of the human diet, as their dietary intake may reach 1 g, as suggested by Kuhnau (1976). Of this total amount of mixed flavonoids the major part is provided by beverages such as tea (one cup of tea contains about 150 mg of catechins) and red wine (one glass of red wine contains about 300 mg of total polyphenols, expressed as gallic acid equivalents) (Simonetti *et al.*, 1997). Flavonols and flavones contribute much less (up to 50 mg per day), as their content in normally consumed veg-

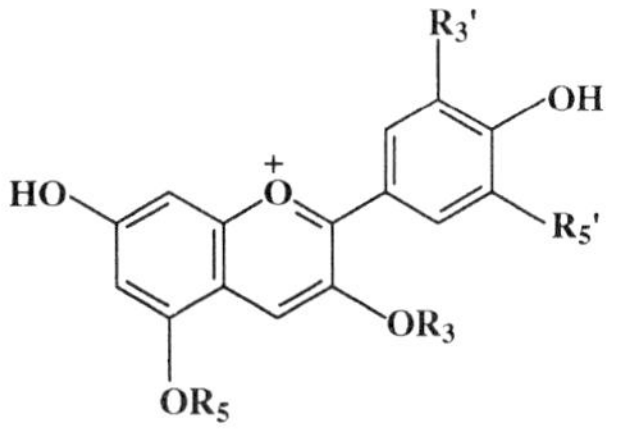

Anthocyanidins (R$_3$= R$_5$ = H)   Anthocyanin (R$_3$, R$_5$ = glycosides)

Compound	R$_3$'	R$_5$'	Compound	R$_3$	R$_5$
Delphinidin (De)	OH	OH	De-3-glucosides	glucose	H
Cyanidin (Cy)	OH	H	De-3,5-diglucosides	glucose	glucose
Pelargonidin (Pe)	H	H			
Peonin (Pn)	OCH$_3$	H			
Malvidin (M)	OCH$_3$	OCH$_3$			
Petunidin (Pt)	OCH$_3$	OH			

Figure 4   Structure of anthocyanidins and related glycosides.

etables and fruits is quite low (few mg/100 g fresh weight) (Table I). Tea and wine provide similar low amounts of flavonols. Thus, considering flavonols and flavones as actual dietary intake of flavonoids may be misleading.

FLAVONOID ACTIVITIES

Biological activities of flavonoids have become well known in recent years. Many studies suggest that flavonoids have beneficial effects on human health due to their antioxidant capacity (Bors *et al.*, 1996) and their ability to (a) modulate the activity of different enzymes (Middleton and Kandaswami, 1994; Melzig, 1996), (b) interact with specific receptors (Xiaoduo *et al.*, 1996; Medina *et al.*, 1997), (c) exert vasodilatory effects (Duarte *et al.*, 1993), and (d) chelate metal ions such as Cu and Fe (Leibovitz and Muller, 1993).

Based on their daily intake, which exceeds largely that of other antioxidants (vitamin C, 70–100 mg/day; vitamin E, 7–10 mg/day; β-carotene, 2–3 mg/day), flavonoids provide a major contribution to ensure the antioxidant potential of the diet. In other words, dietary flavonoids may represent an important exogenous defense against the imbalance between prooxi-

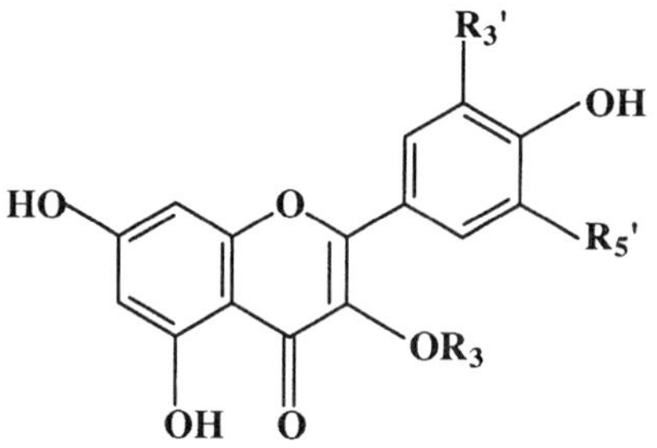

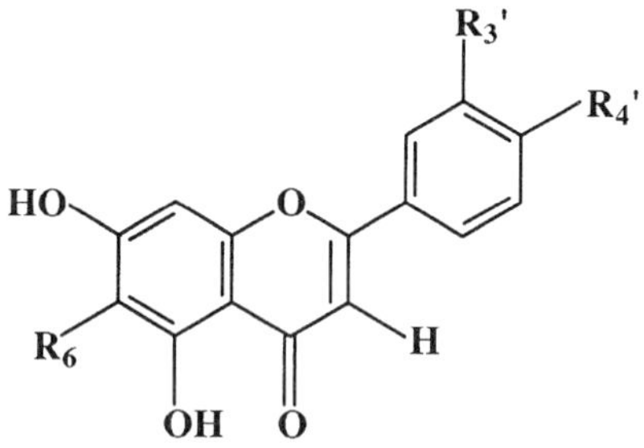

Flavonols aglycones and related glycosides

Aglycones	R_3	$R_{3'}$	$R_{5'}$
Quercetin	H	OH	H
Myricetin	H	OH	OH
Isorhamnetin	H	OCH_3	H
Kaempferol	H	H	H
Glycosides			
Isoquercitrin	glucose	OH	H
Rutin	rutinose	OH	H
Quercitrin	arabinose	OH	H
Hyperoside	galactose	OH	H
Astragalin	glucose	H	H
Myricitrin	glucose	OH	OH

Flavones aglycones

Aglycones	$R_{3'}$	$R_{4'}$	R_6
Chrysin	H	H	H
Apigenin	H	OH	H
Luteolin	OH	OH	H
Baicalein	H	H	OH
Sinensetin	OH	OH	OH

Figure 5 Structures of flavonols, flavones, and related glycosides.

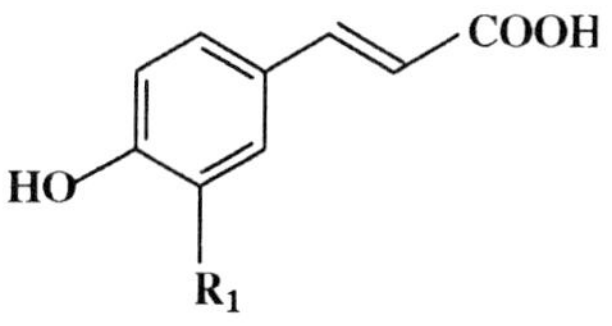

Cinnamic acid and derivatives

Compound	R_1
Caffeic acid	OH
p-Coumaric	H
Ferulic acid	OCH_3

Figure 6 Structure of cinnamic acid and some derivatives.

Table I   Concentration (μg/g) of Flavonols in Legumes, Vegetables, and Fruits

Food name	Rutin	Quercetin	Kaempferol
Potato	nd[a]	nd	23.40
Tomato	0.9	1.58	7.50
Green pepper	21.70	14.10	3.17
Eggplant	nd	1.58	nd
Carrot	4.38	nd	7.20
Parsley	nd	7.03	45.10
Radish	nd	nd	3.36
Cabbage	5.54	nd	7.20
Broccoli	1.17	9.77	16.10
Spinach	nd	nd	nd
Lettuce	nd	4.78	nd
Onion	nd	33.70	14.09
Cucumber	1.89	nd	7.61
Kiwi fruit	nd	2.07	30.60
Watermelon	nd	nd	18.10
Orange	nd	17.50	31.50
Peach	nd	1.08	6.54
Apple	nd	5.27	26.70
Persimmon	nd	nd	nd
Grape	8.24	nd	16.80

[a] Not detected.

dants and antioxidants, i.e., the oxidative stress. Indeed, flavonoids play an active role in the diminished formation of reactive oxygen species (ROS), as they affect enzymes that catalyze oxidation–reduction reactions, including mitochondrial succinoxidase, NADH-oxidase, and enzymes involved in arachidonic acid metabolism (Middleton and Kandaswami, 1994). In addition, highly oxidizing ROS are reduced by flavonoids, which in turn are transformed in less aggressive aroxyl radicals (Jovanovic *et al.*, 1997). Indeed, flavonoids have been reported to prevent LDL peroxidation, which is thought to play an important role in atherosclerosis (De Whalley *et al.*, 1990; Vinson *et al.*, 1995; Serafini *et al.*, 1996).

Unfortunately, most positive evidence of these activities has been reached by performing *in vitro* studies on isolated compounds; however, it is difficult to relate this evidence to *in vivo* situations and to the complex matrix of foods and beverages. Thus, the absorption of dietary flavonoids represents a critical point in evaluating their potential health benefits. In past years, many efforts have been made to overcome this difficulty and now better information on the absorption of flavonoids from regular foods, beverages, and food supplements is available as detailed in this chapter. Limited space, however, restricts us to primarily describing the results obtained for two specific classes, i.e., flavonols and catechins, with focus on our own experience.

FLAVONOL AND CATECHIN ABSORPTION AND METABOLISM

In the early 1970s Das (1971) reported that (+)-catechin was absorbed in the gastrointestinal tract following administration to healthy volunteers (4.2 g in the form of gelatin capsules taken orally with water at breakfast). (+)-Catechin was excreted in the urine together with several unidentified metabolites, and the amount excreted within 24 hr was about 7.5% of the administered dose. Unchanged (+)-catechin (about 19% of the dose) was found in the 48-hr fecal collection, and several phenolic metabolites were detected in plasma, but they were not quantified. Some years later, Gugler *et al.* (1975) described that no measurable concentrations of quercetin or its derivatives were detected in plasma or urine after oral administration of a single dose of 4 g of this flavonol. Approximately 53% of the oral dose was recovered unchanged in the feces. Apart from their discrepancy, both these studies suffer from the limitation that individual flavonoids were administered at doses largely exceeding dietary doses. Indeed, humans consume a range of flavonoids from different sources, and their highest daily intake is approximately 1 g.

Studies performed in rats and humans fed flavonols (from vegetables and phytoceuticals) and catechins (from green tea) have confirmed that these compounds are partly absorbed by the intestine. However, debate is open as to whether glycosylated forms of flavonols (that are prevalent in dietary plants) are absorbed from the digestive tract.

It is generally stated that flavonoid glycosides are hydrolyzed before being absorbed (Kuhnau, 1976). However, this has been questioned by Hollmann *et al.* (1995), who described that quercetin glycosides from onions in ileostomy patients could be absorbed (even if less than quercetin itself) by the small intestine. According to Manach *et al.* (1997), this finding could depend on a modification of digestive processes in ileostomy patients (such as a different site of bacterial colonization) or on the components of the food matrix rather than glycosylation. Indeed, Manach *et al.* (1996) reported that only quercetin is absorbed from the small intestine, whereas rutin (quercetin-3-O-rutinoside) needs to be hydrolyzed by cecal microflora before being absorbed. Thus, these authors conclude that aglycones should be better available than glycosides, as aglycones are absorbed both in the small intestine and in the large bowel. More recently, Hollman *et al.* (1997)

reported on the absorption of quercetin in two healthy subjects after ingestions of fried onions, which are a good source of quercetin glucosides. Unfortunately, no evidence of the presence of these glucosides in plasma was provided, as plasma samples were hydrolyzed with 2 *M* HCl and only the free aglycone could be detected. In conclusion, more work is needed to endorse the assumption that flavonol glycosides are absorbed as such by the gut in healthy normal subjects.

However, it has been proved that flavonols and catechins are largely converted to glucuronyl derivatives in the intestinal mucosa. These first-passage derivatives are transferred to the liver, where methylation, sulfation, and further conjugation with glucuronic acid, sulfate, and glycine take place (Hackett, 1986).

According to Da Silva *et al.* (1997), plasma obtained at 1 hr after administration of (−)-epicatechin (EC) to rats showed the presence of glucuronides, O-methylglucuronide sulfate, and glucuronide sulfate of EC. Glucuronide conjugates were predominant, whereas free EC ranged from 7 to 13% of the total absorbed amount, depending on the intake of (−)-epicatechin (10–50 mg, respectively). Similarly, Manach *et al.* (1996) reported that plasma of rats receiving a single meal containing 0.25% quercetin contained glucurono-sulfo conjugates of quercetin and its methoxy derivative (Fig. 7). The authors detected epigallocatechin gallate and epicatechin gallate in human plasma after ingestion of green tea infusion (Pietta *et al.*, 1997b). The levels of these two free catechins increased significantly after enzymatic hydrolysis with glucuronidase/sulfatase, indicating that they were present in plasma mainly in the conjugated forms.

According to these studies, the percentage of the absorption, determined by measuring plasma levels of flavonols or catechins after enzymatic hydrolysis, does not seem to exceed 2–3% of the ingested dose. It is also likely that, as with other micronutrients, the existence of a steady-state concentration of these compounds could result in diminished absorption. Thus, it is conceivable that the major parts of these flavonoids are either degraded to phenolic acids in the large intestine or excreted in the feces.

Indeed, the colon is a metabolically active compartment, where microorganisms degrade either unabsorbed flavonols and catechins coming from the small intestine or their conjugates secreted into the duodenum via the gall bladder. Colon bacteria are capable of hydrolyzing the conjugated forms (Sheline, 1973) or of splitting the heterocyclic C ring of flavonols and catechins (Griffiths, 1982). The bacterial degradation in the colon depends on the hydroxylation pattern of the compound involved, and a variety of phenolic acids are produced. These are absorbed and subsequently metabolized by enzymes present mainly in the liver, where 3′-O-methylation by catechol-O-methyltransferase (Ting Zhu *et al.*, 1994), dehydroxylation,

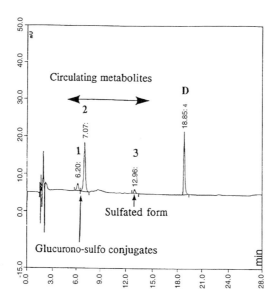

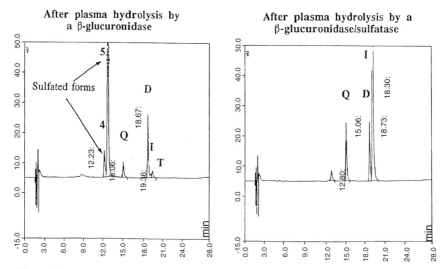

Figure 7   Conjugated derivatives of quercetin in the plasma of rats adapted to a 0.2% quercetin diet. Q, quercetin; D, diosmetin; I, isorhamnetin; T, tamarixetin.

β-oxidation, and conjugation with glucuronic acid, sulfate, and glycine occurs. The resulting metabolites are then excreted into bile, plasma, and urine.

Some of these metabolites were detected in human urine after administration of an extract of *Ginkgo biloba* (Pietta *et al.*, 1997a) or ingestion of green tea infusions (Pietta *et al.*, 1998c).

As far as flavonols are concerned, urine samples of volunteers who received 1 g of ginkgo flavonol glycosides contained different substituted benzoic acids, i.e., 4-hydroxybenzoic acid glucuronide, 4-hydroxyhippuric acid, 3-methoxy-4-hydroxyhippuric acid, 3,4-dihydroxybenzoic acid, 4-hydroxybenzoic acid, hippuric acid, and 3-methoxy-4-hydroxybenzoic acid. The total amount of metabolites found in the 48-hr urine collection was approximately 150 mg (0.6 mM). No phenylacetic or phenylpropionic acid derivatives could be detected in human urine, but their formation cannot be excluded. Some of these metabolites were monitored in rat plasma (Pietta *et al.*, 1995a), whereas their levels in human plasma have not yet been established. More sensitive methods are needed to detect these phenolic acids in the complex plasma matrix.

Similar results were obtained when an infusion of green tea (400 mg of total catechins) was given to healthy volunteers. Urine samples collected at 0–48 hr contained detectable amounts of the final catechin metabolites, including 4-hydroxybenzoic acid, 3,4-dihydroxybenzoic acid, 3-methoxy-4-hydroxyhippuric acid, and 3-methoxy-4-hydroxybenzoic acid (Fig. 8). The total content of these metabolites averaged 60 mg (0.7 mM) (Pietta *et al.*, 1997b).

Due to the limited knowledge on flavonol or catechin absorption across the intestinal membrane (Crevoisier *et al.*, 1975), the catechin absorption was examined by perfusing everted small intestine segments with a 100-ml green tea extract isotonic solution (equivalent to 75 mg of EGCg, 8.2 mg of ECg, 1.3 mg of EGC, 2.3 mg of EC, and 0.8 mg of C) or an isotonic solution. The levels of catechins absorbed across the intestinal membrane were detected after 1 hr of perfusion. EGC was absorbed at a higher extent than EC, whereas the related gallates were absorbed very poorly (Table II). A small part of these catechins was retained by the small intestine tissue [scraped mucosal layer and the residual tissue of the small intestine (Table III)]. Not surprisingly, the initial levels of α-tocopherol either in the scraped mucosal layer or in the residual intestine were maintained by the presence of catechins as compared to their dramatic decrease (-82%) after perfusion with isotonic solution (Table IV). Thus, this *ex vivo* approach may give useful information, and the comparative absorption of specific dietary flavonols is currently under investigation by the authors' group at Milan.

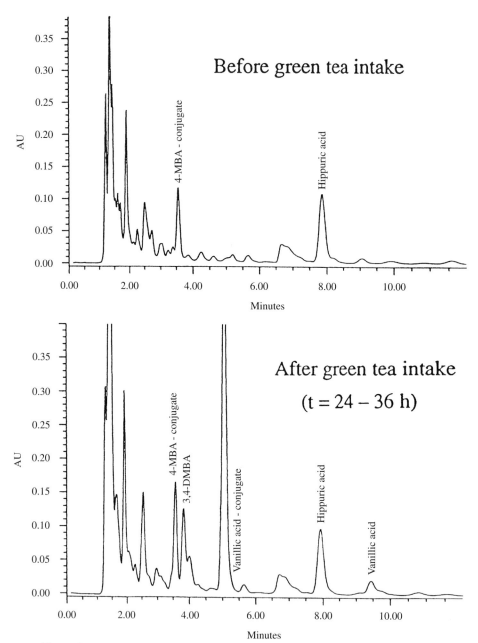

Figure 8   Urine catechin metabolites after intake of green tea infusion (400 mg EGCg).

Table II   Catechin Absorption (Percentage ± SD) in Everted Small Intestine Segments

Intestinal segment	Proximal	Medial	Distal
Gallic acid	0.26 ± 0.06	0.40 ± 0.07	0.55 ± 0.05
Epigallocatechin	0.59 ± 0.07	0.88 ± 0.10	0.99 ± 0.12
Epicatechin	0.47 ± 0.09	0.72 ± 0.08	0.84 ± 0.14
Egallocatechin gallate	0.02 ± 0.01	0.04 ± 0.02	0.08 ± 0.10
Epicatechin gallate	0.01 ± 0.01	0.03 ± 0.01	0.07 ± 0.07

RADICAL-SCAVENGING AND ANTIOXIDANT CAPACITY OF SELECTED FLAVONOL AND CATECHIN METABOLITES

The radical-scavenging capacity of some metabolites from ginkgo flavonol glycosides was studied following the inhibition of chemiluminescence either in human polymorphonuclear cells (PMNs) stimulated by N-formylmethionylleucylphenylalanine (FMLP) or opsonized zymosan (OZ) or in a cell-free system using horseradish peroxidase (Heilmann *et al.,* 1995). Results indicated that metabolites with an o-dihydroxy group in the phenyl group are as effective as the flavonol precursor (quercetin) in quenching free radicals (Merfort *et al.,* 1995). Methylation of one hydroxyl slightly decreases this ability. Metabolites lacking this structural characteristic are poorly active, and the length of the alkyl chain or the position of the hydroxy group in the phenyl ring seems to have some role (Krol *et al.,* 1995).

Concerning the antioxidant capacity, the new approach introduced by Miller and Rice-Evans (1997) was applied to establish the ability of some metabolites from ginkgo flavonols and green tea catechins to scavenge the ABTS$\bullet^+$ radical cation compared with that of Trolox (TEAC) (Table V)

Table III   Catechin Retained (Percentage ± SD) by the Intestinal Wall

Intestine	Scraped mucosal layer	Residual tissue[a]
Gallic acid	0.32 ± 0.02	0.40 ± 0.02
Epigallocatechin	0.76 ± 0.26	0.67 ± 0.27
Epicatechin	1.05 ± 0.36	0.06 ± 0.02
Epigallocatechin gallate	1.83 ± 0.72	0.71 ± 0.27
Epicatechin gallate	3.41 ± 1.21	1.17 ± 0.44

[a] Residual tissue of small intestine after scraping the mucosal layer.

Table IV   Contents of α-tocopherol (μg/mg ± SD) in Freeze-Dried Everted Intestine Wall

Intestine	Scraped mucosal layer	Residual tissue
Baseline	0.113 ± 0.013	0.169 ± 0.044
Green tea infusion	0.097 ± 0.014	0.132 ± 0.017
Isotonic solution	0.019 ± 0.001[a]	0.090 ± 0.003[a]

[a] $p < 0.05$ vs baseline

(Pietta *et al.*, 1998b). Results for these phenolic acids confirmed that the position of the hydroxy groups, their methylation, and the length of the alkyl chain influence the antioxidant potential. An o-hydroxy group (catechol structure) is the main requirement for strong antioxidant capacity, as is evidenced for 3,4-dihydroxyphenylacetic acid and 3,4-dihydroxybenzoic acid. TEAC values increased with lengthening of the chain bearing the carboxyl group, as evidenced for 4-hydroxybenzoic, 4-hydroxyphenylacetic (0.34 mM) (Rice-Evans *et al.*, 1996), and 4-hydroxyphenylpropionic acids.

The carboxyl group makes the resulting phenols less antioxidant in na-

Table V   TEAC Values of Selected Flavonols, Catechin Metabolites, and Other Natural Antioxidants

Compound	Source[a]	TEAC (mM)
4-Hydroxybenzoic acid	Gb, GT	0.07 ± 0.02
4-Hydroxyhippuric acid	Gb	0.20 ± 0.02
3,4-Dihydroxybenzoic acid	Gb, GT	1.01 ± 0.04
3-Methoxy-4-hydroxybenzoic acid (vanillic acid)	Gb, GT	1.19 ± 0.06
3-Methoxy-4-hydroxyhippuric acid	Gb, GT	1.29 ± 0.04
3-Hydroxyphenylacetic acid	Gb	0.97 ± 0.08
3,4-Dihydroxyphenylacetic acid	Gb	2.16 ± 0.10
3-Methoxy-4-hydroxyphenylacetic acid (homovanillic acid)	Gb	1.63 ± 0.11
3-Hydroxyphenylpropionic acid	Gb	1.03 ± 0.05
4-Hydroxyphenylpropionic acid	Gb, GT	1.60 ± 0.09
4-Methylcatechol	GT	1.25 ± 0.03
3,4-Dihydroxycinnamic acid (caffeic acid)	GT	1.26 ± 0.01
Epigallocatechin gallate	GT	4.80 ± 0.06
Quercetin	Gb	4.72 ± 0.10
Rutin	Gb	2.4 ± 0.12
Ascorbic acid		1.0 ± 0.02
α-Tocopherol		1.0 ± 0.03

[a] Gb, *Ginkgo biloba*; GT, green tea.

ture, as is shown by the lower TEAC values of 3,4-dihydroxybenzoic and vanillic acids as compared with 3,4-dihydroxyphenylacetic and homovanillic acids. Conjugation of 4-hydroxybenzoic and vanillic acids with glycine produces a slight increase of TEAC values, as the secondary amino group has a lower electron-withdrawing capacity than the free carboxy group. This behavior satisfies the Hammett's correlation between antioxidant capacity and electron-donating or -withdrawing properties of the substituents (Jovanovic *et al.*, 1991). In fact, better electron-donating groups (such as —OH, —OCH$_3$, alkyl) reduce the redox potential of phenols and increase their antioxidant efficacy. On the other side, electron-withdrawing groups (such as —COOH, —CH=CH-COOH) increase the redox potential, disqualifying these phenols as antioxidants (compare 3,4-dihydroxyphenylacetic acid with 3,4-dihydroxycinnamic acid, i.e., caffeic acid).

Concerning the possible *in vivo* effects of these metabolites, the amounts detected in the 0- to 48-hr urine sample after the intake of green tea (400 mg catechins) and ginkgo flavonols (1 g) were about 60 (0.7 mM) and 150 (0.6 mM) mg, respectively. These levels could account for a possible contribution to the body antioxidant potential.

IN VIVO ANTIOXIDANT MECHANISM

Despite increasing evidence on the potential of flavonoids as dietary antioxidants, little is known about their precise *in vivo* mechanism. In addition to direct radical-quenching activity, flavonoids and their metabolites may act through a sequence involving an interaction with physiologic antioxidants. Indeed, because of their lower redox potentials (E$_7$ = 700–540 mV) (Jovanovic *et al.*, 1994), flavonoids and their metabolites are capable of oxidizing reactive oxygen species, such as OH•, RO•, ROO•, and O$_2$•$^-$, which have redox potentials in the range of 2130–1000 mV (Buettner, 1993). The resulting water-soluble aroxyl radicals (ArO•) are less aggressive than the original ROS, even if they are thermodynamically and kinetically able to oxidize hydrophilic antioxidants, such as ascorbate (E$_7$ = 282 mV) (Bors and Buettner, 1997). Urate (E$_7$ = 590 mV) may also be oxidized by those ArO• with E$_7$ > 590 mV. Thus, a decrease of the plasma levels of ascorbate and, to a lower extent, of urate should be feasible. For its hydrophobic character, α-tocopherol (E$_7$ = 500 mV) is less involved (Niki, 1996) and, if oxidized, is regenerated by both ascorbate and glutathione (Reed, 1993; Palozza and Krinski, 1992). The latter can also contribute to the reduction in the ability of dehydroascorbate to recover ascorbate (Buettner, 1993), and a decrease of total glutathione in plasma should be expected. Concerning β-carotene, it should be consumed less than

α-tocopherol, as described previously (Palozza and Krinski, 1992; Bast *et al.*, 1996). Consequently, the sparing effect on liposoluble vitamin E and β-carotene should improve the defense against the oxidation of membrane polyunsaturated fatty acids (PUFA). To investigate this possible *in vivo* antioxidant mechanism of dietary flavonoids, 15 healthy volunteers with a diet low in flavonoid-containing foods and beverages were enrolled and divided randomly into two groups. The controls ($n = 7$) received no supplement whereas the treated subjects received one tablet of green tea extract (total catechins 100 mg) per day for 4 consecutive weeks. Samples of peripheral blood were taken before and after treatment; plasma levels of ascorbate, urate, total glutathione, and β-carotene were detected. α-Tocopherol was measured in plasma and in red blood cells (RBC). The latter were also analyzed for the content of membrane phosphatidylcholine fatty acids. As shown in Table VI, the experimental results were consistent with the scenario postulated previously. Indeed, water-soluble ascorbate, total glutathione, and urate decreased. Plasma α-tocopherol did not undergo a loss, whereas RBC α-tocopherol and plasma β-carotene increased. The levels of phosphatidylcholine PUFA of RBC membranes increased significantly (Fig. 9), which correlates well with the rise of liposoluble α-tocopherol and β-carotene. Based on these results (Pietta and Simonetti, 1988b), which confirm previous data (Pietta *et al.*, 1996) and are in good agreement with preliminary reports of other authors (Vieira *et al.*, 1997; Castineira *et al.*, 1997), it may be suggested that catechins and their metabolites exert their antioxidant protection *in vivo* through a cascade involving ROS and hydrophilic antioxidants, resulting in an overall protection of liposoluble vitamin E and β-carotene. Both of these vitamins are known as efficient an-

Table VI   Effects of Chronic Ingestion of Green Tea Catechins[a]

	Before receiving supplement (mean ± SE)	After receiving supplement (mean ± SE)
Plasma		
Ascorbic acid (mg/dl)	1.05 ± 0.24	0.87 ± 0.018
Uric acid (mg/liter)	399.7 ± 25.5	363.8 ± 16.7
Total glutathione (μmol/liter)	618.2 ± 32.7	601.3 ± 46.4
α-Tocopherol (μg/ml)	10.6 ± 0.3	9.6 ± 0.4
β-Carotene (μg/dl)	31.1 ± 5.3	37.9 ± 5.3[b]
RBC membranes		
α-Tocopherol (μg/g)	1.89 ± 0.37	1.99 ± 0.26[b]

[a] Involves 15 healthy volunteers.
[b] $p < 0.05$.

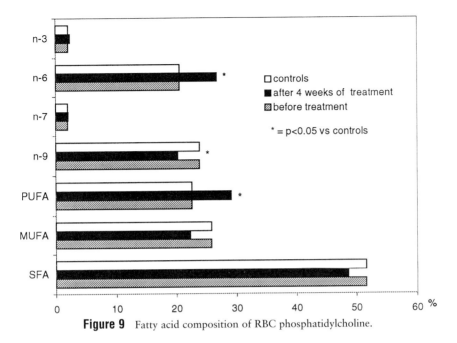

Figure 9   Fatty acid composition of RBC phosphatidylcholine.

tioxidants for low-density lipoprotein (LDL) (Wagner *et al.*, 1996) and their sparing by dietary flavonoids may partly explain the beneficial role of these components of our diet in preserving oxidative modifications of LDL (Zhang *et al.*, 1997).

EVIDENCE OF A RELATIONSHIP BETWEEN RATE AND EXTENT OF ABSORPTION AND MODIFICATION OF PLASMA ANTIOXIDANT STATUS

After having confirmed that a chronic intake of green tea catechins leads to *in vivo* antioxidant activity, it was of interest to investigate a possible relationship between rate and extent of absorption and modification of selected markers of the plasma antioxidant status. To our knowledge, published reports on this issue were lacking, which encouraged us to develop the following study. Healthy volunteers were supplemented with single doses of green tea (Greenselect) or *G. biloba* (Ginkgoselect) extracts. Time course concentrations of plasma epigallocatechin gallate (for green tea) and kaempferol (for ginkgo) were followed and correlated with percentage variations of ascorbate, total glutathione, and total radical antioxidant parameter (TRAP).

Green Tea Catechins

Healthy male volunteers (n = 6, aged 22 ± 2 years) were selected using a self-administered questionnaire to assess the intake of fruits and vegetables. Study subjects had a habitual low intake of flavonoid-rich foods, were not taking antioxidants supplements, and abstained from beverages containing flavonoids for 3 days before the study. Subjects received eight capsules of Greenselect, equivalent to 400 mg EGCg (Fig. 10).

Plasma samples, collected from each participant before and at 1,2,3,4,5, and 6 hr after ingestion, showed the presence of free EGCg and epicatechin gallate (ECg) (Fig. 11). EGCg was chosen as a biomarker of green tea catechin absorption, whereas ECg and other catechin conjugates eluting with the front were not followed.

The time course of EGCg plasma concentration after ingestion of Greenselect is exemplified in Fig. 12. The mean peak levels of EGCg (0.9 ± 0.2 μg/ml, 2 μM) were attained 2 hr after ingestion and then decreased slowly to approximately 0.4 ± 0.1 μg/ml at 6 hr. Percentage absorption (as by measuring EGCg) did not exceed 2% of the given dose. However, in ad-

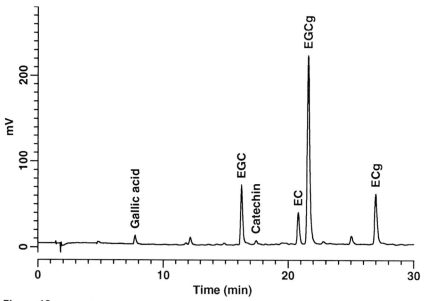

Figure 10   Typical HPLC of Greenselect. Detector: Coulochem II (ESA). Guard cell: −640 mV. E_1: −100 mV. E_2: 640 mV. Gradient: (A) 0.3% phosphoric acid; (B) acetonitrile; 0–25% B in 30 min; flow rate: 2.0 ml/min. Column: Symmetry C_{18}, 220 × 4.6 mm. Sensitivity: 2 ng. EGC, Epigallocatechin; EC, epicatechin; EGCg, epigallocatechin gallate; ECg, epicatechin gallate.

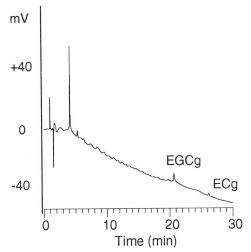

Figure 11 Typical HPLC of plasma EGCg and ECg, Detector: Coulochem II (ESA). Guard cell: −640 mV. E_1: −100 mV. E_2: 640 mV. Gradient: (A) 0.3% phosphoric acid; (B) acetonitrile; 0–25% B in 30 min; flow rate: 2.0 ml/min. Column: Symmetry C_{18}, 220 × 4.6 mm. Sensitivity: 2 ng. EGCg, epigallocatechin gallate; ECg, epicatechin gallate.

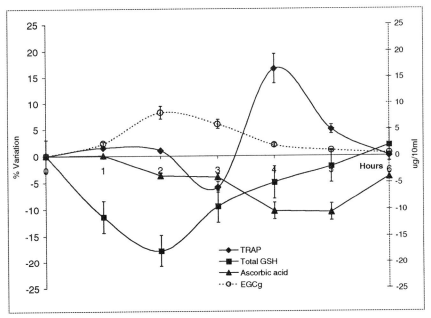

Figure 12 Time course of EGCg plasma concentration and percentage variations of TRAP, total GSH, and ascorbic acid.

dition to EGCg, other catechin conjugates were present in plasma, and their total concentration may be higher, as reported by Wiseman *et al.* (1997). Furthermore, the presence of EGCg in plasma at 6 hr after ingestion suggests that regular consumption of green tea (one cup of tea contains about 150 mg of total catechins) may determine a possible steady-state plasma concentration of EGCg (and related compounds).

Levels of plasma ascorbate and total glutathione and the TRAP value were measured at the same intervals of EGCg detection (Fig. 12). Plasma ascorbate concentration started decreasing at 1 hr, reaching a value of $-10 \pm 2\%$ at 4 hr after ingestion. This decrease was transient, and an almost complete recovery to initial values occurred at 6 hr after supplementation. Concerning total glutathione plasma levels, a negative variation of 15% occurred gradually within the first 2 hr.

Ascorbate and total glutathione variations were well correlated with the EGCg time course concentration. This correlation was less evident for TRAP. There was a significant rise at 4 hr (15 $\pm$ 6%) after an initial decrease, and its variation is due mostly to the partial recovery of glutathione. These data indicate that an intake of green tea equivalent to 400 mg EGCg (approximately 4 cups of tea) has an effect on plasma antioxidant activity and seem to confirm the results obtained by Serafini *et al.* (1996) rather than those reported by Maxwell and Thorpe (1997).

No significant variations of vitamin E and β-carotene could be measured after the intake of Greenselect. It is likely that a single dose administration is not capable of inducing modifications of these liposoluble vitamins and that a regular consumption is required, as evidenced in the midterm study mentioned in the Section on *In Vivo* Antioxidant Mechanism. (Pietta *et al.*, 1998a).

Ginkgoflavonol Glycosides

Ginkgoselect contains approximately 6% terpenoids (ginkgolides) and 24% flavonol glycosides. Of the latter, kaempferol derivatives are an important part, as determined by measuring the kaempferol aglycone after controlled acid hydrolysis of the extract (Pietta, 1997). Thus, it is not surprising that the HPLC trace of plasma after the ingestion of Ginkgoselect (1500 mg equivalent to 360 mg of flavonol glycosides given to four healthy volunteers) shows a major peak with a retention time and an ultraviolet (UV) spectrum that can be ascribed to kaempferol (Fig. 13). Peaks eluting with the front have UV spectra typical of flavonols and possibly are glucuronide or sulfate conjugates. However, the attention was focused on kaempferol as a biomarker of ginkgo flavonol glycoside absorption.

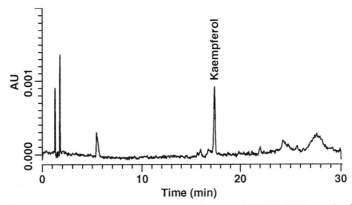

Figure 13 Typical HPLC of plasma kaempferol. Detector: UV-DAD 370 nm. Gradient: (A) 0.3% phosphoric acid; (B) acetonitrile; 15–40% B in 30 min; flow rate: 2.0 ml/min. Column: Symmetry C_{18}, 220 × 4.6 mm. Sensitivity: 0.5 ng.

The time course of kaempferol concentration in plasma is shown in Fig. 14. The highest levels of kaempferol of about 80 ng/ml were detected at 3 hr after ingestion and were equivalent to approximately 0.29 μM. However, this value is limited to the chosen biomarker, and the presence of other flavonols in free or conjugated forms cannot be excluded. Judging from the areas of peaks eluting with the front, a three- to fourfold concentration should be feasible.

As shown in Fig. 14, both ascorbate and total glutathione plasma concentrations decreased at different rates and extents, and their modifications were correlated to the time course concentration of the biomarker. Ascorbate underwent maximum negative variation ($-10 \pm 2\%$) at 1 hr, with an almost complete recovery at 4 hr. There was a rise in total glutathione at 1 hr ($+5 \pm 2\%$), possibly due to other components of Ginkgoselect, and then a decrease to $-18 \pm 3\%$ at 2 hr. Glutathione initial levels were restored at 3 h. TRAP showed two peaks of positive variation, i.e., at 1 ($+8 \pm 1\%$) and 4 ($+18 \pm 2\%$) hr. The first peak may be attributed to the increasing amount of the biomarker and glutathione. The second peak, which occurred at times when the biomarker concentration was quite low, is due to the recovery of total glutathione, as also observed after Greenselect ingestion. Similar to the green tea intake, vitamin E and β-carotene were not affected by the ingestion of a single dose of ginkgo flavonol glycosides.

In both green tea and ginkgo trials, control subjects did not show modifications of plasma ascorbate, total glutathione, and TRAP.

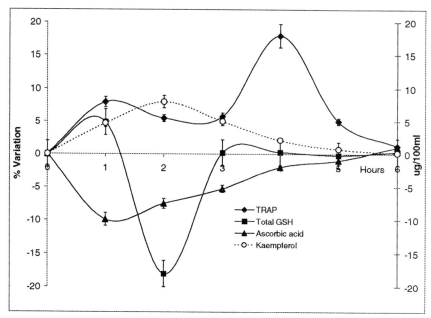

Figure 14   Time course of kaempferol plasma concentration and percentage variations of TRAP, total GSH, and ascorbic acid.

These preliminary results (Pietta *et al.*, 1998a) indicate that the decrease of ascorbate and total glutathione and the increase of TRAP depend on the rate and extent of catechins and ginkgo flavonol glycosides absorption, thereby proving a correlation between dynamic and kinetic data.

SUMMARY

This brief review of some of the current research on dietary flavonoids shows that (a) green tea catechins and ginkgo flavonol glycosides are partly absorbed as intact molecules or their metabolites and transferred into the blood; (b) some metabolites of catechins and flavonols maintain radical-scavenging and antioxidant capacity comparable to that of their precursor and vitamins C and E; (c) dietary flavonoids and their metabolites are likely to play their antioxidant role *in vivo* through a cascade involving reactive oxygen species and hydrophilic ascorbate, urate, and glutathione, with a resulting sparing of liposoluble vitamin E and β-carotene, and subsequent RBC membrane PUFA protection against oxidation; and (d) a correlation exists between dynamic (percentage variations of plasma ascorbate, total glutathione, and TRAP) and kinetic (rate and extent of absorption) data monitored after a single dose intake of green tea catechins or ginkgo flavonol glycosides.

ACKNOWLEDGMENTS

The authors acknowledge the contribution of Claudio Gardana, Antonella Brusamolino, and Daniele Contino to this work. Indena SpA is acknowledged for its support.

REFERENCES

Bast, A., van der Plas, R. M., and Haenen, G. R. M. (1996). β-Carotene as antioxidant. *Eur. J. Clin. Nutr.* **50**, s54–s56.

Blot, B. J., Chow, W. H., and McLanghlin, J. K. (1996). Tea and cancer: A review of the epidemiological evidence. *Eur. J. Cancer Prev.* **5**, 425–438.

Bors, W., Heller, W., Michel, C., and Stettmaier, K. (1996). Flavonoids and polyphenols: Chemistry and biology. *In* "Handbook of Antioxidants" (E. Cadenas and L. Packer, eds.), pp. 409–446. Dekker, New York.

Bors, W., and Buettner, G. R. (1997). The vitamin C radical and its reaction. *In* "Vitamin C in Health and Disease" (L. Packer and J. Fuchs, eds.), pp. 75–94. Dekker, New York.

Buettner, G. R. (1993). The pecking order of free radicals and antioxidants: Lipid peroxidation, α-tocopherol, and ascorbate. *Arch. Biochem. Biophys.* **300**, 535–543.

Castineira, M., Pèrez-de-Luelmo, P., Marin, D., Lasuncion, M. A., and Ortega, H. (1997). Monitoring oxidative stress status in antioxidant diet-therapies. *In* "Proc. 16th International Congress of Nutrition, Montreal, Canada," p. 138.

Crevoisier, C., Buri, P., and Boucherat, J. (1975). Etude du transport de trois flavonoides a travers des membranes artificials et biologiques. *Pharm. Acta Helv.* **59**, 231–236.

Da Silva, E. L., Piskula, M., and Terao, J. (1998). Enhancement of antioxidant ability of rat plasma by oral administration of (−)-epicatechin. *Free Rad. Biol. Med.* **24**, 1209–1216.

Das, N. P. (1971). Studies on flavonoid metabolism: Absorption and metabolism of (+)-catechin in man. *Biochem. Pharmacol.* **20**, 3435–3445.

De Whalley, C. V., Rankin, S., Hoult, J. R. S., Jessup, W., and Leake, D. S. (1990). Flavonoids inhibit the oxidative modification of low density lipoproteins by macrophages. *Biochem. Pharmacol.* **39**, 1743–1750.

Duarte, J., Vizcaino, F. P., Utrilla, P., Jimenez, J., Tamargo, J., and Zarzuelo, A. (1993). Vasodilatory effects of flavonoids in rat aortic smooth muscle: Structure activity relationship. *Biochem. Pharmacol.* **24**, 857–862.

Graf, E. (1992). Antioxidant potential of ferulic acid. *Free Radic. Biol. Med.* **13**, 435–448.

Graham, G. (1992). Green tea composition, consumption and polyphenol chemistry. *Prevent. Med.* **21**, 334–350.

Griffiths, L. (1982). Mammalian metabolism of flavonoids. *In* "The Flavonoids: Advances in research" (J. Harborne and T. Mabry, eds.), pp. 691–718. Chapman and Hall, London.

Gugler, R., Leschik, M., and Dengler, H. J. (1975). Disposition of quercetin in man after single oral and intravenous doses. *Eur. J. Clin. Pharmacol.* **9**, 234–299.

Hackett, A. M. (1986). The metabolism of flavonoid compounds in mammals. *In* "Plant

Flavonoids in Biology and Medicine: Biochemical, Pharmacological and Structure Activity Relationships" (V. Cody, E. Middleton, and J. B. Harborne, eds.), pp. 177–194. Liss, New York.

Haslam, E. (1989). "Plant polyphenols." Cambridge Univ. Press, Cambridge, UK.

Heilmann, J., Merfort, I., and Weiss, M. (1995). Radical scavenger activity of different 3',4'-dihydroxy-flavonols and 1,5-dicaffeoylquinic acid studied by inhibition of chemiluminescence. *Planta Med.* **61**, 435–438.

Hermann, K. (1976). Flavonols and flavones in plants: A review. *J. Food Technol.* **111**, 433–448.

Hermann, K. (1989). Occurrence and content of hydroxycinnamic and hydroxybenzoic acid compounds in foods. *Crit. Rev. Food Sci. Nutr.* **28**, 315–347.

Hollman, P. C. H., de Vries, J. H. M., Leeuwen, S. D., Mengelers, M. J. B., and Katan, M. B. (1995). Absorption of dietary quercetin glycosides and quercetin in healthy ileostomy volunteers. *Am. J. Clin. Nutr.* **62**, 1276–1282.

Hollman, P. C. H., Gaag, M. V. D., Mengelers, M. J. B., van Trjip, J. M. P., de Vries, J. H. M., and Katan, M. B. (1996). Absorption and disposition kinetics of dietary antioxidant quercetin in man. *Free Radic. Biol. Med.* **21**, 703–707.

Jovanovic, S. V., Steenken, S., Simic, M. G., and Hara, Y. (1997). Antioxidant properties of flavonoids: Reduction potentials and electron transfer reactions of flavonoid radicals. *In* "Flavonoids in Health and Disease" (C. A. Rice-Evans and L. Packer, eds.), pp. 137–161. Dekker, New York.

Jovanovic, S. V., Steenken, S., Tosic, M., Marjanovic, B., and Simic, M. C. (1994). Flavonoids as antioxidants. *J. Am. Chem. Soc.* **116**, 4846–4851.

Jovanovic, S. V., Tosic, M., and Simic, M. G. (1991). Use of Hammett correlation and for σ^+ calculation of one-electron redox potentials of antioxidants. *J. Phys. Chem.* **95**, 10824–10827.

Krol, W., Czuba, Z., Sceller, S., Paradowski, Z., and Shani, J. (1994). Structure-activity relationships in the ability of flavonols to inhibit chemiluminescence. *J. Ethnopharmacol.* **41**, 121–126.

Kuhnau, J. (1976). The flavonoids: A class of semi-essential food components: Their role in human nutrition. *World Rev. Nutr. Diet.* **24**, 117–191.

Leibovitz, B., and Muller, J. A. (1993). Bioflavonoids and polyphenols. *J. Optimal Nutr.* **2**, 17–35.

Maffei Facino, R., Carini, M., Aldini, G., Bombardelli, E., Morazzoni, P., and Morelli, R. (1994). Free radical scavenging action and anti-enzyme activities of procyanidins from *Vitis vinifera*. *Arzneim.-Forsch./Drug Res.* **44**, 592–601.

Manach, C., Morand, C., Demigné, C., Texier, O., Régérat, F., and Rèmèsy, C. (1997). Bioavailability of rutin and quercetin in rats. *FEBS Lett.* **409**, 12–16.

Manach, C. (1998). Personal communication.

Masquelier, J. (1965). The leucoanthocyanidins in the dead bark of *Pinus maritima. Bull. Soc. Pharm. Bordeaux* **104**, 33–36.

Maxwell, S., and Thorpe, G. (1996). Tea flavonoids have little short term impact on serum antioxidant activity. *Brit. Med. J.* **313**, 220.

Mazza, G., and Miniati, E. (1993). "Anthocyanins in Fruits, Vegetables and Grains." CRC Press, Boca Raton, FL.

Medina, J. H., Viola, H., Wolfman, C., Marder, M., Wasowski, C., Calvo, D., and Paladini, A. C. (1997). Flavonoids: A new family of benzodiazepine receptor ligands. *Neurochem. Res.* **22**, 419–425.

Melzig, M. F. (1996). Inhibition of adenosine deaminase activity of aortic aendothelial cells by selected flavonoids. *Planta Med.* **62**, 20–21.

Merfort, I., Heilmann, J., Weiss, M., Pietta, P. G., and Gardana, C. (1995). Radical scavenger activity of metabolites from *Ginkgo biloba* flavonoids. *Planta Med.* **62**, 289–292.

Middleton, E. J. R., and Kandaswami, C. (1994). The impact of plant flavonoids on mammalian biology: Implications for immunity, inflammation and cancer. *In* "The Flavonoids: Advances in Research since 1986" (J. B. Harborne, ed.), pp. 619–652. Chapman & Hall, London.

Miller, N. J., and Rice-Evans, C. A. (1997). Factors influencing the antioxidant activity determined by the ABTS[*+] radical cation assay. *Free Radic. Res.* **26**, 195–199.

Niki, E. (1996). α-Tocopherol. *In* "Handbook of Antioxidants" (F. Cadenas and L. Packer, eds.), pp. 3–25. Dekker, New York.

Palozza, P., and Krinsky, N. I. (1992). α-Carotene and β-tocopherol are synergistic antioxidants. *Arch. Biochem. Biophys.* **297**, 184–187.

Pietta, P. G. (1997). Flavonoids in medicinal plants. *In* "Flavonoids in Health and Disease" (C.A. Rice-Evans and L. Packer, eds.), pp. 61–110. Dekker, New York.

Pietta, P. G., Gardana, C., and Mauri, P. L. (1997a). Identification of *Ginkgo biloba* flavonol metabolites after oral administration to humans. *J. Chromatogr. B* **693**, 249–255.

Pietta, P. G., Gardana, C., Mauri, P. L., Maffei Facino, R., and Carini, M. (1995a). Identification of flavonoid metabolites after oral administration to rats of a *Ginkgo biloba* extract. *J. Chromatogr. B* **673**, 75–80.

Pietta, P. G., Mauri, P. L., Simonetti, P., and Testolin, G. (1995b). HPLC and MEKC determination of major flavonoids in selected food pools. *Fresenius J. Anal. Chem.* **352**, 788–792.

Pietta, P. G., and Simonetti, P. (1998). Dietary flavonoids and interaction with endogenous antioxidants. *Biochem. Mol. Biol. Int.* **44**, 1069–1074.

Pietta, P. G., Simonetti, P., Gardana, C., Brusamolino, A., Morazzoni, P., and Bombardelli, E. (1997b). Catechin metabolites after intake of green tea infusions. *Biofactors* **8**, 111–118.

Pietta, P. G., Simonetti, P., Gardana, C., Brusamolino, A., Morazzoni, P., and Bombardelli, E. (1998a). Relationship between rate and extent of catechin and ginkgoflavonol absorption and plasma antioxidant status. *Biochem. Mol. Biol. Int.* in press.

Pietta, P. G., Simonetti, P., and Rice-Evans, C. A. (1998b). Trolox equivalent antioxidant capacity of selected flavonol and catechin metabolites. Submitted for publication.

Pietta, P. G., Simonetti, P., Roggi, C., Brusamolino, A., Maccarini, L., and Testolin, G. (1996). Dietary flavonoids and oxidative stress. *In* "Natural Antioxidants and Food Quality in Atherosclerosis and Cancer Prevention" (J. T. Kumpulainen and J. T. Salonen, eds.), pp. 249–255. The Royal Society of Chemistry, Cambridge.

Reed, D. J. (1993). Interaction of vitamin E, ascorbic acid and gluathione in protection against oxidative damage. *In* "Vitamin E in Health and Disease" (L. Packer and J. Fuchs, eds.), pp. 269–281. Dekker, New York.

Reinlin, K., and Block, G. (1996). Phytoestrogen content of foods: A compendium of literature values. *Nutr. Cancer* **26**, 123–148.

Rice-Evans, C. A., and Miller, N. J. (1995). Antioxidants: The case for fruit and vegetables in the diet. *Br. Food J.* **97**, 35–40.

Rice-Evans, C. A., Miller, N. J., and Paganga, G. (1996). Structure-antioxidant activity relationships of flavonoids and phenolic acids. *Free Radic. Biol. Med.* **20**, 933–956.

Rice-Evans, C. A., and Packer, L. (eds.) (1997). "Flavonoids in Health and Disease." Dekker, New York.

Serafini, M., Ghiselli, A., and Ferro-Luzzi, A. (1996). In vivo antioxidant effect of green and black tea in man. *Eur. J. Clin. Nutr.* **50**, 28–32.

Sheline, R. (1991). "Handbook of Mammalian Metabolism of Plant Compounds." CRC Press, Boca Raton, FL.

Simonetti, P., Pietta, P. G., and Testolin, G. (1997). Polyphenol content and total antioxidant potential of selected italian wines. *J. Agric. Food Chem.* **45**, 1152–1155.

Ting Zhu, B., Ezell, E. L., and Liehr, J. C. (1994). Catechol-O-methyltransferase-catalyzed rapid O-methylation of mutagenic flavonoids. *J. Biol. Chem.* **269**, 292–299.

Vieira, O., Laranjinha, J., Madeira, V., and Almeida, L. (1997). Oxidation of free cholesterol and cholesteryl esters in LDL challenged with ferrylmyoglobin: Concerted antioxidant activity of dietary flavonoids with ascorbate. *In* "Proc. SFFR Europe Summer Meeting," pp. 134–135. Abano Terme, Italy.

Vinson, J. A., Dabbagh, Y. A., Serry, M. M., and Jang, J. (1995). Plant flavonoids, especially tea flavanols, are powerful antioxidants using in vitro oxidation model for heart disease. *J. Agric. Food Chem.* **43**, 2800–2802.

Wagner, B. A., Buettner, G. R., and Burns, C. P. (1996). Vitamin E slows the rate of free radical-mediated lipid peroxidation in cells. *Arch. Biochem. Biophys.* **334**, 261–267.

Wiseman, S., Tijburg, L., and Korver, O. (1997). The food industry and functional foods. *In* "Proc. Intern. Symposium on Antioxidant Food Supplements in Human Health," p. 43. Kaminoyama-city, Japan.

Xiao-duo, J., Melman, N., and Jacobson, K. A. (1996). Interactions of flavonoids and other phytochemicals with adenosine receptors. *J. Med. Chem.* **39**, 781–788.

Zhang, A., Chan, P. T., Lus, Y. S., Ho, W. K. K., and Chen, Z. Y. (1997). Inhibitory effect of jasmine green tea epicatechin isomers on LDL-oxidation. *J. Nutr. Biochem.* **8**, 334–340.

III Natural Source Antioxidants

20 From Ancient Pine Bark Uses to Pycnogenol

G. Drehsen
Basel, Switzerland

INTRODUCTION

In ancient times, pine terminology did not consistently correspond to one particular botanical race. The "Georgics" of Virgil, a manual of agriculture in verse written in 31 B.C., refers to *P. pinaster.* At that time, the Romans used *Pinus* for pine and *Pinaster* for *P. silvestris.* Further terminology confusion follows from the pines mentioned in the "Natural History" of Pliny the elder (A.D. 23–79), the most important writing of antiquity, which deeply influenced the early encyclopedists.

In modern scientific English, as published in 1967 by N. T. Mirov, the terminology variation was essentially reduced to scientific numbering of the then known pine species and subspecies or races.

The phenomenon of races, including climate races through geographical separation and genetic fixation, has been proved by provenance tests and analysis of the genetic variability of proteins of *P. pinaster.* "Pin des Landes," a climatic–geographic race of *P. pinaster* in France, was used principally as a sand binder of the Atlantic south of Bordeux for the last 200 centuries, which resulted in culture separation. This French maritime pine, in France usually called "Pin des Landes," is used to produce origin-guaranteed Pycnogenol, a proprietary, pure natural extract from this bark. Pycnogenol is the registered name by Horphag Research.

The research link between Pycnogenol as a natural product of modern science and the traditional pine bark uses is evident. Traditional medical use of pine bark to reduce inflammation can be followed back to Hippocrates, the "Father of Medicine" ($\approx$ 400 B.C.). In the Middle Ages, the "Thesaurus Medicaminum" of the Zurich pharmacist Hans Minner (1479) refers to wound healing with pine bark. The Maritime Indians of North America in 1535 also reportedly used pine bark as food, for drinks, and for medicine. Early empirical usage of pine bark is now explained from the observation that the natural pine bark antioxidants that protect against free oxygen radicals also possess anti-inflammatory activity.

The pharmacology of the 19th century, with its extensive use of the anti-inflammatory willow bark, followed empirically the modern concept that free radicals are produced at sites of inflammation. The evolution from willow bark to anti-platelet aspirin is synonymous for the successful development of modern science. The same is true for the history of pine bark uses and the subsequent development and usage of contemporary Pycnogenol.

ANCIENT PINE NOMENCLATURE

Although the differentiation of pine species started with Theophrastus (370–285 B.C.), the founder of the scientific study of botany, the irregular usage of definitions and nomenclature has to be accepted as a reality for some pine trees (36). To the middle of the 20th century, the most irritating confusion in the literature concerned *Pinus pinaster* and *Pinus silvestris* (3).

The inconsistent terminology use for pines is traceable to the botany of the Talmud produced in Babylonia and Palestine where *P. silvestris,* this typical pine species of *Northern* Eurasia, is described as the tree used to build the temple and the royal palace of Jerusalem (1K. 6,9). At the same time, the Talmud differentiated between *P. larix* and *P. cedrus,* which indicates that the designated pine tree was *P. pinaster,* available and well known along the Mediterranean Sea, whereas *P. silvestris* as defined today, grows in Scandinavia, Germany, and Russia (16).

In Greek mythology the nomenclature for coniferae was also confused. A prominent example is in the verse "Cranes of Ibycus," written by the Greek poet c540 B.C. If plant geography is considered, the so-called "Poseidon's spruce copse" could not have existed in Greece. Research into the plant kingdom of Greek mythology revealed that this spruce copse would rather be a culture of maritime pines along the coast. Even the holy tree of Poseidon is supposed to be the maritime pine, the adequate pine species for Poseidon, the ancient Greek god of the sea (33, 36).

"Poseidon's spruce copse" was further identified to surround the sanctuary of "Poseidon's Isthmus," which is the "Isthmus of Corinth," where maritime pines were the dominant species. Corinth, one of the wealthiest and most powerful of the ancient Greek cities on the "Isthmus of Corinth," was the site of the "Isthmian Games," one of the great national festivals held every 2 years. The victor of the "Isthmian Games" was celebrated with a victory garland made up of maritime pine branches. According to Plutarch, antique representations often saw Poseidon, god of the sea, with a maritime pine garland. The old Corinthian silver coins showed the cone of the maritime pine as a money sign (33, 36).

The identification of the maritime pine is further supported by Pliny the elder (A.D. 23–79), who was by far the most important writer in the so-called "Dark Ages." He deeply influenced the early encyclopedists with his "Natural History" in which he summarized popular medicine. The Romans at that time, among them Pliny and Virgil, differentiated clearly between *"Pinus"* for *pine,* but used *"Pinaster"* for *P. silvestris.* The "Georgics" of Virgil, written in 31 B.C., is a manual of agriculture in verse that refers to *Pinaster* (1).

The same terminology confusion continues throughout the Middle Ages. In Novi Herbarii by Brunfels (1536), *Pinaster* was considered a genus, whereas others at the same time still considered *Pinaster* a variety of *P. silvestris* (37, 41). This disagreement of terminology and species designation existed at least until the appearance of the reference book "Die Nutzhölzer der Welt" (= the economic woods of the world) in 1942 (3). In 1967, the terminology variation was essentially reduced to scientific numbering known pine species (32). Since then, science has shown that *P. pinaster* exists in five different climate races due to geographic separation and special vegetation conditions (42).

CLIMATE UNIQUENESS OF THE FRENCH MARITIME PINE

P. pinaster grows naturally on sandy coastal soils, but on the drift sands of the Biscay Bay area, these forests were man-made to recover the desert of Les Landes. Already in 1751 this pine forest was described in the literature under the name "landes de Bordeaux" (13). The botanical and environmental reasons for this maritime pine forest were summarized as early as 1910 in the 11th edition of the "Encyclopaedia Britannica": "*P. pinaster* is an important species from its vigorous growth in the sand drifts of the coast where it has been grown more extensively and successfully than any other tree. When once established, the tree is rarely overthrown even on the loosest sand" (54).

Although the phenomenon of races, including climate races, is well known in the plant kingdom, the proof of different climate races of *P. pinaster* belongs to recent research. The Atlantic pine species "Pin des Landes," or "French maritime pine from Landes de Gascogne," is one of the prominent botanical examples for the existence of such a climate race. Historically, "Pin des Landes" is a well-described maritime pine with detailed scientific reports available since 1872 (8, 9, 14, 53). Its major geographical and environmental determinations, as published in 1874 by A. Boireau (8), are the Atlantic ocean, lowland sand dunes with strong hydromorphic drainage conditions, and over 1,000,000 hectares of maritime pine forest. Characteristics of the growth conditions have remained stable for the last 200 years, partly due to state-controlled afforestation. The harsh climate of the Atlantic along the southwest coast of France at the Bay of Biscay attributed to the synonym "Atlantic pine" and finally determined the structure of the genetic and physiological features.

Since the mid-1960s, modern botanical and genetic research was able to differentiate between the genetic origin of (a) French "Landais" maritime pine, (b) Iberian pine, and (c) Moroccan pine. The most vigorous race was found to be the Landais, which is best adapted to cooler, frost-prone seasons with hot summers (25, 28, 30, 39, 48). A trial comparing 10 maritime pine provenances confirmed that the ranking for vigor remained stable throughout the test period (12).

Provenance comparisons undertaken in Australia were also able to demonstrate the inheritance and dominance of such racial characteristics (22). Moreover, the occurrence of different climatic races was proved by the failure of the introduction of pine seeds from the Iberian peninsula into the pine cultures of "Landes de Gascogne," as well as by failed attempts to cultivate the French maritime pine in Korea (21, 28).

The genetic variability of maritime pines from seven geographical origins was proved by the analysis of the genetic variability of proteins. The clustering was in agreement with the Atlantic, Mediterranean, and North African structuration of maritime pines as established from terpene data (2, 34).

For the production of origin-guaranteed Pycnogenol, the sophisticated botanical research findings are an important quality element. The French maritime pine of Landes de Gascogne is the sole source for this proprietary flavonoid extract. Figure 1, photographic views of the forest taken in 1997, shows the typical culture scenery.

TRADITIONAL PINE BARK USES

In ancient cultures, e.g., in Egypt and India, medicine derived its origin from religion. Worship of Asclepius as the god of healing relied on priests, not physicians. The Homeric Greeks more often prescribed dietetic reme-

Figure 1 Photographic views of the French maritime pine forest Landes de Gascogne.

dies than drugs. The Hippocratic school attached great importance to diet, especially in chronic cases (24).

During the years of the government of the Roman emperor Tiberius, Aulus Cornelius Celsus described the medicine of his time in eight books. Then, medicine comprised three parts: (a) regulation of living conditions (dietetics), (b) administration of drugs (pharmacotherapy) and (c) use of hands (surgery) (24). More than 2000 years later, this concept of medicine is still valid. Especially, dietary supplements (DSHEA law, US) or food supplements (EU) are used to maintain health because of their diverse actions on the general state of health. According to modern science, the role of selected diets in disease prevention is to protect against free radical attack.

Many of the dietary and medicinal plants in ancient times were used as tea, often made up from bark due to the ease of plant handling. More than 100 years ago, the green willow bark tea was used to relieve headache, then an established medicine.

Bark-derived traditional medicines strongly influence present-day health care all over the world. Aspirin is the first truly synthetic drug developed to achieve a better efficacy than its natural parent compound salicylic acid from the willow bark. Salicylic acid and the polyphenolic catechin are natural willow bark constituents with free radical-scavenging activity (19). According to modern science, the role of selected diets in disease prevention is to protect against free radical attack.

Due to the ancient custom of "recording cases" for gain of experience, the history of pine bark uses for (a) food and emergency food procurement (17, 29, 56), (b) beverage and tea drinks (10, 18, 43, 44, 49, 51, 52), and (c) treatment of symptoms and diseases (10, 18, 23, 26, 31, 40, 43, 44, 49, 51, 52, 55) can be traced back to several ethnic cultures and geographic parts of the world. Especially in the circumpolar area, e.g., Russia and Scandinavia, pine bark was used predominantly as emergency bread under famine. Today, pine bark bread is still for sale in Finland, as many people remember it from the famine years of World War II.

Some of the ancient medicinal uses of pine bark are cited below.

America

- The inner bark of the white pine tree was pounded as a poultice for any kind of inflamed wound, sore, or ulcer (10, 18).
- White pine bark made a basis for cough syrup (55).

Europe

- Putting pine bark into warm vinegar and sucking on the bark was used to relieve tooth ache pain (31, 40).

- French maritime pine bark was used against bleeding, scurvy, kidney, and bladder symptoms (15).
- Pine bark use suggestions by Hieronymus Bock: for cough and weight loss from disease, together with wine as a diuretic and for dysuria, with liquid for diarrhea, for liver disease, and for skin disease and skin ulcers. (23)

Pharmacology of the 19th century explained these medical effects, although on the acknowledged basis of the 2000-year-old classification of the Roman school (regulation of living conditions with diet) (24, 27).

Willow bark, as well as other barks with catechin polyphenolics, were classified pharmacologically as "astringents" containing tannins. The tannin-like polyphenolic molecules explain pine bark uses for diarrhea and in a broad range of skin diseases and skin ulcers. They were grouped in the class of "tonics" and were defined to increase the energy of organic fibres (27).

Indications for tonic astringents were separated into general nutrition and subclassified diseases. Dilation of vessels was related to scurvy, bleeding, and symptoms of venous insufficiency (27).

With respect to modern flavonoid research and findings, the early pharmacological explanation of tonic astringents is most interesting and as a general concept (polyphenol-protein complexation) still valid today. The mechanism of action of tonic astringents was described as follows: "The mechanical firmness and elasticity of organic fibres is called tonus. Tonic astringents will assure that the tissue of organs will get firm and indirectly better nourished. This action is accompanied by an increase of blood volume through a strong pulse – pulse frequency will rather slow down. The red colour of the periphery will subsequently intensify" (27).

Early pharmacology of the willow bark, as well as of other bark extracts, does not directly mention the anti-inflammatory effect. However, one can assume that the just-cited general mechanisms of action of "tonic astringents" not only include the established effects on microcirculation, but indirectly also cover anti-inflammatory activity.

To summarize, the early description of the pharmacology of "tonic astringents" offers a valid mechanistic link between the polyphenolic bark constituents in the willow tree and the French maritime pine, as well as between the sophisticated research profiles of aspirin and Pycnogenol.

PYCNOGENOL

Pycnogenol is the name for the proprietary extract from the bark of the French maritime pine from Landes de Gascogne with the typical reddish color and bark texture shown in Fig. 2.

Figure 2   Photographic views of the bark of French maritime pine of Landes de Gascogne.

The extraction process uses ethanol and water in a patented (EU 0313441) process-controlled inox chain. The extracted natural bark polyphenolics are procyanidins (e.g., procyanidin B1, B3, B6, B7), the monomers catechin and taxifolin, phenolic acids, and vanillin. The bark origin and its standardized processing and composition define the natural extract as water soluble, astringent, and acidic.

The extract composition reflects the natural equilibrium of the water-soluble flavonoids contained in French maritime pine bark. The monomers and phenolic acids of this bark form particular procyanidin condensation products. Pycnogenol contains procyanidins with chain lengths between 2 (dimer) and 12 monomeric units.

The proprietary research and development of Pycnogenol through Horphag Research started in the mid-1960s and was strongly influenced by the early mechanistic concept of "tonic astringents" as discussed in the previous section. During the first 10–20 years, clinical research was undertaken in Europe to demonstrate the action as a capillary protectant (increase of capillary resistance, decrease of capillary permeability, and fragility). These circulation benefits with the strengthening of blood vessels and better general nourishment of the body corresponded perfectly to the early pharmacological understanding and was rather easy to measure clinically.

The maintenance of vascular integrity and the prevention of microvascular complications are age-related dietetic or therapeutic necessities. Only recently have the mechanisms of cell damage by free radicals been associated with life-style, aging, and some age-related diseases. Because of the polyphenolic structure of the water soluble-flavonoids of Pycnogenol, the antioxidant effects have been intensively researched in the past few years.

The antioxidant action of Pycnogenol includes free radical scavenging (oxygen-free radicals, nitric oxide free radicals) and absorption of ultraviolet light (4, 20, 35, 38, 45–47, 50). The anti-inflammatory activity of Pycnogenol was also proved in different experimental settings (4, 5, 7).

Activity was researched for single extract constituents as well as for the original proprietary extract (e.g., 6, 7). The proprietary combination of antioxidant molecules (water-soluble flavonoids) that form Pycnogenol provides synergistic protection because the different flavonoid molecules operate by different mechanisms (e.g., 4–7, 11, 35, 38, 45–47, 50).

The recent recognition and proof of major Pycnogenol mechanisms of action are all attributed to the extract composition, hence the water-soluble flavonoids. The synergistic and different mechanisms taken together explain why Pycnogenol is a capillary protectant with all the benefits of circulation enhancement. The early pharmacological wording and meaning for the actions of bark-derived "tonic astringents" (27) were confirmed through international, sophisticated research of the past few years, an important phenomenon per se, considering the facts of traditional pine bark uses for centuries.

The development from willow bark to aspirin is similar to the efforts documented for pine bark uses leading to Pycnogenol. The antiplatelet activity of Pycnogenol was established in the latest preclinical and clinical research initiated and headed by Dr. P. Rohdewald, University of Münster, Germany, and was then confirmed by Dr. R. R. Watson, University of Arizona, Tucson, Arizona. In a two-armed clinical study, Pycnogenol was administered versus aspirin to smokers. Smoking caused stress to the healthy volunteers, which was detectable in relevant biochemical parameters.

In summary, results of the clinical studies showed that 100 mg oral Pycnogenol reduced blood clotting *without* prolonging bleeding time, whereas the same blood clotting reduction was obtained with 500 mg oral aspirin, but with the unwanted effect of the prolongation of the bleeding time.

SUMMARY

This chapter examined the relationship between the ancient knowledge of pine bark and the latest research trends and development. It is astonishing to learn from the comparison of history to modern science that astringent tonics had a justified role in antiquity which continues to be valid for the future.

REFERENCES

1. Abbe, E. (1965). "The Plants of Virgil's Georgics: Commentary and Woodcuts." pp. 7–11. Cornell Univ. Press, Ithaca, NY.
2. Bahrmann, N., Zivy, M., Damerval, C., and Baradat, P. (1994). Organisation of the variability of abundant proteins in seven geographical origins of maritime pine (*Pinus pinaster* Ait.). *Theor. Appl. Gene.* **88**(3–4), 407–411.
3. Bärner, J., and Müller, J. F. (1942). "Die Nutzhölzer der Welt," Vol. 1, pp. 83. J. Neumann, Neudamm.
4. Blazso, G., Gabor, M., Sibbel, R., and Rohdewald, P. (1994). Antiinflammatory and superoxide radical scavenging activities of a procyanidins containing extract from the bark of *Pinus pinaster* Sol. and its fractions. *Pharm. Pharmacol. Lett.* **3**, 217–220.
5. Blazso, G., Rohdewald, P., Sibbel, R., and Gabor, M. (1995). Antiinflammatory activities of procyanidin-containing extracts from *Pinus pinaster* Sol. *In* "Flavonoids and Bioflavonoids" (S. Antus, M. Gabor, and K. Vetschera, eds.). Proceedings of the International Bioflavonoid Symposium, Vienna, Austria.
6. Blazso, G., Gaspar, R., Gabor, M., Rüve, H.-J., and Rohdewald, P. (1996). ACE inhibition and hypotensive effect of a procyanidins containing extract from the bark of *Pinus pinaster* Sol. *Pharm. Pharmacol. Lett.* **6**, 8–11.
7. Blazso, G., Gabor, M., and Rohdewald, P. (1997). Anti-inflammatory activities of procyanidin-containing extracts from *Pinus pinaster* Ait. after oral and cutaneous application. *Pharmazie* **52**, 380–382.
8. Boireau, A. (1874). Essai sur le pin des Landes es ses produits. 4°. Ecole de pharmacie. A. Derenne, Paris.
9. Buffault, P. (1937). Sur la végétation du pin maritime dans les dunes de Gascogne. *Revue Eaux Forêts*, 401–407.
10. Chandler, F. R., Freeman, L. and Hooper, S. N. (1979). Herbal remedies of the Maritime Indians. *J. Ethnopharmacol.* **1**, 49–68.
11. Cheshier, J. E., Ardestani-Kaboudanian, S., Liang, B., Araghiniknam, M., Chung, S., Lane, L., Castro, A., and Watson, R. R. (1996). Immunomodulation by Pycnogenol in retrovirus-infected or ethanol-fed mice. *Life Sci.* **58**, 87–96.
12. Danion, F. (1994). Stand features and height growth in a 36-year-old maritime pine (*Pinus pinaster* Ait.) Provenance test. *Silvae Genet.* **43**(1), 52–62.
13. Diderot, D. (1751). "Encyclopédie ou Dictionnaire raisonné des sciences, des arts et des métiers, par une société de gens de lettres," Vol. XII, pp. 629–636. Samuel Faulche, Neufchastel.
14. Dive, P. (1872). Essai sur une arbre du genre Pinus qui croit spontanément dans les landes de Gascogne. Ecole de pharmacie, Paris.

15. Dragendorff, G. (1898). "Die Heilpflanzen der verschiedenen Völker und Zeiten. Ihre Anwendung, wesentlichen Bestandteile und Geschichte," p. 66. Ferd. Enke, Stuttgart.

16. Duschak, M. (1871). "Zur Botanik des Talmud," pp. 55–59. Published by the rabbi of Gaya, Pest.

17. Eidlitz, K. (1969). "Food and Emergency Food in the Circumpolar Area." Inaugural Ph.D. Dissertation, University of Uppsala.

18. Fielder, M. (1975). "Plant Medicine and Folklore," pp. 55–195. Winchester Press, New York.

19. Forlines, D. R., Tavenner, T., Malan, J. C. S., and Karchesy, J. J. (1992). Plants of the olympic coastal forests: Ancient knowledge of materials and medicines and future heritage. *In* "Plant Polyphenols" (R. W. Hemingway and P. E. Laks, eds.), pp. 767–782. Plenum Press, New York.

20. Guochang, T. (1993). "Ultraviolet Radiation-Induced Oxidative Stress in Cultured Human Skin Fibroblasts and Antioxidant Protection. Inaugural Ph.D. Dissertation, University of Jyväskylä, Finland.

21. Han, Y. C., Ryu, K. O., Lee, K. Y., Chang, S. O., and Youn, K. S. (1987). Adaptation trial of maritime pine-*Pinus pinaster* Ait. in Korea. *Res. Rep. Instit. For. Genet.* **23**, 107–111.

22. Hopkins, E. R., and Butcher, T. B. (1993). Provenance comparisons of *Pinus pinaster* Ait. in western Australia. *CALM Sci.* **1**(1), 55–105.

23. Hoppe, B. (1969). Das Kräuterbuch des Hieronymus Bock. Wissenschafts-historische Untersuchung, pp. 377–378. Anton Hiersemann, Stuttgart.

24. Kollesch, J., and Nickel, D. (1994). Antike Heilkunst. Ausgewählte Texte aus den medizinischen Schriften der Griechen und Römer. Reclam, Stuttgart.

25. Lemoine, B., Gelpe, J., Ranger, J., and Nys, C. (1986). Biomass and growth of maritime pine *Pinus pinaster*: A study of variability in a 16-year-old stand. *Ann. Sci. For. (Paris)* **43**(1), 67–84.

26. Lescarbot, M. (1911). History of New France. *In* "The Publications of the Champlain Society," Vol. 2, pp. 149–154. The Champlain Society, Toronto.

27. Lessing, M. B. (1866). Kurzer Abriss der Materia Medica. Ein Repetitorium. 2nd ed., pp. 1–50, 95–103, 125–136. Arthur Felix, Leizig.

28. Le Tacon, F., Bonneau, M., Gelpe, J., Boisseau, T., and Baradat, P. (1994). Dieback of maritime pine in the Landes of Gascogne following the introduction of seed from the Iberian Peninsula and the excessively cold weather in 1962–1963 and 1985. *Rev. For. Francaise (Nancy)* **46**(5), 474–484.

29. Lindgren, J. (1918). Läkemedelsnamn. Ordförklaring och Historik, pp. 286–288. Berlingska Boktryckeriet, Lund.

30. Loustau, D., Crepeau, S., Guye, M. G., Sartore, M., and Saur, E. (1995). Growth and water relations of three geographically separate origins of maritime pine (*Pinus pinaster*) under saline conditions. *Tree Physiol.* **15**(9), 569–576.

31. Minner, H. (1479). "Thesaurus Medicaminum." Codex 81 University Library, Marburg.

32. Mirov, N. T. (1967). "The Genus Pinus." Ronald Press Company, New York.

33. Murr, J. (1890). Die Pflanzenwelt in der Griechischen Mythologie, pp. 110–122. Reprint by University Publishers, Innsbruck.

34. Nguyen, A., and Lamant, A. (1988). Pinitol and myo inositol accumulation in water-stressed seedlings of maritime pine. *Phytochem. (Oxford)* **27**(11), 3423–3428.

35. Noda, Y., Anzai, K., Mori, A., Kohno, M., Shinmei, M., and Packer, L. (1997). Hydroxyl and superoxide anion radical scavenging activities of natural source antioxidants using the computerized JES-FR30 ESR spectrometer system. *Biochem. Mol. Biol. Int.* **42**, 35–44.

36. Peters, H. (1922). Aus der Geschichte der Pflanzenwelt in Wort und Bild. (Society of Pharmacy History ed.), pp. 161–168. Arthur Nemayer Verlag, Mittenwald.
37. Pritzel, G., and Jessen, C. (1882). "Die deutschen Volksnamen der Pflanzen," pp. 277–281. Philipp Cohen, Hannover.
38. Rong, Y., Li, L., Shah, V., and Lau, B. (1995). Pycnogenol protects vascular endothelial cells from t-butyl hydroperoxide induced oxidant injury. *Biotechnol. Ther.* **5**, 117–126.
39. Saur, E., Rotival, N., Lambrot, C., and Trichet P. (1993). Maritime pine die-back on the West Coast of France: Growth response to sodium chloride of 3 geographic races in various edaphic conditions. *Ann. Sci. Forest. (Paris)* **50**(4), 389–399.
40. Schmitz, U. (1974). Edition and commentary on the "Thesaurus medicaminum" of Minner. Inaugural Ph.D. Dissertation, Philipps University, Marburg.
41. Schneider, W. (1974). Pflanzliche Drogen. Sachwörterbuch zur Geschichte der pharmazeutischen Botanik. *In* "Lexikon zur Arzneimittelgeschichte," pp. 69–79. Govi-Verlag, Frankfurt a.M.
42. Schütt, P., Schuck, H. J., and Stimm, B. (1992). "Lexikon der Forstbotanik. Morphologie, Pathologie, Ökologie und Systematik wichtiger Baum- und Straucharten," pp. 372–373. Ecomed, Landsberg.
43. Speck, F. G., and Dexter, R. W. (1951). Utilization of animals and plants by the Micmac Indians of New Brunswick. *J. Wash. Acad. Sci.* **41**, 250–259.
44. Speck, F. G., and Dexter, R. W. (1951). Utilization of animals and plants by the Malecite Indians of New Brunswick. *J. Wash. Acad. Sci.* **42**, 1–7.
45. Ueda, T. Ueda T., and Armstrong D. (1996). Preventive effect of natural and synthetic antioxidants on lipid peroxidation in the mammalian eye. *Ophthalmic Res.* **28**, 184–192.
46. Virgili, F., Kobuchi, H., and Packer, L. (1997). Nitrogen monoxide (NO) metabolism: Antioxidant properties and modulation of inducible NO synthase activity in activated macrophages by procyanidins extracted from *Pinus maritima* (Pycnogenol). Submitted for publication.
47. Virgili, F., Kobuchi, H., and Packer, L. (1997). Procyanidins extracted from *Pinus maritima* (Pycnogenol): Scavengers of free radical species and modulators of nitrogen monoxide metabolism in activated murine RAW 264.7 macrophages. Submitted for publication.
48. Von Euler F., Baradat, P., and Lemoine, B. (1992). Effects of plantation density and spacing on competitive interactions among half-sib families of maritime pine. *Can. J. For. Res.* **22**(4), 482–489.
49. Wallis, W. D., and Wallis, R. S. (1955). "The Micmac Indians of Eastern Canada." University of Minnesota Press, Minneapolis.
50. Wei, Z. H., Peng, Q. L., and Lau, B. H. S. (1997). Pycnogenol enhances endothelial cell antioxidant defenses. *Redox Rep.* **3**(4), 219–224.
51. Youngken, H. W. (1924). The drugs of the North American Indian. *Am. J. Pharm.* **96**, 485–502.
52. Youngken, H. W. (1925). The drugs of the North American Indian II. *Am. J. Pharm.* **97**, 257–271.
53. La Grande Encyclopédie (1901). Volume XXVI, pp. 939–941. Paris
54. The Encyclopaedia Britannica (1910/1911). Eleventh Edition, Vol. XXI, pp. 621–625.
55. The Dispensatory of the United States of America (1955). 25th Edition. Lippincott, Philadelphia.
56. Scientific Consultative Committee of National Defence. (1979). Report Series A. 5/A/79 On the possibilities of relying on wild plants and animals as a source of nutrition. Part IV OSA. Helsinki.

21 Procyanidins from *Pinus marittima* Bark: Antioxidant Activity, Effects on the Immune System, and Modulation of Nitrogen Monoxide Metabolism

F. Virgili, H. Kobuchi,† Y. Noda,† E. Cossins,†
and L. Packer†*

*National Institute of Nutrition
Rome, Italy

† Department of Molecular and Cell Biology
University of California
Berkeley, California 94720

INTRODUCTION

Polyphenols are attracting increasing interest as nutritional compounds recognized to positively affect human health. Several different epidemiological reports have indicated that consumption of foods rich in polyphenols is associated with a lower incidence of degenerative diseases (Editorial, 1994; Steinmetz *et al.*, 1996), and experimental data are accumulating regarding phenolic compounds as natural phytochemical antioxidants important for human health (Halliwell, 1996b). Flavonoids constitute a large class of polyphenolic compounds, ubiquitous in plants, containing a number of phenolic hydroxyl groups attached to ring structures conferring a strong antioxidant activity. Flavonoids often occur in nature as glycosidic derivatives, which have been reported to be more readily absorbed than the simple form (Hollman *et al.*, 1995). Flavonoids are normally ingested by consumption of fruits and vegetables but some flavonoids, such as the flavanols, catechin or epicatechin and the flavonol quercetin, are constituents of plant-derived beverages such as tea or red wine. Estimates of the total daily intake range from about 20 mg up to 1 g depending on the characteristics of the diet (Hertog *et al.*, 1993). Complex mixtures of flavonoids have frequently been used in the past as the only possible remedy for various diseases, and currently are widely used, even in affluent countries, both as an alternative and as a complement to "official" medical therapy (i.e., phytomedicine).

Study of the biological effects of phytochemicals is important for at least two different reasons: first to provide a biochemical and molecular background to the safe and proper use of these compounds. Second, results obtained with these groups of compounds could be applied to a broader family of compounds, thus providing a basis for the better understanding of the biological activity of polyphenolic compounds as therapeutic agents in various human disorders.

PYCNOGENOL

Pycnogenol is a proprietary extract of the bark of the French pine tree *Pinus marittima* obtained by water extraction followed by ethyl acetate in order to eliminate nonwater-soluble substances. Even though its chemical composition is still not totally known, the main constituents of Pycnogenol are known to be phenolic compounds, broadly divided into monomers (catechin, epicatechin, and taxifolin) and condensed flavonoids classified as procyanidins/proanthocyanidins. These condensed polyphenols are mainly constituted by "bricks" of the flavan-3-ols catechin and epicatechin, linked together, from dimers up to heptamers (see Fig. 1). Pycnogenol also contains phenolcarbonic acids (such as caffeic, ferulic, and *p*-hydroxybenzoic acids) as minor constituents and glycosylation products, i.e., glucopyranosid derivatives of either flavanols or phenolcarbonic acids as lesser constituents. Chapter 20 of this book has presented historical landmarks on the utilization of pine bark extract through time. In a more recent window of time, the biological activity of Pycnogenol has been studied using rigorous scientific approaches, and more data are now available. In this chapter, information on the effect of Pycnogenol both as an antioxidant and on immune system responses that have appeared in the peer-reviewed literature in recent years will be presented, together with some results obtained in the authors' laboratory on the effect of Pycnogenol in modulating nitrogen monoxide (nitric oxide, NO) metabolism in macrophages.

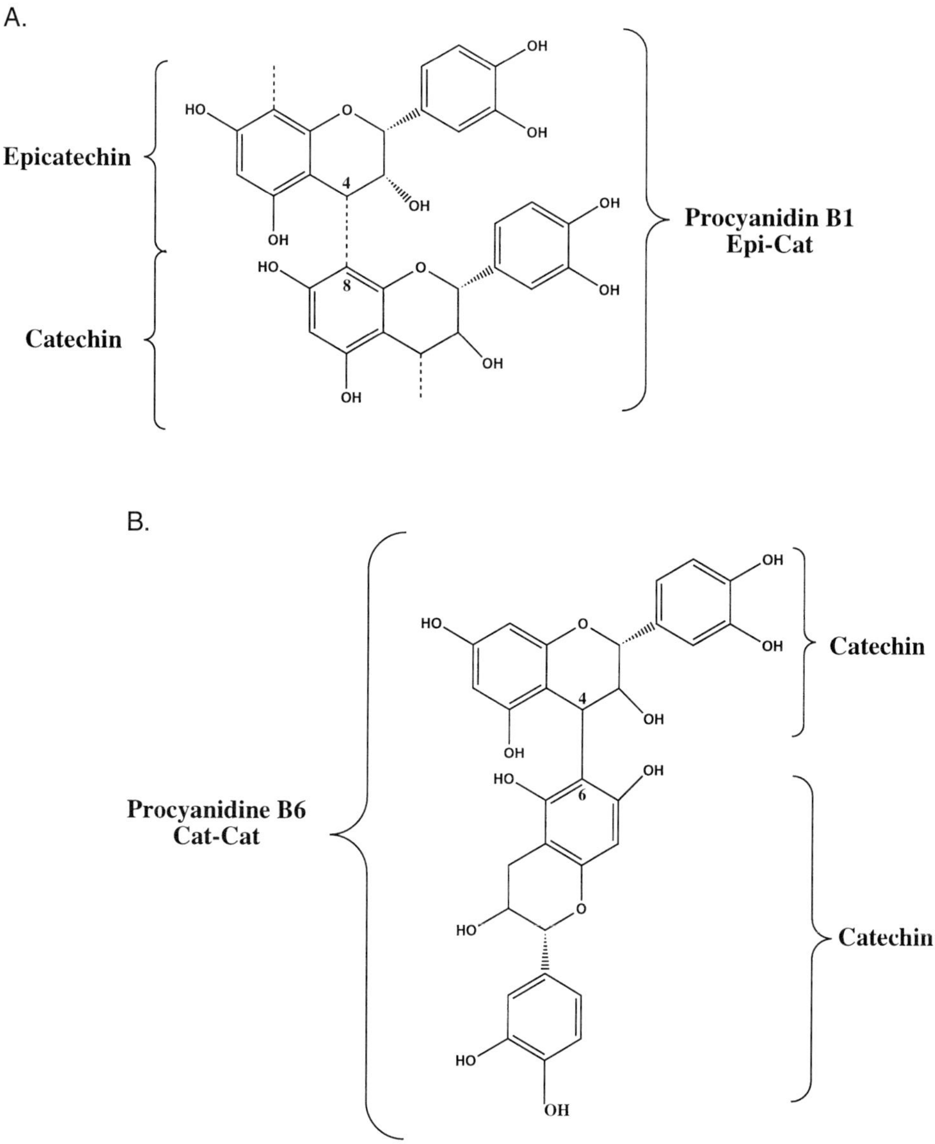

A.

Epicatechin

Catechin

Procyanidin B1
Épi-Cat

B.

Procyanidine B6
Cat-Cat

Catechin

Catechin

Figure 1 Chemical structures of procyanidine-like molecules: Procyanidine (A) B1 and (B) B6 are shown as examples.

PYCNOGENOL AS AN ANTIOXIDANT

Polyphenols, particularly flavonoids, are effective antioxidant molecules. They act as free radical scavengers because of their ability to exchange an hydrogen electron, with appropriate redox potential with respect to the radical species to be scavenged, the resulting radical formed being stabilized by delocalization (Rice-Evans *et al.*, 1997). Due to the multiple phenolic groups in their basic molecular structure, flavonoids and flavonoid-containing plant extracts are potentially able to quench free radicals by forming more stable oxidized products. In addition to this "quenching" effect, some flavonoids may act as preventive antioxidants by chelating transition metals, thus precluding the formation of hypervalent metal forms, which initiate the peroxidative process (Ursini *et al.*, 1989). Research on antioxidant defenses has focused on the potential benefits of both purified phytochemicals and plant extracts such as the pine bark extract Pycnogenol. A number of different studies have addressed the antioxidant capacity of Pycnogenol using either simplified assay systems *in vitro* or by testing its activity in cultured cell models and in perfused organs. Some of the most important free radicals in the biological environment, namely reactive oxygen species lipid peroxides, $O_2^{\bullet-}$ and HO•, and the nitrogen reactive species NO•, have been investigated *in vitro*.

Studies *in Vitro*

Pycnogenol has been reported to act as a strong antioxidant in 1,1 diphenyl-2-pycrylhydrazyl (DPPH) radical quenching (van Jaarsveld *et al.*, 1996). In this *in vitro* model, Pycnogenol was able to quench the stable radical DPPH with an activity comparable to that of both purified catechin and α-tocopherol. Other investigators have reported that Pycnogenol is an efficient scavenger of both the superoxide anion radical (Blazso *et al.*, 1994; Elstner *et al.*, 1990; Virgili *et al.*, 1998a) and the hydroxyl radical (Virgili *et al.*, 1998a). The superoxide anion scavenging activity was studied by Virgili *et al.*, (1998a) utilizing ESR techniques, and Pycnogenol activity was reported as superoxide dismutase (SOD) equivalents in the order of hundreds of units of SOD per milligram of Pycnogenol. In the same investigation, the specific scavenging activity of Pycnogenol toward the hydroxyl radical, generated by the iron/ascorbate redox system, was also studied by means of ESR spectroscopy and was reported to be in the order of micromoles EPC-K1 (a water-soluble synthetic reference antioxidant) per milligram of Pycnogenol in solution. Both OH• and $O_2^{\bullet-}$ scavenging activity were maintained after treatment with ascorbate oxidase, indicating that ascorbate, possibly present in the mixture, was not responsible for the antioxidant ac-

Table I   Synopsis of Results on Pycnogenol as an Antioxidant

Experimental system	Effects of pycnogenol	Reference
Superoxide-generating system or iron/ascorbate	Dose-dependent scavenging effect by ESR; best antioxidant activity among other plant extracts (such as *Ginkgo biloba* and others)	Virgili *et al.* (1998) Noda *et al.* (1997)
Sodium nitroprusside decomposition	Dose-dependent decrease of nitrite accumulation	Virgili *et al.* (1998)
Superoxide-generating system	Different fractions from Sephadex chromatography were tested; all showed a strong antioxidant activity, with the one corresponding to oligomers being the most active	Blazso *et al.* (1994)
Ascorbyl radical generated by ascorbate oxidase	Extension of ascorbyl radical lifetime by ESR	Cossins *et al.* (1998)
Copper-induced LDL oxidation	Regeneration of tocopheryl radical	Cossins *et al.* (unpublished observations)
Iron-induced lipid peroxidation of retina-purified rod outer segments and pigment epithelium	Protective activity and enhances α-tocopherol protection	Ueda *et al.* (1996)
Xanthine–xanthine oxidase, sugar autoxidation, or light-dependent oxidation of Rose Bengal	Dose-dependent inhibition in all systems	Elstner *et al.* (1990)
Structural damage of DNA induced by iron/ascorbate	Decrease of single- and double-strand formation	Nelson *et al.* (1998)
Zymosan-activated J774 macrophages	Dose-dependent inhibition of oxidative burst	Nelson *et al.* (1998)
Cu-induced oxidation of LDL	Dose-dependent decrease of TBA-RS formation	Nelson *et al.* (1998)
Cultured human endothelial cells (ECV line)	Increase of α-tocopherol levels either in basal condition or after oxidative stress	Virgili *et al.* (1998b)
Normal endothelial cell (PAEC) in steady state or challenged by either $O_2\bullet^-$ or H_2O_2	Increased GSH levels and increased activity of GSH peroxidase, GSH reductase, SOD, and catalase	Wei *et al.* (1997)
Normal endothelial cell (PAEC) challenged by *t*-BOOH	Decrease of TBA-RS and enhancement of cell viability	Rong *et al.* (1994–95)
In vivo administration to rats (up to 50 mg Pycnogenol/day)	Increase of α-tocopherol levels in the heart, decrease of ascorbate levels in the lungs, no effects on other organs; lack of protection in ischemia/reperfusion-induced damage	van Jaarsveld *et al.* (1996)

A.

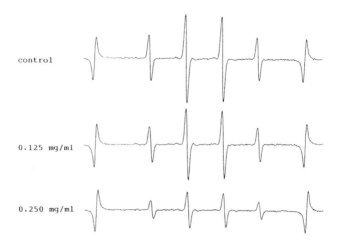

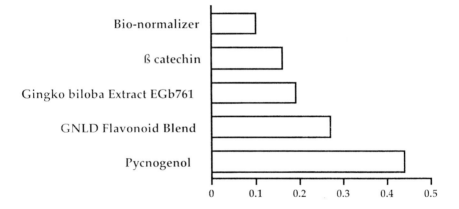

EPC-K equivalents (μmoles/mg)

Figure 2 (A) OH• scavenging activity of Pycnogenol measured by ESR. (Top) Spectra of DMPO-OH spin adducts and the effect of Pycnogenol (one of the experiments giving similar results is shown). (Bottom) Comparison of OH• scavenging activity of different plant extracts. Values are reported as EPC-K equivalents per milligram of extract.

tivity. However, $O_2^{\bullet-}$ scavenging activity was in part affected by ultrafiltration, implying the contribution to the antioxidant activity of high molecular weight compounds present in the mixture (see Fig. 2). When compared to other phytochemicals and plant extracts, Pycnogenol has been demon-

B.

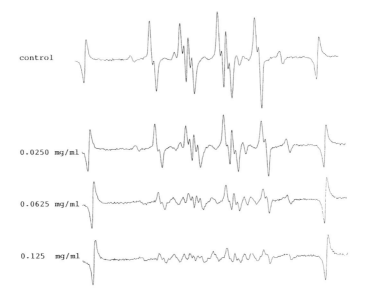

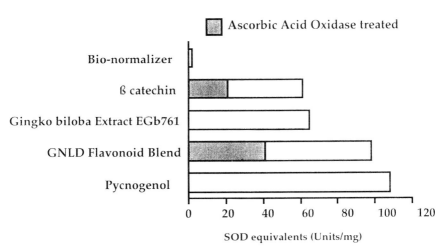

SOD equivalents (Units/mg)

Figure 2 *continued* (B) O_2^- • scavenging activity of Pycnogenol measured by ESR and the effect of ascorbic acid oxidase treatment on scavenging activity. (Top) Spectra of DMPO-OO$^-$•· spin adducts and the effect of Pycnogenol. (Bottom) Comparison of O_2^- • scavenging activity of different plant extracts; shaded bars indicate O_2^- • scavenging activity after ascorbic acid oxidase treatment. Values are reported as SOD equivalents per milligram of extract.

strated to be the highest ranking oxygen free radical scavenger (Noda *et al.,* 1997) and its activity was not affected by ascorbate oxidase treatment, suggesting that the observed antioxidant activity is not to be ascribed to ascorbic acid possibly contained in the bark (See Fig. 2).

Blazso and co-workers (1994) tested *in vitro* the antioxidant activity of three different chromatographic fractions (separated by the dimensional sieve Sephadex LH) of Pycnogenol against superoxide radical anion. The whole extract and each of its fractions were reported to inhibit the superoxide-induced reduction of Nitroblue tetrazolium (NBT) to formazan in a dose-dependent fashion. The most active fraction was the one containing oligomeric procyanidins. Interestingly, a strong correlation between *in vitro* antioxidant activity and *in vivo* anti-inflammatory effect (Pycnogenol and the immune system) was reported (Blazso *et al.,* 1994).

Elstner and Kleber (1990) studied the scavenging activity of Pycnogenol in different models, with the aim of mimicking *in vitro* different physiological conditions that could lead to oxidative stress (namely xanthine/xanthine oxidase, NADH/diaphorase, copper/dihydroxyfumarate, and rose bengal activation by light). They reported that pine bark procyanidins react preferentially with both hydroxyl radical and singlet oxygen and therefore suggested a possible role of Pycnogenol flavonoids in the therapy of pathologies related to the formation of these aggressive oxygen species.

Nelson and collaborators (1998) studied the capacity of Pycnogenol to protect the low-density fraction of human plasma lipoproteins (LDL) from copper-induced oxidation and reported a dose-dependent decrease in lipid peroxide generation, starting with Pycnogenol concentrations as low as 2 μg/ml. In this model, in order to achieve a comparable degree of protection, about 100 μg/ml of tocopherol succinate had to be added. Some criticism may be raised about the validity of such a comparison, as pine bark extract is composed mainly of water-soluble molecules whereas tocopherol succinate is likely to be localized within the lipid core of LDL. The generation of radical species induced by copper would challenge LDL in the water phase, where Pycnogenol may exert its tremendous protective effect (Nelson *et al.,* 1998). In the same paper, the ability to protect DNA from iron/ascorbate-induced damage was also studied. A minimization of single- and double-strand DNA breaks was reported, suggesting a potentially important role of Pycnogenol in the protection against free radical-induced injury to genetic material and possibly to the consequences of gene mutation (Emerit, 1994).

Nitrogen monoxide (nitric oxide) is attracting increasing interest as a free radical molecule that becomes potentially harmful once its concentration overwhelms its neurotransmitter and second messenger functions. NO toxicity will be discussed later, but first a study conducted in the authors' laboratory has demonstrated that Pycnogenol significantly decreased, in a dose-dependent fashion, the accumulation of nitrite after the spontaneous decomposition of sodium nitroprusside, acting as a nitric oxide radical scavenger (Virgili *et al.*, 1998a).

Studies on Cultured Cells

Pycnogenol has also shown a significant antioxidant capacity in cultured endothelial cells and macrophages. In cultured normal endothelial cells (pulmonary artery endothelial cells, PAEC), protection from both lipid peroxidation and cell damage induced by *tert*-butylhydroperoxide (*t*-BOOH) has been reported after preincubation with 20 to 80 μg/ml Pycnogenol (Rong *et al.*, 1994–1995). In the same system, Pycnogenol was reported to induce a dose-dependent decrease in the steady-state production of both superoxide and hydrogen peroxide and to decrease the rate of hydrogen peroxide accumulation following treatment with a superoxide-generating system (Wei *et al.*, 1997). In this study, the effect on superoxide and hydrogen peroxide levels was attributed to the effects of Pycnogenol on glutathione (GSH) redox status and on the enhancement of the activity of the enzymatic machinery that regulates GSH redox status. In fact, a significant increase in GSH levels, an increased activity of the GSH redox enzymes (GSH reductase and GSH peroxidases), and, finally, an increase in the enzymatic activity of both SOD and catalase were reported. The effect of Pycnogenol on enzyme activity was proposed by Wei and collaborators (1997) to be mediated by an increase of protein. The possible effect of Pycnogenol on antioxidant enzyme expression deserves further studies, as the induction of the synthesis of the most important antioxidant enzyme machineries would actually amplify the organism antioxidant defenses. Observations obtained on endothelial cells are of interest in that free radical-induced endothelial dysfunction is considered one the major causative factors in the development of atheroma. These results provide information relevant to understanding the potential beneficial effects of Pycnogenol in circulatory diseases (Halliwell, 1996a). From a similar perspective, the observation that preincubation with Pycnogenol significantly decreased the extent of oxidative burst of rat macrophages induced by zymosan treatment in a dose-dependent fashion (Nelson *et al.*, 1998) substantiates a possible therapeutic utilization of Pycnogenol and other related substances in chronic inflammatory and cardiovascular diseases. It is of interest to note that some polyphenolic molecules present in red wine, and considered responsible for the beneficial effects of the so-called "French paradox" (Kanner *et al.*, 1994), also belong to the same family of Pycnogenol flavonoid components.

Protective activity has also been reported for another tissue naturally exposed to photooxidative stress, the eye. Pycnogenol was found to protect both purified rod outer segments and the pigment epithelium of the retina from lipid peroxidation induced by ferric iron and to enhance the protective effect of α-tocopherol (Ueda *et al.*, 1996).

Studies *in Vivo*

Studies *in vivo* are fundamental in assessing the real bioavailability and efficacy of ingested flavonoids as antioxidants. *In vivo*, using rats as animal models, supplementation with Pycnogenol in the diet was found to be associated with a specific and significant increase in α-tocopherol levels in rat hearts. No effects, with the exception of a significant decrease of ascorbic acid level in the lungs, were observed in other organs (van Jaarsveld *et al.*, 1996). In the same study, when isolated hearts were subjected to ischemia/reperfusion injury, neither dietary Pycnogenol nor addition of Pycnogenol to the perfusate significantly decreased tissue damage, assessed both as release of low molecular weight iron and mitochondrial oxidative phosphorylation (van Jaarsveld *et al.*, 1996). However, catechin, which is a major component of procyanidins contained in Pycnogenol, was found to affect all these parameters, significantly protecting the heart from ischemia/reperfusion injury. These data indicate that the antioxidant potential of Pycnogenol is partially available *in vivo*, but also suggest that the pattern of absorption and distribution of Pycnogenol flavonoid components is complex. Pycnogenol components may exert important bioactivity in selected target organs and tissues. A fuller understanding of this aspect of pine bark extract behavior and, in general, of polyphenols deserves further studies, utilizing either other animal models or clinical studies on human subjects.

Another interesting possibility is that Pycnogenol flavonoids, as the result of their intermediate redox potential, may exert their activity within the cellular antioxidant network. Studies (not yet published) conducted on different cell lines (macrophages, endothelium) by the authors' laboratory showed that incubation with Pycnogenol is associated with higher tocopherol levels both in normal conditions and following oxidative stress. These observations could confirm in intact cells other ESR studies that Pycnogenol may prolong the ascorbyl radical lifetime and regenerate the tocopheryl radical in copper-induced LDL oxidation.

Finally, although not related directly to antioxidant activity, but with relevance to cardiovascular pathologies Blazso and co-workers (1996) reported a moderate although significant hypotensive activity following IV injections of Pycnogenol through inhibition of the angiotensin-converting enzyme (ACE). Rohdewald (1998) observed the inhibition of smoking-induced aggregation of platelets, similar to that due to aspirin after the ingestion of 125 mg Pycnogenol.

It may be mentioned further that, despite the relatively few studies that have appeared in peer-reviewed literature, Pycnogenol has attracted the attention of a large number of investigators, and a substantial amount of potentially interesting data has been reported in sources not so readily available. Many of those interesting reports have been reviewed by Rohdewald (1997).

Taken together, data suggest that the pine bark extract Pycnogenol may play an important role in the prevention of vascular disorders associated with oxidative injury.

PYCNOGENOL AND THE IMMUNE SYSTEM

Several studies have addressed the effects of Pycnogenol on the immune system. Retrovirus infection and chronic ethanol consumption are known to induce abnormalities in the function and structure of different cells involved in humoral and cellular immunity. In a study conducted in either LP-BM5 retrovirus-infected or ethanol-fed mice, Pycnogenol has been reported to significantly enhance interleukin-2 production in mitogen-activated isolated splenocytes and to increase natural killer cytotoxicity (Cheshier *et al.*, 1996). In agreement with van Jaarsveld *et al.* (1996), the same study reports no effect on hepatic α-tocopherol levels after Pycnogenol supplementation.

Blazso and co-workers (1994) reported that intraperitonal injections, either of Pycnogenol or its chromatographic fractions, significantly inhibit the formation of croton oil-induced edema in the rat ear. The most active fraction, as for the antioxidant activity, was the one containing oligomeric procyanidins (Blazso *et al.*, 1994), suggesting that anti-inflammatory and antioxidant activities are related.

The same authors have also reported that Pycnogenol supplementation in the diet inhibits the formation of edema induced either by croton oil in the ear or by Compound 48/80 in the rat paw model. The topical application of Pycnogenol, but not dietary supplementation, also significantly protected rat skin from UV-induced irritation (Blazso *et al.*, 1997).

NITRIC OXIDE METABOLISM IN PHYSIOLOGY AND DISEASE

Nitric oxide is synthesized in mammalian cells using L-arginine and oxygen as substrates by a family of enzymes referred to as nitric oxide synthases (NOS) (Nathan, 1992). NO biosynthesis is regulated in excitable tissues by an increase in intracellular calcium, which activates NOS due to the dependence of the constitutive enzyme on calmodulin. In other tissues, e.g., in macrophages, NO synthesis by an inducible form of NOS (iNOS) is mainly regulated at the transcriptional level following different extracellular stimuli, such as bacterial wall components or various cytokines (Nathan *et al.*, 1994). Because NO displays a remarkable number of different roles acting as an intracellular messenger, as a transcellular signal, or as a cytotoxic species, NOS may therefore be implicated in a wide spectrum of different functions. Thus, a correlation exists between the toxic and homeostatic functions of NO and its production in large and or limited amounts, respectively.

Physiological vs Pathological Functions of Nitric Oxide in the Central and Peripheral Nervous System

In neuronal systems, NO has been proposed to be implicated in the central regulation of blood flow, respiratory rate, circadian rhythm, and sleep cycle and in various neuroendocrine responses (Schmidt *et al.*, 1994). NO plays an important role in normal brain functions, such as memory, the learning process, modulation of wakefulness (Yamada *et al.*, 1995), and for the regulation of both noradrenaline and dopamine release and uptake (Pogun *et al.*, 1994). In this context, NO has been proposed to participate in the long-term potentiation (LTP) of synapses by traveling backward across the synapse and enhancing the release of neurotransmitter in the presynaptic neuron (Schmidt and Walter, 1994). In fact, various reports have indicated that NO may have a role in the mechanisms of storage and retrieving information (i.e., the basis of learning and memory processes) in neuronal cells. In the rat, both spatial learning tasks and LTP can be blocked by the administration of NO synthase inhibitors and N-methy-D-aspartate (NMDA) receptor antagonists. The effect of NO on other types of learning has also been examined with conflicting results (Ingram *et al.*, 1996).

However, the derangement of NO metabolism has been suggested to be associated with various brain pathologies, such as Alzheimer's disease, cerebral ischemia, stroke, and other disorders (Gutteridge, 1995). Moreover, NO-generating compounds have been reported to have a biphasic effect on dopamine release (i.e., an initial enhancement followed by inhibition), corroborating the hypothesis of the involvement of NO in dopamine-related disorders. Finally, because of the free radical nature of NO, it is also to be considered that oxidative stress may affect neuronal physiology by altering membrane integrity and in turn the responsiveness of different receptor systems leading to a deficit in memory or in other brain malfunctions.

Physiological Functions of Nitric Oxide in the Circulatory System

Among its various physiological functions, NO constitutively synthesized by endothelial cells stimulates the soluble guanylyl cyclase in smooth muscle cells by binding to iron in the heme-containing active site of the enzyme, therefore acting as a vasorelaxing agent (Christopherson *et al.*, 1997). NO reduces platelet aggregation and adhesion (Radomski *et al.*, 1990) and inhibits smooth muscle cell proliferation (Nakaki *et al.*, 1990). Other antiatherogenic NO activities have also been reported related to the inhibition of leukocyte recruitment and adhesion to the endothelium (Tsao *et al.*, 1994) and monocyte chemotaxis (Bath *et al.*, 1991). Finally, NO and NO donors have been demonstrated to limit the cytokine-induced activation of endothelial cells and therefore negatively affect the expression of the adhesion molecules VCAM-1, ICAM-1, and P-selectin by repressing their gene expression at the level of NF-κB transcription factor binding (De Caterina *et al.*, 1995). All of these observations contribute to compose a picture of NO as a molecule with both strong anti-inflammatory and antiatherogenic properties within the vessel walls, but even in the same environment, i.e., in the circulation, the picture may be different if the concentration of NO arises above sustainable levels and when associated with a simultaneous production of superoxide radical, as described later.

Physiological vs Pathological Functions of Nitric Oxide in the Immune System

In macrophages, the high output metabolic pathway is due to the inducible form of NOS (iNOS), which generates NO at a micromolar concentration to exert a nonspecific immune response. In fact, NO has been demonstrated to be cytotoxic against various foreign or infectious materials, such as bacteria, parasites, helmints, and viruses, and also against tumor cells (Moncada *et al.*, 1991). However, NO has a suppressive effect on T-lymphocyte proliferation, leading to increased sensitivity to certain pathogens during chronic stages of the immune response (Krenger *et al.*, 1996; Sternberg *et al.*, 1996). During inflammation associated with different pathologies, such as arthritis or Crohn's disease, the production of NO increases significantly and may become "autodestructive" as is known to occur in chronic inflammatory diseases. In fact, NO overproduction has been reported in autoimmune disease, transplanted organ rejection, and sepsis (Hooper *et al.*, 1995; Schmidt and Walter, 1994). Moreover, the free radical nature and the high reactivity with $O_2^{\cdot-}$ with the subsequent generation of $ONOO^-$ renders NO a potent prooxidant molecule able to induce oxidative stress potentially harmful toward virtually all cellular targets (Epe *et al.*, 1996; Luperchio *et al.*, 1996). NO and reactive nitrogen species formed during the reaction of NO with superoxide or with oxygen have been reported to modify free and protein-bound amino acid residues, to inhibit enzymatic activities, to induce lipid peroxidation, and to deplete cellular antioxidant levels. All these features may be associated with the development of different pathologies (Halliwell, 1996a; Liu *et al.*, 1995; Rubbo *et al.*, 1996).

In summary, NO production, both by constitutive NOSs or iNOS, may be considered a biological process potentially leading to opposite outcomes, either physiological or pathological, depending on the ability of the system in controlling both the expression of iNOS activity and the nonspecific effects of NO. On this basis, in particular the inducible enzyme, iNOS, has been proposed to play an important role in the pathogenesis associated with states of shock, inflammation, and autoimmune disease; the constitutive form of NOS, referred to as "neuronal NOS" (nNOS), is also probably involved in ischemic brain damage, epilepsy, and nociception. The therapeutic utilization of specific NOS inhibitors or modulators may therefore be useful in different pathological conditions.

MODULATION OF NITRIC OXIDE METABOLISM IN MACROPHAGES BY PYCNOGENOL

It has been demonstrated that the complex mixture of flavonoids extracted from the bark of *P. marittima* significantly affected iNOS expression and activity in macrophages (the murine RAW 264.7 cell line) challenged by lipopolysaccaride (LPS) and interferon (IFN)-γ (Virgili *et al.*, 1998). *In vitro*, it was found that Pycnogenol had a remarkable modulatory effect on iNOS enzyme activity (see Fig. 3), producing a slight stimulatory effect at a low concentration (10 μg/ml) and acting as a powerful inhibitor of iNOS activity at higher, although still physiologically achievable, concentrations (50–100 μg/ml). This biphasic effect on iNOS activity was confirmed in cultured cells after preincubation with Pycnogenol and subsequent activation with LPS and IFN-γ, where a similar biphasic modulation of the generation of NO_2^- and NO_3^- has also been observed (see Fig. 4).

In addition to the effect on iNOS enzyme activity, the effect of treatment with Pycnogenol on the expression of the iNOS gene at the transcriptional level was investigated. Antioxidants have been reported to modulate gene expression by directly affecting cellular redox status or transcription factors (such as NF-κB and AP-1) binding to DNA and transactivation (Mizuno *et al.*, 1996; Suzuki *et al.*, 1992, 1993). No effects of Pycnogenol pretreatment were observed on the activation of either NF-κB or IRF-1, the transcription factors involved in LPS and IFN-γ induced expression

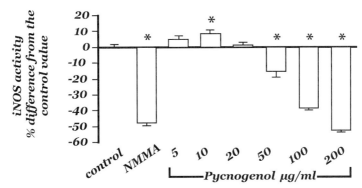

Figure 3   Pycnogenol effect on iNOS enzyme activity. The iNOS enzyme was obtained from RAW 274.7 monocyte–macrophage cells following 24 hr of activation with LPS and IFN-γ as described by Kobuchi *et al.* (1997). The enzyme assay was performed in the presence of increasing concentrations of Pycnogenol. The histogram shows either the inhibition or the activation of iNOS enzyme activity expressed as the percentage difference from the activity in the absence of Pycnogenol. The effect of 50 μM NMMA (N-monomethyl arginine) is shown for comparison. Values represent the mean of at least three different experiments (ISE). The asterisk indicates that the difference from control is significant by ANOVA ($p < 0.05$).

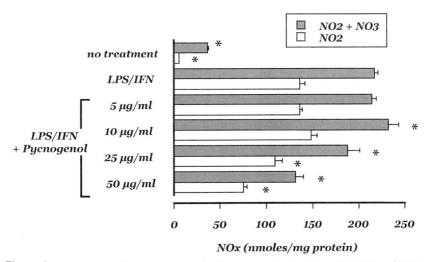

Figure 4   Pycnogenol effect on NO_2^- and NO_3^- accumulation induced by LPS and IFN-γ in RAW 264.7 monocyte–macrophages. Cells were preincubated with different concentrations of Pycnogenol and then treated with LPS and INF-γ (10 ng/ml and 5 U/ml, respectively). After 24 hr the concentration of NO_2^- and NO_3^- in the medium was assessed as described by Virgili *et al.* (1998). NOx indicates NO_2^- plus NO_3^- values. Data (mean $\pm$ SE) refer to three or more experiments. The asterisk indicates that the difference from LPS and IFN-γ treatment is significant by ANOVA ($p < 0.05$).

of iNOS (Nathan and Xie, 1994). However, in activated macrophages, a significant dose-dependent decrease of iNOS mRNA expression, associated with preincubation with the pine bark extract, was observed (see Fig. 5).

The mechanism by which pine bark extract acts on these different cellular targets has not yet been investigated. iNOS inhibitory activity is likely to be aspecific. The high affinity of flavonoids to proteins may be associated with a noncompetitive or an in-competitive activity. However, when compared to purified flavonoids or to other plant extracts, Pycnogenol inhibitory activity is in the highest rank (Virgili *et al.*, 1998). The effect on mRNA expression may be due to a specific effect on mRNA stability or possibly mediated by an effect on some still unknown modulatory step in the cellular regulation of gene expression. Finally, the antioxidant properties of

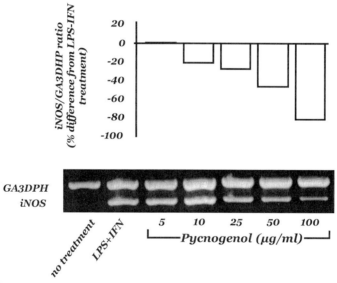

Figure 5 Effect of overnight preincubation with different concentrations of Pycnogenol on iNOS mRNA expression on RAW 264.7 monocyte–macrophage cells treated with LPS and INF-γ (10 ng/ml and 5 U/ml respectively). Six hours after the induction, total RNA was extracted and iNOS mRNA was assayed by RT-PCR as described by Virgili *et al.* (1998). For quantification, PCR bands on the photograph of the gel were scanned using a laser densitometer linked to a computer analysis system. The relative iNOS signal was normalized against the corresponding G3PDH signal obtained from the same sample, and data were expressed as the iNOS/G3PDH ratio. (Bottom) iNOS and glyceraldehyde-3-phosphate dehydrogenase (GA3DPH) electrophoretic signals. (Top) Percentage difference from LPS and IFN-γ treatment of iNOS/G3PDH ratio. One out of at least three experiments providing similar results is shown.

procyanidins may indirectly affect the cellular redox status and therefore the cell response to stimulation (Suzuki *et al.*, 1997).

On the basis of these results and previous data, Pycnogenol and plant polyphenols in general appear promising tools for the nonpharmacological control of NO overflow during chronic inflammation and as preventive treatment against different pathologies that have been proposed to be associated with a dysregulation of NO production, such as arteriosclerosis, cardiovascular disease, arthritis, and others. In particular the mixture of procyanidins extracted from pine bark, composed mainly of catechin as monomers or condensed polymers, is a powerful scavenger of reactive oxygen and nitrogen species and appears able to efficiently and specifically modulate NO metabolism in activated macrophages. The findings reported here provide a background for a better understanding of the mechanism of the biological activities not only of Pycnogenol, but also of other related flavonoids.

FUTURE STUDIES AND POSSIBLE FIELDS OF APPLICATION OF PYCNOGENOL

There are several indications for future studies. The modulatory effect of Pycnogenol on NO metabolism renders it possibly suitable for the treatment of peripheral and central neuropathies. Some behavioral diseases, such as attention deficit disorder, might also be other fields of application.

The iNOS inhibitory activity of Pycnogenol suggests different possible therapeutic applications, such as in Alzheimer's disease, multiple sclerosis, rehumathoid arthritis, and other diseases characterized by a localized high generation of NO.

The strong antioxidant capacity, together with the ability to play a central role in the cellular antioxidant network, suggests that Pycnogenol can favorably affect the cell response to oxidative stress conditions. In the endothelium, this pivotal role may prevent cellular disfunctions and those cell responses, such as the activation of NF-κB transcription factor and the subsequent expression of different genes (TNF, IL1, adhesion molecules), that have been proposed to participate in the formation of the atheromatous plaque. In other tissues, such as in the smooth muscle cells of the vascular intima, control of the cell redox status following oxidative challenge may regulate the cell proliferative response, thus participating in the protection of cardiovascular diseases.

In conclusion, if future studies will confirm and corroborate the importance of flavonoids for the health, then new avenues will be explored to understand the complex relationship between the human body and these plant components.

REFERENCES

Bath, P. M. W., Hassal, D. G., Gladwin, A. M., Palmer, M. R. J., and Martin, F. J. (1991). Nitric oxide and prostacyclin: Divergence of inhibitory effects on chemotaxis and adhesion to endothelium in vitro. *Artherioscler. Thromb.* **11**, 254–260.

Blazso, G., Gabor, M., and Rohdewald, P. (1997). Anti-inflammatory activities of procyanidin-containing extract from *Pinus pinaster* Ait after oral and cutaneous application. *Pharmazie* **52**, 380–382.

Blazso, G., Gabor, M., Sibbel, R., and Rohdewald, P. (1994). Anti-inflammatory and superoxide radical scavenging activities of procyanidins containing extract from the bark of *Pinus pinaster* Sol. and its fractions. *Pharm. Pharmacol* **3**, 217–220.

Blazso, G., Gaspar, R., Gabor, M., Ruve, H.-J., and Rohdewald, P. (1996). ACE inhibition and hypotensive effect of a procyanidins containing extract from the bark of *Pinus pinaster* Sol. *Pharm. Pharmacol. Lett.* **6**, 8–11.

Cheshier, J. E., Ardestani-Kaboudanian, S., Liang, B., Araghiniknam, M., Chung, S., Lane, S., Castro, A., and Watson, R. R. (1996). Immunomodulation by pycnogenol in retrovirus infected or ethanol fed mice. *Life Sci.* **58**, 87–96.

Christopherson, K. S., and Bredt, D. S. (1997). Nitric oxide in exitable tissues: Physiological roles and disease. *J. Clin. Invest.* **100**, 2424–2429.

De Caterina, R., Libby, P., Peng, H.-B., Thannickal, W. J., Rajavashisth, T. B., Gimbrone, M. A., Shin, W. S., and Liao, J. K. (1995). Nitric oxide decreases cytokine induced endothelial activation. *J. Clin. Invest.* **96**, 60–68.

Editorial. (1994). Dietary flavonoids and risk of coronary heart disease. *Nutr. Rev.* **52**, 59–61.

Elstner, E. F., and Kleber, E. (1990). Radical scavenger properties of leucocyanidine. *In* "*Flavonoids in Biology and Medicine. III. Current Issues in Flavonoid Research*" (N. P. Das, ed.), pp. 227–235. Natl. Univ of Singapore Press, Singapore.

Emerit, I. (1994). Reactive oxygen species, chromosome mutation, and cancer: Possible role of clastogenic factors in carcinigenesis. *Free Radic. Biol. Med.* **16**, 99–109.

Epe, B., Ballmaier, D., Roussyn, I., Brivida, K., and Sies, H. (1996). DNA damage by peroxynitrite characterized with DNA repair enzymes. *Nucleic Acids Res.* **24**, 4105–4110.

Gutteridge, J. M. C. (1995). Hydroxyl radicals, iron, oxidative stress and neurodegeneration. *Ann. N.Y. Acad. Sci.* **738**, 201–213.

Halliwell, B. (1996a). Antioxidants in human health and disease. *Annu. Rev. Nutri.* **16**, 33–50.

Halliwell, B. (1996b). Oxidative stress, nutrition and health: Experimental strategies for optimization of nutritional antioxidant intake in humans. *Free Radic. Res.* **25**, 57–74.

Hertog, M. G. L., Fesrens, E. J. M., Hollman, P. C. K., Katan, M. B., and Kromhout, D. (1993). Dietary antioxidant flavonoids and risk of coronary heart disease: The Zutphen elderly study. *Lancet* **342**, 1007–1011.

Hollman, P. C. H., de Vries, J. H. M., van Leeuven, S. D., Mengelers, M. J. B., and Katan, M. B. (1995). Absorbtion of dietary quercetin glycosides and quercetin in healthy ileostomy volunteers. *Am. J. Clin. Nutr.* **62**, 1276–1282.

Hooper, D. C., Ohnishi, S. T., Kean, R., Numagami, Y., Dietzschold, B., and Koprowski, H. (1995). Local nitric oxide production in viral and autoimmune diseases of the central nervous system. *Proc. Natl. Acad. Sci. USA* **92**, 5312–5316.

Ingram, D. K., Shimada, A., Spangler, E. L., Ikari, H., Hengemihle, J., Kuo, H., and Greig, N. (1996). Cognitive enhancement: New strategies for stimulating cholinergic, glutamatergic, and nitric oxide system. *Ann. N.Y. Acad. Sci.* **786**, 348–361.

Kanner, J., Frankel, E., Granit, R., German, B., and Kinsella, J. E. (1994). Natural antioxidants in grapes and wines. *J. Agric. Food Chem.* **42**, 64–69.

Kobuchi, H., Droy-Lefaix, M. T., Christen, Y., and Packer, L. (1997). Ginkgo biloba extract (EGb761): Inhibitory effect on nitric oxide production in the macrophage cell line RAW 264.7. *Biochem. Pharmacol.* **53**, 897–903.

Krenger, W., Falzarano, G., Delmonte, J. J., Snyder, K. M., Byon, J. C., and Ferrara, J. L. (1996). Interferon-gamma suppresses T-cell proliferation to mitogen via the nitric oxide pathway during experimental acute graft-versus-host disease. *Blood* **88**, 1113–1121.

Liu, R. H., and Hotchkiss, J. H. (1995). Potential genotoxicity of chronically elevated nitric oxide: A review. *Mutat. Res.* **339**, 73–89.

Luperchio, S., Tamir, S., and Tannembaum, S. R. (1996). NO induced oxidative stress and glutathione metabolism in rodent and human cells. *Free Radic. Biol. Med.* **21**, 513–519.

Mizuno, M., Droy-Lefaix, M. T., and Packer, L. (1996). *Gingko biloba* extract EGb 761 is a suppressor of AP-1 transcription factor stimulated by phorbol 12-myristate 13- acetate. *Biochem. Mol. Biol. Int.* **39**, 395–401.

Moncada, S., Palmer, R. M. J., and Higgs, E. A. (1991). Nitric oxide: Physiology, pathophysiology and pharmacology. *Pharmacol. Rev.* **43**, 109–142.

Nakaki, T., Nakayama, M., and Kato, R. (1990). Inhibition by nitric oxide and nitric oxide producing vasodilators of DNA synthesis in vascular smooth muscle cells. *Eur. J. Pharmacol.* **189**, 347–353.

Nathan, C. (1992). Nitric oxide as a secretory product of mammalian cells. *FASEB J.* **6**, 3051–3064.

Nathan, C., and Xie, Q. W. (1994). Regulation of biosynthesis of nitric oxide. *J. Biol. Chem.* **269**, 13725–13728.

Nelson, A. B., Lau, B. H. S., Ide, N., and Rong, Y. (1998). Pycnogenol inhibits macrophage oxidative burst, lipoprotein oxidation and hydroxil radical induced DNA damage. *Drug Dev. Indust. Med.* **24**, 1–6.

Noda, Y., Anzai, K., Mori, A., Kohno, M., Shinmei, M., and Packer, L. (1997). Hydroxyl and superoxide anion radical scavenging activities of natural source of antioxidants using the computerized JES-FR30 ESR spectrometeer system. *Biochem. Mol. Biol. Int.* **42**, 35–44.

Pogun, S., and Kumar, M. J. (1994). Regulation of neurotransmitter reuptake by nitric oxide. *Ann. N.Y. Acad. Sci.* **738**, 305–315.

Radomski, M. W., Palmer, R. M., and Moncada, S. (1990). An L-arginine/nitric oxide pathway present in human platelets regulates aggregation. *Proc. Natl. Acad. Sci. USA* **85**, 5193–5197.

Rice-Evans, C. A., Miller, N. J., and Pananga, G. (1997). Structure-antioxidant activity relationship of flavonoids and phenolic acids. *In* "*Flavonoids in Health and Disease*" (L. Packer and C. A. Rice-Evans, eds.), pp. 199–209. Dekker, New York.

Rohdewald, P. (1997). Pycnogenol. *In* "Flavonoids in Health and Disease" (C. Rice-Evans and L. Packer, eds.), pp. 405–419. Dekker, New York.

Rohdewald, P. (1998). Method for controlling the reactivity of human blood platelets by oral administration of the extract of the maritime pine (Pycnogenol). U.S. Patent No. 5,720,956, US.

Rong, Y., Li, L., and Lau, B. H. (1994–1995). Pycnogenol protects vascular endothelial cells from t-butyl hydroperoxide induced oxidant injury. *Biotechnol. Ther.* **5**, 117–126.

Rubbo, H., Darley-Usmar, V., and Freeman, B. A. (1996). Nitric oxide regulation of tissue free radical injury. *Chem. Res. Toxicol.* **9**, 809–820.

Schmidt, H. H. H. W., and Walter, U. (1994). NO at work. *Cell* **78**, 919–925.

Steinmetz, K. A., and Potter, J. D. (1996). Vegetable, fruit and cancer prevention: A review. *J. Am. Diet Assoc.* **96**, 1027–1039.

Sternberg, M. J., and Mabbott, N. A. (1996). Nitric oxide-mediated suppression of T-cell responses during *Trypanosoma brucei* infection: Soluble trypanosoma products and interferon-gamma are synergistic inducers of nitric oxide synthase. *Eur. J. Immunol.* **26,** 539–543.

Suzuki, Y. J., Aggarwal, B. B., and Packer, L. (1992). α-Lipoic acid is a potent inhibitor of NF-κB activation in human T cells. *Biochem. Biophys. Res. Commun.* **189,** 1709–1715.

Suzuki, Y. J., Forman, H. J., and Sevanian, A. (1997). Oxidants as stimulators of signal transduction. *Free Radic. Biol. Med.* **22,** 269–285.

Suzuki, Y. J., and Packer, L. (1993). Inhibition of NF-κB activation by vitamin E derivatives. *Biochem. Biophys. Res. Commun.* **193,** 277–283.

Tsao, P. S., McEvoy, L. M., Drexler, H., Butcher, E. C., and Cooke, J. P. (1994). Enhanced endothelial adhesiveness is attenuated by L-arginine. *Circulation* **89,** 2176–2182.

Ueda, T., Ueda, T., and Armstrong, D. (1996). Preventive effect of natural and synthetic antioxidants on lipid peroxidation in the mammalian eye. *Ophthalmol. Res.* **28,** 184–192.

Ursini, F., Maiorino, M., Hochstein, P., and Ernster, L. (1989). Microsomal lipid peroxidation: Mechanism of initiation. *Free Radic. Biol. Med.* **6,** 31–36.

van Jaarsveld, H., Kuyl, J. M., Schulemburg, D. H., and Wiid, N. M. (1996). Effect of flavonoids in the outcome of myocardial mitochondrial ischemia/reperfusion injury. *Res. Commun. Mol. Pathol. Pharmacol.* **91,** 65–75.

Virgili, F., Kobuchi, H., and Packer, L. (1998a). Procyanidins extracted from *Pinus marittima* (Pycnogenol): Scavengers of free radical species and modulators of nitrogen monoxide metabolism in activated murine RAW 264.7 macrophages. *Free Radic. Biol. Med.* **24,** 1120–1129.

Virgili, F., Kim, D., Packer, L. (1998b). Procyanidins extracted from pine bark protect α-tocopherol in ECV 304 endothelial cells challenged by activated RAW 264.7 macrophages: role of nitric oxide and peroxynitrite. *FEBS Letters* **431,** 315–318.

Wei, Z., Peng, Q., and Lau, B. H. S. (1997). Pycnogenol enhances endothelial cell antioxidant defences. *Redox Rep.* **3,** 147–155.

Yamada, K., Noda, Y., Nakayama, S., Komori, Y., Sugihara, H., Haegawa, T., and Nabeshima, T. (1995). Role of nitric oxide in learning and memory in monoamine metabolism in the rat brain. *Br. J. Pharmacol.* **115,** 852–858.

22 Antioxidant Properties of *Ginkgo biloba Extract: EGb 761*

Marie-Thérèse Droy-Lefaix* and Lester Packer†

* Department of Pharmacology
IPSEN Institute
75781 Paris cedex 16, France

† Department of Molecular and Cell Biology
University of California
Berkeley, California 94720

INTRODUCTION

Oxygen radicals from endogenous and exogenous sources are demonstrated to be implicated in many diseases. They have been shown to be a major cause of endogenous damage to DNA, proteins, lipids, and other macromolecules. Their release into tissues results in various disorders, such as ischemia, aging, and numerous degenerative diseases, including cardiovascular diseases, cancer, and neurodegenerative diseases (Ames *et al.,* 1993; De Groot, 1994; Halliwell, 1994; Packer, 1995).

However, because antioxidants are demonstrated to be potential protective substances, the *Ginkgo biloba* extract EGb 761 (Ipsen France) was tested as a therapeutic agent. This product is extracted from the dried leaves of the tree, the *G. biloba,* believed to be the oldest living tree species. EGb 761 is a standardized extract containing 24% flavonoids (ginkgo flavone glycosides), 6% terpen lactones such as ginkgolides A, B, C, J, and bilobalides, some organic acids, and other constituents (De Feudis, 1991).

The purpose of this chapter is to summarize data relating to the antioxidant functions of EGb 761 and to speculate on the potential relationship of pathological conditions in brain, sensorial tissues such as eye, olfactory bulb, and ear, and the vascular system.

RADICAL-SCAVENGING ACTIVITIES OF *Ginkgo biloba* EXTRACT

Effect on Superoxide Anion

The univalent reduction of oxygen produces the superoxide anion radical ($O_2^{\cdot-}$), characterized by its unpaired electron (Halliwell and Gutteridge, 1989). In the presence of metal ions such as copper or iron, all aqueous systems can produce the hydrogen peroxide (H_2O_2) by a dismutation reaction at the origin of the generation of hydroxyl radical (OH•). As demonstrated in *in vitro* experiments, EGb 761 can scavenge $O_2^{\cdot-}$ when it is generated by the irradiation of γ rays (Gardès-Albert *et al.*, 1993; Marcocci *et al.*, 1994) or by phenazine methosulfate-NADH (Haramaki *et al.*, 1996; Packer *et al.*, 1997). Inhibition of xanthine oxidase activity is also noted in a dose-dependent manner that can prevent the release of $O_2^{\cdot-}$ (Marcocci *et al.*, 1994).

Effect on Hydroxyl Radical

The hydroxyl radical is a powerful oxidant, reacting even with the most stable structure. This radical is extremely reactive and toxic for cell membranes. The *G. biloba* extract, *in vitro*, is determined to be effective on the OH• generation produced by irradiation with γ rays (Gardès-Albert *et al.*, 1993; Haramaki *et al.*, 1996) or using dihydroxybenzoic acid on the Fenton reaction (Pincemail and Deby, 1986). It inactivates the formation of radicals by a scavenging effect that seems correlated with the presence of flavonoids in the extract (Roback and Gryglewski, 1988). Hydroxyl radicals are strongly scavenged with EGb 761, even at the lowest concentration of 25 μg/ml (Packer *et al.*, 1997).

Consequently, *ex vivo,* a partial inhibition of the hydroxyl radical DMPO spin adduct DMPO-OH in rats is observed in coronary effluents after ischemia–reperfusion in isolated hearts of rats treated with a dose of 60 mg/kg, given orally, for 24 days (Pietri *et al.*, 1993).

Effect on Peroxyl Radicals

EGb 761 totally inhibits membrane lipid peroxidation, as demonstrated by measuring the formation of malondialdehyde in rat liver microsome (Pincemail and Deby, 1986). It is also effective (7.5 μg/ml) against the UV-C irradiation-induced formation of malondialdehyde and the polyunsaturated fatty acid damage noted in rat liver microsomes (Dumont *et al.*, 1992). EGb 761 also quenches, in a dose-dependent manner, lipoperoxides generated by the production of chemoluminescence from luminol in the presence of DOPC liposomes in the reaction with an azo initiator 2,2'-azo-bis(2,4-dimethylvaleronitrile) (AMVN) (Marcocci *et al.*, 1994; Maitra *et al.*, 1995). This EGb 761 effect on peroxide formation may be due to the presence of flavonoids in the extract, contributing to the inhibition of lipoperoxidation (Bors *et al.*, 1990; Yuting *et al.*, 1990).

Effect on Nitric Oxide

The sustained production of high amounts of nitric oxide (NO), following the induction of iNOS, can react with superoxide anion to induce the formation of peroxylnitrite (ONOO-), which is very toxic for tissues, as demonstrated in many pathologies. *In vitro*, EGb 761 is found to suppress the level of iNOS mRNA in LPS/IFN-γ-activated macrophages by decreasing the stability of iNOS mRNA or by affecting its translational regulation (Kobushi *et al.*, 1997). Furthermore, EGb 761 also inhibits NOS enzyme activity and, by concomitant inhibitory actions, the production of NO, with an initial effect as soon as the concentration of 20 μg/ml is reached (Kobushi *et al.*, 1997).

BIOLOGICAL APPROACH OF *Ginkgo biloba* EXTRACT

Oxidant attack is at the origin of destructive processes at different cell levels. By scavenging the free radicals, EGb 761 may protect the cells against DNA damage and membrane disruption and seems to have beneficial effects on the brain, the cardiovascular system, and on sensorial tissues such as the olfactive bulb, the eye, and the ear. Special effects are also noted in aging disorders (Droy-Lefaix, 1997).

EGb 761 and Cell Components

Mitochondria and Anoxia–Reoxygenation

In the presence of an excess of free radicals, several mitochondrial alterations may occur. Lipid peroxidation can disturb lipid bilayer permeability essential for the oxidative phosphorylation. Protein thiol oxidation can modify enzyme conformation and the activity and the base oxidation in mitochondrial DNA that can alter the products of mitochondrial DNA (mtDNA). Consequently, these oxidant effects seem to induce permanent damage at the level of oxidative phosphorylation.

On rat liver mitochondria, anoxia–reoxygenation experimentation leads to an alteration in oxide–reductase activities and to a decrease of the

ADP/O ratio, resulting in a decrease in oxidative phosphorylation efficiency. Acute administration of EGb 761 (50–100 μg/ml) in the incubation medium is responsible for an improvement in oxidative phosphorylation, preventing the production of superoxide anion after reperfusion (Sluze *et al.*, 1998; Willet *et al.*, 1997).

Mitochondria and Aging

However, the increased generation of oxygenated free radicals may be responsible for the age-associated oxidative damage that occurs in the mitochondria (Benzi and Moretti, 1995). A significant factor contributing to the aging process is the decline of mitochondrial oxidative phosphorylation due to the accumulation of mutations in mtDNA (Linanne *et al.*, 1989; Wallace, 1992). Lipid peroxidation may lead to mtDNA degradation (Harman, 1956; Miquel, 1991), and mitochondrial-free radical production is shown to increase with age (Bandy and Davison, 1990; Ames *et al.*, 1993). The importance of oxidative mtDNA lesions may be appreciated, measuring the amount of oxidized base, as demonstrated with 8-oxo-7,8-dihydro 2'-deoxyguanosine (oxo 8 and 6) (Hruszkewycz, 1992; Sastre *et al.*, 1997). In brain and in liver of 27-month-old rats, as compared to young 4-month-old rats, the oxidative damage to mtDNA increases significantly with aging. Treatment with EGb 761, given orally for 3 months before the mtDNA evaluation, at a dose of 100 mg/kg, significantly prevents mtDNA damage (Sastre *et al.*, 1997).

The measurement of peroxide generation on young and old rats confirms that the mtDNA protective effect of EGb 761 in brain and liver is due to the antioxidant properties of the extract. The strong increase of peroxides, due to aging, is reduced significantly if the rats are treated for 3 months with EGb 761 (100 mg/kg/po) (Sastre *et al.*, 1997). Furthermore, the oxidative damage to the mtDNA that occurs during aging seems, in relation to the oxidation of the mitochondria glutathione, to be linked to an increase of the GSSG:GSH ratio noted in untreated rats. These results confirm the key role of mitochondrial gluthatione in the protection against free radical damage (Sastre *et al.*, 1997).

In addition, treatment with EGb 761 prevents age-associated impairments in respiratory function, protecting the mitochondria against a decrease of the respiratory parameters, especially in state 4, which corresponds to the resting mitochondria respiration state. In state 4, the stimulation of the membrane potential in brain mitochondria from 27-month-old rats is around 30% of that found in state 4 of young rats. Treatment with EGb 761 (100 mg/kg/po/3 months) is effective, preventing a decrease in energy status (Sastre *et al.*, 1997).

These experiments confirm that oxidative damage is responsible for the decline of mitochondrial activity causing an inhibition of mitochondrial respiration. EGb 761 may be beneficial, preventing, in trapping oxygenated free radicals, mitochondrial disorders.

EGb 761 and Membranes

Neurotransmitters

The accumulation of lipoperoxides can lead to the degradation of polyunsaturated fatty acids, resulting in changes in membrane structure characterized by a decrease in fluidity and an increase in viscosity, inducing the rigidity of the membranes. In brain, these membrane alterations can induce disorders in neurotransmitter uptakes (Pastusko *et al.*, 1983). EGb 761 can totally inhibit PUFA degradation and the appearance of thiobarbituric acid-reactive substances (TBARS) in the microsomes exposed to UV-C irradiation (Dumont *et al.*, 1993). With this effect, EGb 761 may provide effective and persistent protection to the membrane and thereby to the neurotransmitter uptake, such as demonstrated on adrenergic and serotoninergic receptors in cognitive brain areas. In hippocampus, using rauwolcine, a specific ligand for α_2-adrenoreceptors, on 24-month-old rats, there is a decrease of the number of binding sites when compared to young 4-month-old rats. After treatment with EGb761, given before measurements (5 mg/ip/21 days), the binding of rauwolcine is increased significantly. The same results are noted with 5HT1 receptors in rat hippocampus and in the cerebral cortex. In 24-month-old rats, the specific (3H)8-OH-DPAT binding to these receptors is lower when compared to young rats. A significant increase occurs after treatment with EGb 761 (5 mg/kg/ip) (Huguet *et al.*, 1995).

Ginkgo biloba EXTRACT AND PATHOLOGICAL CONDITIONS

Cerebral Ischemic Damage

During ischemic injury, major changes are demonstrated to generate cerebral function disturbances. Consequences of ischemia are widely associated with alterations in the membrane Na,K-ATPase function (Maixent and Lelièvre, 1987; Ames, 1991). In the nervous system, NA,K-ATPase plays a critical role during ischemia, as it is largely involved in the maintenance of ionic gradients, membrane potential, and cell volume (Lees, 1991). Due to the accumulation of fatty acids and their metabolites (Panetta *et al.*, 1987), cerebral ischemia is reported to decrease the activity of this enzyme as early as 30 min postocclusion (Jamme *et al.*, 1995). A prevention of this early reduction in Na,K-ATPase activity is noted with EGb 761 given to mice at a dose of 100 mg/kg/po/10 days. After 1 and 6 hr of focal ischemia after middle cerebral artery occlusion, Na,K-ATPase activity is decreased 43 and 62%, respectively, in ipsilateral and controlateral cortex. No significant effect of 1 and 6 hr of ischemia is observed if the mice are pretreated with EGb 761, preserving a cerebral protection essential to the neuronal activity (Maixent *et al.*, 1998). A positive correlation is found between the variations in malondialdehyde (MDA) content and the activity of the Na,K-ATPase, confirming that alterations of its activity following ischemia occur through the lipid peroxidation and disruption of the Na,K-ATPase membrane environment (Maixent *et al.*, 1998).

Stress

Some brain areas are particularly sensitive to free radical damage such as hippocampus, especially affected by Alzheimer' disease. EGb 761 can protect hippocampal neurons against toxicity induced by hydrogen peroxide (H_2O_2) and sodium nitroprussiate (SNP), two generators of reactive oxygen species. Exposure of rat primary mixed or enriched neuronal hippocampal cell cultures to H_2O_2 or SNP results in a concentration-dependent decrease in cell survival. Concomitant incubation of cells with EGb 761 (10–100 μg/ml) significantly protects, depending on the concentration against oxygenated free radical toxicities induced by H_2O_2 or SNP (Bastianetto *et al.*, 1998).

Apoptosis

Apoptosis, a form of programmed cell death, is controlled by an elaborate network of signaling pathways through receptors, protein kinases, second messengers, phosphorylated protein intermediates, and factors that can regulate the expression of individual genes or groups of genes (Lavin *et al.*, 1993). Oxidative stress was reported to induce the apoptosis of neurons in the cortex and in the neurosensorial tissues such as retina and olfactive bulb (Berkelaar *et al.*, 1994; Filipkowski *et al.*, 1994; Ratan *et al.*, 1994; Nitatori *et al.*, 1995). In this way, olfactive neurons are known to be very sensitive to apoptotic effects because of their frequent exposure to aggressive, physical, chemical, or biological factors vehiculated by inspired air. Olfactory neurons seem a good target for studying apoptosis mechanisms in the nervous system. As demonstrated in rats, bulbectomy induces neuronal olfactive apoptosis, with major DNA degradation, 24 to 48 hr following lesion formation. When rats are pretreated with EGb 761, given orally at doses of 50 and 100 mg/kg for 10 days, depending on the dose, there is a reduction of DNA fragmentation and a significant decrease of the expression of C-fos at the dose of 100 mg/kg (Jourdan *et al.*, 1997).

Reactive Oxygen Species and Retina

Light Retinal Degeneration

In sensorial tissues, EGb 761 is always demonstrated to have antioxidant effects in the eye, specially at the level of the retina, known to have a remarkable capacity for morphological and biochemical adaptation to different environmental luminances. Nevertheless, it is well established that prolonged light induces oxidative reactions, resulting in damage to the photoreceptor cells (Noëll, 1987), supporting the idea that lipid peroxidation occurs in the retina in response to light. In fact, the retina contains a high level of polyunsaturated fatty acids, which are a preferred substrate for peroxidation (Fliesler and Anderson, 1983). In albinos rats, a 24-hr fluorescent light exposure at 1700 lux induces retinal degeneration linked to a significant loss of the retinal function (Ranchon *et al.*, 1998b), as demonstrated by electroretinographic parameters that give specific responses of the retina to a light stimulus and require cell membrane integrity. There is a significant decrease of the maximal b wave (B_{max}) (60%) and PIII ($PIII_{max}$) (40%) amplitudes. These functional disorders are accompanied by morphological alterations that are characterized by a reduction of the outer nuclear layer. The photoreceptors become significantly shorter in the inferior and superior retina. Treatment of rats with EGb 761 for 2 weeks before light exposure and for 4 weeks after exposure, at a dose of 100 mg/kg/po/day, protects the photoreceptors against light-induced damage. The saturated amplitudes B_{max} and $PIII_{max}$ do not vary significantly and rods appear normal, similar to those noted with 1,3-dimethylthiourea given at 500 mg/kg/ip 24 hr before light exposure and just before exposure (Ranchon *et al.*, 1998a).

Ischemia–Reperfusion and Retina

The major role of free radicals in the pathology of the retina is also demonstrated in a model of ischemia–reperfusion generated in rats by clamping the retinal artery for 90 min followed with a reperfusion of from 4 to 24 hr. In such a model, histological analysis shows degenerative changes with major edema and neutrophil invasion in the different layers of the retina. In the presence of EGb 761, given orally at a dose of 100 mg/kg/10 days before the experiment, the free radical degenerative changes are reduced significantly (Szabo *et al.*, 1991a,b).

As alterations in the redox state can alter membrane permeability, retinal ion contents are measured by atomic absorption spectrophotometry using various wavelengths after washing out the blood and extracellular fluid and then drying and washing the retina. In untreated rats, homeostasis is destroyed with an enhancement in Na^+ and Ca^{2+} ions and a decrease in K^+ and Mg^{2+} ions. Pretreatment with EGb 761 (100 mg/kg/po/day) given for 10 days and until the rats are sacrificed significantly reduces this ion shift imbalance (Szabo *et al.*, 1993).

As ischemia–reperfusion has consequences on visual function, electroretinogram records were done. After ischemia induced by hypertony with a 110-mg Hg pressure for 1 hr, b wave ERG amplitudes were recorded 1, 4, 24, and 48 hr after reperfusion. In rats treated with EGb 761 (100 mg/kg/po/10 days), results showed that the b wave ratio was significantly higher when compared to untreated rats, indicating a protection against the deleterious effect of ischemia. Scanning electron microscopy observations confirm this antioxidant effect of the extract. In nontreated animals, photoreceptors are totally disorganized with junction rupture, necrosis of the inner and outer segments, and edema (Menerath *et al.*, 1998). After EGb 761 treatment, the photoreceptor layer retains a normal appearance (Menerath *et al.*, 1995a,b).

Ischemia–Reperfusion and Ear

Other potential applications of EGb 761 include the inner ear, especially at the cochlea. This organ is very sensitive to ischemic disorders, whose function depends strongly on ionic gradient maintenance and on the endolymphatic potential controlled by the vascular stria (Billett *et al.*, 1989; Pujol *et al.*, 1992; Widick *et al.*, 1994; Cazals *et al.*, 1993; Ren *et al.*, 1995).

Complete blockage of the cochlear artery lasting from 5 to 8 min was achieved in gerbils, and cochlear function was monitored with distortion produced by otoacoustic emissions (DPOAE-primary levels 60 dB). Ears show a typical profile of DPOAE changes with time, with rapid decay starting about 20 sec after the interruption of cochlear blood flow, then a long-lasting plateau around 30 dB with respect to the initial DPOAE level. Reperfusion induces an almost immediate but incomplete recovery of DPOAE, followed by a rapid secondary decay lasting 200–250 sec, and a last stage of difference of monotonous recovery so that DPOAE levels are within 1 DB of their initial level at 600 sec. Pretreatment of gerbils 2 weeks before the experiment, at a dose of 100 mg/kg/po/day, shows no difference in the time course of DPOAE during ischemia. The treated animals recover significantly more rapidly than the controls (Avan *et al.*, 1996).

Ginkgo biloba EXTRACT AND CARDIOVASCULAR SYSTEM

Human Low-Density Lipoprotein Oxidation

Among the factors that may influence the progress of artery sclerosis is the oxidative alterations of low-density lipoproteins (LDL). In oxidative LDLs, as demonstrated on human LDL incubated in phosphate-buffered saline at 37°C with the hydrophilic peroxyl radical initiator 2,2′-azobis(2-amidinopropane) hydrochloride (AAPH), there is a decrease in vitamin E and β-carotene levels and an increase in the cholesterol level. If the LDL incubation is done with EGb 761, an effect is seen with a concentration of 10 μg/ml with total inhibition at 100 μg/ml. The loss of vitamin E and β-carotene is corrected in a dose-dependent manner and the cholesterol is decreased (Haramaki *et al.*, 1996). Furthermore, EGb 761 completely inhibits LDL peroxidation, as shown by a decrease of trytophan fluorescence induced by incubation with AAPH (Haramaki *et al.*, 1996). These results suggest that EGb 761 may prevent LDL lipoperoxidation in both lipolytic and hydrophylic phases and may be useful in the prevention of atherosclerosis (Haramaki *et al.*, 1996).

Inflammatory Reaction

Reactive oxygen species also play a major role in initiation, duration, and breakdown of inflammation during which polymorphonuclear neutrophils (PMN) are recruited in tissues and activated as shown in the ischemia–reperfusion syndrome. Free radicals can increase the adhesion of neutrophils to endothelial cells. Human umbilical cord vein endothelial cells are exposed to H_2O_2 and $O_2^{\bullet-}$, released by the oxidation of hypoxanthine–xanthine oxidase. The adhesion of remaining PMN is measured by the myeloperoxidase level. EGb 761 incubation at 100 μg/ml totally inhibits this adhesion, as well as dimethylthiourea (7.5 mM) and iron chelators such as desferioxamine (1 mM), suggesting the involvement of the hydroxyl radical (OH•) in such a reaction (Gozin *et al.*, 1997).

Cardiac Ischemia–Reperfusion

The uncontrolled formation of reactive oxygen species, including free radicals, should provide reliable explanations of myocardial ischemia–reperfusion. Cardioprotective effects of EGb 761 against these pathological processes are shown using isolated rat hearts (Guillon *et al.*, 1988; Tosaki *et al.*, 1993; Haramaki *et al.*, 1994). Oral administration or perfusion of EGb 761 significantly reduces the hydroxyl formation as shown by the spin-trapping method using ESR (Tosaki *et al.*, 1993). Furthermore, perfusion with EGb 761 suppresses the decrease in tissue total ascorbate and the oxidation of tissue ascorbate due to cardiac ischemia–reperfusion (Haramaki *et al.*, 1994).

To assess the development of oxidative stress in cardiac ischemia–reperfusion that occurs during cardiac surgery in humans, an electron-spin resonance study was controlled by double-blinding the samples. Noninvasive spin trapping with 5,5-dimethyl-1-pyrroline 1-oxide (DMPO) and direct detection of stable ascorbyl free radical (AFR) were achieved in blood samples collected during cardiac arrest and in the course of aortic declamping. The induction of ischemia by aortic clampage is associated with a significant increase of DMPO-OH signal strength in peripheral and coronary sinus blood samples at T0 in relation with an increase in oxidative stress. After declamping, DMPO-OH levels are enhanced significantly and remain higher after reperfusion, whereas a decrease appears in AFR levels (Culcasi *et al.*, 1993). Beneficial effects of 5-day preoperative oral treatment with EGb 761 (320 mg/day) are shown with significant improvement. The DMPO-OH blood content is reduced and the decrease of AFR is strongly prevented (Culcasi *et al.*, 1993).

CONCLUSION

As radical-mediated processes are implicated in a number of chronic degenerative diseases and aging, there is evidence for a significant role of antioxidants in health maintenance and disease protection. Among these defense mechanisms, natural substances such as *G. biloba* extract EGb 761 (Ipsen France) can be effective compounds, protecting the tissues from damage and functional disturbances. EGb 761 scavenges superoxide, hydroxyl, peroxyl radicals, and nitric oxide and can protect cellular functions in different pathological conditions.

EGb 761 may directly improve the mitochondria respiratory chain function by preventing oxidative damage to mtDNA, oxidation of mitochondrial GSH, and peroxide formation in the mitochondria. During aging and ischemic injury, this antioxidant also protects the brain, which is very sensitive to the presence of oxygenated free radicals that directly interfere with the polyunsaturated fatty acids (PUFAs) in the membranes. EGb 761 can totally inhibit PUFAs degradation. Thus, EGb 761 may be effective to neurotransmitter uptake. It also preserves the activity of the NA,K-ATPase membrane-bound enzyme, which is linked to the maintenance of ionic gradients and membrane potential.

In sensorial tissues, EGb 761 can reduce apoptosis in the olfactive bulb by decreasing DNA fragmentation. Concerning vision, degenerative lesions of the retina due to light exposure, ischemic disorders, or aging are prevented. At the inner ear level, it can also inhibit alterations of cochlear sensitivity due to ischemia–reperfusion.

In the cardiovascular system, by a direct effect on oxidative low-density lipoproteins, EGb 761 results in a decrease in atherosclerosis evolution. It prevents lipid peroxidation and the loss of both vitamin E and β-carotene in LDL oxidation. It also accelerates cardiac mechanical recovery after ischemia–reperfusion, preventing ascorbate leakage and oxidation.

In conclusion, with its antioxidant properties, EGb 761, a free radical scavenger, may provide protection from the increasingly high free radicals generated from the external or from endogenous biological reactions. As there is evidence that free radical processes have a higher incidence in degenerative diseases with consequences in cardiovascular disease, brain disorders such as Alzheimer's dementia and Parkinson disease, diabetes, retinopathies, and ear disorders, EGb 761 can lead to new approaches in therapy in cases of oxidative stress.

REFERENCES

Ames, A. (1991). Energy requirement of CNS cells as related to their function and to their vulnerability to ischemia: A commentary based on studies on retina. *Can. J. Physiol. Pharmacol.* **70**, 158–164.

Ames, B. N, Shigenaga, M. K., and Hagen, T. M (1993). Oxidants, antioxidants and the degenerative diseases of aging. *Proc. Natl. Acad. Sci. USA* **90**, 7915–7922.

Avan, P., Mom T., Gilain L., and Droy-Lefaix, M. T. (1996). Effet de l'EGb 761 sur un modèle de surdité brusque par clampage réversible de l'artère labyrinthique chez la gerbille. Journées Multirégionales d'Oto-rhino-laryngologie, Poitiers, France.

Bandy, B., and Davison, A. J. (1990). Mitochondrial mutations may increase oxidative stress: Implications for carcinogenesis and aging. *Free Radic. Biol. Med.* **8**, 523–539.

Bastianetto, S., Ramassamy, C., Christen, Y., Poirier, J., and Quirion, R. (1998). *Ginkgo biloba* extract (EGb 761) prevents cell death induced by oxidative stress in hippocampal neuronal cell cultures. *In* "Advances in *Ginkgo biloba* Extract Research" (L. Packer and Y. Christen, eds.), Vol. 7. Elsevier, Paris, pp. 85–99.

Benzi, G., and Moretti, A. (1995). Age and peroxidative stress-related modifications of the cerebral, enzymatic activities linked to mitochondria and the glutathione system. *Free Radic. Biol. Med.* **19**, 77–101.

Berkelaar, M., Clarke, D. B., Wang, Y. C., Bray, G. M., and Aguayo, A. J. (1994). Axotomy results in delayed death and apoptosis of retinal ganglion cells in adult rats. *J. Neurosci.* **14**, 4368–4374.

Billett, T. E., Thorne, P. R., and Gavin, J. B. (1989). The nature and progression of injury in the organ of Corti during ischemia. *Hear. Res.* **41**, 143–156.

Bors, X., Heller, W., Michel, C., and Saran, M. (1990). Flavonoids as antioxidants: Determination of radical-scavenging efficiencies. *Methods Enzymol.* **189**, 343–355.

Cazals, Y., Horner, K., and Didier, A. (1993). Experimental models of cochlear pathologies associated with ischemia. *In* "Advances in *Ginkgo biloba* Extract Research" (C. Ferradini, M. T. Droy-Lefaix, and Y. Christen, eds.), Vol. 2, pp. 115–122. Elsevier, Paris.

Culcasi, M., Pietri, S., Carrière, I., d'Arbigny, P., and Drieu, K. (1993). Electron-spin-resonance study of the protective effects of *Ginkgo biloba* extract (EGb 761) on reperfusion-induced free-radical generation associated with plasma ascorbate consumption during open-heart surgery in man. *In* "Advances in *Ginkgo biloba* Extract Research" (C. Ferradini, M. T. Droy-Lefaix, and Y. Christen, eds.), Vol. 2, pp. 153–162. Elsevier, Paris.

De Feudis, F. V. (1991). "*Ginkgo biloba* Extract (EGb 761): Pharmacological Activities and Clinical Applications." Elsevier, Paris.

De Groot, H. (1994). Reactive oxygen species in tissue injury. *Hepato-gastroenterol.* **41,** 328–332.

Droy-Lefaix, M. T. (1997). Effect of the antioxidant action of *Ginkgo biloba* extract (EGb 761) on aging and oxidative stress. *Age* **20,** 141–149.

Dumont, E., Petit, E., d'Arbigny, P., Tarrade, T., and Nouvelot, A. (1993). *Gingko biloba* extract (EGb 761) protects membrane polyunsaturated fatty acids and proteins against degradation induced by UV-C irradiation. *In* "Advances in *Ginkgo biloba* Extract Research" (C. Ferradini, M. T. Droy-Lefaix, and Y. Christen, eds.), Vol. 2, pp. 173–184. Elsevier, Paris.

Dumont, E., Petit, E., Tarrade, T., and Nouvelot, A. (1992). UV-C irradiation induced-peroxidative degradation of microsomal fatty acid and proteins: Protection by an extract of *Ginkgo biloba* (EGb 761). *Free Radic. Biol. Med.* **13,** 197–203.

Filipkowski, R. K., Hetman, M., Kaminska, B., and Kaczmarek, L. (1994). DNA fragmentation in rat brain after intraperitoneal administration of kaïnate. *NeuroReport* **5,** 1538–1540.

Fliesler, W. K., and Anderson, R. E. (1983). Chemistry and metabolism of lipids in the vertebrate retina. *Prog. Lipid. Res.* **22,** 79–131.

Gardès-Albert, M., Ferradini, C., Sekaki, A., and Droy-Lefaix, M. T. (1993). Oxygen-centered free radicals and their interaction with EGb 761 or CP202. *In* "Advances in *Ginkgo biloba* Extract Research" (C. Ferradini, M. T. Droy-Lefaix, and Y. Christen, eds.), Vol. 2, pp. 1–11. Elsevier, Paris.

Gozin, A., Da Costa, L., Andrieu, V., Droy-Lefaix, M. T., and Pasquier, C. (1997). Oxygen radicals activate protein phosphorylation in endothelial cells: Effect of the antioxidant *Gingko biloba* extract, EGb 761. SFFR-Asia/Costam/Unesco-MCBN workshop: Biological Oxidants and Antioxidants: Molecular Mechanisms and Health Effects, Penang, Malaysia.

Guillon, J. M., Rochette, L., and Baranès J. (1988). Effects of *Ginkgo biloba* extract on two models of experimental myocardial ischemia. *In* "Rökan (*Ginkgo biloba*): Recent Results in Pharmacology and Clinic" (E. W. Fünfgeld, ed.), pp. 153–161. Springer-Verlag, Berlin.

Halliwell, B. (1994). Free radicals, antioxidants and human disease: Curiosity, cause, or consequence? *Lancet* **344,** 721–724.

Halliwell, B., and Gutteridge, J. M. C. (1989). "Free Radicals in Biology and Medicine," 2nd Ed. Clarendon Press, Oxford.

Haramaki, N., Aggarwal, S., Kawabata, T., Droy-Lefaix, M. T., and Packer, L. (1994). Effects of natural antioxidant *Ginkgo biloba* extract (EGb 761) on myocardial ischemia–reperfusion injury. *Free Radic. Biol. Med.* **16,** 789–794.

Haramaki, N., Packer, L., Droy-Lefaix, M. T., and Christen, Y. (1996). Antioxidant actions and health implications of *Ginkgo biloba* extract. *In* "Handbook of Antioxidants" (E. Cadenas and L. Packer, eds.), pp. 487–510. Dekker, New York.

Harman, D. (1956). Aging: A theory based on free radical and radiation chemistry. *J. Gerontol.* **11**, 298–300.

Hruszkewycz, A. M. (1992). Lipid peroxidation and mtDNA degeneration: A hypothesis. *Mutat. Res.* **275**, 243–248.

Huguet, F., Drieu, K., Tarrade, T., and Piriou A. (1995). Cerebral adrenergic and serotoninergic receptor loss in rats during aging can be reversed by *Ginkgo biloba* extract (EGb 761). *In* "Advances in *Ginkgo biloba* Extract Research" (Y. Christen, Y. Courtois, and M. T. Droy-Lefaix, eds.), Vol. 4, pp. 77–88. Elsevier, Paris.

Jamme, I., Petit, E., Divoux, D., Gerbi, A., Maixent, J. M., and Nouvelot, A. (1995). Modulation of mouse cerebral Na,K-ATPase activity by oxygen radicals. *NeuroReport* **7**, 333–337.

Jourdan, F., Didier, A., Coronas, V., Moyse, E., and Rouiller, D. (1997). Apoptose neuronale et neurogénèse chez l'adulte: Le modèle du système olfactif périphérique. *In* "Rencontres IPSEN en ORL, tome 1" (Y. Christen, L. Collet, and M. T. Droy-Lefaix, eds.), pp. 121–129. Irvinn, Paris.

Kobushi, H., Droy-Lefaix, M. T., Christen, Y., and Packer, L. (1997). *Ginkgo biloba* extract (EGb 761): Inhibitory effect on nitric oxide production in the macrophage cell line RAW 264.7. *Biochem. Pharmacol.* **53**, 897–903.

Lavin, M. F., Baxter, G. D., Song, Q., Fidik, D., and Koxacs, E. (1993). Protein modification in apoptosis. *In* "Programmed Cell Death, the Cellular and Molecular Biology of Apoptosis" (M. Lavin and D. Watters, eds.), pp. 45–57. Harwood Academic, Brisbane.

Lees, G. J. (1991). Inhibition of sodium-potassium-ATPase: A potentially ubiquitous mechanism contribution to central nervous system neuropathology. *Brain. Res.* **16**, 283–300.

Linanne, A. W., Marzuki, S., Ozawa, T., and Tanaka, M. (1989). Mitochondrial DNA mutation as an important contributor to aging and degenerative diseases. *Lancet* **3**, 642–645.

Maitra, I., Marcocci, L., Droy-Lefaix, M. T., and Packer, L. (1992). Peroxyl radical scavenging activity of *Ginkgo biloba* EGb 761. *Free Radic. Biol. Med.* **49**, 1649–1655.

Maixent, J. M., and Lelièvre L. G. (1987). Differential inactivation of inotropic and toxic digitalis receptors in ischemic dogs heart: Molecular basis of the deleterious effect of digitalis. *J. Biol. Chem.* **262**, 12458–12462.

Maixent, J. M., Pierre, S., Jamme, I., Gerbi, A., Robert, K., Droy-Lefaix, M. T., and Nouvelot, A. (1998). Protective effects of EGb 761 on the alterations of Na,K-ATPase isoenzyme activities during cerebral ischemia in mouse. *In* "Advances in *Ginkgo biloba* Extract Research" (L. Packer, and Y. Christen, eds.), Vol. 7. Elsevier, Paris, 45–55.

Marcocci, L., Packer, L., Droy-Lefaix, M. T., Sekaki, A. H., and Gardès-Albert, M. (1994). Antioxidant action of *Gingko biloba* extract EGb 761. *In* "Methods in Enzymology" (L. Packer, ed.), Vol. 234, pp. 462–475. Academic Press, San Diego.

Menerath, J. M., Cluzel, J., Droy-Lefaix, M. T., and Doly, M. (1995a). Electroretinographic assessment of a free radical scavenger (EGb 761) treatment on rat retinal ischemia. *Invest. Ophthalmol. Vis. Sci.* **36**, S394.

Menerath, J. M., Doly, M., Droy-Lefaix, M. T., Besse, G., and Vennat, J. (1995b). Evaluation électroretinographique et histologique de l'ischémie rétinienne chez le rat par hypertonie intra-oculaire. *Ophthalmologie* **9**, 291–300.

Menerath, J. M., Droy-Lefaix, M. T., and Doly, M. (1998). Scanning electron microscopy after cryofracture of normal and ischemic rat retinas. *Ophthalmic Res.* **30**, 216–220.

Miquel, J. (1991). An integrated theory as the result of mitochondrial-DNA mutation in differentiated cells. *Arch. Gerontol. Geriatr.* **12**, 99–117.

Nitatori, T., Sato, N., Waguri, S., Karasawa, Y., Araki, H., Shibanai, K., Kominami, E., and Uchiyama, Y. (1995). Delayed neuronal death in the CA1 pyramidal cell layer of the gerbil hippocampus following transient ischemia is apoptosis. *J. Neurosci.* **15**, 1001–1011.

Noëll, W. K. (1987). Retinal damage by light damage to the retina varies with time of day of bright light exposure. *Physiol. Behav.* **39**, 607–613.

Packer, L. (1995). Oxidative stress, antioxidants, aging and disease. *In* "Oxidative Stress and Aging" (R. G. Cutler, L. Packer, J. Bertram, and A. Mori, eds.), pp. 1–14. Birkauser Verlag, Basel.

Packer, L., Saliou, C., Droy-Lefaix, M. T., and Christen, Y. (1997). *Ginkgo biloba* extract EGb761: Biological actions, antioxidant activity, and regulation of nitric oxide synthase. *In* "Flavonoids in Health and Disease" (C. A. Rice-Evans and L. Packer, eds.), pp. 303–341. Dekker, New York.

Panetta, T., Marcheselli, V. L., Braquet, P., Spinnewyn, B., and Bazan, N. G. (1987). Effects of platelet activating factor (BN 52021) on free fatty acids, polyphosphoinositides and blood flow in the gerbil brain: Inhibition of ischemia-reperfusion induced cerebral injury. *Biochem. Biophys. Res. Commun.* **149**, 580–587.

Pastuszko, A., Gordon-Majozak, W., and Dabrowiecki, Z. (1983). Dopamine uptake in striatal synaptosomes exposes to perixidation "in vitro." *Biochem. Pharmacol.* **32**, 141–146.

Pietri, S., Culcasi, M., Vernière, E., d'Arbigny, P., and Drieu, K. (1993). Effect of *Ginkgo biloba* extract (EGb 761) on free radical-induced ischemia-reperfusion injury in isolated rat hearts: A hemodynamic and electron-spin-resonance investigation. *In* "Advances in *Ginkgo biloba* Extract Research" (C. Ferradini, M. T. Droy-Lefaix, and Y. Christen, eds.), Vol. 2, pp. 163–171. Elsevier, Paris.

Pincemail, J., and Deby, C. (1986). Propriétés antiradicalaires de l'extrait de *Ginkgo biloba*. *Presse. Med.* **15**, 1475–1479.

Pujol, R., Puel, J. L., Eybalin, M. (1992). Implication of non-NMDA and NMDA receptors in cochlear ischemia. *NeuroReport* **3**, 299–302.

Ranchon, I., Gorrand, J. M., Cluzel, J., Droy-Lefaix, M. T., and Doly, M. (1998a). Functional rescue of photoreceptors from light-induced damage by DMTU and EGb 761. *Invest. Opthalmol. Vis. Sci.*, in press.

Ranchon, I., Gorrand, J. M., Cluzel, J., Vennat, J. C., and Doly, M. (1998b). Light induced variation on retina sensitivity in rats. *Curr. Eye Res.*, in press.

Ratan, R. R., Murphy, T. H., and Baraban, J. M. (1994). Oxidative stress induces apoptosis in embryonic cortical neurons. *J. Neurochem.* **62**, 376–379.

Ren, T. Y., Brown, J. N., Zhang, M. S., Nuttall, A. L., and Miller, J. M. (1995). An ischemia-reperfusion model in gerbil cochlea. *Abstr. Ass. Res. Otol.* **566**.

Robak, J., and Gryglewski, R. J. (1988). Flavonoids are scavengers of superoxide anions. *Biochem. Pharmacol.* **37**, 837–841.

Sastre, J., Milan, A., de La Asuncion, J. G., Pallardo, F. V., Droy-Lefaix, M. T., and Vina, J. (1997). Prevention of age-associated mitochondrial DNA damage by *Ginkgo biloba* extract Egb 761. *In* "Proceedings of the VIII Biennal meeting, International Society For Free Radical Research." J. R. Prous, Barcelona, in press.

Sluse, F. F., Du, G. H., Willet, K., Detry, O., Mouithys-Mickalad, A., and Droy-Lefaix, M. T. (1998). Protective effects of *Ginkgo biloba* (EGb 761) on functional impairments of mitochondria induced by anoxia-reoxygenation in situ and in vitro. *In* "Advances in *Ginkgo biloba* Extract Research" (L. Packer, and Y. Christen, eds.), Vol. 7. Elsevier, Paris, pp. 139–148.

Szabo, M. E., Droy-Lefaix, M. T., Doly, M., and Braquet, P. (1991a). Free radical-mediated effects in reperfusion injury: A histologic study with superoxide dismutase and EGb 761 in rat retina. *Ophthalmic Res.* **23**, 225–234.

Szabo, M. E., Droy-Lefaix M. T., Doly, M., Carré C., and Braquet P. (1991b). Ischemia and reperfusion induced histological changes in the rat retina: Demonstration of a free-radical mediated mechanism. *Invest. Opthalmol. Vis. Sci.* **32,** 1471–1478.

Szabo, M. E., Droy-Lefaix, M. T., Doly, M., and Braquet, P. (1993). Modification of ischemia-reperfusion induced ion shifts (NA^+, K^+, CA^{2+}, MG^{2+}). *Ophthalmic. Res.* **25,** 1–9.

Tosaki, A., Droy-Lefaix, M. T., Pali, T., and Das, D. K. (1993). Effects of SOD, catalase, and a novel antiarrhythmic drug, EGb 761, on reperfusion-induced arrhythmias in isolated rat hearts. *Free Radic. Biol. Med.* **14,** 361–370.

Wallace, D. C. (1992). Diseases of the mitochondrial DNA. *Annu. Rev. Biochem.* **61,** 1175–1212.

Widick, M. P., Telischi, F. F., Lonsbury-Martin, B. L., and Stagner, B. B. (1994). Early effects of cerebellopontine angle compression on rabbit distortion-product otoacoustic emissions: A model for monitoring cochlear function during acoustic neuroma surgery. *Otolaryngol. Head Neck Surg.* **111,** 407–416.

Willet, K., Detry, O., Droy-Lefaix, M. T., and Sluse, F. E. (1997). Effects of ischemia on the mitochondrial oxidative-phosphorylation: In situ protective effects of EGb 761. *In* "Proceedings of the VIIIth biennal Meeting of International Society for Free Radical Research." J. R. Prous, Barcelona, in press.

Yuting, C., Rongliang, Z., Zhongjian, J., and Yong, J. (1990). Flavonoids as superoxide scavengers and antioxidants. *Free Radic. Biol. Med.* **9,** 19–21.

23 Downregulation of Agonist-Induced Nitric Oxide Synthesis and Cell–Cell Adhesion by *Ginkgo biloba* Extract EGb 761: Therapeutic Efficacy in Circulatory Disorders

Sashwati Roy,† Hirotsugu Kobuchi,† Chandan K. Sen,† Marie-Thérèse Droy-Lefaix, and Lester Packer†*

† Membrane Bioenergetics Group
Department of Molecular and Cell Biology
University of California
Berkeley, California 94720

* Department of Pharmacology
IPSEN Institute
75016 Paris, France

INTRODUCTION

Extracts from the leaves or fruits of *Ginkgo biloba* (Ginkgoaceae) have been used therapeutically for centuries in traditional Chinese medicine. In recent years, the use of extracts from *Ginkgo* leaves has been used extensively in Europe for the treatment of vascular and cognition disorders. The demand for *Ginkgo* is increasing at 26% a year, one of the highest for any medicinal plant (Masood, 1997). Almost 2000 tons of this product are used each year, mostly in Germany (Masood, 1997). EGb 761 (IPSEN, Paris) is a standardized extract prepared from the leaves of *G. biloba*. Two major classes of compounds that constitute the active components of EGb 761 are 24% flavonoid glycosides and 6% terpenoids (Drieu, 1986; DeFeudis, 1991). The flavonoid fraction is composed mainly of three flavonols: quercetin, kaempferol, and isorhamnetin, which are linked to a sugar (De-Feudis, 1991). The terpenoid fraction is composed of ginkgolides and bilobalides (Drieu, 1988).

EGb 761 has been shown to have cardioprotective and neuroprotective effects. The mechanisms underlying the beneficial effects of EGb 761 are only partially understood. One of the main mechanisms seems related to the antioxidant properties of EGb 761, which appear to be due to the synergistic action of its major constituents such as flavonoids, terpenoids, and organic acids (Packer *et al.*, 1996). EGb 761 directly scavenges hydroxyl, superoxide, peroxyl, and nitric oxide radicals *in vitro* (Marcocci *et al.*, 1994; Maitra *et al.*, 1995). EGb 761 also protects against free radical-mediated damage in biological model systems, including ischemia–reperfusion injury of organs (Haramaki *et al.*, 1994) and oxidative modification of low-density lipoprotein (Maitra *et al.*, 1995; Yan *et al.*, 1995). To elucidate the molecular mechanisms underlying the therapeutic efficacy of EGb 761 in cerebral and vascular disorders, the effect of this extract on agonist-induced nitric oxide synthesis and cell–cell adhesion was studied.

NITRIC OXIDE METABOLISM

Nitric oxide (NO), first identified as an endothelium-derived relaxation factor (EDRF), plays a critical role in the regulation of cerebral and vascular functions. NO is synthesized in mammalian cells via the oxidation of L-arginine by a family of NO synthases (NOS). NOS are either constitutive (cNOS) or inducible (iNOS) (Nathan and Xie, 1994). cNOS is calcium dependent and produces small amount of NO over several minutes in response to stimuli that elevate intracellular Ca^{2+} (Nathan and Xie, 1994). Nanomolar concentrations of NO produced by cNOS are sufficient for intracellular signaling. Inducible NOS is activated with stimuli, including lipopolysaccharide (LPS), interferon-γ (IFN-γ), interleukin (IL)-1, or tumor necrosis factor-α (TNF-α) in macrophages, smooth muscle cells, endothelial cells, hepatocytes, and other cells. Following activation, iNOS produces high levels of NO that are sustained for long periods (Curran *et al.*, 1989; Busse and Mulsch, 1990; Steurer *et al.*, 1990; Gross *et al.*, 1991; Schulz *et al.*, 1992). Macrophages can greatly increase their production of both NO and superoxide anion simultaneously, resulting in the formation of peroxynitrite, which, through further reaction, can exert even stronger oxidant effects (Ischiropoulos *et al.*, 1992). Therefore, high amounts of NO generated by the activation of iNOS are potentially cytotoxic, capable of injuring host tissue indiscriminately by itself or by the formation of peroxynitrite.

It has been reported previously that EGb 761 can scavenge nitric oxide directly (Marcocci *et al.*, 1994). Regulation of cellular nitric oxide metabolism by EGb 761 in macrophages has been characterized (Kobuchi *et al.*, 1997).

Direct Interactions of EGb 761 with NO

Oxyhemoglobin is a well-known scavenger of NO. NO oxidizes oxyhemoglobin to the methemoglobin form (Doyle and Hoekstra, 1981). Scavengers of NO compete with hemoglobin for NO and affect the rate of hemoglobin oxidation depending on their concentration. EGb 761 inhibited the oxidation of hemoglobin to methemoglobin in a dose-dependent manner, and the inhibition was dependent on the concentration of hemoglobin (Fig. 1A). In these experiments, NO was produced by the reaction of hydroxylamine with complex I of catalase. Sodium nitroprusside decomposes in aqueous solution at physiological pH to generate NO. A similar inhibition of NO production by EGb 761 was also observed when NO was generated from sodium nitroprusside (Fig. 1B) (Marcocci *et al.*, 1994).

Effect of EGb 761 on NO Production in Macrophages

The ability of EGb 761 to influence the production of NO in macrophages was investigated. The mouse monocyte/macrophage cell line RAW 264.7 was treated with EGb 761. This was followed by the activation of cells with LPS (10 ng/ml) and IFN-γ (10 U/ml). Nitrite production by cultured macrophages in response to LPS-IFN-γ was used as an index for NO synthesis. EGb 761 inhibited nitrite accumulation in the culture medium from macrophages activated with LPS and IFN-γ. The inhibition of nitrite production by EGb 761 was dose dependent (Fig. 2).

EGb 761 Downregulates NOS Activity and iNOS mRNA Expression in Macrophages

To characterize the mechanism underlying the inhibition of nitrite production by EGb 761, the possible effect of EGb 761 on NOS activity has been investigated. NOS activity was determined by the conversion of radiolabeled arginine to citrulline using a cell-free cytosolic preparation from LPS/IFN-γ-activated RAW 264.7 cells. Figure 3 illustrates the inhibitory ef-

A.

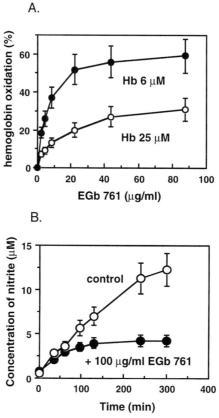

B.

Figure 1   Effect of EGb 761 on (A) hemoglobin oxidation by NO generated from the reaction of complex I of catalase with hydroxylamine and (B) the amount of nitrite generated during incubation with 5 mM sodium nitroprusside. Each point represents the mean ± SD of four experiments (Reprinted from Marcocci *et al.*, 1994 with permission).

fect of EGb 761 on NOS activity. Further studies were performed to evaluate whether the effect of EGb 761 on NOS activity was due to altered expression of the iNOS gene in macrophages. When macrophages were stimulated with LPS/IFN-γ, marked elevation of the iNOS mRNA level was observed. Induced iNOS mRNA expression reached a maximum 6–8 hr following stimulation. Treatment of cells with EGb 761 significantly downregulated LPS/IFN-γ-induced iNOS mRNA expression (Fig. 4). Maximal suppression (41%) of the expression of iNOS mRNA was observed at a concentration of 200 μg/ml. These results indicate that the inhibitory effect of EGb 761 on NO production in macrophages is mediated by the downregulation of iNOS gene expression (Kobuchi *et al.*, 1997).

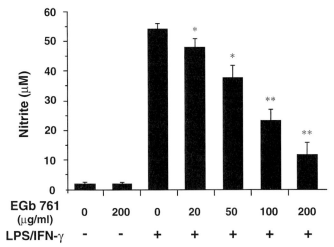

Figure 2 Effect of EGb 761 on NO production by macrophages. RAW 264.7 macrophages
(2.5×10^5 cells/0.5 ml/well) were stimulated with LPS (10 ng/ml) plus IFN-γ (10 U/ml) in the
presence of varying concentrations of EGb 761. EGb 761 was added immediately after stimulation. After incubation for 20 hr, nitrite in the conditioned medium was measured. Control
indicates untreated cells. Values represent the mean ± SD of triplicate samples (Reprinted from
Kobuchi *et al.*, 1997 with permission). * $p < 0.05$ and ** $p < 0.01$ compared with the LPS/
IFN-γ-stimulated group.

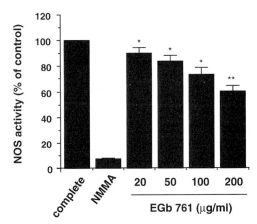

Figure 3 Effect of EGb 761 on NOS activity. NOS activity was determined by the conversion of radiolabeled arginine to citrulline using a cytosolic preparation from macrophages.
N^G-Monomethylarginine (NMMA), an inhibitor of NOS activity, was present in the assay at
100 μM. EGb 761 was added into the "complete" assay mixture. All values are expressed as
a percentage of control and represent the mean ± SD of three independent experiments * $p <$
0.05 and ** $p < 0.01$ compared with the LPS/IFN-γ-stimulated group (Reprinted from
Kobuchi *et al.*, 1997 with permission).

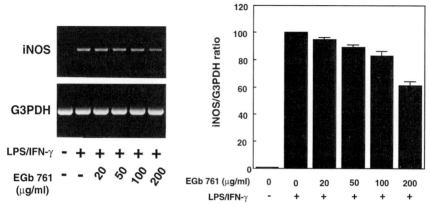

Figure 4   Effect of EGb 761 on the expression of iNOS mRNA in macrophages. (Left) Gel photograph of PCR-amplified cDNA derived from iNOS and G3PDH mRNA. Macrophages (2.5×10^6 cells/5 ml/28 cm^2 well) were incubated with LPS (10 ng/ml) plus IFN-γ (10 U/ml) in the presence of various concentrations of EGb 761. After incubation for 7 hr, cells were harvested and total RNA was extracted. (Right) Dose and time response of iNOS mRNA to EGb 761. iNOS mRNA was quantified by densitometry and expressed as 100% control for the iNOS/G3PDH ratio (ratio of iNOS to G3PDH) of mRNA after cells were stimulated with LPS (10 ng/ml) plus IFN-γ (10 U/ml). Values represent the mean ± SD of triplicate samples (Reprinted from Kobuchi *et al.*, 1997 with permission).

The NO metabolism modulatory properties of EGb 761 expand the role of EGb 761 as an antioxidant agent, and the fact that EGb 761 affects a substance that regulates the vascular tone may help explain its effectiveness in a variety of conditions in which blood flow is affected.

CELL ADHESION

Cell–cell adhesion is critical to the generation of effective immune responses and is dependent on the expression of a variety of cell surface receptors (Bevilacqua, 1993). Increased expression of adhesion molecules and altered cell adhesion properties play a critical role in a number of circulation-related pathologies, including inflammation, atherosclerosis, postischemic vascular disorders, and diabetes (Dosquet *et al.*, 1992; Munro, 1993; Thiery *et al.*, 1996). Intercellular adhesion molecule-1 (ICAM-1, CD54) and vascular cell adhesion molecule-1 (VCAM-1, CD 106) are inducible cell surface glycoproteins (Albelda *et al.*, 1994). The expression of these surface proteins is known to be induced in response to activators such as cytokines (TNF-α, IL-1 α and β), phorbol 12-myristate 13-acetate (PMA), LPS, and oxidants (Albelda *et al.*, 1994). Ligands for ICAM-1 and

VCAM-1 on lymphocyte are lymphocyte function-associated antigen-1 (LFA-1, CD11a/CD18) and very late antigens-4 (VLA-4), respectively (Albelda *et al.*, 1994).

Regulation of both ICAM-1 and VCAM-1 gene expression has been related to oxidative stress through specific reduction–oxidation (redox)-sensitive transcriptional or posttranscriptional mechanisms (Marui *et al.*, 1993; Ikeda *et al.*, 1994). Various antioxidants, including flavonoids, have been reported to affect the expression of cell adhesion molecules.

Regulation of Cell Adhesion by Flavonoids

Treatment of human endothelial cells with certain hydroxyflavones and flavanols has been reported to inhibit cytokine-induced ICAM-1, VCAM-1, or E-selectin expression in human endothelial cells (Gerritsen *et al.*, 1995). Apigenin is a flavone that inhibits adhesion molecule expression in endothelial cells in a dose- and time-dependent manner. An effect of this flavone at the transcriptional level has been demonstrated (Gerritsen *et al.*, 1995). Apigenin also inhibits TNF-α-induced ICAM-1 expression *in vivo* (Panes *et al.*, 1996). Cell adhesion regulatory effects of flavonoids are also consistently evident from other independent studies. The flavonoid delphinidin chloride (CAS 528-53-0, IdB 1056) inhibited acetylcholine (Ach) and sodium nitroprusside (SNP)-induced adherence of leukocytes to the venular endothelium in diabetic hamsters (Bertuglia *et al.*, 1995). The flavonoids 5-methoxyflavanone and, more potently, 5-methoxyflavone downregulated indomethacin-induced leukocyte adherence to mesenteric venules (Blank *et al.*, 1997). The flavonoid 2-(3-amino-phenyl)-8-methoxychromene-4-one (PD 098063) selectively blocks TNF-α induced VCAM-1 expression in endothelial cells in a concentration-dependent manner with half-maximal inhibition at 19 μM but had no effect on ICAM-1 expression. This selective inhibition of agonist-induced VCAM-1 protein and gene expression by PD 098063 was through an NF-κB-independent mechanism(s) (Wolle *et al.*, 1996). Because of the high flavonoid content in EGb 761, the authors investigated whether this extract has any regulatory functions in cell–cell adhesion.

Effect of EGb 761 on Agonist-Induced ICAM-1 Expression

The expression of ICAM-1 was assayed in EGb 761-pretreated human endothelial (ECV) cells following activation with PMA (100 nM) for 24 hr. The expression was markedly induced in response to PMA treatment for 24 hr (Fig. 5). Pretreatment of ECV cells with EGb 761 (100 μg/ml) for 72 hr downregulated the PMA-induced expression of ICAM-1 (Fig. 5).

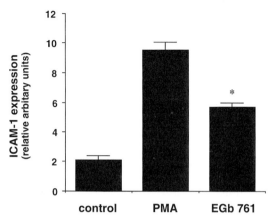

Figure 5 Effect of EGb 761 pretreatment on PMA-induced ICAM-1 expression in human vascular endothelial (ECV) cells. ECV cells were pretreated with EGb 761 (100 μg/ml) for 72 hr and were then activated with 100 nM PMA for 24 hr. Data are mean ± SD of at least three experiments. * $p < 0.01$ (student's t test) when compared with PMA-induced ICAM-1 expression.

Inhibition of PMA-Stimulated Adhesion of Human Jurkat T Lymphocytes to Endothelial Cells

A lymphocyte adhesion assay was performed to verify whether the inhibitory effect of EGb 761 on the expression of ICAM-1 was also effective in the downregulation of actual cell-to-cell adhesion. A significant decrease in PMA-induced adherence of Jurkat T cells to endothelial cells was observed following the pretreatment of endothelial cells with EGb 761 (50–100 μg/ml) for 72 hr (Fig. 6). Such downregulation of PMA-induced adherence of Jurkat T cells to endothelial cells by EGb 761 was dose dependent (25–100 μg/ml). Based on these observations, it may be concluded that EGb 761 may be expected to have important therapeutic potential for the treatment of a variety of inflammatory diseases involving an increase in leukocyte adhesion and trafficking.

THERAPEUTIC POTENTIAL

Several clinical trials have been conducted to assess the efficacy of EGb 761 circulatory and brain disorders. A multicenter 52-week-long placebo-controlled and double-blind study was conducted in the United States to investigate the efficacy and safety of EGb 761 in Alzheimer's disease and multi-infarct dementia by a standard assessment of cognition and behavior.

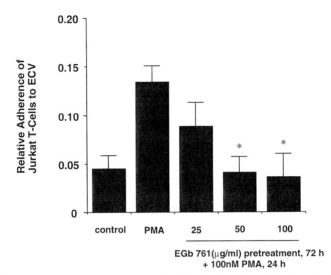

Figure 6 Adhesion of human Jurkat T cells to PMA-activated endothelial (ECV) cells is inhibited by EGb 761 pretreatment. ECV cells were pretreated with EGb 761 (25–100 μg/ml) for 72 hr and were then activated with 100 nM PMA for 24 hr. Cells were washed three times with PBS and were then cocultured with calcein-AM-labeled Jurkat T cells for 1 hr. Data are mean $\pm$ SD of at least five experiments. * $p < 0.01$ (student's t test) when compared with PMA-induced adhesion of Jurkat T cells to ECV.

EGb 761 significantly stabilized and improved cognitive performances and the social functioning of demented patients (Le Bars *et al.*, 1997). Most controlled clinical trials that have investigated the efficacy of EGb 761 in cerebral insufficiencies have reported beneficial effects (Kleijnen and Knipschild, 1992; Letzel and Schoop, 1992; Kanowski *et al.*, 1996; Letzel *et al.*, 1996). Symptoms of cerebral insufficiency have been associated with impaired cerebral circulation (DeFeudis, 1991). Therefore, the beneficial effects of EGb 761 in cerebral insufficiencies may partly be by improving cerebral circulation.

SUMMARY

EGb 761 has been shown to directly scavenge many of the reactive oxygen and nitrogen species, such as nitric oxide, hydroxyl, superoxide, and peroxyl radicals. Some of the protective effects of EGB 761 thus may be attributed to its direct radical scavenging activity. Regulatory effects of EGb 761 on two major cellular systems, nitric oxide metabolism and cell adhesion processes, have also been observed. Both nitric oxide and cell adhesion processes play critical roles in the regulation of vascular functions. EGb 761 inhibited lipopolysaccharide and IFN-γ-induced nitric oxide production in the macrophage cell line. Both activity and mRNA expression of iNOS were dose dependently inhibited by EGb 761. Pretreatment of human endothelial cells with EGb 761 downregulated the phorbol ester-induced expression of ICAM-1 and actual adhesion of lymphocytes to endothelial cells. Thus regulation of these important cell regulatory systems by *G. biloba* extract EGb 761 may be implicated in its beneficial effects in circulatory disorders.

REFERENCES

Albelda, S. M., Smith, C. W., and Ward, P. A. (1994). Adhesion molecules and inflammatory injury. *FASEB J.* **8,** 504–512.

Bertuglia, S., Malandrino, S., and Colantuoni, A. (1995). Effects of the natural flavonoid delphinidin on diabetic microangiopathy. *Arzneimittelforschung* **45,** 481–485.

Bevilacqua, M. P. (1993). Endothelial-leukocyte adhesion molecules. *Annu. Rev. Immunol.* **11,** 767–804.

Blank, M. A., Ems, B. L., O'Brien, L. M., Weisshaar, P. S., Ares, J. J., Abel, P. W., McCafferty, D. M., and Wallace, J. L. (1997). Flavonoid-induced gastroprotection in rats: Role of blood flow and leukocyte adherence. *Digestion* **58,** 147–154.

Busse, R., and Mulsch, A. (1990). Induction of nitric oxide synthase by cytokines in vascular smooth muscle cells. *FEBS Lett.* **275,** 87–90.

Curran, R. D., Billiar, T. R., Stuehr, D. J., Hofmann, K., and Simmons, R. L. (1989). Hepatocytes produce nitrogen oxides from L-arginine in response to inflammatory products of Kupffer cells. *J. Exp. Med.* **170,** 1769–1774.

DeFeudis, F. V. (1991). "*Ginkgo biloba* Extract (EGb 761): Pharmacological Activities and Clinical Applications." Series Elsevier, Paris.

Dosquet, C., Weill, D., and Wautier, J. L. (1992). Molecular mechanism of blood monocyte adhesion to vascular endothelial cells. *Nouv. Rev. Fr. Hematol.* **34**(Suppl.), S55–59.

Doyle, M. P., and Hoekstra, J. W. (1981). Oxidation of nitrogen oxides by bound dioxygen in hemoproteins. *J. Inorg. Biochem.* **14,** 351–358.

Drieu, K. (1986). Preparation et definition de l'extrait de Ginkgo biloba. *Presse. Med.* **15,** 1455–1457.

Drieu, K. (1988). "*Ginkgo biloba:* Recent Results in Pharmacology and Clinic." Series Springer-Verlag, Berlin.

Gerritsen, M. E., Carley, W. W., Ranges, G. E., Shen, C. P., Phan, S. A., Ligon, G. F., and Perry, C. A. (1995). Flavonoids inhibit cytokine-induced endothelial cell adhesion protein gene expression. *Am. J. Pathol.* **147,** 278–292.

Gross, S. S., Jaffe, E. A., Levi, R., and Kilbourn, R. G. (1991). Cytokine-activated endothelial cells express an isotype of nitric oxide synthase which is tetrahydrobiopterin-dependent, calmodulin-independent and inhibited by arginine analogs with a rank-order of potency characteristic of activated macrophages. *Biochem. Biophys. Res. Commun.* **178,** 823–829.

Haramaki, N., Aggarwal, S., Kawabata, T., Droy-Lefaix, M. T. T., and Packer, L. (1994). Effects of natural antioxidant *Ginkgo biloba* extract (EGB 761) on myocardial ischemia-reperfusion injury. *Free Radic. Biol. Med.* **16.**

Ikeda, M., Schroeder, K. K., Mosher, L. B., Woods, C. W., and Akeson, A. L. (1994). Suppressive effect of antioxidants on intercellular adhesion molecule-1 (ICAM-1) expression in human epidermal keratinocytes. *J. Invest. Dermatol.* **103,** 791–796.

Ischiropoulos, H., Zhu, L., and Beckman, J. S. (1992). Peroxynitrite formation from macrophage-derived nitric oxide. *Arch. Biochem. Biophys.* **298,** 446–451.

Kanowski, S., Herrmann, W. M., Stephan, K., Wierich, W., and Hoerr, R. (1996). Proof of efficacy of the *Ginkgo biloba* extract EGb 761 in outpatients suffering from mild to moderate primary degenerative dementia of the Alzheimer type or multi-infract dementia. *Pharmacopsychiatry* **29,** 47–56.

Kleijnen, J., and Knipschild, P. (1992). *Ginkgo biloba* for cerebral insufficiency. *Br. J. Clin. Pharmacol.* **34,** 352–358.

Kobuchi, H., Droy-Lefaix, M. T., Christen, Y., and Packer, L. (1997). *Ginkgo biloba* extract (EGb 761): Inhibitory effect on nitric oxide production in the macrophage cell line RAW 264.7. *Biochem. Pharmacol.* **53,** 897–903.

Le Bars, P. L., Katz, M. M., Berman, N., Itil, T. M., Freedman, A. M., and Schatzberg, A. F. (1997). A placebo-controlled, double-blind, randomized trial of an extract of gingko biloba for dementia. *J. Am. Med. Assoc.* **278,** 1327–1332.

Letzel, H., Haan, J., and Feil, W. B. (1996). Nootropics. *J. Drug Dev. Clin. Pract.* **8,** 77–94.

Letzel, H., and Schoop, W. (1992). *Gingko biloba* extract EGb 761 and pentoxifylline in intermittent claudication: Secondary analysis of the clinical effectiveness. *Vasa* **21,** 403–410.

Maitra, I., Marcocci, L., Droy-Lefaix, M. T., and Packer, L. (1995). Peroxyl radical scavenging activity of *Ginkgo biloba* extract EGb 761. *Biochem. Pharmacol.* **49,** 1649–1655.

Marcocci, L., Maguire, J. J., Droy-Lefaix, M. T., and Packer, L. (1994). The nitric oxide-scavenging properties of *Ginkgo biloba* extract EGb 761. *Biochem. Biophys. Res. Commun.* **201,** 748–755.

Marcocci, L., Packer, L., Droy-Lefaix, M. T., Sekaki, A., and Gardes-Albert, M. (1994). Antioxidant action of *Ginkgo biloba* extract EGb 761. *Methods Enzymol.* **234,** 462–475.

Marui, N., Offermann, M. K., Swerlick, R., Kunsch, C., Rosen, C. A., Ahmad, M., Alexander, R. W., and Medford, R. M. (1993). Vascular cell adhesion molecule-1 (VCAM-1) gene transcription and expression are regulated through an antioxidant-sensitive mechanism in human vascular endothelial cells. *J. Clin. Invest.* **92,** 1866–1874.

Masood, E. (1997). Medicinal plants threatened by over-use. *Nature* **385,** 570.

Munro, J. M. (1993). Endothelial-leukocyte adhesive interactions in inflammatory diseases. *Eur. Heart. J.* **14**(Suppl K.), 72–77.

Nathan, C., and Xie, Q. W. (1994). Regulation of biosynthesis of nitric oxide. *J. Biol. Chem.* **269,** 13725–13728.

Packer, L., Marcocci, L., Haramaki, N., Kobuchi, H., Christen, Y., and Droy-Lefaix, M. T. (1996). Antioxidant properties of *Ginkgo biloba* extract EGb 761 and clinical applications. *In* "Antioxidant Properties of *Ginkgo biloba* Extract EGb 761 and Clinical Applications". (L. Packer, M. G. Traber, and W. Xin, eds.). AOCS Press, Champaign, IL.

Panes, J., Gerritsen, M. E., Anderson, D. C., Miyasaka, M., and Granger, D. N. (1996). Apigenin inhibits tumor necrosis factor-induced intercellular adhesion molecule-1 upregulation in vivo. *Microcirculation* **3,** 279–286.

Schulz, R., Nava, E., and Moncada, S. (1992). Induction and potential biological relevance of a Ca(2+)-independent nitric oxide synthase in the myocardium. *Br. J. Pharmacol.* **105,** 575–580.

Steurer, J., Siegenthaler-Zuber, G., Siegenthaler, W., Suter, S., Kessler, F. J., Vahlensieck, M.,

Streuli, R., and Lingg, G. (1990). Paroxysmal non-hereditary angioedema. *Dtsch. Med. Wochenschr.* **115**, 1586–1590.

Thiery, J., Teupser, D., Walli, A. K., Ivandic, B., Nebendahl, K., Stein, O., Stein, Y., and Seidel, D. (1996). Study of causes underlying the low atherosclerotic response to dietary hypercholesterolemia in a selected strain of rabbits. *Atherosclerosis* **121**, 63–73.

Wolle, J., Hill, R. R., Ferguson, E., Devall, L. J., Trivedi, B. K., Newton, R. S., and Saxena, U. (1996). Selective inhibition of tumor necrosis factor-induced vascular cell adhesion molecule-1 gene expression by a novel flavonoid: Lack of effect on transcription factor NF-kappa B. *Arterioscler. Thromb. Vasc. Biol.* **16**, 1501–1508.

Yan, L. J., Droy-Lefaix, M. T., and Packer, L. (1995). *Ginkgo biloba* extract (EGb 761) protects human low density lipoproteins against oxidative modification mediated by copper. *Biochem. Biophys. Res. Commun.* **212**, 360–366.

24 Reactive Oxygen Species Increase Neutrophil Adherence to Endothelial Cells and Activate Tyrosine Phosphorylation of Cytoskeleton Proteins

Alexia Gozin, Hassan Sellak, Elisabeth Franzini, and Catherine Pasquier

INSERM U479
CHU Xavier Bichat
Paris, France

INTRODUCTION

Oxidative stress, defined as an increase in the production of reactive oxygen species (ROS) such as superoxide anion ($O_2^{\cdot-}$), hydrogen peroxide (H_2O_2), and hydroxyl radical (HO•), has been related to reperfusion injury in the heart and other organs (1). Parks and Granger (2) and McCord (3) suggested that, in reperfusion injury, xanthine oxidase (XO) derived from xanthine dehydrogenase (XD) conversion during ischemia was a source of ROS during reperfusion of ischemic tissues; other cellular sources of ROS have also been suggested (4–6). Several laboratories have shown that reperfusion of ischemic tissues leads to a marked local increase in the number of polymorphonuclear neutrophils (PMN) adhering to endothelial cells (7–10). The adherence of PMN to endothelial cells is related to many factors, especially adhesion molecules (11); the latter include intercellular adhesion molecule (ICAM-1), E- and P-selectins on the endothelium, and L-selectin, β_2 integrins, and carbohydrates such as sialyl-Lewis^x on PMN. Endothelial cells also express platelet-activating factor (PAF), which binds to specific receptors on PMN.

A large part of our knowledge on the interaction of the endothelium with PMN is derived from *in vitro* studies using cultured endothelial cells isolated from human umbilical cord vein (HUVEC) and isolated PMN. Although these models are not strictly identical to *in vivo* conditions, they have shown that after exposure to histamine or thrombin, endothelial cells express P-selectin (GMP-140) and PAF within minutes (12–14); in contrast, inflammatory mediators, including tumor necrosis factor α (TNF α), interferon-γ, interleukin-1 (IL-1), and lipopolysaccharide (LPS), cause strong expression and synthesis of ICAM-1 and E-selectin (15), but only after a few hours. It has also been reported that endothelial cells exposed *in vitro* to H_2O_2 for 1 to 4 hr bind a significant number of PMN and that this binding is related to the expression of PAF and P-selectin (16).

The authors and others (17–19) have shown that ROS, *in vitro,* stimulated PMN adhesion, to HUVEC, suggesting an important role of ROS in PMN recruitment during inflammation; this adhesion seemed to be due to the intracellular hydroxyl radical (HO•) produced by a Fenton reaction inside HUVEC (17, 20) as it has been shown that hydroxyl radical scavengers decreased it. In order to look for an activity of HO• as an intracellular or extracellular mediator in the mechanism involved in the increased adherence promoted by H_2O_2, the authors used compounds capable of abolishing the initial step of its metabolism such as HO• scavengers, entering cells or not, and iron chelators unable to penetrate cells.

The intracellular mechanism by which HO• mediates the adhesiveness of HUVEC is unclear. A growing body of evidence suggests that ROS, produced directly or by γ-irradiation in other cell types, may function as intracellular signaling molecules by activating different various tyrosine kinases (21–24). Tyrosine-phosphorylated proteins play an important role in the regulation of cell-adhesion molecules (21, 25–27). Indeed, both genistein and herbimycin A attenuated the adhesion of neutrophils and monocytes to endothelial cells in response to both interleukin 1 and TNFα (28).

This chapter focuses on the involvement of adhesion molecules in the increased adherence of neutrophils to ROS-stimulated endothelial cells, on the oxygen species responsible for this adhesion, and on the intracellular signaling pathway leading to cytoskeleton modification by ROS.

IMPLICATION OF ADHESION MOLECULES IN ROS-STIMULATED NEUTROPHIL ADHERENCE

The PMN–endothelial interaction induced by ROS is a critical event leading to localization and extravasation of PMN. HUVEC treated with nontoxic amounts of ROS showed increased binding to PMN (17). The number of PMN adhering was about 2- to 2.5-fold higher than with untreated HUVEC. The increased adherence appeared to be due to ICAM-1, without its upregulation, and to the involvement of a carbohydrate ligand on HUVEC. In experimental conditions, the oxidative stress did not damage HUVEC, as cell monolayers treated for 15 min with HX-XO in HBSS buffer (pH 7.4, 37°C) were viable at 4 and 24 hr (their morphology did not differ from that of controls, LDH was not released into the medium, and cells remained attached to the support).

HX-XO-exposed endothelial cell adhesion to PMN was not P-selectin dependent; this proadhesive molecule has been reported to mediate PMN adherence to endothelial cells treated with H_2O_2 for long periods (16). Other agonists have also been reported to induce P-selectin upregulation (29). The authors found that the HUVEC preparation expressed no P-selectin on the cell surface before or after stimulation with ROS. A large amount of P-selectin was, however, found in permeabilized HUVEC, as previously described, in Weibel-Palade bodies (21). Another argument against the functional involvement of P-selectin in the authors' system was the absence of effect of anti-P-selectin monoclonal antibodies (MoAb) on the adherence of PMN to HX-XO-exposed HUVEC. This MoAb, however, clearly inhibited PMN adherence to PMA- and thrombin-stimulated HUVEC (by about 50 and 40%, respectively). PMA- and thrombin-induced adherence are known to be partly dependent on P-selectin (14, 29). In contrast to these results, Patel *et al.* (16) observed an increased expression of P-selectin on HUVEC after H_2O_2 treatment. However, these authors used a much larger amount of H_2O_2; the HX-XO concentrations used in this study led to the generation of about 30–40 μmol L^{-1} H_2O_2.

Another proadhesive molecule that might be expressed after the short activation of endothelial cells is the platelet-activating factor. It is synthetized rapidly by hydrolysis of the PAF precursor phospholipid (30) and can bind to its receptor on the PMN (31), serving as a signal for the upregulation of CD11/CD18 glycoproteins on the PMN plasma membrane. In the authors' system, a PAF receptor antagonist failed to antagonize PMN adherence after HUVEC treatment with HX-XO. In contrast, after HUVEC stimulation with PMA or thrombin, the PAF receptor antagonist inhibited PMN adherence (by about 20 and 40%, respectively), suggesting that the inhibitor was effective. In contrast, Lewis *et al.* (32) reported that this phospholipid mediates PMN adhesion to HUVEC treated with H_2O_2; however, the amount of H_2O_2 generated by the authors' system was again lower than that in the study by Lewis and colleagues.

In another study, Gasic *et al.* (33) reported that perfusion of vessels with H_2O_2 (1 mM), with no evidence of endothelial cell injury, induced PAF-dependent adherence of PMN to vessels by a mechanism that involved CD18 on PMN and ICAM-1 on the endothelium. This adherence was transient, mediated by PAF, and CD18 (ICAM-1) dependent. The cell model (vessels and canine jugular venous endothelial cells) and H_2O_2 concentration (1 mM) may explain the observed differences compared to the authors' model.

PMN adherence to HX-XO-treated HUVEC was not inhibited by 4-bromophenacyl bromide, a phospholipase A2 inhibitor, whereas PMN adherence to PMA- and thrombin-stimulated HUVEC was inhibited (by about 26 and 36%, respectively); these degrees of inhibition were similar to those obtained with the PAF receptor antagonist, showing that PAF is not involved in PMN adherence to HX-XO-treated HUVEC, whereas it is involved when HUVEC are stimulated with thrombin and PMA.

Arachidonic acid metabolites can be produced by HX-XO-exposed endothelial cells, and their release stimulates PMN (34). Arachidonic acid-derived products were not involved in the authors' system, as indomethacin did not inhibit PMN adhesion to HX-XO-, PMA-, or thrombin-stimulated HUVEC. By washing monolayers before the adhesion assay, the potential effect of HUVEC-derived soluble products was eliminated, which could have accumulated during HX-XO treatment.

Unlike P-selectin and E-selectin, ICAM-1 is expressed constitutively on the endothelial cell surface (35), although it can be markedly upregulated on stimulation by cytokines, including TNF α and IL-1β. Earlier studies indicated that upregulation of ICAM-1 by cytokines requires protein synthesis and takes 3 to 4 hr (24); ICAM-1 is a ligand for CD11a/CD18 and CD11b/CD18 on PMN. These receptors are expressed on resting PMN isolated from blood (36). Inhibition of PMN adherence to ROS-treated HUVEC by anti-ICAM-1 MoAb demonstrates the contribution of ICAM-1 to the increased adherence. Another argument for the involvement of constitutive ICAM-1 in the authors' model is that anti-CD11/CD18 MoAbs also inhibited PMN adherence to HX-XO-treated HUVEC. It has been reported that durable adhesion of PMN to thrombin-stimulated endothelial cells is dependent on ICAM-1 upregulation and that this ICAM-1 activation induced by thrombin occurs through a protein synthesis-independent mechanism (37). In the authors' model, no quantitative increase of constitutive ICAM-1 on endothelial cells was observed, as short exposure to XO, PMA, and thrombin failed to up-regulate ICAM-1. The difference between these results and ours is difficult to explain, but no evidence of an intracellular ICAM-1 pool (resembling that of P-selectin) when endothelial cells are permeabilized has been found by us or others (33).

Another point of interest is that mannose-6-P, and *N*-acetylneuraminic acid (sialic acid) inhibited the binding interaction between ROS-exposed HUVEC and PMN and that these carbohydrates appear to act on PMN but not on HUVEC. It is likely that mannose-6-P and *N*-acetylneuraminic acid interact at an active site of a ligand involved in PMN adherence to HUVEC. These data suggest that HUVEC surface carbohydrates and carbohydrate-binding molecules on PMN may contribute to the adherence of PMN to HUVEC stimulated by ROS, thrombin, and PMA.

The partial inhibition of adherence in the presence of anti-ICAM-1 and/or anti-CD18 MoAbs suggests the involvement of another non-ICAM-1-dependent mechanism in this adherence; this was supported by the total inhibition obtained when these antibodies were combined with mannose-6-P and sialic acid, suggesting that L-selectin on PMN recognizes a carbohydrate ligand on HUVEC. This effect, observed in the absence of shear stress, might be more important in flow conditions (33).

It has been reported that only Mel-14 equivalent to human TQ-1/Leu-8 (i.e., anti-L-selectin) (39, 40) significantly blocks PMN binding at 4°C (41) and that leukocyte integrins do not operate efficiently at cold temperatures (42, 43). In the experimental conditions used in this chapter, residual PMN adherence at 4°C to HX-XO-treated HUVEC may have been due to the contribution of L-selectin binding to carbohydrates expressed by HUVEC. Indeed, L-selectin was found on PMN after isolation at 4 and 37°C, and its binding activity on ROS-treated HUVEC was inhibited by an anti-L-selectin MoAb at 4°C. Anti-L-selectin did not inhibit PMN adherence at 37°C; on the contrary, it increased it, possibly by stimulating PMN. Whether L-selectin association to carbohydrate structure expressed by endothelial cells stimulated by ROS induces an increased expression of β_2 integrins needs further studies. Mannose-6-P and sialic acid inhibited PMN adherence to HX-XO-exposed HUVEC at 4°C; at this temperature only the adherence due to L-selectin persisted, with the role of ICAM-1 (or CD18) being negligible.

In conclusion, the increased binding of PMN to HX-XO-treated HUVEC observed here involved ICAM-1, but was independent of its up-regulation, and another non-ICAM-dependent mechanism in which carbohydrates expressed on HUVEC recognize L-selectin on PMN.

ROS INVOLVED IN PMN–ENDOTHELIAL CELLS ADHESION

The increase in PMN adherence to HUVEC seemed to be dependent on the production of H_2O_2 by the HX-XO system, as it was abolished by catalase. The same results were obtained with glucose oxidase but not with H_2O_2 added directly to the cells, although the same conditions of culture (confluent cells) were used. H_2O_2 enters cells, however, it is produced, and the fact that it does not have the same effect when added to cultures only once can be explained by its rapid metabolization by glutathion peroxydase and catalase, which cannot eliminate a permanent accumulation of H_2O_2 (released by XO or GO). It should thus be effective only when it is provided continuously to the cells by an enzymatic process. In contrast, extracellular $O_2^{\cdot -}$ did not seem to be involved (apart from its role as a precursor of H_2O_2), as it was not inhibited by SOD. Neither catalase nor SOD affected the increased adherence induced by PMA or thrombin, showing that PMN adherence to HX-XO-treated HUVEC is ROS specific.

As H_2O_2 is a moderately reactive oxidant species but can give rise to highly reactive HO• by accepting electrons from transition metals such as ferrous iron (Fe^{2+}), its effect was tested by using the HO• scavengers dimethylthiourea (DMTU), pentoxifylline (Ptx), and N-acetylcysteine (NAC) and the iron chelators deferioxamine (DF) and hydroxybenzylethylene diamide (HBED) (Fig. 1). These scavengers and iron chelators effectively inhibited the increase in PMN adherence to HX-XO-treated HUVEC, but did not affect the adherence of control HUVEC or HUVEC stimulated with PMA or thrombin. These findings suggest that HO• radicals are involved in promoting adherence through a different mechanism than that induced by PMA and thrombin, which involves cell receptors. The action of HO• scavengers and iron chelators suggests that HO• could be produced inside the cells by a Fenton reaction. Indeed, the source of catalytic transition metals ions inside the cells is well documented; an increase in this source could be due to the activity of heme oxygenase induced by the oxidative stress (44) and also to the potential presence of $O_2^{\cdot -}$ inside the cells (45), which is able to release iron from iron-binding proteins such as ferritin and transferrin (46). It is thus conceivable that catalytic transition metals ions are present inside the cells and contribute to the formation of HO•.

As HO• can also be located outside the cell, its extracellular activity was studied by adding the iron chelator transferrin, which does not enter the cells. This compound had no effect on PMN adherence to HX-XO-treated HUVEC, suggesting that the activity of HO• is not acting on the outer membrane of the cells. These results are consistent with work suggesting that HO• is an intracellular messenger; indeed, HO• was an intermediate in the production of interleukin-8 (IL-8) in response to LPS in a whole blood model. Scavengers of HO• strongly inhibited IL-8 production induced by LPS, TNF, IL-1, phytohemagglutinin, and aggregated immune complexes (47, 48).

It can also be assumed that the biomolecular targets of HO• are cell sur-

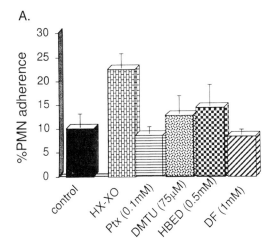

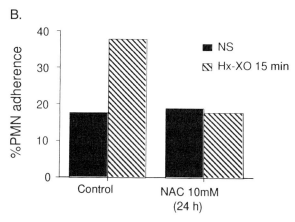

Figure 1 PMN adherence to ROS-stimulated HUVEC. Endothelial cells were stimulated with HX-XO (2×10^{-4} M, 4.5 mU ml^{-1}) in 24-well gelatin-coated plates for 15 min in serum-free HBSS, pH 7.4, at 37°C or with HBSS buffer alone, as indicated. Neutrophils were added (10^6/ml/well) to the wells and left for 15 min at 37°C. Adherence was determined by MPO measurement. (A) Before activation, HUVEC were incubated with 0.1 mM pentoxifylline (Ptx), 75 μM DMTU, 0.5 mM hydroxybenzylethylene diamide (HBED), or 1 mM desferioxamine (DF) for 24 hr, Values are the mean ± SD of at least three determinations. $p < 0.001$ versus HX-XO-stimulated HUVEC. (B) Before activation, HUVEC were incubated for 24 hr with 10 mM N-acetylcysteine (NAC) and adherence was measured.

face components, as trace iron in HBSS could bind to the plasma membrane; HO• could then activate membrane phospholipids, which could be involved in the increased adherence.

INTRACELLULAR MECHANISMS INVOLVED IN PMN ADHESION TO ROS-STIMULATED HUVEC

As ROS are able to activate tyrosine kinases, the authors looked for the effect of tyrosine kinase inhibitors on PMN adhesion to ROS-stimulated HUVEC. Herbimycin A, genistein, and erbstatin have been shown to inhibit this adherence (Fig. 2), suggesting that the intracellular mechanism of adherence involves tyrosine kinases.

P125^{FAK} is a member of a nonreceptor protein tyrosine kinase family that plays a role in regulating the changes in actin cytoskeleton organization. p125^{FAK} in fibroblasts is associated with focal adhesions; its tyrosine phosphorylation and that of the cytoskeleton-associated protein paxillin (PAX) and p130cas can be activated by $\beta 1$ and $\beta 3$ integrins, by a variety of regulatory peptides and lipids mediating cell growth and differentiation and acting through G-protein-coupled receptors, and by several other compounds (25, 26, 49–51). Vascular endothelial growth factor (VEGF) has been shown in endothelial cells to rapidly tyrosine phosphorylate p125^{FAK} and paxillin; this event is potentially involved in the migratory cell response to this factor (52).

ROS and possibly hydroxyl radicals HO• are able to stimulate the phosphorylation of p125^{FAK}, paxillin, and p130cas in endothelial cells (Fig. 3). As cytochalasin D was able to inhibit tyrosine phosphorylation and PMN adhesion at the same time (personal results), it was postulated that a correlation may exist among the phosphorylation of cytoskeleton pro-

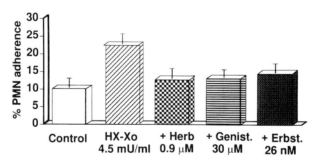

Figure 2   Effect of tyrosine kinase inhibitors on PMN adherence to ROS-stimulated HU-VEC. Endothelial cells were stimulated with HX-XO as described in Fig. 1. Tyrosine kinase inhibitors were incubated with cells before stimulation for 10 min: 30 μM genistein, 900 nM herbimycin, or 26 nM erbstatin. Adherence was then measured as indicated and expressed as a percentage (10% of adherent PMN represents 10^5 PMN). Values are the mean ± SD of at least three determinations. $p < 0.001$ versus HX-XO-stimulated HUVEC.

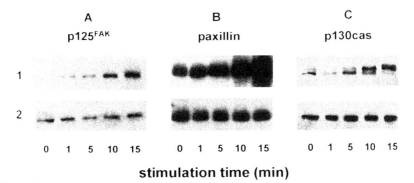

stimulation time (min)

Figure 3   Effect of HO• stimulation on p125^{FAK}, paxillin (PAX), and p130cas phosphorylation. HUVEC cultured to confluence in 6-well gelatin-coated plates were treated with HX-XO (HX, 2×10^{-4} M; XO, 4.5 mU ml^{-1}) for 1 to 15 min and were immunoprecipitated with (A) PAX, (B) p125^{FAK}, and (C) p130cas mAbs, respectively. Tyrosine phosphorylation was analyzed by anti-Tyr(P) Western blotting (1: A, B, C). The amount of proteins in each lane was checked by Western blotting with the respective antibodies: anti-p125^{FAK}, antipaxillin, and anti-p130cas (2: A, B, C). This profile is representative of at least three different experiments.

teins, modification of cell membrane structure and cytoskeleton organization which may play a role in the increase neutrophil adhesion mediated by ROS.

Antioxidants, DMTU, Ptx, NAC, and *Ginkgo biloba* extract and iron chelators (DF and HBED) inhibited the tyrosine phosphorylation of p125^{FAK} and paxillin (Figs. 4 and 5), suggesting a possible correlation between ROS-triggered adherence and tyrosine phosphorylation. Tyrosine-phosphorylated proteins play an important role in the regulation of cell adhesion molecules (25–27), as has been shown for the adherence of neutrophils and monocytes to endothelial cells in response to both interleukin-1 and tumor necrosis factor (28). A number of interesting findings have been made regarding p125^{FAK} and its cellular function. P125^{FAK} seems to be responsible for phosphorylating the components of focal adhesion (tensin, paxillin, or talin) and regulating the interactions of integrins with the cytoskeleton and/or the extracellular matrix. Hydroxyl radicals mediating tyrosine phosphorylation of the cytosolic focal adhesion protein kinase and of the cytoskeleton-associated proteins paxillin and p130cas can be considered as signaling molecules, involved in the modification of a cellular function such as adhesion, through this phosphorylation.

Thus the ROS-induced increase in PMN adhesion to HUVEC could be due to cytoskeleton reorganization through the tyrosine phosphory-

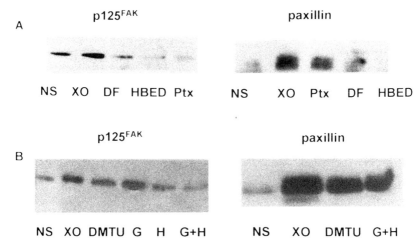

Figure 4 Effect of tyrosine kinase inhibitors, antioxidants, and iron chelators on p125^{FAK} and PAX phosphorylation. HUVEC were preincubated (A) for 24 hr with the iron chelators DF (1 mM) and HBED (0.5 mM) and for 10 min with Ptx (0.1 mM) and (B) for 10 min with DMTU (0.75 mM) or for 30 min with the tyrosine kinase inhibitors genistein (G, 30 μM) and herbimycin A (H, 0.9 μM) or with a mixture of both (G + H) at the same concentrations and then treated with HX-XO (HX, 2 $\times$ 10^{-4} M; XO, 4.5 mU ml^{-1}) for 15 min in serum-free HBSS, pH 7.4, at 37°C in a CO$_2$ incubator. p125^{FAK} and paxillin were immunoprecipitated with their respective antibodies, and Western blots were analyzed by the antiphosphotyrosine antibody.

lation of cytoskeleton proteins. The luminal surface of endothelial cells bears most of the adhesion molecules that can bind leukocytes, such as ICAM-1 and ICAM-2 (53–55). This preferential localization may be regulated by the normal intracellular organization of the actin cytoskeleton.

Tyrosine phosphorylation in endothelial cells is sensitive to exogenous-oxidizing agents and provides evidence that ROS may function as signaling molecules regulating tyrosine phosphorylation in protein cytoskeleton rearrangement. The prime candidate for the intracellular oxidant ultimately responsible for the induction of phosphotyrosine is the hydroxyl radical. A correlation between the tyrosine phosphorylation of p125^{FAK}, paxillin, and p130cas and the increase in PMN adherence to ROS-stimulated HUVEC was observed, but a direct relationship remains to be demonstrated. Future studies will examine the translocation of phosphorylated proteins to the membrane and their link with adhesion molecules on endothelial cells exposed to ROS.

A

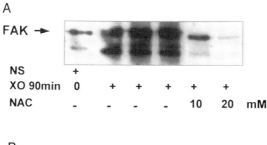

FAK →

NS   +
XO 90min   0   +   +   +   +   +
NAC   -   -   -   -   10   20   mM

B

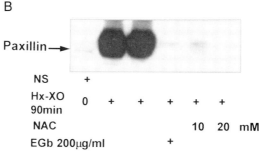

Paxillin →

NS   +
Hx-XO   0   +   +   +   +   +
90min
NAC   10   20   mM
EGb 200μg/ml   +

Figure 5   Effect of antioxidants on tyrosine phosphorylation of p125^{FAK} and paxillin. HUVEC were preincubated (A) for 24 hr with NAC (10 and 20 mM) and (B) with 10 and 20 mM of NAC and 200 μg/ml of *Ginkgo biloba* extract; they were then treated with HX-XO (HX, 2×10^{-4} M; XO, 2.5 mU ml^{-1}) for 90 min in serum-free HBSS, pH 7.4, at 37°C in a CO_2 incubator. p125^{FAK} and paxillin were immunoprecipitated with their respective antibodies, and Western blots of p125^{FAK} (A) and paxillin (B) were probed with the antiphosphotyrosine antibody.

REFERENCES

1. Parks, D. A., Bulkley, G. B., Granger, D. N., Hamilton, S. R., and McCord, J. M. (1982). Ischemic injury in the cat small intestine: Role of superoxide radicals. *Gastroenterology* **82,** 9–15.
2. Parks, D. A., and Granger, D. N. (1983). Ischemia-induced vascular changes: Role of xanthine oxidase and hydroxyl radicals. *Am. J. Physiol.* **245,** G285–G289.
3. McCord, J. M. (1985). Oxygen derived free radicals in postischemic tissue injury. *N. Engl. Med.* **312,** 159–163.
4. Tate, R. M., and Repine, J. E. (1983). Neutrophil and the adult respiratory distress syndrome. *Am. Rev. Respir. Dis.* **128,** 552–559.
5. Weiss, S. J., Young, J., Lobuglio, A. F., Slivka, A., and Nimeh, N. F. (1981). Role of hydrogen peroxide in neutrophil-mediated destruction of endothelial cells. *J. Clin. Invest.* **68,** 714–721.
6. Babior, B. M., Curnutte, J. T., and McMurrich, B. J. (1976). The particulate superoxide forming system from human neutrophils. *J. Clin. Invest.* **58,** 989–996.
7. Suzuki, M., Inaunen, W., Kvietys, P. R., Grisham, M. B., Meininger, C., Schelling, M. E.,

Granger, H. J., and Granger, D. N. (1989). Superoxide mediates reperfusion-induced leukocyte-endothelial cell interactions. *Am. J. Physiol.* **257,** H1740–H745.

8. Granger, D. N., Hollwarth, M. E., and Parks, D. A. (1986). Ischemia-reperfusion injury: Role of oxygen-derived free radicals. *Acta Physiol. Scand. (Suppl).* **548,** 47–63.

9. Suzuki, M., Grisham, M. B., and Granger, D. N. (1991). Leukocyte-endothelial cell adhesive interactions: Role of xanthine oxidase derived oxidants. *J. Leukocyte Biol.* **50,** 488–494.

10. Zimmerman, B. J., Grisham, M. B., and Granger, D. N. (1990). Role of oxidants in ischemia/reperfusion-induced granulocyte infiltration. *Am. J. Physiol.* **258,** G185–G190.

11. Zimmerman, G. A., Prescott, S. M., and McIntyre, T. M. (1992). Endothelial cell interactions with granulocytes: Tethering and signaling molecules. *Immunol. Today* **13,** 93–100.

12. Geng, J. G., Bevilacqua, M. P., Moore, K. L., McIntyre, T. M., Prescott, S. M., Kim, J. M., Bliss, G. A., Zimmerman, G. A., and McEver, R. P. (1990). Rapid neutrophil adhesion to activated endothelium mediated by GMP-140. *Nature* **343,** 757–760.

13. Prescott, S. M., Zimmerman, G. A., and McIntyre, T. M. (1984). Human endothelial cells in culture produce platelet-activating factor (1-alkyl-2-acetyl-sn-glycero-3-phosphocholine) when stimulated with thrombin. *Proc. Natl. Acad. Sci. USA* **81,** 3534–3538.

14. Lorant, D. E., Patel, K. D., McIntyre, T. M., McEver, R. P., Prescott, S. M., and Zimmerman, G. A. (1991). Coexpression of GMP-140 and PAF by endothelium stimulated by histamine or thrombin: A juxtacrine system for adhesion and activation of neutrophils. *J. Cell. Biol.* **115,** 223–234.

15. Pohlman, T. H., Stanness, K. A., Beatty, P. G., Ochs, H. D., and Harlan, J. M. (1986). An endothelial cell surface factor(s) induced in vitro by lipopolysaccharide, interleukine 1, and tumor necrosis factor-alpha increases neutrophil adherence by CDw18-dependent mechanism. *J. Immunol.* **136,** 4548–4553.

16. Patel, K. D., Zimmerman, G. A., Prescott, S. M., McEver, R. P., and McIntyre, T. M. (1991). Oxygen radicals induce human endothelial cells to express GMP-140 and bind neutrophils. *J. Cell. Biol.* **112,** 749–759.

17. Sellak, H., Franzini, E., Hakim, J., and Pasquier, C. (1994). Reactive oxygen species rapidly increase endothelial ICAM-1 ability to bind neutrophils without detectable upregulation. *Blood* **83,** 2669–2677.

18. Marui, N., Offermann, M. K., Swerlick, R., Kunsch, C., Rosen, C. A., Ahmad, M., Alexander, R. W., and Medford, R. M. (1993). Vascular cell adhesion molecule-1 (VCAM-1) gene transcription and expression are regulated through an antioxidant-sensitive mechanism in human vascular endothelial cells. *J. Clin. Invest.* **92**(4), 1866–1874.

19. Aoki, T., Suzuki, Y., Suzuki, K., Miyata, A., Oyamada, Y., Takasugi, T., Mori, M., Fujita, H., and Yamaguchi, K. (1996). Modulation of ICAM-1 expression by extracellular glutathione in hyperoxia-exposed human pulmonary artery endothelial cells. *Am. J. Respir. Cell Mol. Biol.* **15**(3), 319–327.

20. Franzini, E., Sellak, H., Marquetty, C., Babin-Chevaye, C., Hakim, J., and Pasquier, C. (1996). Inhibition of human neutrophil binding to hydroxyl peroxide-treated endothelial cells by cAMP and hydroxyl radical scavengers. *Free Radic. Biol. Med.* **21,** 15–23.

21. Bauskin, A. R., Alkalay, I., and Ben-Neriah, Y. (1991). Redox regulation of a protein tyrosine kinase in the endoplasmic reticulum. *Cell* **66,** 685–696.

22. Nakamura, K., Hori, T., Sato, N., Sugle, K., Kavakami, T., and Yodoi, J. (1993). Redox regulation of a src family protein tyrosine kinase $p56^{lck}$ in T cells. *Oncogene* **8,** 3133–3139.

23. Schieven, G. L., Mittler, R. S., Nadler, S. G., Kirihara, J. M., Bolen, J. B., Kanner, S. B., and Ledbetter, J. A. (1994). ZAP-70 tyrosine kinase, CD45, and T cell receptor involve-

ment in UV- and H_2O_2-induced T cell signal transduction. *J. Biol. Chem.* **269**, 20718–20726.

24. Brumell, J. H., Burkhardt, A. L., Bolen, J. B., and Grinstein, S. (1996). Endogenous reactive oxygen intermediates activate tyrosine kinases in human neutrophils. *J. Biol. Chem.* **271**, 1455–1461.

25. Seufferlein, T., and Rozengurt, E. (1994). Sphingosine induces p125[FAK] and paxillin tyrosine phosphorylation, actin stress fiber formation, and focal contact assembly in Swiss 3T3 cells. *J. Biol. Chem.* **269**, 27610–27617.

26. Kornberg, L., Earp, H. S., Parsons, T., Schaller, M., and Juliano, R. L. (1992). Cell adhesion or integrin clustering increases phosphorylation of a focal adhesion-associated tyrosine kinase. *J. Biol. Chem.* **267**, 23439–23442.

27. Bockholt, S. M., and Burridge, K. (1993). Cell spreading on extracellular matrix proteins induces tyrosine phosphorylation of tensin. *J. Biol. Chem.* **268**, 14565–14567.

28. McGregor, P. E., Agrawal, K., and Edwards, J. D. (1994). Attenuation of human leukocyte adherence to endothelial cell monolayers by tyrosine kinase inhibitors. *Biochem. Biophys. Res. Commun.* **198**, 359–365.

29. Hattori, R., Hamilton, K. K., Fugate, R. D., McEver, R. P., and Sims, P. J. (1989). Stimulated secretion of endothelial von Willebrand factor is accompanied by rapid redistribution to the cell surface of the intracellular granule membrane protein GMP-140. *J. Biol. Chem.* **264**, 7768–7771.

30. Whatley, R. E., Nelson, P., Zimmerman, G. A., Stevens, D. L., Parker, C. J., McIntyre, T. M., and Prescott, S. M. (1989). The regulation of platelet-activating factor production in endothelial cells, the role of calcium and protein kinase C. *J. Biol. Chem.* **264**, 6325–6333.

31. Tonnesen, M. G., Anderson, D. C., Springer, T. A., Knedler, A., Ardi, N., and Hensen, P. M. (1989). Adherence of neutrophils to cultured human endothelial cells: Stimulation by chemotactic peptides and lipid mediators and dependence upon the MAC-1, LFA-1, p150, 95 glycoprotein family. *J. Clin. Invest.* **83**, 637–646.

32. Lewis, M. S., Whatley, R. E., Cain, P., McIntyre, T. M., Prescott, S. M., and Zimmerman, G. A. (1988). Hydrogen peroxide stimulates the synthesis of platelet-activating factor by endothelium and induces endothelial cell-dependent neutrophil adhesion. *J. Clin. Invest.* **82**, 2045–2055.

33. Gasic, A. C., McGuire, G., Krater, S., Farhood, A. I., Golstein, M. A., Smith, C. W., Entman, M. L., and Taylor, A. A. (1991). Hydrogen peroxide pretreatment of perfused canine vessels induces ICAM-1 and CD18-dependent neutrophil adherence. *Circulation* **84**, 2154–2166.

34. Hoover, K. L., Karnovsky, M. J., Avsten, K. F., and Levis, R. A. (1984). Leukotriene B4 action on endothelium mediates augmented neutrophil-endothelial adhesion. *Proc. Natl. Acad. Sci. USA* **81**, 2191–2193.

35. Luscinskas, F. W., Cybulsky, N. I., Kiely, J., Peckins, C. S., Davis, V. M., and Gimbrone, M. A. (1991). Cytokine-activated human endothelial monolayers support enhanced neutrophil transmigration via a mechanism involving both endothelial-leukocyte adhesion molecule-1 and intercellular adhesion molecule-1. *J. Immunol.* **146**, 1617–1625.

36. Pober, J. S., Gimbrone, M. A., Lapierre, L. A., and Dendrick, D. L. (1986). Overlapping patterns of activation of human endothelial cell by interleukin 1, tumor necrosis factor and immune interferon. *J. Immunol.* **137**, 1893–1896.

37. Kuijpers, T. W., Tool, A. T. J., Van der Schoot, C. E., Ginsel, L. A., Onderwater, D. R., Roos, D., and Verhoeven, A. J. (1991). Membrane surface antigen expression on neutrophils: A reappraisal of the use of surface markers for neutrophil activation. *Blood* **78**, 1105–1111.

38. Sugama, Y., Tirupathi, C., Janakidevi, K., Andersen, T. T., Fenton, J. W., and Malik, A. B. (1992). Thrombin-induced expression of endothelial P-selectin and intercellular adhesion molecule-1: A mechanism for stabilizing neutrophil adhesion. *J. Cell. Biol.* **119**, 935–944.

39. Camerini, D., James, S. P., Stamenkovic, I., and Seed, B. (1989). Leu-8/TQ-1 is the human equivalent of the Mel-14 lymph node homing receptor. *Nature (London)* **342**, 78–82.

40. Tedder, T. F., Penta, A. C., Levine, H. B., and Freedman, A. S. (1990). Expression of the human leukocyte adhesion molecule, Lam1 identity with the TQ1 and Leu-8 differentiation antigens. *J. Immunol.* **144**, 532–540.

41. Jutila, M. A., Rott, L., Berg, E. L., and Butcher, E. C. (1989). Function and regulation of the neutrophil Mel-14 antigen in vivo: Comparison with LFA-1 and MAC-1. *J. Immunol.* **143**, 3318–3324.

42. Shaw, H., Ginther-Luce, G. E., Quinones, R., Gress, R. E., Springer, T. A., and Sanders, M. E. (1986). Two antigen-independent adhesion pathways used by human cytotoxic T-cell clones. *Nature* **323**, 262–264.

43. Marlin, S. D., and Springer, T. A. (1987). Purified intercellular adhesion molecule-1 (ICAM-1) is a ligand for lymphocyte function associated antigen (LFA-1). *Cell* **51**, 813–818.

44. Vile, G. F., Basu-Modak, S., Waltner, C., and Tyrrel, R. M. (1994). Heme oxygenase 1 mediates an adaptive response to oxidative stress in human skin fibroblasts. *Proc. Natl. Acad. Sci. USA* **91**, 2607–2610.

45. Wayne, O. W., Narayanan, P. K., and Robinson, J. P. (1994). Intracellular hydrogen peroxide and superoxide detection in endothelial cells. *J. Leukocyte Biol.* **55**, 253–258.

46. Bolann, B. J., and Ulvik, R. J. (1990). On the limited ability of superoxide to release iron from ferritin. *Eur. J. Biochem.* **193**, 899–904.

47. De Forge, L. E., Preston, A. M., Takeuchi, E., Kenny, J., Boxer, L. A., and Remick, D. G. (1993). Regulation of interleukin 8 gene expression by oxidant stress. *J. Biol. Chem.* **268**, 25568–25576.

48. De Forge, L. E., Fantone, J. C., Kenny, J. S., and Remick, D. G. (1992). Oxygen radical scavengers selectively inhibit interleukin 8 production in human whole blood. *J. Clin. Invest.* **90**, 2123–2129.

49. Schaller, M. D., Borgman, C. A., Cobb, B. S., Vines, R. R., Reynolds, A. B., and Parsons, J. T. (1992). pp125[FAK], a structurally distinctive protein-tyrosine kinase associated with focal adhesions. *Proc. Natl. Acad. Sci. USA* **89**, 5192–5196.

50. Hanks, S. K., Calalb, M. B., Harper, M. C., and Patel, S. K. (1992). Focal adhesion protein-tyrosine kinase phosphorylated in response to cell attachment to fibronectin. *Proc. Natl. Acad. Sci. USA* **89**, 8487–8491.

51. Huang, M. M., Lipfert, L., Cunningham, M., Brugge, J. S., Ginsberg, M. H., and Shattil, S. J. (1993). Adhesive ligand binding to integrin $\alpha_{IIb}\beta_3$ stimulates tyrosine phosphorylation of novel protein substrates before phosphorylation of pp125[FAK]. *J. Cell. Biol.* **122**, 473–483.

52. Abedi, H., and Zachary, I. (1997). Vascular endothelial growth factor stimulates tyrosine phosphorylation and recruitment to new focal adhesions of focal adhesion kinase and paxillin in endothelial cells. *J. Biol. Chem.* **272**, 15442–15451.

53. Springer, T. A. (1990). Adhesion of the immune system. *Nature* **346**, 425–434.

54. Diamond, M. S., Stauton, D. E., de Fougerolles, A. R., Stacker, S. A., Garcia-Aguilar, J., Hibbs, M. L., and Springer, T. A. (1990). ICAM-1 (CD54): A counter receptor for Mac-1 (CD11b/CD18). *J. Cell. Biol.* **111**, 3129–3139.

55. de Fougerolles, A. R., Stacker, S. A., Schwarting, R., and Springer, T. A. Characterization of ICAM-2 and evidence for a third counter receptor for LFA-1. *J. Exp. Med.* **174**, 253–267.

25 Natural Phenolic Antioxidants and Their Impact on Health

Edwin N. Frankel

Department of Food Science and Technology
University of California
Davis, California 95616

INTRODUCTION

Flavonoids are an important part of the diet because they can modulate lipid peroxidation involved in atherogenesis, thrombosis, and carcinogenesis (Kinsella *et al.*, 1993). Known properties of flavonoids include free radical scavenging, strong antioxidant activities in preventing the oxidation of low-density lipoproteins (LDL), inhibition of hydrolytic and oxidative enzymes (phospholipase A_2, cyclooxygenase, lipoxygenase), and anti-inflammatory actions.

Diets rich in fruits and vegetables are associated with lower risks of coronary heart disease and cancer (Block *et al.*, 1992). Fruits and vegetables are considered to be the main source of dietary antioxidants. Although vitamin E, vitamin C, and carotenoids have received the most attention as anticarcinogens and as defenses against degenerative diseases of aging (Ames *et al.*, 1993), the nutritional role of flavonoid antioxidants in fruits and vegetables has generally been overlooked. These flavonoids are found in many fruits and vegetables in a great variety of structures and specificities. As active antioxidants, flavonoids may be especially important in protecting against human degenerative diseases. However, little is known about their absorption and excretion in humans and very little is known about their bioavailability.

This chapter reviews some of the chemical and biological properties of plant phenolic antioxidants in foods and possible mechanisms for their protective effects in biological systems. The aim is to relate the extensive *in vitro* studies on flavonoids to their potential health effects in foods and beverages.

CHEMICAL AND BIOLOGICAL PROPERTIES

Two important properties of flavonoids include their antioxidant activity and their metal chelation properties. They behave as antioxidants by donating electrons to radicals and breaking the radical chains. Several antioxidant mechanisms may be considered, including scavenging radicals (ROO•, RO•) and activated oxygen species (HO•, $O_2^•$, $^1O_2$), inactivating metal ions by chelation, complexing with proteins (enzymes, apoprotein-B of LDL) and metal-binding sites of enzymes, exhibiting synergism by reducing radicals from oxidized α-tocopherol and ascorbic acid, and partitioning according to their polarity to become distributed at different oxidation sites. Flavonoids can also bind one or two copper atoms by chelation, thus inhibiting copper-activating enzymes.

Oxidation of LDL plays a key role in the initial phases of atherosclerosis (Steinberg *et al.*, 1989; Esterbauer *et al.*, 1992). When polyunsaturated lipids (LH) in LDL or tissue membranes become oxidized in the presence of metals and hydroperoxides, an alkyl radical L• is formed (Fig. 1). The alkyl radical reacts with oxygen to form peroxyl radicals (LOO•), which react again with LH to produce hydroperoxides (LOOH). These hydroperoxides decompose readily in the presence of metals to produce aldehydes. These aldehydes react with the apoprotein-B of LDL to produce oxidized LDL. Phenolic antioxidants (AH) can inhibit two important steps in the free radical chain reaction by reacting with (1) peroxyl radicals and inhibiting hydroperoxide formation and (2) alkoxyl radicals and inhibiting aldehyde formation. To properly evaluate the effectiveness of phenolic antioxidants, it is not only necessary to measure the formation of hydroperoxides, but also the formation of aldehydes, which cause biological damage. The inhibition of aldehyde formation is an important function of phenolic antioxidants that has been poorly appreciated in the literature.

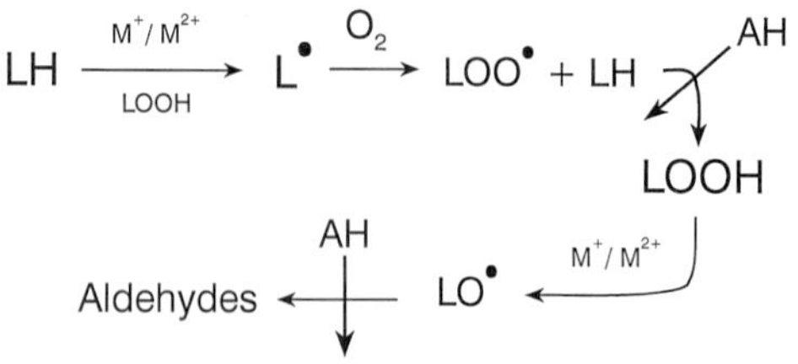

Figure 1   Oxidation of polyunsaturated lipids in LDL and mechanism of antioxidant action.

Many plant flavonoids inhibit enzymatic reactions, especially those that produce and respond to oxidized lipids. These enzymes are highly sensitive to the presence of oxidants and antioxidants. Phenolic antioxidants may reduce the peroxidation of polyunsaturated lipids and LDL and decrease macrophage foam cell formation.

PROTECTIVE ROLE OF PLANT FLAVONOIDS

Flavonoids are known to inhibit cyclooxygenase and lipoxygenase (Moroney *et al.*, 1988; Kinsella *et al.*, 1993). These enzymes play a key role in platelet aggregation and in the formation of macrophages, prostaglandins, and leukotrienes. By downregulating these enzymes, phenolic compounds may reduce thrombosis and chronic inflammatory reactions involved in immune suppressions by reducing hydroperoxides, downregulating the arachidonic acid cascade, and reducing platelet aggregation and thrombotic tendencies. These physiological events tend to increase chronic diseases such as heart diseases and cancer. Three possible mechanisms for the protective role of plant flavonoids include (a) inhibiting LDL oxidation, protecting cholesterol from forming cytotoxic compounds and preventing *atherosclerosis*, (b) interfering with the *immune response* by reducing monocytes from vessel walls, and (c) interfering with platelet aggregation involved in the clotting process and preventing *thrombosis*.

The susceptibility of lipoprotein particles to oxidation *in vitro* was shown to increase with prolonged time spent in circulation, or aging, *in vivo* in an avian model (Walzem *et al.*, 1995). These results imply that plasma LDL from hypercholesterimic individuals is older than LDL from individuals with normal plasma cholesterol. Accordingly, LDL particles from hypercholesterimic individuals are subjected to elevated oxidant stress by prolonging their exposure to intravascular oxidation (German *et al.*, 1997). Therefore, individuals with high LDL cholesterol may require a higher antioxidant intake for protection against the damaging effects of higher circulating levels of oxidatively susceptible LDL.

NATURAL PHENOLIC ANTIOXIDANTS

Wines

Considerable research has been published on the flavonoid compounds of grapes and wine. Grapes are the world's largest fruit crop, and about 80% of the total crop is used in wine making (Mazza, 1995). Dealcoholized red wine and red grapes were found to be important sources of flavonoid antioxidants that inhibit the oxidation of LDL *in vitro* (Frankel *et al.*, 1993). The antioxidant activity of phenolics in wines was evaluated in a number of *in vitro* systems using either copper or biological catalysts (myoglobin, cytochrome c, and iron ascorbate) (Kanner *et al.*, 1994). Lipid peroxidation inhibition of 50% (I_{50}) by wine phenolics was achieved at concentrations of 0.2 to 0.9 μg/ml phenolics. The relative antioxidant activity of 20 commercial California wines in inhibiting LDL oxidation *in vitro* varied from 46 to 100% with red wines and from 3 to 6% with white wines (Frankel *et al.*, 1995). The activity of wines to protect LDL from oxidation was distributed widely among the principal phenolic components, including catechin, gallic acid, epicatechin, malvidin 3-glucoside, rutin, myricetin, quercetin, caffeic acid, and cyanidin.

Grapes and Grape Juice

The antioxidant activities of extracts of different table and wine grapes were tested in lecithin liposomes and in human LDL *in vitro*. In lecithin liposomes oxidized with copper acetate, grape extracts inhibited conjugated diene formation by 25–68% and hexanal formation by 49–98% (Yi *et al.*, 1997). Anthocyanins were the most abundant phenolic compounds in extracts of red grapes, and flavonols were most abundant in extracts of white grapes. In LDL oxidized with copper sulfate, hexanal formation was inhibited by 62–91% (Meyer *et al.*, 1997). The relative antioxidant activity of grape extracts toward LDL oxidation correlated with the concentration of total phenols, anthocyanins, and flavonols. Extracts of different grape varieties showed significantly different trends in antioxidant activity between liposome and LDL systems. Commercial white and red grape juices also significantly inhibited the oxidation of LDL (66–73%) (Frankel *et al.*, 1997). Therefore, grape extracts and juices, especially red grape juices, contribute a significant source of phenolic antioxidants that, like wine, may have potential beneficial health effects in protecting against atherosclerosis.

Pure Phenolic Compounds

The more active fractions purified from a Petite Syrah wine containing components of the catechin family were evaluated for their activity in inhibiting human LDL oxidation by copper (Teissedre *et al.*, 1996). The highest activity was found with the procyanidin dimers, the trimer, and the monomers catechin, epicatechin and myricetin, and lower activity for the gallic acid, quercetin, caffeic acid, and rutin. The relative antioxidant activity of pure phenolic compounds tested at two concentrations decreased in the following order: catechin > myricetin = epicatechin = rutin > gallic acid > quercetin > cyanidin. α-Tocopherol had relatively weak activity. Some dimers and trimers separated from red grape seeds had the same activity as catechin, but other dimers and trimers were less active. These oligomers may have different protein-binding capacity than the monomeric catechin. Thus, the numerous phenolic compounds in wines, grapes, and plant flavonoids found in the diet are potent antioxidants in various physiologically relevant systems. Several anthocyanins were tested *in vitro* on human LDL and on lecithin liposomes (Satué-Gracia *et al.*, 1997). In LDL oxidized with 10 μM copper, malvidin was the best antioxidant, followed by delphinidin, cyanidin, and pelargonin. In the liposome system oxidized with 10 μM copper, pelargonin followed by malvidin in antioxidant activity, but cyanidin and delphinidin were prooxidants.

The properties of various phenolic compounds are thus very system dependent. Caution is required in interpreting the antioxidant activities determined with nonspecific assays in the absence of suitable substrate targets.

Rosemary Extracts

Rosemary extracts provide a major source of natural antioxidants used commercially in foods. Carnosic acid and carnosol are the most important active components of rosemary extracts and rosmarinic acid is a minor constituent. Rosemary extracts, carnosic acid, and rosmarinic acid were more active in bulk corn oil than carnosol (Frankel *et al.*, 1996). However, rosemary compounds were less active in corn oil emulsions than in bulk corn oil, and rosemary extracts, carnosic acid, and carnosol were more active than rosmarinic acid. The polar hydrophilic components of rosemary extracts were less active in the emulsion system because they partitioned into the water phase and became less protective than in the bulk oil system.

Green Tea Catechins

Green tea catechins have attracted much attention because of their physiological effects, including antimutagenic and antitumorigenic activities. Studies showed that the hydrophilic catechol-type flavonoids of green tea were more effective in inhibiting the oxidation of liposomes composed of egg phosphatidylcholine than the lipophilic α-tocopherol (Terao *et al.*, 1994). However, kinetic studies showed that flavonoids had much smaller rate constants for the inhibition of oxidation of methyl linoleate than α-tocopherol. The marked differences in activity between hydrophilic catechins and lipohilic α-tocopherol may be due to their relative partition between the water phase and the surface environment of the phospholipid bilayers. This interfacial phenomenon was invoked to explain the contrasting properties of polar and nonpolar antioxidants in bulk oil versus oil-in-water emulsion systems (Frankel *et al.*, 1994).

The antioxidant activities of pure components of green tea and different commercial green teas were evaluated in different lipid systems (Huang and Frankel, 1997; Frankel *et al.*, 1997). Green teas were active antioxidants in bulk corn oil but were prooxidant in the corresponding oil-in-water emulsions oxidized. Green tea catechins were also active antioxidants in lecithin liposomes oxidized in the presence of cupric acetate as the catalyst. The marked variation in activity of green teas may be attributed in part to differences in their relative partition between phases in various lipid systems. The improved antioxidant activity observed for green teas in lecithin liposomes compared to corn oil emulsions can be explained by the greater affinity of the polar catechin gallates for the polar surface environment of the lecithin bilayers, thus affording better protection against oxidation.

CONCLUSIONS

Dealcoholized red wine and grapes are important sources of flavonoid antioxidants that are potent inhibitors of *in vitro* oxidation of LDL. The activity of commercial wines to protect LDL from oxidation was compared with the activity of pure phenolic compounds found in wine. The antioxidant activities of grape extracts and commercial grape juices were comparable to that of commercial wines. The extracts of various grape varieties showed different trends in antioxidant activity between liposome and LDL systems. Rosemary extracts and components, tea catechins, and commercial green teas had strong antioxidant activities in corn oil and lecithin liposomes, but not in corn oil emulsions. Therefore, the relative activity of natural phenolic antioxidants is affected greatly by the test system used and the biological target to be protected.

Natural phenolic compounds are found in many fruits and vegetables in a great variety of structures and specificities. A 140-ml glass of red wine or tea is estimated to contain approximately 0.3 g of phenolic materials consisting largely of flavonoids. As potent antioxidants, flavonoids may be especially important in protecting against human disease and represent a positive potential in our diet that requires further research to improve our understanding of their mechanism of action.

Natural phenolic antioxidants inhibit biological oxidation by a complex multistep mechanism (Fig. 1). Unfortunately, many contemporary studies tend to use nonspecific indicators of oxidation and antioxidation to determine "antioxidant capacity" or "antioxidant status." Phenolic compounds can participate in several antioxidant defenses, including preventing

oxidant formation, scavenging activated oxidants, reducing reactive intermediates, and inducing repair systems. To improve our understanding of
these complex interactions in different systems, the use of nonspecific assays
for antioxidant capacity would be risky because they do not provide information on the biological target(s) protected. The better approach is to measure specific products of oxidation in both relevant *in vitro* and *in vivo* biological systems.

REFERENCES

Ames, B. N., Shigenaga, M. K., and Hagen, T. M. (1993). Oxidants, antioxidants, and the degenerative diseases of aging. *Proc. Natl. Acad. Sci. USA* **90**, 7915–7922.

Block, G., Patterson, B., and Subar, A. (1992). Fruits, vegetables, and cancer prevention: A review of the epidemiological evidence. *Nutr. Cancer* **18**, 1–29.

Esterbauer, H., Gebicki, J., Puhl, H., and Jurgens, G. (1992). The role of lipid peroxidation
and antioxidants in oxidative modification of low density lipoproteins. *Free Radic. Biol.
Med.* **13**, 341–390.

Frankel, E. N., Bosanek, C. A., Meyer, A. S., Silliman, K., and Kirk, L. L. (1998). Commercial
grape juices inhibit the in vitro oxidation of human low-density lipoproteins. Submitted
for publication.

Frankel, E. N., Huang, S.-W., and Aeschbach, R. (1997). Antioxidant activity of green teas in
different lipid systems. *J. Am. Oil Chem. Soc.* **74**, 1309–1315.

Frankel, E. N., Huang, S.-W., Aeschbach, R., and Prior, E. (1996). Antioxidant activity of a
rosemary extract and its constituents, carnosic acid, carnosol, and rosmarinic acid, in
bulk oil and oil-in-water emulsion. *J. Agric. Food Chem.* **44**, 131–135.

Frankel, E. N., Huang, S.-W., Kanner, J., and German, J. B. (1994). Interfacial phenomena in
the evaluation of antioxidants: Bulk oils vs emulsions. *J. Agric. Food Chem.* **42**,
1054–1059.

Frankel, E. N., Kanner, J., German, J. B., Parks, E., and Kinsella, J. E. (1993). Inhibition of in
vitro oxidation of human low-density lipoprotein with phenolic substances in red wine.
Lancet **341**, 454–457.

Frankel, E. N., Waterhouse, A. L., and Teissedre, P. L. (1995). Principal phenolic phytochemicals in selected California wines and their antioxidant activity in inhibiting oxidation of
human low-density lipoproteins. *J. Agric. Food Chem.* **43**, 890–894.

German, J. B., Frankel, E. N., Waterhouse, A. L., and Walzem, R. L. (1997). Phenolics and targets of chronic disease. *In* "Wine: Nutritional and Therapeutic Benefits" (T. R. Watkins,
ed.), Symposium Series No. 661, pp. 196–214. American Chem. Society, Washington,
DC.

Huang, S.-W., and Frankel, E. N. (1997). Antioxidant activity of tea catechins in different lipid
systems. *J. Agric. Food Chem.* **43**, 3033–3038.

Kanner, J., Frankel, E. N., Granit, R., German, J. B., and Kinsella, J. E. (1994). Natural antioxidants in grapes and wines. *J. Agric. Food Chem.* **42**, 64–69.

Kinsella, J. E., Frankel, E. N., German, J. B., and Kanner, J. (1993). Possible mechanisms for
the protective role of antioxidants in wine and plant foods. *Food Technol.* **4**, 85–89.

Mazza, G. (1995). Anthocyanins in grapes and grape products. *Crit. Rev. Food Sci. Nutr.* **35**,
341–371.

Meyer, A., Yi, O.-S., Pearson, D. A., Waterhouse, A. L., and Frankel, E. N. (1997). Inhibition of human low-density lipoprotein oxidation in relation to composition of phenolic antioxidants in grapes (*Vitis vinifera*). *J. Agric. Food Chem.* **45**, 1638–1643.

Moroney, M.-A., Alcaraz, M. J., Forder, R. A., Carey, F., and Hoult, J. R. S. (1988). Selectivity of neutrophyl 5-lipoxygenase and cyclo-oxygenase inhibition by an anti-inflammatory flavonoid glycoside and related aglycone flavonoids. *J. Pharm. Pharmacodyn.* **278**, 4–12.

Satué-Gracia, M. T., Heinonen, M., and Frankel, E. N. (1997). Anthocyanins as antioxidants on human low-density lipoprotein and lecithin-liposome systems. *J. Agric. Food Chem.* **45**, 3362–3367.

Steinberg, D., Parthasarathy, S., Carew, T. E., Khoo, J. C., and Witztum, J. L. (1989). Beyond cholesterol: Modification of low-density lipoprotein that increase its atherogenicity. *N. Engl. J. Med.* **320**, 915–924.

Teissedre, P. L., Frankel, E. N., Waterhouse, A. L., Peleg, H., and German, J. B. (1996). Inhibition of in vitro human LDL oxidation by phenolic antioxidants from grapes and wines. *J. Sci. Food Agric.* **70**, 55–61.

Terao, J., Piskula, M., and Yao, Q. (1994). Protective effect of epicatechin, epicatechin gallate and quercetin on lipid peroxidation in phospholipid bilayers. *Arch. Biochem. Biophys.* **308**, 278–284.

Walzem, R. L., Watkins, S., Frankel, E. N., Hansen, R. J., and German, J. B. (1995). Older plasma lipoproteins are more susceptible to oxidation: A linking mechanism for the lipid and oxidation theories of atherosclerotic cardiovascular disease. *Proc. Natl. Acad. Sci. USA* **92**, 7460–7464.

Yi, O.-S., Meyer, A. S., and Frankel, E. N. (1997). Antioxidant activity of grape extracts in a lecithin liposome system. *J. Am. Oil Chem. Soc.* **74**, 1301–1307.

26 Antioxidants in Herbs: Polyphenols

Takuo Okuda

Okayama University
Tsushima, Okayama 700-8530, Japan

INTRODUCTION

The name "herb" is interpreted in several ways. The "herb" in Japan often means some plant rich in aroma, mostly introduced via Europe and used for perfume and cooking, differentiating it from the other medicinal plants. However, "medicinal plants" comprise, in addition to those of mild effects usable for home treatment, those with acute toxicity, such as aconitum, poppy, and digitalis.

The herbs discussed in this chapter are medicinal plants of mild activity. They cover a wide range of plants of various origins, applicable to both medicine and food; they exclude those of acute toxicity.

ANTIOXIDANT COMPONENTS DISTRIBUTED WIDELY AND ABUNDANTLY IN HERBS

Among the most widely and abundantly distributed antioxidant components in this type of herb are a variety of phenolic compounds represented by (−)-epigallocatechin gallate (EGCG), α-tocopherol, and sesaminol (a lignan), although there are many other antioxidants in herbs. Polyphenols of numerous chemical structures contained in herbs (1,2) are the main subjects of this chapter.

VOLATILE AND NONVOLATILE COMPONENTS IN HERBS

Volatile compounds composing essential oils are easily discerned components of herbs and are often regarded as the characteristic components of each herb. However, experience in Kampo medicine (traditional Chinese medicine practiced in Japan) in recent years has shown that most of the essential oils contained in the crude drugs disappear within an hour when extracted with boiling water in a kettle. It is improbable that these essential oils can be retained in the extract and its preparation after concentration and drying of the extract *in vacuo,* has been done for practically all of the herb preparations.

Precautions Against Loss of Plant Polyphenols in Herb Preparations

Polyphenols in the nonvolatile fractions of herbs also cannot be safeguarded against loss, as from decomposition occurring during drying in the sunshine, extraction, and concentration, as shown in Figs. 1 and 2. The herb in Fig. 1 is *Perilla frutescens* (Shiso) in which rosmarinic acid is the main component (3): that in Fig. 2 is *Geranium thunbergii* (Gen-no-shoko) in which geraniin is the main component (4).

Remarkable differences in behavior are observed between rosmarinic acid and geraniin. When drying fresh plants of these species under direct sunshine or in an oven, rosmarinic acid in *P. frutescens* is destroyed quickly

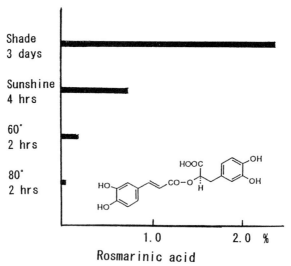

Figure 1   Decrease of rosmarinic acid content in dried *Perilla frutescens* leaf.

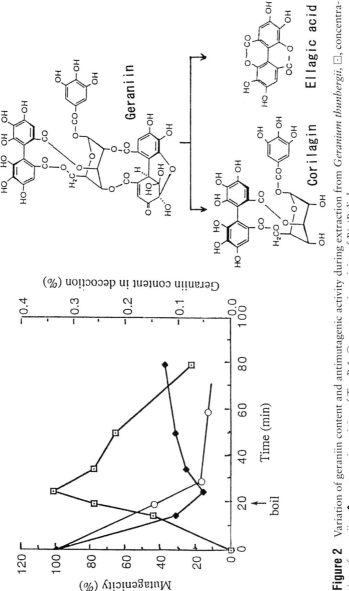

Figure 2 Variation of geraniin content and antimutagenic activity during extraction from *Geranium thunbergii*, □, concentration of geraniin; ◆, mutagenic activity of Trp-P-1; ○, mutagenic activity of B(a)P-diol.

(Fig. 1), whereas geraniin in G. *thunbergii* is not destroyed by this exposure to sunshine for a day. However, rosmarinic acid in *P. frutescens* is extracted without decomposition in boiling water, whereas extensive hydrolysis of geraniin occurs on its extraction from the herb in boiling water (5). This hydrolysis of geraniin induces marked variation in the antioxidant, in the antimutagenic (Fig. 2), and in other activities of the extract. It should be noted that, unlike geraniin in the herb, crystalline geraniin once isolated from the herb is fairly stable in boiling water.

Care therefore must be taken regarding the differences in the quality of the herb preparations due to the method of preparation. The chemical analysis of the preparations, such as that by HPLC combined with spectra, often gives reliable answers under these circumstances (6).

POLYPHENOLS AS ANTIOXIDANT COMPONENTS IN HERBS

Modern classification of the main plant polyphenols and some examples of polyphenols of each type and herbs that are rich in these polyphenols (2) are listed in Table I.

The name "plant polyphenols" covers a wide range of compounds belonging to hydrolyzable tannins (metabolites of gallic acid) (2), condensed tannins (proanthocyanidins = condensates of flavans) (1), caffeates (3,7), lignans (8), flavonoids, and compounds of several other types having a polyphenolic structure in their molecules. Essentially all of these com-

Table I   Polyphenol Classification and Polyphenol-Rich Plants

Type of polyphenol	Polyphenol (example)	Plant, drug
Metabolites of gallic acid		
Small molecule polyphenols	Gallic acid	*Quercus* sp.
Hydrolyzable tannins		
Gallotannin monomers	Pentagalloylglucose	Chinese gall
Ellagitannin and oxidized	Corilagin, geraniin	*Geranium* sp.
ellagitannin monomers		
Ellagitannin oligomers	Agrimoniin	*Agrimonia* sp.
Metabolites of catechins		
Monomeric catechins	EGCG, ECG, EGC	Green tea, gambir
Condensed tannins		*Diospyros kaki*, cinnamon
(proanthocyanidins)		
Caffeates		
Caffetannins	Diccafeoylquinic acid	*Artemisia* sp.
Labiataetannins	Rosmarinic acid	Rosemary

pounds are polyphenols, but lignans and flavonoids are usually classified separately. "Tannins" in herbs involve hydrolyzable and condensed tannins, along with several other types of polyphenols, such as caffeic acid derivatives, represented by rosmarinic acid and its analogs, called labiataetannins, which are the main component in many labiate herbs, and mono- and oligo-caffeoylquinic acids called caffetannins, which are contained in the plants of *Artemisia* species and coffee.

ANTIOXIDANT ACTIVITIES OF FLAVONOIDS

Flavonoids are compounds of a large family, distributed widely in herbs, and have antioxidant activity if their molecules possess some polyphenolic structures. Among these flavonoids, quercetin and some other flavonols are distributed widely, often as their glycosides, in plants used as vegetables, fruits, or medicines and are potent antioxidants.

An example of a popular herb rich in polyphenols of flavonoidal structure is licorice. Although glycyrrhizic acid (a triterpenoid glucuronide) is well known as a component of licorice, licorice is also rich in the total amount and structural variety of flavonoids. The structures of four flavonoids isolated from licorice, with higher superoxide-suppressing activities than other licorice phenolics, along with their effects are shown in Fig. 3.

These activities were measured in three ways, combining different methods of generating and detecting superoxide. Licochalcone B and several others exhibit the most potent effects of all of the screened 20 licorice flavonoids (9). These effects and other antioxidant effects are regarded as based on their radical scavenging activity (10).

Some of these flavonoids from licorice inhibited the products from lipoxygenase in arachidonic acid metabolism. An example is the inhibition of the formation of leukotrienes B_4 and C_4 in human polymorphonuclear neutrophils by licochalcones A and B (11).

These flavonoidal compounds showing such strong antioxidant activity possess phenolic hydroxyl groups located adjacent to each other. Licochalcone B from licorice, having antioxidant activity higher than that of quercetin, has three oxygen functions vicinal to each other on a phenyl ring in its molecule.

The extract of *Silybum marianum* (Compositae), a European antihepatotoxic drug, contains silymarin, a mixture containing silybin (=silibinin), which is a conjugate of a dihydroflavonol and coniferyl alcohol. This silybin, an antioxidant whose molecular size is about twice that of a flavonoid monomer, was found as the main active component in the extract (12) (Fig. 4).

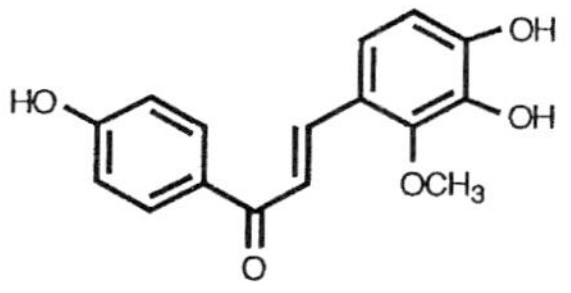

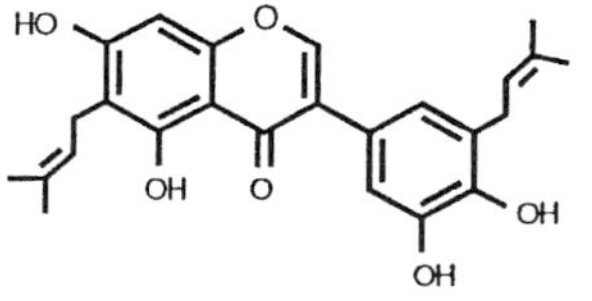

Isoangustone A

Xanthine-XOD-NBT	20μM
PMS-NADH-NBT	61μM
Xanthine-XOD cytochrome c	11μM

Licochalcone B

Xanthine-XOD-NBT	1.6μM
PMS-NADH-NBT	3.6μM
Xanthine-XOD cytochrome c	2.0μM

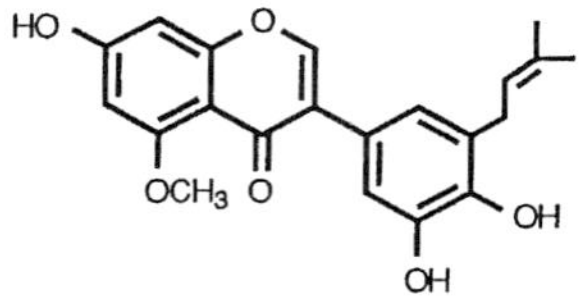

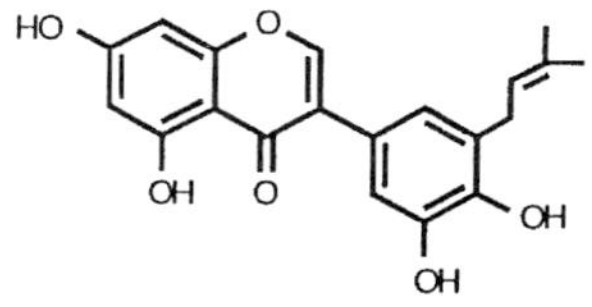

Glycyrrhisoflavone

Xanthine-XOD-NBT	4.2μM
PMS-NADH-NBT	11μM
Xanthine-XOD cytochrome c	8.0μM

Glisoflavone

Xanthine-XOD-NBT	14μM
PMS-NADH-NBT	20μM
Xanthlne-XOD cytochrome c	15μM

Figure 3   Examples of flavonoids from licorice and their superoxide suppression.

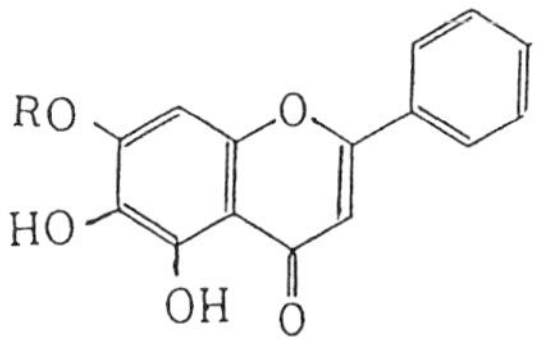

Baicalein (R=H)
Baicalin (R=glucuronyl)

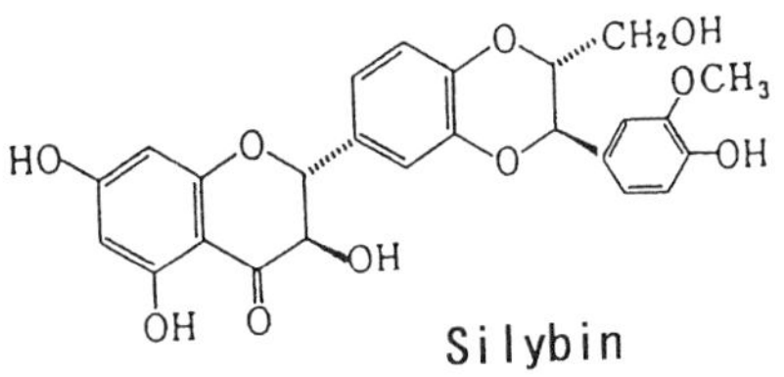

Silybin

Figure 4 Structures of baicalin, baicalein, and silybin.

COMPARISON OF ANTIOXIDANT ACTIVITIES OF FLAVONOIDS AND TANNINS

Data in Fig. 5 show that the superoxide-suppressing activity (13) of pedunculagin (14), which is one of the ellagitannins, extensively surpasses that of quercetin. Pedunculagin has two HHDP (hexahydroxydiphenoyl) groups in its molecule and is found in medicinal plants of *Quercus* species and in some others.

It is also observed that the superoxide-suppressing activity of pedunculagin is more than 10 times that of allopurinol in the xanthine-XOD-NBT system and over 20 times that in the xanthine-XOD-cytochrome c system, and that pedunculagin inhibits the PMS-NADH-NBT system, which is not inhibited by allopurinol (9).

These potent activities are often found among ellagitannins and gallotannins of various structures and are attributable to the presence of the HHDP and/or galloyl groups in their molecules.

STRUCTURE, MOLECULAR SIZE, AND ANTIOXIDANT ACTIVITY OF POLYPHENOLS

The number of phenolic hydroxyl groups on a benzene ring, particularly those adjacent to each other, and their total number in a molecule induce large differences in the antioxidant and in other activities of polyphe-

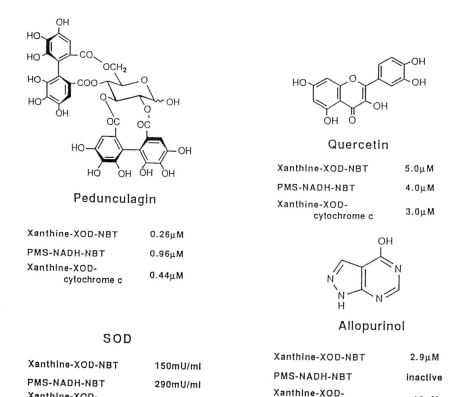

Figure 5   Comparison of superoxide suppression by an ellagitannin and a flavonol.

nols. Molecules having a pyrogallol portion(s) generally have remarkably higher activities than those having catechol. For instance, the superoxide-suppressing activities of gallic acid and EGCG, having three phenolic hydroxyl groups adjacent to each other, are 5 to 10 times that of their analogs having only two hydroxyl groups vicinal to each other (13) (Fig. 6).

Caffeic acid and its derivatives comprising caffetannins and labiata-etannins belong to the latter type of polyphenols. However, compounds having catechol portions at the two ends of the molecule, such as rosmarinic acid, inhibit lipid peroxidation more potently than other catechol derivatives (15) (Fig. 7). Activities of the analogs of these compounds having alkoxyl group in place of hydroxyl groups are generally low.

The increase of galloyl groups on a glucopyranose ring in a molecule highly elevates the antioxidant potencies. The biosynthetic formation of the HHDP group by intramolecular coupling between the galloyl groups in

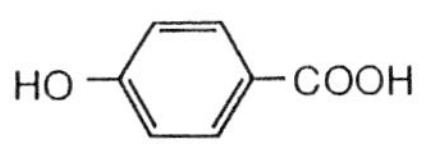

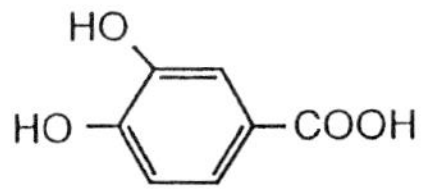

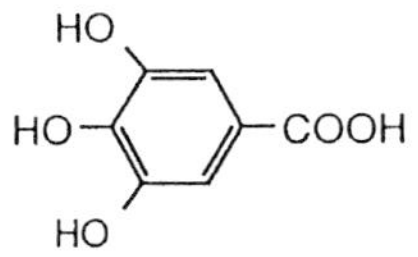

p-Hydroxy-benzoic acid

Xanthine-XOD-NBT	> 100μM
PMS-NADH-NBT	> 100μM
Xanthine-XOD-cytochrome c	————

Protocatechuic acid

Xanthine-XOD-NBT	15μM
PMS-NADH-NBT	9.0μM
Xanthine-XOD-cytochrome c	26μM

Gallic acid

Xanthine-XOD-NBT	1.8μM
PMS-NADH-NBT	9.2μM
Xanthine-XOD-cytochrome c	2.3μM

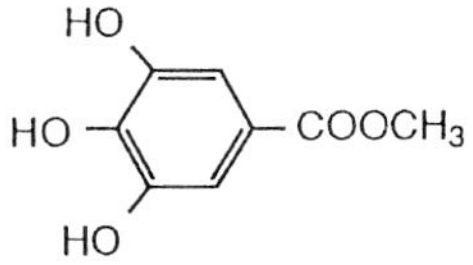

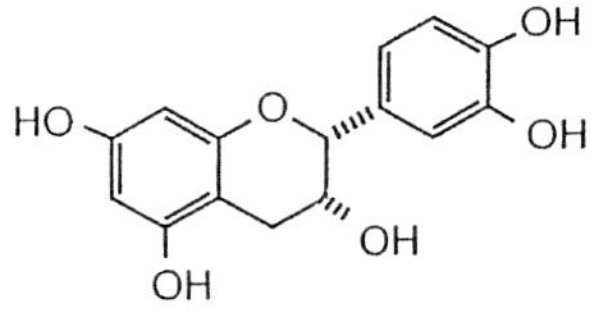

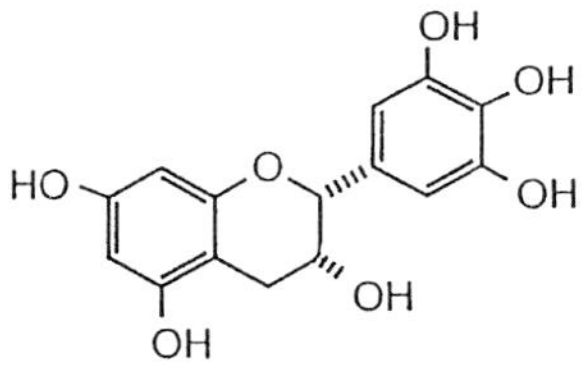

Methyl gallate

Xanthine-XOD-NBT	1.9μM
PMS-NADH-NBT	2.1μM
Xanthine-XOD-cytochrome c	3.2μM

(-)-Epicatechin

Xanthine-XOD-NBT	4.8μM
PMS-NADH-NBT	38μM
Xanthine-XOD-cytochrome c	16μM

(-)-Epigallocatechin

Xanthine-XOD-NBT	1.0μM
PMS-NADH-NBT	5.0μM
Xanthine-XOD-cytochrome c	1.4μM

Figure 6 Small molecule plant polyphenols and superoxide suppression.

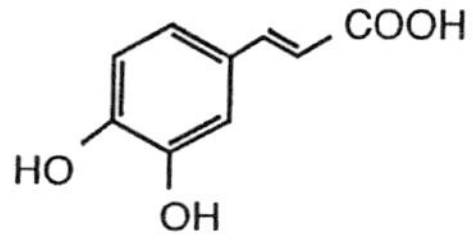

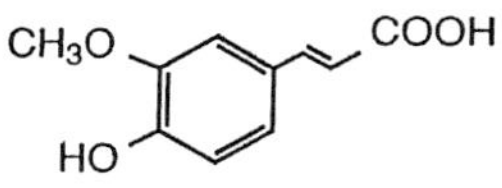

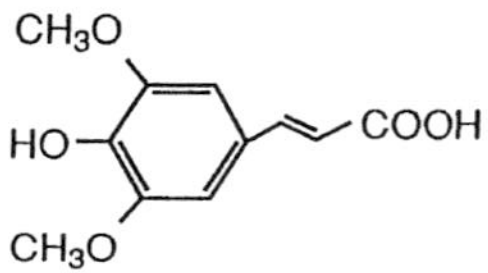

Caffeic acid

Xanthine-XOD-NBT	3.8μM
PMS-NADH-NBT	30μM
Xanthine-XOD-cytochrome c	24μM

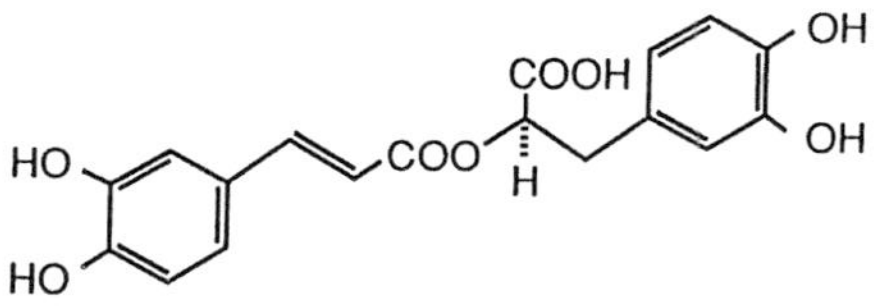

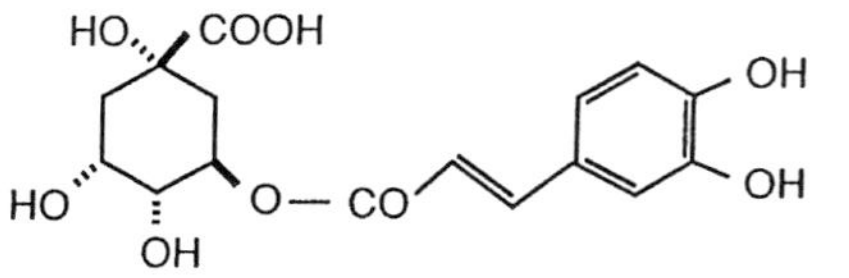

Ferulic acid

> 100μM
> 100μM
—

Sinapic acid

> 100μM
> 100μM
—

Rosmarinic acid

Xanthine-XOD-NBT	1.9μM
PMS-NADH-NBT	11μM
Xanthine-XOD-cytochrome c	56μM

Chlorogenic acid

4.8μM
30μM
8.6μM

Figure 7 Caffeates and superoxide suppression.

polygalloylglucose to form an ellagitannin molecule generally elevates the inhibitory potencies against lipid peroxidation (Fig. 8).

It is noticeable that the effects of ellagitannins such as geraniin in polar solvents are long lasting compared with those of ascorbic acid and α-tocopherol (16). The stability of the free radical generated from the HHDP group is higher than that from gallates, as found by the comparison of ESR spectra (16). The ESR spectra of the gallates show only the signal of the HHDP group produced by the radical coupling between gallates. Products from this coupling were isolated from the solution in high yields (17).

Some examples of ellagitannins of molecular weight 800–1000 are shown in Fig. 9. Their superoxide-suppressing activity is more potent than most of the other types of natural polyphenols (13).

Further biogenetic oxidative transformations of the HHDP group to DHHDP, chebuloyl (2), and several other groups (Fig. 8), however, do not increase the antioxidant potency, as shown by comparisons of geraniin, tellimagrandin II, isoterchebin, and pedunculagin (Fig. 9). Oxidation at these stages of biosynthesis sometimes lower the activity, as shown by that of furosinin (Fig. 9).

The oligomerization of hydrolyzable tannin in plants, represented by the formation of coriariin A and rugosin D from tellimagrandin II as shown in Fig. 10, doubles the activity expressed by molar concentration, but the activities expressed by mass concentration of these dimers are similar to those of monomers (13).

Antioxidant activities of the flavan monomers are generally poor. However, these activities are enhanced greatly by galloylation at O-3 of the flavan skeleton, as represented by EGCG and (−)-epicatechin gallate (ECG) (13) (Fig. 11).

Oligomerization of these flavans producing condensed tannins (proanthocyanidins) also enhances antioxidant activities and other activities. Enhancement of the activity by the presence of a galloyl or pyrogallol group in a molecule, however, is more intense than that by the oligomerization of the flavan skeleton (13) (Fig. 11).

A similar trend in the structure–activity relationship was observed on the measurement of radical-scavenging activities of these polyphenols on DPPH (13).

REVERSE ACTIVITIES OF ANTIOXIDANTS

Although plant polyphenols are generally antioxidants under ordinary environments, they can also be oxidants under some particular circum-

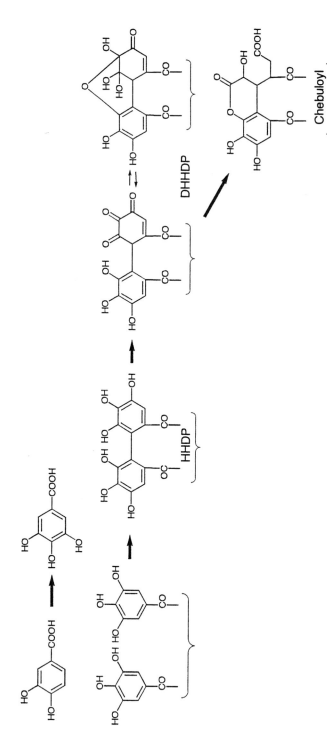

Figure 8 Formation of HHDP, DHHDP, and chebuloyl groups in ellagitannin molecules.

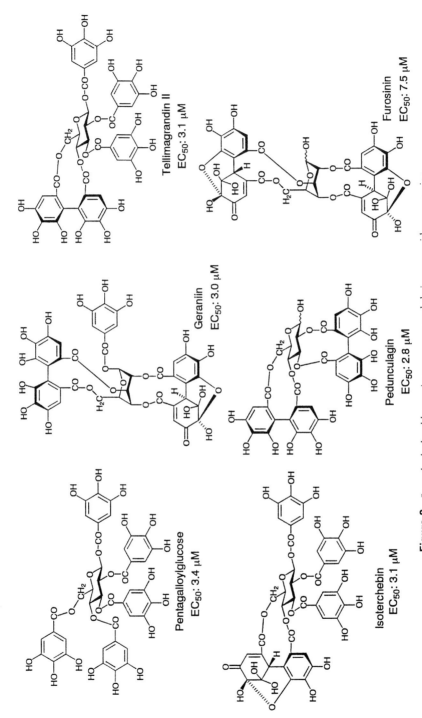

Figure 9 Some hydrolyzable tannin monomers and their superoxide suppression.

Tellimagrandin II
EC$_{50}$: 3.1 μM

Furosinin
EC$_{50}$: 7.5 μM

Geraniin
EC$_{50}$: 3.0 μM

Pedunculagin
EC$_{50}$: 2.8 μM

Pentagalloylglucose
EC$_{50}$: 3.4 μM

Isoterchebin
EC$_{50}$: 3.1 μM

405

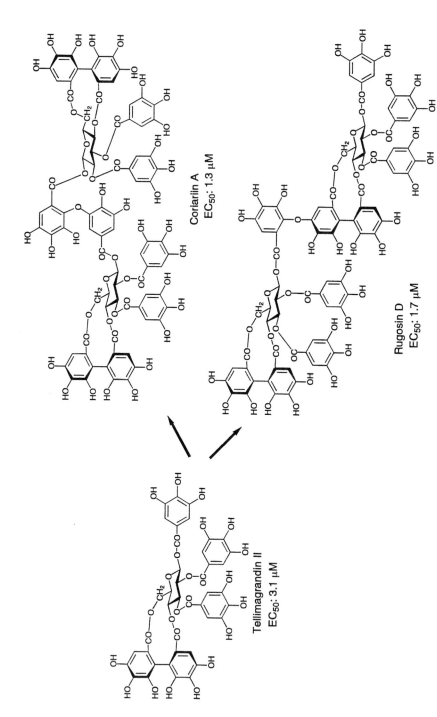

Figure 10 Dimerization of hydrolyzable tannin and superoxide suppression.

Coriariin A
EC$_{50}$: 1.3 μM

Rugosin D
EC$_{50}$: 1.7 μM

Tellimagrandin II
EC$_{50}$: 3.1 μM

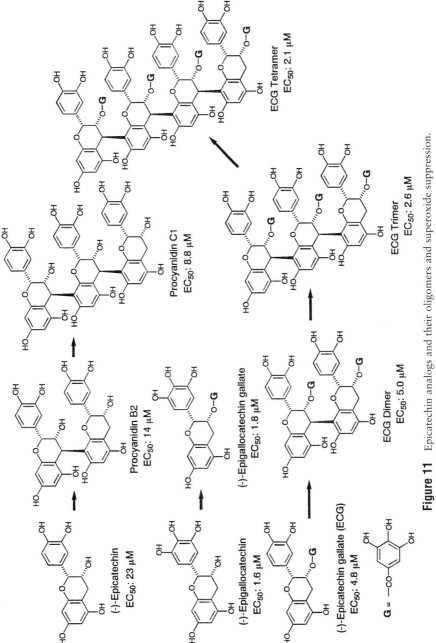

Figure 11 Epicatechin analogs and their oligomers and superoxide suppression.

stances. An example is the promotion of oxidation in the presence of high concentrations of transient metallic ion.

On Cu(II)-catalyzed peroxidation of methyl linoleate, geraniin inhibits peroxidation at the concentration ratio of geraniin/Cu(II) 150–750 [concentration of Cu(II): 6.6×10^{-4} mM]. However, at extraordinarily high concentrations of Cu(II) with the geraniin/Cu(II) ratio 0.1–0.5 [concentration of Cu(II): 1.0 mM], peroxidation is accelerated (18).

The induction of DNA fragmentation is a biochemical hallmark of apoptosis. As shown in Table II, gallic acid is remarkably active among about 80 plant polyphenols in screening systems using human leukemic cell lines and in several other systems (19). The extent of their activities parallels approximately the cytotoxicity of these polyphenols. Contrary to the structure–activity relationship in antioxidant activities, DNA fragmentation by polyphenols of larger molecule (tannins) is generally low (19).

In the *in vivo* test, gallic acid induces the selective death of cancer cells (20). Gallic acid and propyl gallate also inhibit trypanosoma, and propyl gallate inhibits the respiratory system in trypanosoma. These activities are attributable to the elicitation of reactive oxygen by gallic acid.

Iodination of human peripheral blood monocytes is stimulated by polyphenols to various extents. Iodination is the incorporation of radioactive iodine into an acid-insoluble fraction and its stimulation is attributable to the enhancement of the H_2O_2 production of myeloperoxidase. Among the screened polyphenols, the ECG dimer, trimer, and tetramer stimulated iodination most powerfully. Stimulation by smaller polyphenols and hydrolyzable tannins was low (21).

Table II   Induction of DNA Fragmentation

	MW	DNA fragmentation	Cytotoxicity
Gallic acid	170	0–5	5
Pentagalloylglucose	941	50–100	52
Tellimagrandin-II	939	50–100	49
Nobotanin A	1721	50–100	82
Rugosin D	1875	50–100	72
(-)-Epicatechin	290	>200	>200
(-)-Epicatechin gallate	442	50–100	81
(-)-Epicatechin gallate dimer	883	100–150	105

SAFETY OF POLYPHONIES IN FOOD AND HERBS

The old obscure concept of plant polyphenols or tannins, which lacked chemical evidence, should now be discarded. The chemical structures of each polyphenolic compound belonging to tannins contained in many plants, which were unknown in the mid-1970s, have now been determined and their activities possibly contributing to the human health have been investigated extensively. However, the long experience with humans shows that the health effects of many tannin-rich plants should not be overlooked. Evidence concerning EGCG, the main component of "tea tannin," may be one of the steps in the turnover of the concept of tannin.

ACKNOWLEDGMENTS

The author expresses gratitude for helpful contributions to Drs. T. Yoshida, T. Hatano, Y. Fujita, H. Hayatsu, K. Mori, A. Mori, and other colleagues at Okayama University, Dr. H. Fujiki of the Saitama Cancer Center Research Institute, Dr. H. Sakagami of Meikai University, Drs. H. Okuda and Y. Kimura of Ehime University, the late Dr. S. Arichi of Kinki University, and Dr. M. Hiramatsu of Yamagata Technopolis Foundation.

REFERENCES

1. Okuda, T., Yoshida, T., and Hatano, T. (1991). Chemistry and biological activity of tannins in medicinal plants. *In* "Economic and Medicinal Plant Research" (H. Wagner and N. R. Farnsworth, eds.), Vol. 5, pp. 129–165. Academic Press, London.
2. Okuda, T., Yoshida, T., and Hatano, T. (1995). Hydrolyzable tannins and related polyphenols. *In* "Progress in the Chemistry of Organic Natural Products 66" (W. Hertz *et al.*, eds.), pp. 1–117. Springer-Verlag, Vienna.
3. Okuda, T., Hatano, T., Agata, I., and Nishibe, S. (1986). *Yakugaku Zasshi* **106**, 1108.
4. Okuda, T., Yoshida, T., and Hatano, T. (1992). *J. Chem. Soc. Perkin Trans. 1*, 9.
5. Okuda, T., Mori, K., and Hayatsu, H. (1984). *Chem. Pharm. Bull.* **32**, 3755.
6. Okuda, T., Yoshida, T., and Hatano, T. (1989). *J. Natl. Prod.* **52**, 1.
7. Agata, I., Goto, S., Hatano, T., Nishibe, S., and Okuda, T. (1993). *Phytochemistry,* **33**, 508.
8. Agata, I., Hatano, T., Nishibe, S., and Okuda, T. (1989). *Phytochemistry* **28**, 2447.
9. Okuda, T. (1997). Phenolic antioxidants. *In* "Food and Free Radicals" (M. Hiramatsu, T. Yoshikawa, and M. Inoue, eds.). pp. 31–48. Plenum, New York.
10. Hatano, T., Kagawa, H., Yasuhara, T., and Okuda, T. (1988). *Chem. Pharm. Bull.* **36**, 2090.
11. Kimura, Y., Okuda, H., Okuda, T., and Arichi, S. (1988). *Phytother. Res.* **2**, 140.
12. Lorenz, D., Mennicke, W. H., and Behrendt, W. (1982). *Planta Med.* **45**, 216.
13. Hatano, T., Edamatsu, R., Hiramatsu, M., Mori, A., Fujita, Y., Yasuhara, T., Yoshida, T., and Okuda, T. (1989). *Chem. Pharm. Bull.* **37**, 2016.

14. Okuda, T., Yoshida, T., Ashida, M., and Yazaki, K., (1983). *J. Chem. Soc. Perkin Trans. 1*, 1765.

15. Fujita, Y., Uehara, I., Morimoto, Y., Nakashima, M., Hatano, T., and Okuda, T. (1988). *Yakugaku Zasshi* **118**, 129.

16. Fujita, Y., Komagoe, K., Niwa, Y., Uehara, I., Hara, R., Mori, H., Okuda, T., and Yoshida, T. (1988). *Yakugaku Zasshi* **108**, 528.

17. Yoshida, T., Mori, K., Hatano, T., Okumura, T., Uehara, I., Komagoe, K., Fujita, Y., and Okuda, T. (1989). *Chem. Pharm. Bull.* **37**, 1919.

18. Fujita, Y., Komagoe, K., Uehara, I., Morimoto, Y., and Hara, R. (1988). *Yakugaku Zasshi* **108**, 625.

19. Sakagami, H., Satoh, K., Hatano, T., Yoshida, T., and Okuda, T. (1997). *Anticancer Res.* **17**, 377.

20. Inoue, M., Suzuki, R., Sakaguchi, N., Li, Z., Takeda, T., Ogihara, Y., Jiang, B. Y., and Chen, Y. (1995). *Biol. Pharm. Bull.* **18**, 1526.

21. Sakagami, H., Asano, K., Tanuma, S., Hatano, T., Yoshida, T., and Okuda, T. (1992). *Anticancer Res.* **12**, 377

27 Mixed Japanese Herbs and Age-Related Neuronal Functions

Midori Hiramatsu and Makiko Komatsu

Institute for Life Support Technology
Yamagata Technopolis Foundation
Yamagata 990-2473, Japan

INTRODUCTION

Many neurological diseases such as Parkinson's, Alzheimer's, brain ischemia, amyotrophic lateral sclerosis (ALS), brain injury, and posttraumatic epilepsy are associated with free radicals (Packer *et al.*, 1996). These diseases are involved in neuronal cell death through glutamate-stimulated peroxinitrite and hydroxyl radicals. For this reason, it is desirable to keep a constant level of antioxidant components in the brain tissues to scavenge temporal free radical generation. Trials of the natural plant extract of *Ginkgo biloba*, which has free radical scavenging activity, showed efficacy on dementia (Bars *et al.*, 1997). Amyloid β-protein toxicity in cultured rat brain cells was inhibited after the addition of an extract of pine bark (Pycnogenol), which also has free radical scavenging activity (Rohdwald, 1998). These findings suggest that some natural plant extracts have a potential role for the treatment or the prevention of dementia. This chapter describes efficacy of mixed herb extracts of Sho-saiko-to-go-keishi-ka-shakuyaku-to and Toki-shakuyaku-san on aged neuronal functions.

Antioxidant Food Supplements in Human Health
Copyright © 1999 by Academic Press. All rights of reproduction in any form reserved.

SHO-SAIKO-TO-GO-KEISHI-KA-SHAKUYAKU-TO

Sho-saiko-to-go-keishi-ka-shakuyaku-to (Tsumura & Co., Tokyo, Japan) is a vacuum-concentrated boiled extract of nine herbs. A 7.5-g aliquot of Sho-saiko-to-go-keishi-ka-shakuyaku-to includes the extracts of 7 g of Bupleuri radix (*Bupleurum falcatum* L.), 5 g of Pinelliae tuber (*Pinellia ternata* Breitenbach), 3 g of Scutellariae radix (*Scutearia baicalensis* Georgi), 4 g of Zyzyphi fructus (*Zyzyphus vulgaris* Lamarck var. inermis Bunge), 3 g of Ginseng radix (*Panax ginseng* C.A. Meyer), 2 g of Glychyrrhizae radix (*Glycyrrhiza glabra* L. var. glandulifera Regel et Herder; *G. uralensis* Fisher), 1 g of Zingiiberis rhizoma (*Zingiber officinale* Roscoe), 6 g of Paeoniae radix (*Paeonia albiflora* Pallas var. trichocarpa Bunge), and 4 g of Cinnamomi cortex (*Cinnamomum cassia* Blume). Chemical structures of the main components in each herb are shown in Fig. 1.

Antioxidant Action

Because Sho-saiko-to-go-keishi-ka-shakuyaku-to was developed as an antiepileptic, some of the efficacy of treatment with it was shown in epileptic patients (Aimi, 1962; Mukawa, 1982; Nakane *et al.*, 1985). Sho-saiko-to-go-keishi-ka-shakuyaku-to increased the incidence of convulsions induced by increasing stimulation in El mice (Hiramatsu *et al.*, 1985). Injection of an iron solution into the sensory motor cortex of rats induces epielptogenic discharges (Willmore *et al.*, 1978), and this seizure mechanism is thought to be due to the generation of free radicals with iron (Mori, 1996). Sho-saiko-to-go-keishi-ka-shakuyaku-to inhibited the appearance of epileptic-like discharges after iron solution injection into the sensory motor cortex of rats (Mori and Hiramatsu, 1983). It is therefore suggested that Sho-saiko-to-go-keishi-ka-shakuyaku-to has a free radical scavenging activity. Using x-band electron spin resonance (ESR) spectrometry, it was found that Sho-saiko-to-go-keishi-ka-shakuyaku-to had free radical scavenging activities for superoxide induced by hypoxanthine–xanthine oxidase, hydroxyl radicals generated by α-guanidinoglutarate, and 1,1-diphenyl-2-picrylhydrazyl radicals for carbon-centered radicals induced by $FeCl_2$ and ascorbate in the rat brain homogenate (Hiramatsu *et al.*, 1988b) and for the α-tocopheroxyl radical (Hiramatsu *et al.*, 1990). Although the concentration of lipid peroxide as a marker for thiobarbiturate substance (TBARS) formation in the ipsilateral cortex was increased after the iron solution injection into the left cortex of rats, a 3-month treatment of Sho-saiko-to-go-keishi-ka-shakuyaku-to decreased these concentrations of TBARS and carbon-centered radicals 5 min after the injection of the iron solution (Hiramatsu *et al.*, 1987). In addition, the increased half-life of nitroxide radicals in rat head 1 week after injection of the iron solution was found following the intraperitoneal injection of carbamoyl-PROXYL using an L-band ESR spectrometer. Similarly, a 1-week treatment of Sho-saiko-to-go-keishi-ka-shakuyaku-to inhibited the increase (Komatsu *et al.*, 1996). These *in vivo* techniques showed that Sho-saiko-to-go-keishi-ka-shakuyaku-to has a free radical scavenging activity. Free radicals play a role in the aging process (Harman, 1981) and in Alzheimer's disease (Harman, 1995; Markesbery, 1997). The authors examined the effect of a 1-month administration of Sho-saiko-to-go-keishi-ka-shakuyaku-to on carbon-centered radicals and lipid peroxide formation as a marker of TBARS and found that Sho-saiko-to-go-keishi-ka-shakuyaku-to decreased both in the cerebrum of aged rat brain and increased superoxide dismutase (SOD) activity in the cytosol fraction of hippocampus and hypothalamus of aged rats (Hiramatsu *et al.*, 1988a).

Aged Neuronal Functions

A 1-month-treatment of Sho-saiko-to-go-keishi-ka-shakuyaku-to increased concentrations of norepinephrine in the hypothalamus (Hiramatsu *et al.*, 1986) and 5-hydroxytryptamine (5-HT) in the cerebellum, decreased the 5-HT concentration in the hypothalamus, and had no effect on dopamine concentration in the brain of aged rats (Hiramatsu *et al.*, 1988a). Although choline acetyltransferase activities in the hippocampus, striatum, and pons medulla oblongata were lower than those of adult rats, a 1-month treatment of Sho-saiko-to-go-keishi-ka-shakuyaku-to increased choline acetyltransferase activity in the hippocampus and striatum of aged rats (Hiramatsu *et al.*, 1989). Haba *et al.* (1990) found that Sho-saiko-to-go-keishi-ka-shakuyaku-to had a protective effect on changes in concentrations of acetylcholine and monoamine following cerebral ischemia in Mongolian gerbils. Nagakubo *et al.* (1993) found improvement of the cognitive function in epileptic patients with treatment of Sho-saiko-to-go-keishi-ka-shakuyaku-to, which seems to be due to the enhancement of cholinergic neuronal functions by Sho-saiko-to-go-keishi-ka-shakuyaku-to. Hiramatsu *et al.* (1996) found that treatment of Sho-saiko-to-go-keishi-ka-shakuyaku-to prolonged the life span of senescence accelerated mice (SAMP8), which have a 25% shorter life and are a model animal for aging as described by Takeda *et al.* (1991). Edamatsu *et al.* (1997) observed that N-*tert*-butyl-α-

Bupleuri radix

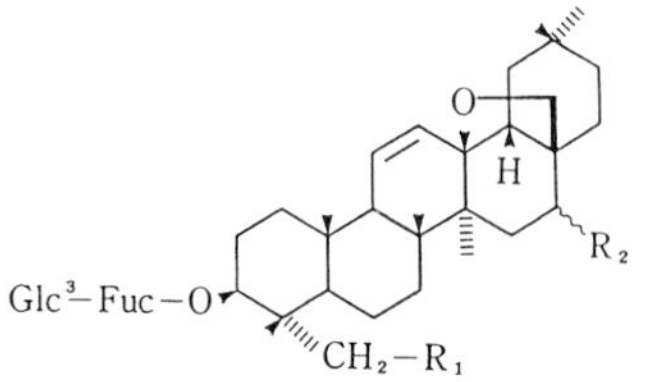
Glc³–Fuc–O
CH₂–R₁
R₂
H
O

	R_1	R_2
saikosaponin a	OH	β-OH
saikosaponin d	OH	α-OH
saikosaponin e	H	β-OH

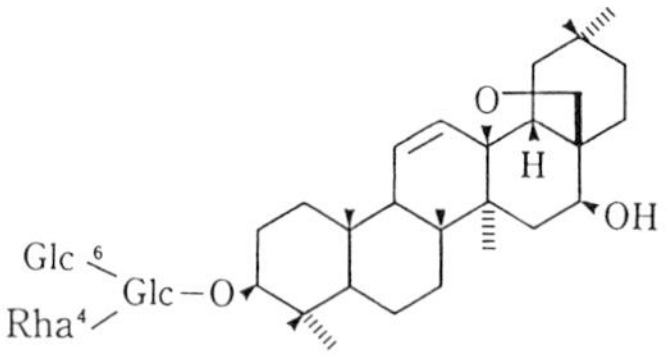
Glc 6
Glc–O
Rha 4
O
H
OH

saikosaponin c

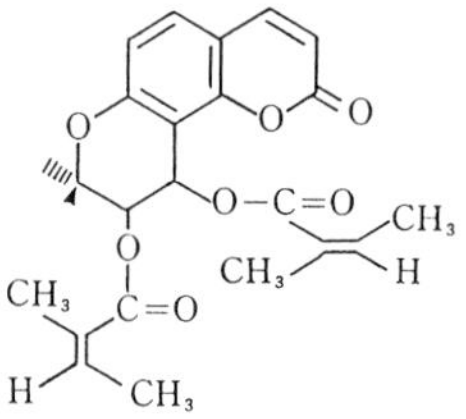

l-anomalin

Pinelliae tuber

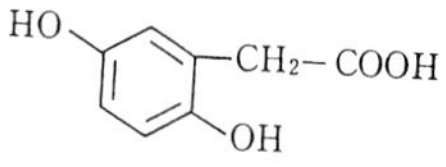
HO
CH2– COOH
OH

homogentisic acid

Scutellariae radix

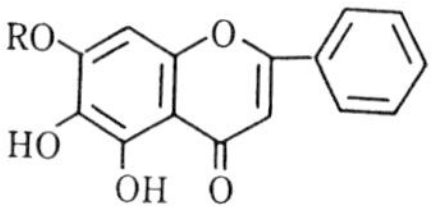
RO
HO
OH
O

baicalein R=H

baicalin R=Glc A

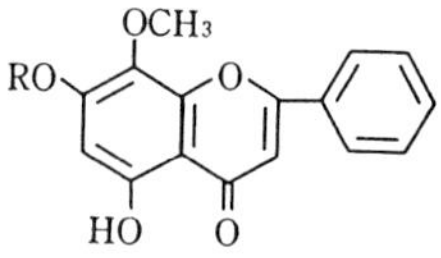
OCH3
RO
HO
O

wogonin R=H

wogonin glucuronide R=Glc A

Zizyphi fructus

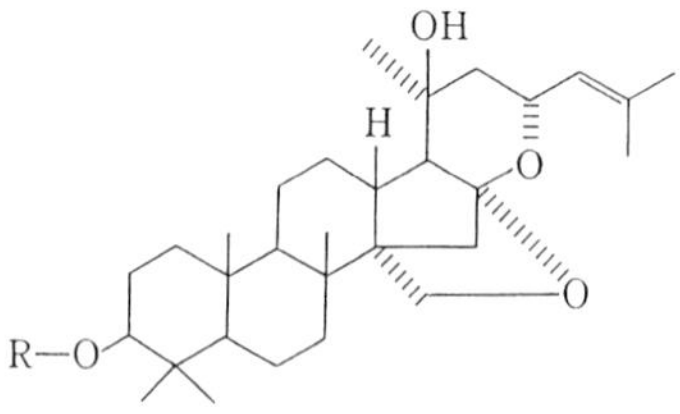
OH
H
O
R—O

zizyphus saponin

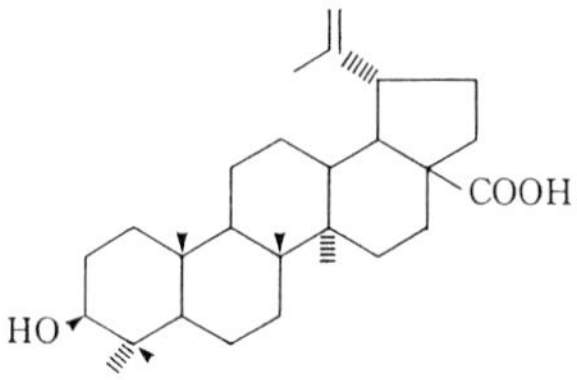
COOH
HO

betulinic acid

Figure 1 Chemical structures of main components in each herb of Sho-saiko-to-go-keishi-ka-shakuyaku-to.

Glycyrrhizae radix

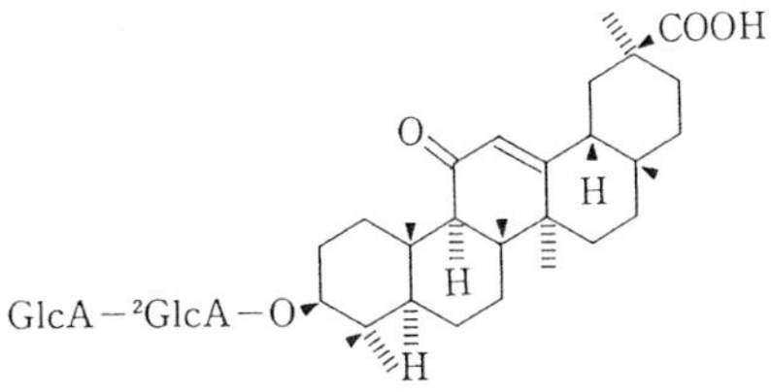

glycyrrhizin

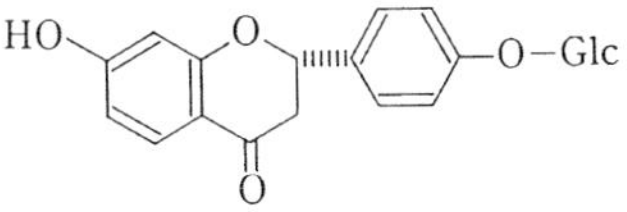

liquiritin

Zingiberis rhizoma

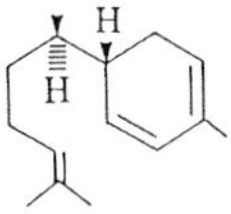

zingiberen

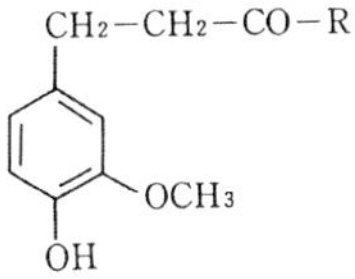

zingerone
$R = -CH_3$
gingerol
$R = -CH_2-CH-(CH_2)_4-CH_3$
$\qquad\qquad |$
$\qquad\qquad OH$
shogaol
$R = -CH=CH-(CH_2)_4-CH_3$

Paeoniae radix

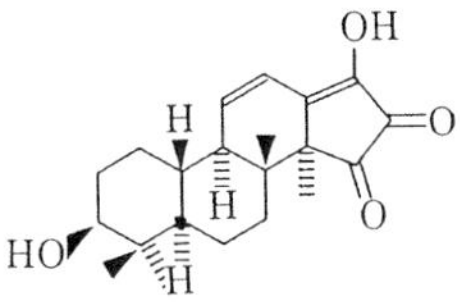

palbinon

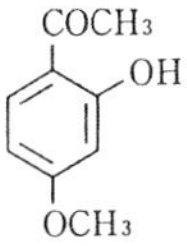

paeonol

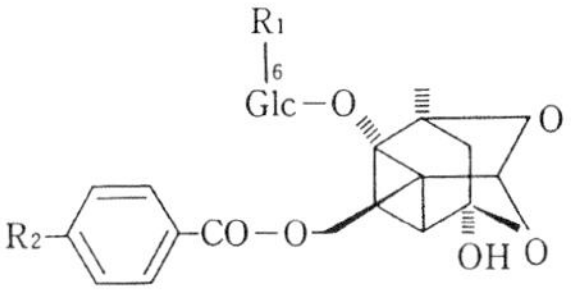

	R_1
paeoniflorin	H
oxypaeoniflorin	H
benzoylpaeoniflorin	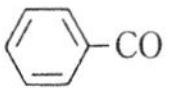 ⟨⟩—CO

Cinnamomi cortex

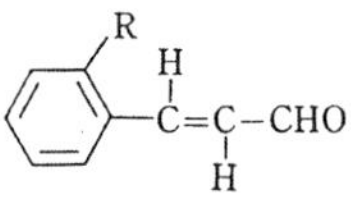

cinnamic aldehyde   R=H
o-methoxycinnamic aldehyde   R=OCH₃

Figure 1   (*Continued*)

phenylnitrone (PBN), a free radical spin-trapping agent, increased the life span of SAMP8. These results suggest that aging is associated with free radicals and that Sho-saiko-to-go-keishi-ka-shakuyaku-to is effective for keeping neuronal functions of aging.

Inhibitory Effect on Neuronal Cell Death

CA1 pyramidal cells of hippocampus in rats were decreased 4 days after a 20-min occlusion of both carotid arteries. Treatment of Sho-saiko-to-go-keishi-ka-shakuyaku-to inhibited cell loss (Sugimoto *et al.*, 1988). Baicalein, which is a flavonoid component of Scutellariae radix, showed more potential activities for free radical scavenging against superoxide, hydroxyl, and 1,1-diphenyl-2-picrylhydrazyl radicals than saikosaponin b_2 of Bupleuri radix, baicalin, and baicalein of Scutellariae radix; ginsenoside Rg_1 of Ginseng radix, paeoniflorin, albiflorin, and paeonol of Peoniae radixt; shogaol and gingerol of Zingiiberis rhizoma; or glycyrrhizin and glycyrrheitnic acid of Glychyrrhizae radix (Mori *et al.*, 1996). Baicalein inhibited the neuronal cell death induced by 5 min of cerebral ischemia in gerbils (Hamada *et al.*, 1993). These results showed that the main active component in Sho-saiko-to-go-keishi-ka-shakuyaku-to for antioxidant action was baicalein.

TOKI-SHAKUYAKU-SAN

Toki-shakuyaku-san is a vacuum-concentrated boiled extract of six herbs. A 7.5-g aliquot of Toki-shakuyaku-san includes the extracts of 4 g of Paeoniae radix (*Paeonia lactiflora* Pallas), 4 g of Atractylodis lanceae rhizoma (*Atractylodes lancea* De Candolle or *Atractylodes chinensis* Koidzumi), 4 g of Alismatis rhizoma (*Alisma orientale* Juzepczuk), 4 g of Hoelen (*Poria cocos* Wolf), 3 g of Cnidii rhizoma (*Cnidium officinale* Makino), and 3 g of Angelicae radix (*Angelica acurtiloba* Kitagawa). Chemical structures of the main components in each herb are shown in Fig. 2.

Toki-shakuyaku-san is used at the Department of Obstetrics and Gynecology for postmenopausal disorders, irregular menstruation, and infertility, as Toki-shakuyaku-san secretes the female hormone estrogen. Alzheimer's disease is high during postmenopausal disorders, in which estrogen secretion is low, suggesting that there is a relationship between dementia and estrogen. This section describes the possibility of Toki-shakuyaku-san as a prophylactic for Alzheimer's dementia in women and for short memory disorder in aging.

Antioxidant Action

Toki-shakuyaku-san has free radical scavenging activities for superoxide, hydroxyl, and 1,1-diphenyl-2-picrylhydrazyl radicals. It inhibited lipid peroxide formation as a marker of TBARS induced by auto-oxidation in the rat cortex homogenate. A 1-month treatment of Toki-shakuyaku-san inhibited TBARS formation and increased SOD activity in the cortex, hippocampus, and striatum of aged rat brain (Ueda *et al.*, 1996).

Aged Neuronal Function

A 1-month treatment of Toki-shakuyaku-san decreased concentrations of dihydroxyphenylacetic acid, homovanilic acid, and 5-hydroxyindoleacetic acid in the cortex, hippocampus, and striatum of aged rats (Ueda *et al.*, 1996) and in the cortex, hippocampus, and striatum of male SAMP8 (unpublished data). These results were not obtained in those of female SAMP8. Concentrations of γ-aminobutyric acid (GABA), glycine, and alanine were elevated in the cortex, hippocampus, and striatum of the female SAMP8, but not in the male SAMP8. GABA, glycine, and alanine are biosynthesized from glucose. Because estrogen augments cerebral glucose utilization (Bishop and Simpkins, 1992), estrogen secreted by Toki-shakuyaku-san might increase concentrations of GABA, glycine, and alanine in these tissues of SAMP8. However, the treatment of estrogen increased GABA binding in the brain (Halbreich, 1997); the treatment may elevate GABA level in these tissues of female but not male SAMP8. Hagino *et al.* (1988, 1990) showed an increase in nicotinic acetylcholine receptor binding and choline acetyltransferase (CAT) activity in the cortex and hippocampus of rats administered Toki-shakuyaku-san. Kishikawa *et al.* (1993) suggested from the experiment of passive avoidance task that Toki-shakuyaku-san alleviates learning disturbances. Fujiwara *et al.* (1989) reported that recognition disorders induced by scopolamine, which is a model for Alzheimer's disease, were improved by treatment with Toki-shakuyaku-san and suggested that the efficacy was dependent on the increase of CAT activity and nicotinic acetylcholine receptor binding. It has been found that CAT activity, muscarinic receptor binding, and acetylcholine esterase activ-

Paeoniae radix

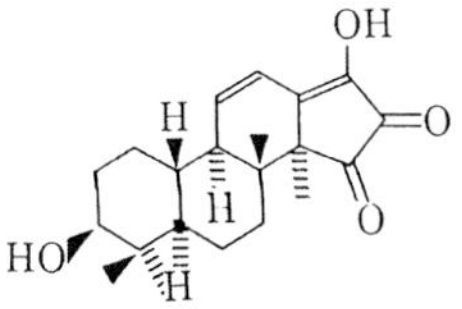

palbinon

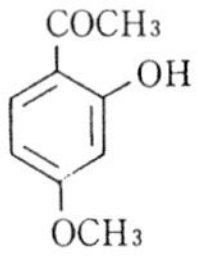

paeonol

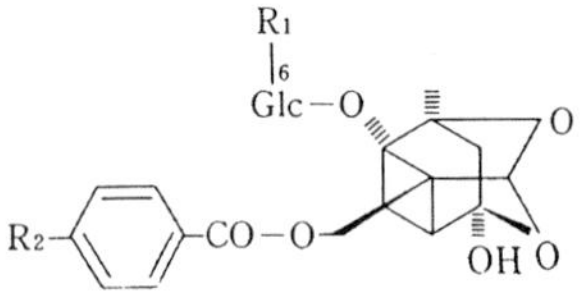

	R_1	R_2
paeoniflorin	H	H
oxypaeoniflorin	H	OH
benzoylpaeoniflorin	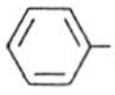-CO	H

Atractylodis Lanceae Rhizoma

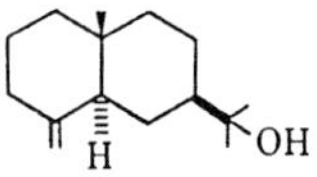

β-eudesmol

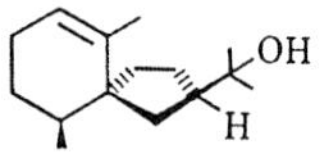

hinesol

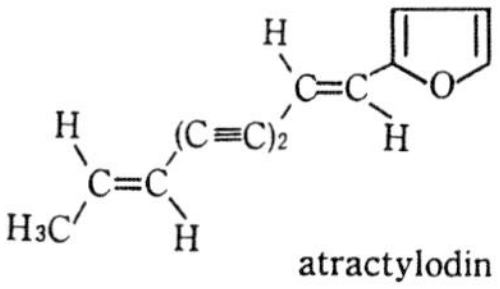

atractylodin

Alismatis rhizoma

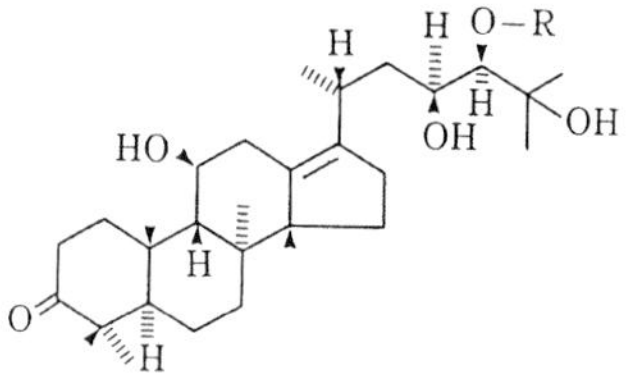

alisol A	R=H
alisol A monoacetate	R=Ac

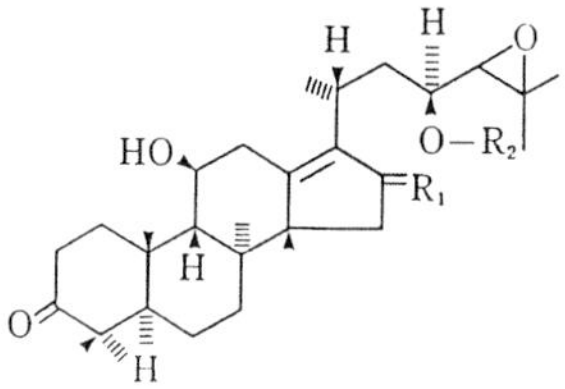

	R_1	R_2
alisol B	<H, H	H
alisol B monoacetate	<H, H	Ac
alisol C	O	H
alisol C monoacetate	O	Ac

$CH_2-O-CO-R$
$|$
$CH-O-CO-R$
$|$
$CH_2-O-PO_3^{\ominus}-CH_2-CH_2-N^{\oplus}(CH_3)_3$

lecithin R=mixture of stearic, palmitic and oleic acids

Figure 2 Chemical structures of main components in each herb of Toki-shakuyaku-san.

Hoelen

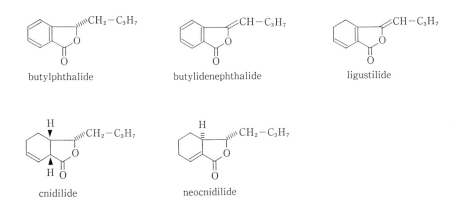

eburicoic acid

Cnidii rhizoma

butylphthalide

butylidenephthalide

ligustilide

cnidilide

neocnidilide

Angelicae radix

ligustilide

$CH_2=CH-C-(C\equiv C)_2-CH-CH=CH-(CH_2)_6-CH_3$

with R_1 and R_2

falcarinol $R_1 = <^{OH}_{H}$ $R_2 = H$

falcarindiol $R_1 = <^{OH}_{H}$ $R_2 = OH$

Figure 2 *(Continued)*

ity are elevated in the striatum of aged rats after a 1-month administration of Toki-shakuyaku-san. These results were not found in adult rats (M. Hiramatsu *et al.*, unpublished data). In addition, it was found that paeoniflorin, a component of Paeoniae radix, improved the impairment of non-spatial learning performance caused by cholinergic dysfunction in rats

(Matsumoto *et al.*, 1997). In addition to these functions on cholinergic neurons, there are some effective reports on senile dementia and Alzheimer's disease with Toki-shakuyaku-san (Mizushima *et al.*, 1989; Yamamoto and Kondo, 1989; Kudo and Sugiura, 1992; Itakura, 1990; Inanaga *et al.*, 1996); these effects have been summarized elsewhere (Hiramatsu *et al.*, 1998).

Neuronal Cell Death

Estrogen attenuated oxidative impairment of synaptic Na^+/K^+-ATPase activity, glucose transport, and glutamate transport induced by amyloid β-peptide and iron (Keller *et al.*, 1997). It also attenuated neuronal injury due to hemoglobin, chemical hypoxia, and excitatory amino acids in murine cortical cultures (Regan and Guo, 1997) and alleviated cognitive dysfunction following transient brain ischemia in ovariectomized gerbils (Kondo *et al.*, 1997). These results suggest that Toki-shakuyaku-san may have a protective effect on neuronal cell death, as Toki-shakuyaku-san secretes estrogen. Watanabe *et al.* (1995) found that Toki-shakuyaku-san inhibited glutamate dehydrogenase activity from cultured cerebellar granule cells induced with glutamate. Neuronal cell death is thought to be induced by peroxinitrite induced by NO and superoxide. In consideration of these facts, Toki-shakuyaku-san may have an inhibitory effect on neuronal cell death by its free radical scavenging activity.

HACHIMI-JIO-GAN

Hachimi-jio-gan is a vacuum-concentrated boiled extract of eight herbs. A 7.5-g aliquot of Hachimi-jio-gan includes the extracts of 6 g of Rehmanniae radix (*Rehmannia glutinosa* Liboschitz var. *purpurea* makino or *R. glutinosa* Liboschitz), 3 g of Corni fructus (*Cornus officinalis* Siebold et Zuccarini), 3 g of Discoreae rhizoma (*Dioscorea japonica* Thunberg or *D. batatas* Decaisne), 3 g of Alismatis rhizoma (*Alisma orientale* Juzepczuk), 3 g of Hoelen (*Poria cocos* Wolf), 2.5 g of Moutan cortex [*Paeonia suffruticosa* Andrews (*Paeonia moutan* Sims)], 1 g of Cinnamomi cortex (*Cinnamomum cassia* Blume), and Aconiti tuber (*Aconitum carmichaeli* Debeaux). Chemical structures of the main components are shown in Fig. 3.

Hachimi-jio-gan is used for diabetes, prostatic hypertrophy, urinary disturbance, sterility, and kidney disturbance. Hachimi-jio-gan has free radical scavenging activities for superoxide and 1,1-diphenyl-2-picrylhydrazyl radicals (Ohnishi *et al.*, 1991). Oral administration inhibited scopolamine-induced memory impairment and restored acetylcholine content in rat brain (Hirokawa *et al.*, 1996). This result suggested that Hachimi-jio-gan has the potential to enhance cholinergic neuronal activities and is effective on dementia.

COGNITIVE ENHANCEMENT THERAPY FOR ALZHEIMER'S DISEASE

At present there is no definitive treatment for Alzheimer's disease. The choline esterase inhibitor of tacrine, enhancers of energy metabolism, and accelerators of vessel circulation have been used in the treatment of Alzheimer's disease. Thus, some trials have been considered in the treatment of Alzheimer's disease; i.e., cholinergic therapy, nootropic of piracetam, neuronotrophic agent of gangliosides, antiamyloid strategy, antipaired helical filaments approach, antioxidant therapy, antialuminium accumulation, and anti-inflammatory alternative were shown as a trial for Alzheimer's disease (Parnetti *et al.*, 1997). The ratio of occurrence of Alzheimer's disease in women is high during postmenopausal disorders, when estrogen secretion is low, suggesting that there is a relationship between dementia and estrogen. For this reason, estrogen treatment in Alzheimer's disease in women has begun (Henderson, 1997; Buckwalter *et al.*, 1997). Toki-shakuyaku-san may have possible potential in Alzheimer's disease as a prophylactic, as it excretes estrogen, has free radical scavenging activity, and enhances cholinegic function (Table I). These functions were also found in Sho-saiko-to-go-keishi-ka-shakuyaku-to and Hachimi-jio-gan. It is not known which component and why mixed herbs selectively enhance cholinergic neuronal functions. However, it is better to consider Toki-shakuyaku-san and other mixed herbs as a prophyalctic in Alzheimer's disease and in other dementia.

CONCLUSION

There is no suitable medicine for the dementia of Alzheimer's disease, vascular dementia, or senile dementia. In addition, the mechanism for de-

Rehmanniae radix

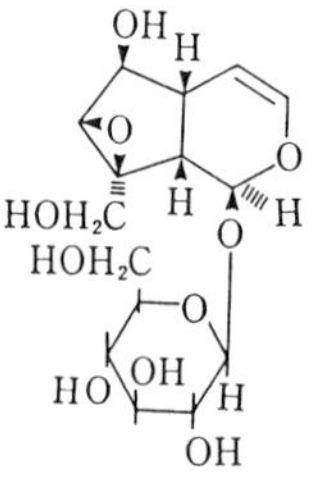

Corni Fructus

catalpol

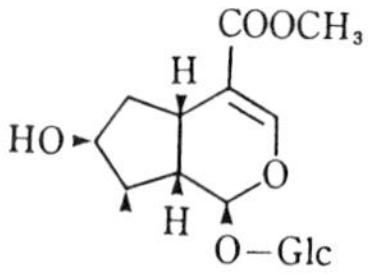

loganin

Alismatis rhizoma

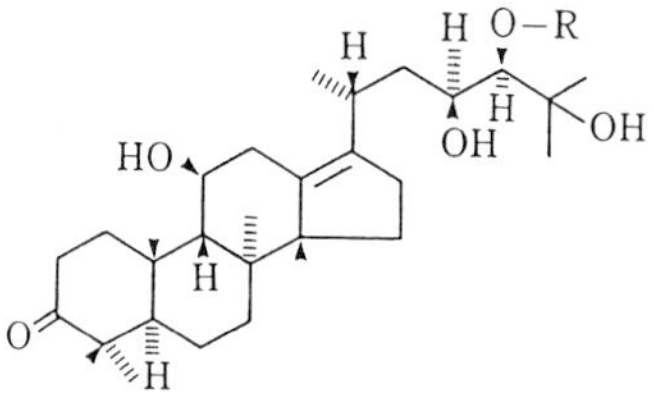

alisol A   R=H
alisol A monoacetate   R=Ac

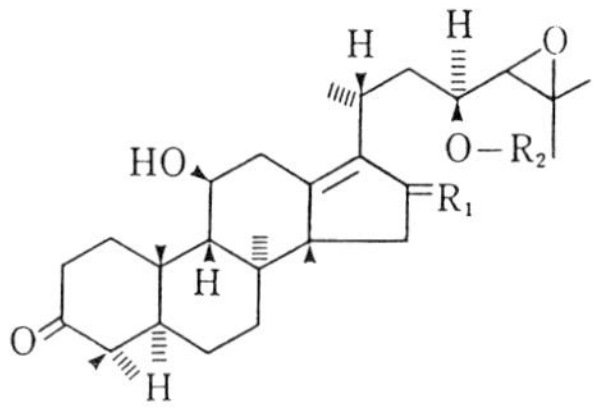

$CH_2-O-CO-R$
$CH-O-CO-R$
$CH_2-O-PO_3-CH_2-CH_2-\overset{\oplus}{N}(CH_3)_3$

lecithin   R=mixture of stearic, palmitic and oleic acids

	R_1	R_2
alisol B	$\langle$ H / H	H
alisol B monoacetate	$\langle$ H / H	Ac
alisol C	O	H
alisol C monoacetate	O	Ac

Hoelen

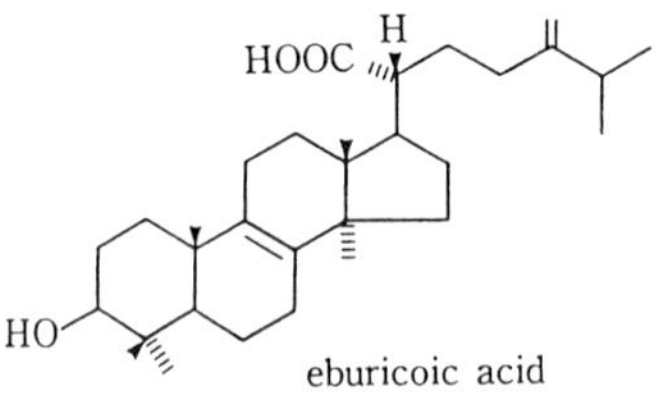

eburicoic acid

Figure 3   Chemical structures of main components in each herb of Hachimi-jio-gan.

Moutan cortex

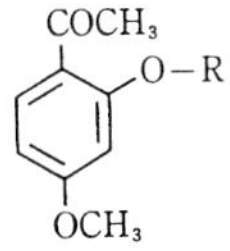

Cinnamomi cortex

paeonol   R=H
paeonoside R=Glc
paeonolide R=Rha−Glc−

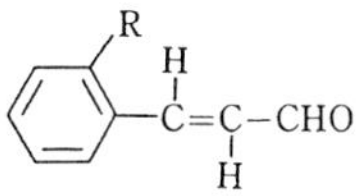

cinnamic aldehyde   R=H
o-methoxycinnamic aldehyde   R=OCH₃

Aconiti tuber

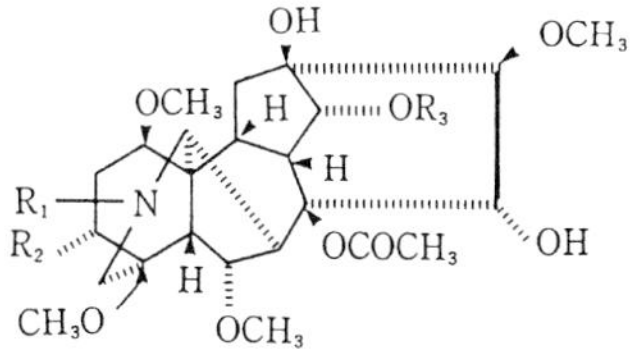

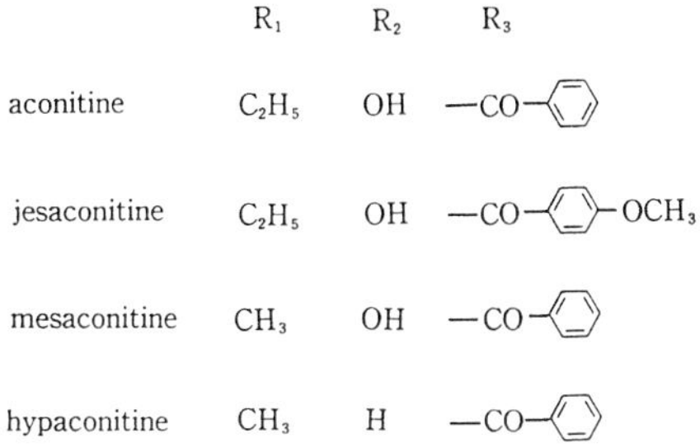

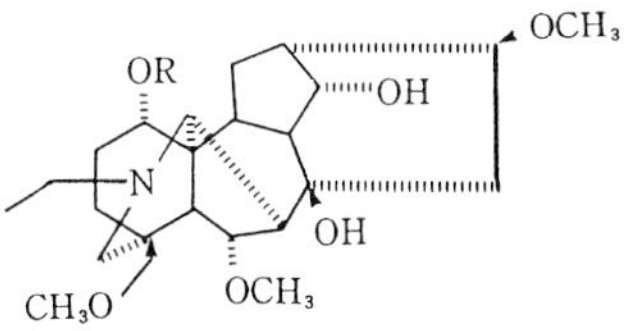

chasmanine R=CH₃
neoline   R=H

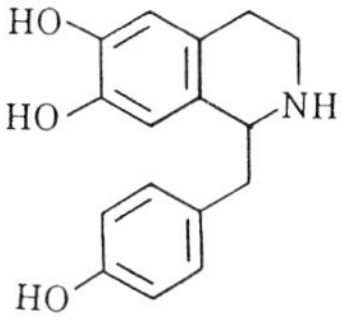

	R₁	R₂	R₃
aconitine	C₂H₅	OH	—CO—
jesaconitine	C₂H₅	OH	—CO—OCH₃
mesaconitine	CH₃	OH	—CO—
hypaconitine	CH₃	H	—CO—

higenamine

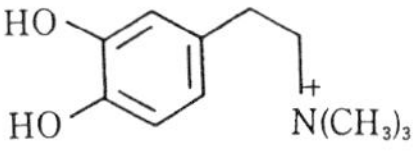

coryneine

Figure 3   (*Continued*)

mentia associated with free radicals has not yet been found. Western medicine, mainly tacrine, has only one pharmacological action: the inhibition of choline esterase activity. However, the authors' findings on mixed herbs showed that they had not only free radical scavenging actions but also that they enhance cholinergic function (Fig. 4). In addition, mixed herbs have estrogen-like functions in the acceleration of microcirculation and enhancement of energy metabolism, as Toki-shakuyaku-san secretes estrogen (Table II). At present, the main trial medicines for Alzheimer's disease are tacrine, agents having accelerators of microcirculation and energy metabolism. For this reason, Toki-shakuyaku-san is

Table I   Clinical Trials for Alzheimer's Disease

Current approach
  Microcirculation enhancer
  Energy production accelerator
  Cholinergic enhancer (choline esterase inhibitor)
    Tacrine
  Nootropics
    Piracetam
Current trials
  Antioxidants (free radical scavenger)
    Estrogens
    Natural plant extracts
      Ginko biloba
      Pycnogenol (pine bark)
    Mixed herbs
      Toki-shakuyaku-san [accelerator for choline acetyltransferase (CAT) activity
        and acetylcholine binding]
      Sho-saiko-to-go-keishi-ka-shakuyaku-to (accelerator for CAT activity)
        Hachimi-jio-gan (accelerator for CAT activity)

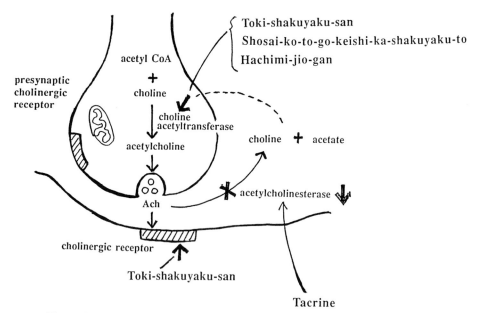

Figure 4   Action site of current trials for Alzheimer's disease on cholinergic neurons.

Table II   Pharmacological Actions of Current Trials for Alzheimer's Disease[a]

	Recognition disorder recovery	Cholinergic actions	Free radical scavenging actions	Neuronal cell death protection	Circulation	Energy production
Tacrine	x	x				
Estrogen	x		x	x	x	x
Toki-shakuyaku-san	x	x	x	x	x	
Sho-saiko-to-go-keishi-ka-shakuyaku-to	x	x	x	x		
Hachimi-jio-gan	x	x	x			
Ginkgo biloba	x		x	x		
Pycnogenol			x	x		

[a] x, effective.

very attractive not only as a treatment, but also as a prophylactic for dementia.

REFERENCES

Aimi, S. (1962). The trial of Sho-saiko-to-go-keishi-ka-shakuyaku-to in epileptic patients. *J. Jpn. Oriental Med.* **13**, 115–118. [in Japanese]

Bishop, J., and Simpkins, J. W. (1992). Role of estrogen in peripheral and cerebral glucose utilization. *Rev. Neurosci.* **3**, 121–137.

Buckwalter, J. G., Schneider, L. S., Wilshire, T. W., Dunn, M. E. and Henderson, V. W. (1997). *Arc. Gerontol. Geriatr.* **24**, 261–267.

Edamatsu, R., Mori, A., and Packer, L. (1995). The spin-trap N-*tert*-alpha-phenylbutylnitrone prolongs the life span of the senescence accelerated mouse. *Biochem. Biophys. Res. Commun.* **211**, 847–849.

Fujiwara, M., and Ueki, S. (1989). Toki-shakuyaku-san (TJ-23) improves scopolamine-induced spatial disruption. *In* "Medicines of Plant Origin in Modern Therapy." A Symposium Report of 4th World Conference on Clinical Pharmacology.

Haba, K., Ogawa, N., and Mori, A. (1990). The effects of sho-saiko-to-go-keishi-ka-shakuyaku-to (TJ-960) on ischemia-induced changes of brain acetylcholine and monoamine levels in gerbils. *Neurochem. Res.* **15**, 487–493.

Hagino, N., and Koyama, T. (1988). Stimulation of nicotinic acetylcholine receptor synthesis in the brain of Toki-shakuyaku-san (TJ-23). *In* "Recent Advances in the Pharmacology of KAMPO (Japanese Herbal) Medicines" (E. Hosoya and Y. Yamamura, eds.), pp. 144–149. Excerpta Medica, Tokyo.

Hagino, N., Sakamoto, S., and Toriizuka, K. (1990). Effect of Toki-shakuyakusan on choline acetyltransferase activity and nicotine acetylcholine receptors in rat brain. *J. Trad. Med.* **7**, 340–341. [in Japanese]

Halbreich, U. (1997). Role of estrogen in postmenopausal depression. *Neurology* **48**, S16–S20.

Hamada, H., Hiramatsu, M., Edamatsu, R., and Mori, A. (1993). Free radical scavenging action of baicalein. *Arch. Biochem. Biophys.* **306**, 261–266.

Harman, D. (1981). The aging process. *Proc. Natl. Acad. Sci. USA* **78**, 7124–7128.

Harman, D. (1995). Free radical theory of aging: Alzheimer's disease pathogenesis. *Age* **18**, 97–119.

Hiramatsu, M., Kabuto, H., and Mori, A. (1985). Effect of Shosaikoto-go-keishikashakuyakuto on convulsions and brain 5-hydroxytryptamine level in El mice. *Neurosciences* **11**, 17–21.

Hiramatsu, M., Kabuto, H., and Mori, A. (1986). Effects of Shosaikoto-go-keishikashakuyakuto (TJ-960) on brain catecholamine level of aged rats. *IRCS Med. Sci.* **14**, 189–190.

Hiramatsu, M., Edamatsu, R., Kabuto, H., and Mori, A. (1988a). Effect of Sho-saiko-to-go-keishi-ka-shakuyaku-to (TJ-960) on monoamines, amino acids, lipid peroxides, and superoxide dismutase in brains of aged rats. *In* "Recent Advances in the Pharmacology of KAMPO (Japanese Herbal) Medicines" (E. Hosoya and Y. Yamamura, eds.), pp. 120–127. Excerpta Medica, Tokyo.

Hiramatsu, M., Edamatsu, R., Kohno, M., and Mori, A. (1988b). Scavenging of free radicals by Sho-saiko-to-go-keishi-ka-shakuyaku-to (TH-960). *In* "Recent Advances in the Pharmacology of KAMPO (Japanese Herbal) Medicines" (E. Hosoya and Y. Yamamura, eds.), pp. 120–127. Excerpta Medica, Tokyo.

Hiramatsu, M., Haba, K., Edamatsu, R., Hamada, H., and Mori, A. (1989). Increased choline acetyltransferase activity by Chinese herbal medicine Sho-saiko-to-go-keishi-ka-shakuyaku-to in aged rat brain. *Neurochem. Res.* **14**, 249–251.

Hiramatsu, M., and Komatsu, M. (1998). Herbal antioxidants and age-related neuronal functions. *In* "Biological Oxidants and Antioxidants: Molecular Mechanisms and Health Effects" (L. Packer and S. H. Ong, eds.), pp. 317–326. AOCS Press, Champaign, IL.

Hiramatsu, M., Komatsu, M., and Ueda, Y. (1996). Aging and herbal antioxidants. *In* "Proceedings of the International Symposium on Natural Antioxidants: Molecular Mechanisms and Health Effects" (L. Packer, M. G. Traber, and W. Xin, eds.), pp. 45–53. AOCS Press, Champaign, IL.

Hiramatsu, M., Velasco, R. D., and Packer, L. (1990). Vitamin E radical reaction with antioxidants in rat liver membranes. *Free Radic. Biol. Med.* **9**, 459–464.

Inanaga, K., Dainoson, K., Ninomiya, Y., Takii, O., Omaru, M., Tanaka, T., Futamata, Y., Tomimatsu, M., Kojima, H., Koga, T., Nishijima, H., Nishikawa, T., Koga, I., and Uchida, Y. (1996). Effects of Toki-shakuyaku-san in patients with cognitive disorders. *Prog. Med.* **16**, 293–300. [in Japanese]

Itakura, M. (1990). Parkinson disease with dementia. *Nihon-Ishikai-Zasshi* **103**, 23–24.

Kishikawa, M., Nishimura, M., Sakae, M., and Iseki, M. (1993). The learning ability and motility of senescence accelerated mice (SAM-P/1) treated with Toki-shakuyaku-san. *Phytother. Res.* **7**, S63–S66.

Komatsu, M., Hiramatsu, M., Yokoyama, H., and Willmore, J. (1996). Effect of TJ-960 on free radical changes within an iron-induced focal epileptogenic region in rat brain measured by in vivo. *Neurosci. Lett.* **205**, 189–192.

Kudo, C., and Sugiura, K. (1992). Senile dementia and Toki-shakuyakusan. *Iyaku J.* **28**, 35–38. [in Japanese]

Le Bars, P. L., Katz, M. M., Berman, N., Itil, T. M., Freedman, A. M., and Schatzberg, A. F. (1997). A placebo-controlled, double-blind, randomized trial of an extract of *Ginkgo biloba* for dementia. *JAMA* **278**, 1327–1332.

Matsumoto, K., Akazawa, K., Murakami, Y., Shimizu, M., Ichiki, H., Maruno, M., and Watanabe, H. (1997). Paeoniflorin ameliorates acquisition impairment of a simple operant discrimination performance caused by unilateral nucleus basalis magnocellularis lesion in rats. *J. Trad. Med.* **14**, 163–168.

Mizushima, Y., and Ikeshita, T. (1989). Effect of Toki-shakuyakusan on senile dementia. *J. Trad. Med.* **6**, 456–457. [in Japanese]

Mori, A., Hamada, H., Ohyama, H., Hiramatsu, M., and Shinohara, S. (1996). Antioxidant effect of TJ-960, a Japanese herbal medicine, against free radical-induced neuronal cell damage. *In* "Natural Antioxidants: Molecular Mechanisms and Health Effects" (L. Packer, M. G. Traber, and W. Xin, eds.), pp. 45–53. AOCS Press, Champaign, IL.

Mori, A., and Hiramatsu, M. (1983). A novel model for post-traumatic epilepsy: Inhibitory effect of sho-saiko-to-go-keishi-ka-shakuyaku-to on iron induced epileptic discharges in rats. *Kampo-Igaku* **7**, 12–16.

Mukawa, J. (1982). Clinical effect of Sho-saiko-to-go-keishi-ka-shakuyaku-to on epileptic patients. *Shinryo-to-Shindan* **19**, 113–117. [in Japanese]

Nagakubo, S., Niwa, S., Kumagai, N., Fukuda, M., Anzai, N., Yamauchi, T., Aikawa, H., Toyoshima, R., Kojima, T., Matsuura, M., Ookubo, Y., and Ootaka, T. (1993). Effects of TJ-960 on Sternberg's paradigm results in epileptic patients. *Jpn. J. Psychiatr. Neurol.* **47**, 609–620.

Nakane, H., Tsuiki, D., Nonaka, K., and Moriyama, S. (1985). Effect of Sho-saiko-to-go-keishi-ka-shakuyaku-to extract on intractable epilepsy. *Rinsho to Kenkyu* **62**, 1914–1923. [in Japanese]

Ohnishi, M., Toda, S., Sugata, R., *et al.* (1991). Effect of diabetes agents on active oxygen species. *Igaku-no-Ayumi* **158**, 447–448. [in Japanese]

Packer, L., Hiramatsu, M., and Yoshikawa, T. (eds) (1996). "Free Radicals in Brain Physiology and Disorders." Academic Press, San Diego.

Parnetti, L., Senin, U., and Mecocci, P. (1997). Cognitive enhancement therapy for Alzheimer's disease, the way forward. *Drugs* **53**, 752–768.

Rohdewald, P. (1998) Pycnogenol. *In* "Flavonoids in Health and Disease" (C. A. Rice-Evans and L. Packer, Eds.), pp. 405–419. Dekker, New York.

Sugimoto, A., Ishige, A., Sudo, K., Sekiguchi, K., Iizuka, S., Itoh, K., Yuzurihara, M., Aburada, M., Hosoya, E., and Sugaya, E. (1988). Protective effect of Sho-saiko-to-go-keishi-ka-shakuyaku-to (TJ-960) against cerebral ischemia. *In* "Recent Advances in the Pharmacology of KAMPO (Japanese Herbal) Medicines" (E. Hosoya and Y. Yamamura, eds.), pp. 112–119. Excerpta Medica, Tokyo.

Takeda, T., Hosokawa, M., Takeshita, S., Irino, M., Higuchi, K., Matsushita, T., Tomita, Y., Yasuhira, K., Hamamoto, H., and Shimizu, K. (1991). A new murine model of accelerated senescence. *Mech. Aging Dev.* **17**, 183–194.

Ueda, Y., Komatsu, M., and Hiramatsu, M. (1996). Free radical scavenging activity of the Japanese herbal medicine Toki-Shakuyaku-San (TJ-23) and its effect on superoxide dismutase activity, lipid peroxides, glutamate, and monoamine metabolites in aged rat brain. *Neurochem. Res.* **21**, 909–914.

Watanabe, Y., Zhang, X. Q., Liu, J. S., Guo, Z., Ohnishi, M., and Shibuya, T. (1995). Protection of glutamate induced neuronal damages in cultured cerebellar granule cells by Chinese herbal medicine: Toki-shakuyaku-san and its comprised six medicinal herbs. *J. Trad. Med.* **12**, 93–101.

Willmore, L. J., Sypert, G. W., and Mouson, J. B. (1978). Chronic focal epileptiform discharges induced by injection into rat and cat cortex. *Science* **200**, 1501–1503.

Yamamoto, T., and Kondo, K. (1989). The treatment of herbs on dementia of the Alzheimer type. *J. Trad. Med.* **6**, 454–455. [in Japanese]

28 Actions of Tea Polyphenols in Oral Hygiene

Yukihiko Hara

Food Research Institute
Mitsui Norin Co., Ltd.
Fujieda City, 426-0133 Japan

INTRODUCTION

Tea polyphenols show potent antioxidative and radical scavenging actions as well as antimicrobial actions, which are perhaps due to the protein-binding property of tea polyphenols. The antioxidative actions of tea polyphenols in relation to lipid peroxidation in various systems have been reported (1). Nanjo *et al.* (2) also reported on the radical scavenging action of catechins and their derivatives on DPPH radicals, demonstrating the structural importance of the *o*-trihydroxyl group in the B ring and the galloyl moiety at the 3 position of flavan-3-ol skeleton in exerting their effects. This chapter reviews some of the actions that aid in the maintenance of oral hygiene.

TEA CATECHINS AND THEAFLAVINS

Tea catechins were extracted from green tea and theaflavins were isolated from black tea. Their structural formulas are shown in Figs. 1 and 2. There are usually four kinds of catechins in green tea, of which (−)-epigallocatechin gallate (EGCg) is predominant, constituting more than 50% of the total catechins. Theaflavins are dimers of catechins that are produced from catechins in the process of black tea manufacturing and constitute 1

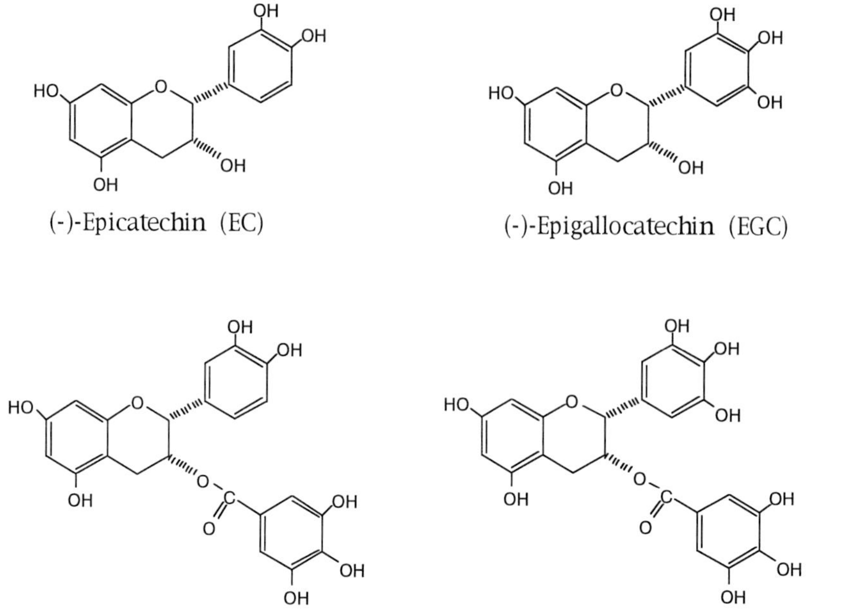

(-)-Epicatechin (EC)

(-)-Epigallocatechin (EGC)

(-)-Epicatechin gallate (ECg)

(-)-Epigallocatechin gallate (EGCg)

Figure 1   Structural formulas of catechins.

to 2% of black tea. Theaflavin digallate (TF3) is the major compound among the four kinds of theaflavins.

INHIBITION OF INFECTIVITY OF INFLUENZA VIRUS

Experiment in Cultured Cells

Tea polyphenols (catechins and theaflavins) were found to be effective inhibitors of influenza viruses in Mardin–Darby canine kidney (MDCK) cells *in vitro*. In order to test the capacity of tea polyphenols to inhibit the infection of influenza A or B viruses to the cells, viruses were mixed with EGCg and TF3 for either 5 or 60 min in phosphate-buffered saline (PBS) and then inoculated on a monolayer of MDCK cells. Without tea polyphenols, viruses will infect the cells and produce approximately 200 plaque-forming units (PFU) over 4 days of incubation in a tissue culture plate under conditions of 33.5°C and 5% CO_2 in air. Plaque formation was inhibited by viruses that were mixed with either EGCg or TF3 as shown in Fig. 3. As is apparent from Fig. 3, in the case of 60 min of contact, com-

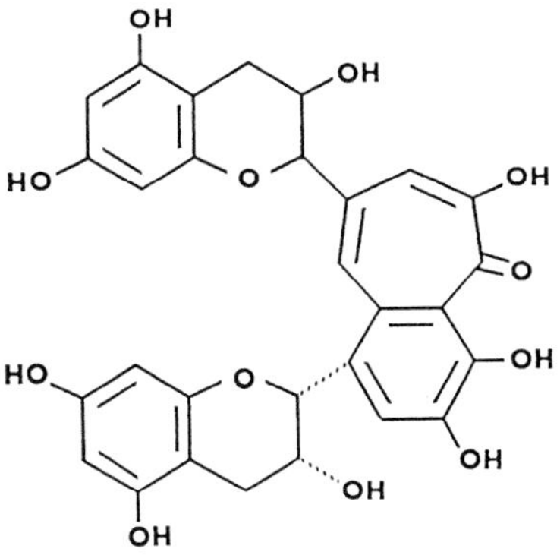

Theaflavin (TF1)

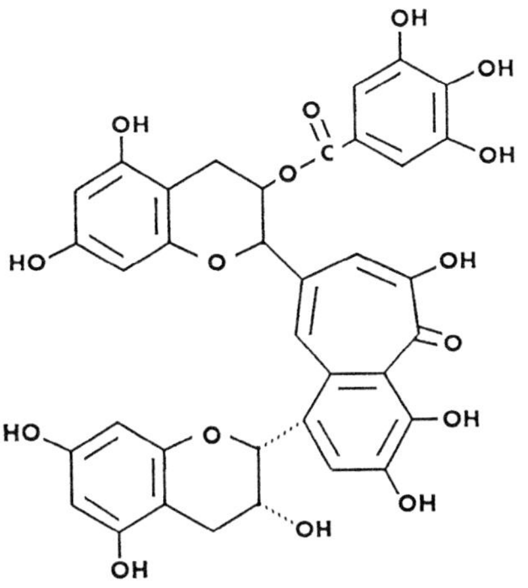

Theaflavin monogallate A (TF2A)

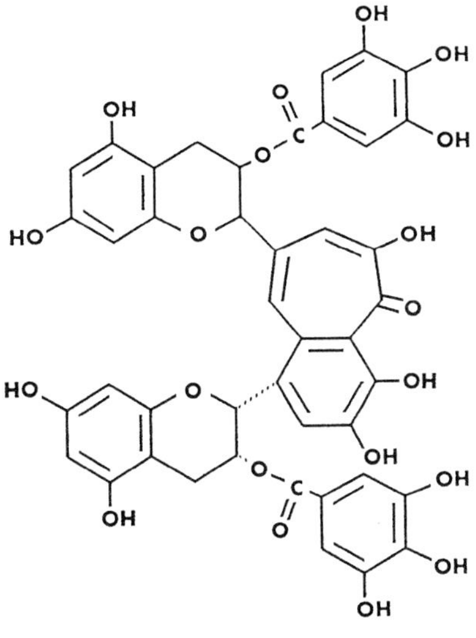

Theaflavin digallate (TF3)

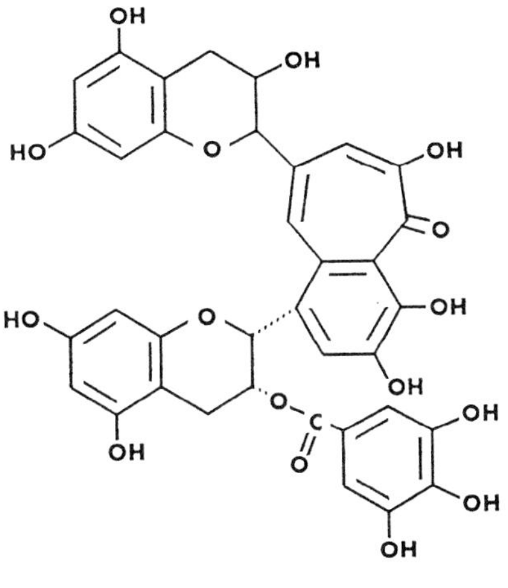

Theaflavin monogallate B (TF2B)

Figure 2 Structural formulas of theaflavins.

plete inhibition of the virus infection was attained by the presence of as low as 1 ppm tea polyphenols in the viral solution (MW: EGCg, 458; TF3, 868). In the case of 5 min of contact, concentrations for complete inhibition were slightly higher, as shown in Fig. 3. These results imply practical utility, as

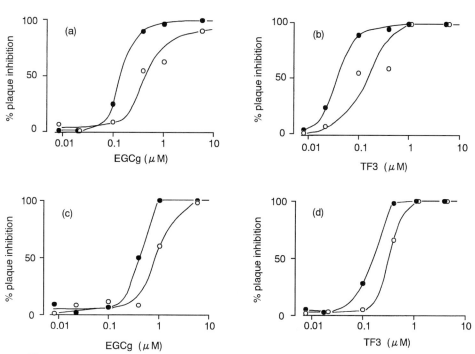

Figure 3 Inhibitory effects of EGCg and TF3 on plaque formation by influenza A virus (a and b) and B virus (c and d). Influenza virus stocks were diluted to 2×10^3 PFU ml^{-1} and incubated with various concentrations of EGCg (a,c) or TF3 (b,d) for 5 min (O) or 60 min (●) at 37°C before virus exposure to MDCK cells. The inhibition of plaque count was scored by the mean of triplicate cultures for each group after assay. Mean PFU ± SD was 78.8 ± 22.1 of control of eight experiments.

the concentration of tea polyphenols in the daily brew ranges from 500 to 1000 ppm. Electron microscopic observation revealed that EGCg and TF3 (1 mM) agglutinated virus particles as well as the antibody during short-time contact (Fig. 4). It was assumed that tea polyphenols may bind to surface glycoproteins of the hemagglutinin antigens of the influenza virus and prevent virus adsorption to the cells (3). Advantages of tea polyphenols in antiflu applications are that they are effective at extremely low concentrations and they are effective regardless of the mutation of the antigen of the virus, whereas the antibody has to be developed in accordance with the mutating antigen. One disadvantage of tea polyphenols is that they cannot act on cells that have already been infected by the virus, i.e., tea polyphenols are effective only in the preventive stage before cellular infection is established.

A

B

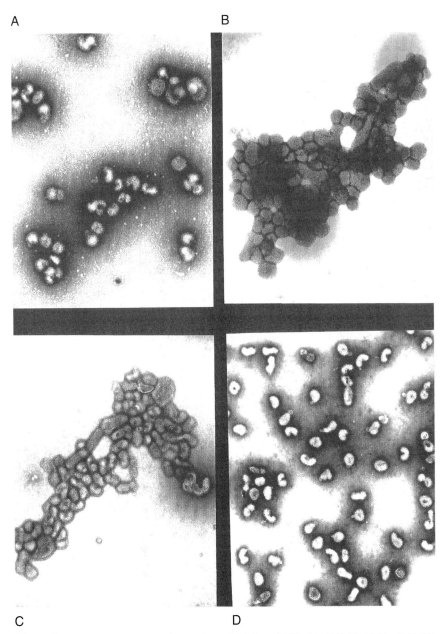

C

D

Figure 4 Electron microscopic observations of EGCg and TF3. (A) PBS, (B) EGCg, (C) TF3, (D) IgG.

Animal Experiments

The antiflu action of tea polyphenols, proved *in vitro*, was tested in animals. Mice were divided into four groups of 10 mice. The number of viruses to be inoculated on each mouse was adjusted to $10^{5.3}$ PFU/mouse ($10^{1.3}$ LD$_{50}$) as follows. Black tea (leaves) was extracted by PBS to a concentration twice that of the normal brew. This extraction was mixed with the same volume of a viral solution in a concentration twice that intended so that the final concentration was $10^{5.3}$ PFU/mouse and the final concentration of tea was that of a normal brew. The following four different solutions (30 μl) were prepared and inoculated intranasally under ether anesthesia: (1) virus and MEM (Eagle's minimum essential medium plus 10% fetal bovine serum); (2) virus and black tea; (3) black tea and MEM, and (4) MEM alone. Body weights and survival rates were observed for 14 days, as shown in Figs. 5 and 6. It is evident that the group of mice receiving virus but not black tea lost weight from day 3, half died by day 6, and all died by day 10. Mice who received a mixed solution of the virus and black tea

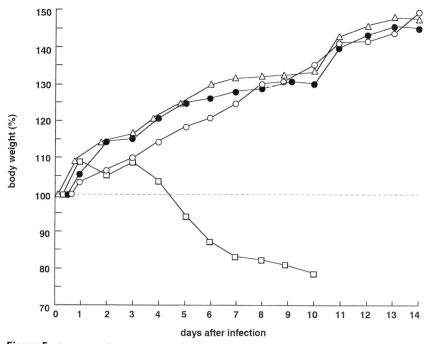

Figure 5   Protection by tea extract of mice against influenza infection. Mice were administered $10^{5.3}$ PFU/mice ($10^{1.3}$LD$_{50}$) of WSN virus. Body weight was observed for 14 days. ○, MEM; □, virus; △, 2% black tea; ●, 2% black tea and virus.

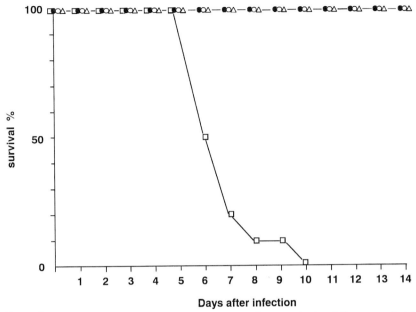

Figure 6   Protection by tea extracts of mice against influenza infection. Mice were administered $10^{5.3}$ PFU/mice ($10^{1.3}$LD$_{50}$) of WSN virus. Survival was observed for 14 days. $\bigcirc$, MEM; $\square$, virus; $\triangle$, 2% black tea; $\bullet$, 2% black tea and virus.

gained weight and the survival rate was as good as the group without the virus. Thus, it was confirmed that infectivity of the flu viruses to the nasal mucous membrane cells was inhibited by prior contact of the viruses with a daily concentration of black tea brew and that there were no recurrences of infectivity of the virus after the absorption inside the body (4).

Because pigs are prone to the epidemic influenza virus in colder seasons, antiflu field tests were conducted in pig pens in northern Japan. From October to March, tea catechins were mixed in the drinking water (0.03% catechins, i.e. 0.1% Polyphenon G) or sprayed from above the pens. Their blood was sampled and antibody titers were determined every month. It was confirmed that titers of the tea catechin group showed much lower figures than those of the noncatechin group. Data will be published later.

PREVENTION OF DENTAL CARIES

Streptococcus mutans, a cariogenic bacteria, produces insoluble sticky glucan on the surface of enamel from sucrose by way of an extracellular enzyme, glucosyltransferase (GTF). Water-insoluble glucan facilitates the accumulation of microorganisms on smooth tooth surfaces (formation of dental plaque) and the development of subsequent dental caries. Therefore, by inhibiting either the growth of *S. mutans* or the action of GTF, chances of developing dental caries will be reduced. Tea polyphenols were found to be effective in both ways.

Inhibition of Glucan Formation

In the [^{14}C]sucrose solution, GTF isolated from *S. mutans* was mixed and incubated for 60 min at 37°C with or without the presence of tea polyphenols. Insoluble glucan was separated from soluble glucan by centrifugation and the degree of incorporation of sucrose into insoluble glucan was determined (5). The percentage of inhibition of insoluble glucan formation by tea polyphenols is shown in Table I. It seems that tea polyphenols influence the formation of glucan synthesis at 1 mg/ml (a heavy drinking concentration) and almost completely inhibit the glucan formation at

Table I   Percentage of Inhibition of Insoluble Glucan Formation by Tea Polyphenols

Tea polyphenols	Concentration	Inhibition (%)
Crude catechins	1 mg/ml	28
	10 mg/ml	93
EC	1 mM	1
	10 mM	42
EGC	1 mM	13
	10 mM	25
ECg	1 mM	35
	10 mM	82
EGCg	1 mM	42
	10 mM	75
Crude theaflavins	1 mg/ml	21
	10 mg/ml	90
TF1	1 mM	57
	10 mM	98
TF2A	1 mM	64
	10 mM	97
TF2B	1 mM	47
	10 mM	97
TF3	1 mM	56
	10 mM	98

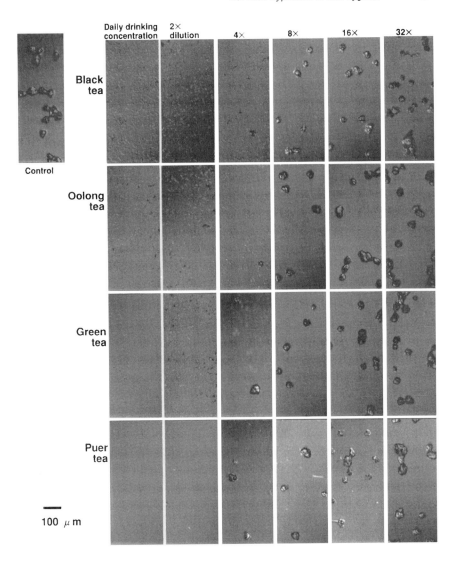

Anti-dental plaque effect of tea beverages

Figure 7   Antidental plaque effect of tea beverages.

10 mg/ml. In the following experiment, using a somewhat more practical model, the inhibition of plaque formation by tea was confirmed. In a petri dish, a brewed tea extract of various concentrations was mixed with 1% sucrose solution plus BHI bouillon. A slide glass was immersed into this solution and S. *mutans* was inoculated. After 3 days of incubation, the num-

ber and the volume of plaques, consisting of insoluble glucan, were observed. As is evident in Fig. 7, there was no growth of plaques in the tea brew of normal concentration or twice diluted concentration. Even at four times the dilution of normal brew, only a few plaques were counted. At more than eight times dilution, plaques grew to the same extent as the control (no tea polyphenols). There was almost no difference in these effects among the varieties of teas: green tea, black tea, oolong tea, or puer tea (6).

Growth Inhibition of *S. mutans*

Growth inhibition curves of *S. mutans* by four kinds of tea are shown in Fig. 8. Viable counts of inoculated *S. mutans* decreased to zero within an hour in the normal brew of green tea. In oolong tea and black tea, *S. mutans* was not affected much. Accordingly, the antibacterial potency of green tea catechins against *S. mutans* was tested. The addition of 500 ppm of green tea catechins (equivalent to the amount found in the normal brew) in physiological saline inhibited the growth of *S. mutans* to zero in about 4 hr, as shown in Fig. 9. It could be surmised from Fig. 9 that there are certain elements in green tea that have a bactericidal effect against *S. mutans*. Re-

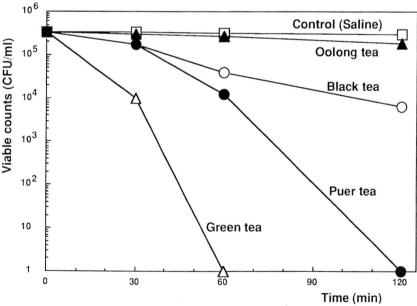

Figure 8   Antibacterial effect of tea beverages against cariogenic bacterium *S. mutans*.

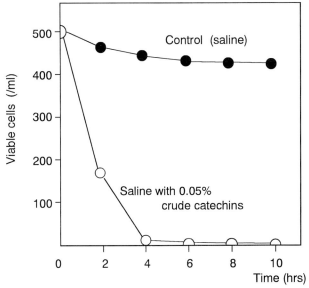

Figure 9   Change in *S. mutans* viable cell numbers after inoculation into saline with 0.05% crude catechins.

search is underway to identify the principal component in green tea in this context (unpublished data).

SUPPRESSION OF BAD BREATH

Representative compounds such as methyl mercaptan, allyl sulfide, or trimethylamine, which are abhorred for causing bad breath, are well suppressed in the presence of tea polyphenols. A 12.5 ppm methyl mercaptan solution was mixed with crude catechins and the head space gas was analyzed by gas chromatography (Fig. 10). It was shown that 500 ppm of tea catechins (equivalent to the concentration of a normal brew) suppressed 12.5 ppm methyl mercaptan, which is a high enough concentration since 0.3–0.5 ppm methyl mercaptan is known to produce bad breath (unpublished data).

Similarly, garlic odor (allyl sulfide) and trimethylamine (fishy smell) were suppressed by adding tea catechins as shown in Figs. 11 and 12. The mode of reaction between odorous compounds and catechins is not elucidated fully. Reaction compounds of catechins with ammonia and amines are proposed in Fig. 13 (7).

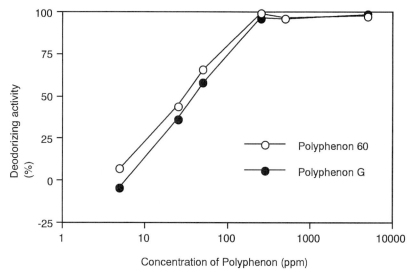

Figure 10 Deodorizing activity of Polyphenon against methyl mercaptan (after 30 min).

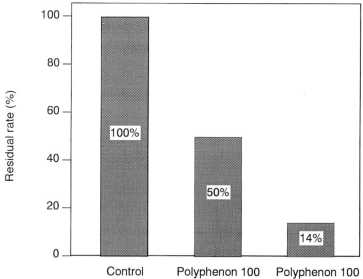

Figure 11 Deodorizing effect of Polyphenon 100 against allyl sulfide (garlic odor).

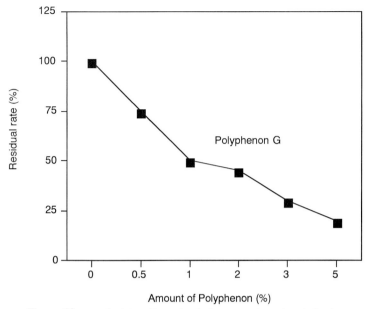

Figure 12 Deodorizing effect of Polyphenon against trimethylamine.

CONCLUSION

Tea polyphenols or tea catechins taken orally remain in the oral cavity for a certain period of time. The interaction of catechins with the oral membrane gives us a sense of astringency. The threshold of EGCg, which is the most astringent, is somewhere around 180 ppm in plain water. It is most probable that tea polyphenols remain in the oral cavity at very low concentrations without us being aware of their presence.

In exerting the potency of radical scavenging actions of tea catechins, the OH group at the 4′ position in the B ring was confirmed to be essential (2). It was also confirmed that nitrogen in amines binds to the 4′ carbon in the B ring to deodorize the smell. It is of interest to find a situation where catechins work as antioxidants and then as deodorizers.

This chapter reviewed several features of tea polyphenols in relation to oral hygiene. It has been confirmed separately that tea catechins interfere with *H. pylori* in the stomach (8), α-amylase (9) or lipids (10) in the small intestine, and bacterial flora in the large intestine (11), all of them in very favorable ways for our good health. The amount of catechins absorbed from the intestine, their metabolism, and fate inside the body are as yet only

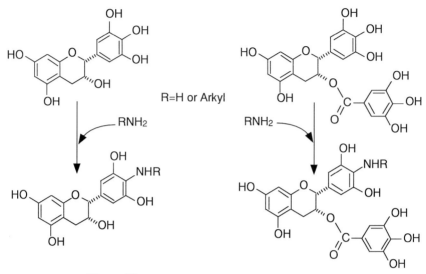

Figure 13   Proposed reaction of catechins with amine.

partly elucidated (12, 13). Detailed studies on the health benefits of tea polyphenols are still needed.

REFERENCES

1. Hara, Y. (1997). Antioxidants in tea and their physiological functions. *In* "Food and Free Radicals" (M. Hiramatsu *et al.*, eds.), pp. 49–65. Plenum Press, New York.
2. Nanjo, F., Goto, K., Seto, R., Suzuki, M. Sakai, M., and Hara, Y. (1996). Scavenging effects of tea catechins and their derivatives on 1,1-diphenyl-2-picrylhydrazyl radical. *Free Radic. Biol. Med.* **21**(6), 895–902.
3. Nakayama, M., Suzuki, K., Toda, M., Okubo, S., Hara, Y., and Shimamura, T. (1993). Inhibition of the infectivity of influenza virus by tea polyphenols. *Antivir. Res.* **21**, 289–299.
4. Nakayama, M., Toda, M., Okubo, S., Hara, Y., and Shimamura, T. (1994). Inhibition of the infectivity of influenza virus by black tea extract. *J. Jpn. Assoc. Infect. Dis.* **68**(7), 824–829.
5. Hattori, M., Kusumoto, I. T., Namba, T., Ishigami, T., and Hara, Y. (1990). Effect of tea polyphenols on glucan synthesis by glucosyltransferase from *Streptococcus mutans. Chem. Pharmacol. Bull.* **38**(3), 717–720.
6. Ishigami, T., and Hara, Y. (1993). Anticarious and bowel modulating actions of tea. Proc. of TEATECH 1993: 125–132, Tea Research Assoc. of India.
7. Nanjo, F., *et al.* (1997). The practice on deodorizing mechanism of catechins against amine odor (Japanese). *Nippon Nogeikagaku Kaishi* **71**, 60. [Abstract]
8. Yamada, M., Murohisa, B., Oguni, I., Harada, N., and Hara, Y. (1997). Effects of tea

polyphenols against *Helicobacter pylori.* Abstract of Papers, Part 1 (069), 213th ACS National Meeting, San Francisco.

9. Hara, Y., and Honda, M. (1990). The inhibition of α-amylase by tea polyphenols. *Agric. Biol. Chem.* **54**(8), 1939–1945.

10. Muramatsu, K., Fukuyo, M., and Hara, Y. (1986). Effect of green tea catechins on plasma cholesterol level in cholesterol-fed rats. *J. Nutr. Sci. Vitaminol.* **32**, 613–622.

11. Goto, K., Kanaya, S., Nishikawa, T., Hara, H., Terada, A., Ishigami, T., Hara, Y. (1998). The influence of tea catechins on fecal flora of elderly residents in long-term care facilities. *Annals of Long-Term Care* **6**(2), 43.

12. Okushio, K., Matsumoto, N., Kohri, T., Suzuki, M., Nanjo, F., and Hara, Y. (1996). Absorption of tea catechins into rat portal vein. *Biol. Pharmacol. Bull.* **19**(2), 326–329.

13. Lee, M.-J., Wang, Z.-Y., Li, H., Chen, L., Sun, Y., Gobbo, S., Balentine, D., and Yang, C.-S. (1995). Analysis of plasma and urinary tea polyphenols in human subjects. *Cancer Epidemiol. Biomark. Prevent.* **4**, 393–399.

29 The Food Industry and Functional Foods: Tea Antioxidants and Cardiovascular Disease

Sheila Wiseman, Ute Weisgerber, Lilian Tijburg, and Onno Korver

Unilever Nutrition Centre
Unilever Research Vlaardingen
The Netherlands

INTRODUCTION

Tea beverage has been considered to be a "functional food" in China for centuries. One of the legends surrounding the origins of tea concerns the Chinese emperor Shen-Nung, who lived around 2700 B.C. While heating some water to allay stomach pains due to overindulgence at dinner, the emperor did not notice that some leaves from a nearby shrub had dropped into the boiling water. He drank the brew and found that it not only tasted delicious but that it also caused the spontaneous disappearance of his stomach pains, thereby demonstrating for the first time beneficial health effects of tea. Nowadays, to define a particular food or food ingredient as "functional" requires rigorous scientific evidence that conclusively demonstrates a physiological benefit. Such evidence is assembled from epidemiological studies, *in vitro* studies, animal studies, and clinical trials with hard or (generally) soft biomarker end points.

This chapter considers evidence related to the effects of tea on cardiovascular disease (CVD) and focuses primarily on the antioxidant properties of tea and tea constituents. CVD is recognized as a multifactorial disease,

Antioxidant Food Supplements in Human Health

and the most established risk factors are elevated blood low-density lipoprotein (LDL) cholesterol levels, high blood pressure, smoking, diabetes, and family history. Oxidation of LDL is now thought to be an important contributory factor in atherosclerotic disease development (Berliner and Heinecke, 1996), and the possible involvement of oxidative processes in the early stages of atherogenesis has stimulated the concept that dietary antioxidants may have a protective function in CVD (Gey, 1995). Data from epidemiological studies and randomized clinical trials indicate protective effects of dietary antioxidants, particularly vitamin E (Kushi *et al.,* 1996; Stephens *et al.,* 1996) and β-carotene (Gey, 1995). In addition, antioxidants such as probucol and vitamin E decrease experimental atherosclerosis in animal models (Frei, 1995). Tea and tea constituents possess potent antioxidant properties and may provide protection against CVD via a number of mechanisms, including inhibition of LDL oxidation.

COMPOSITION OF TEA

Tea is prepared from the leaves of two well-defined varieties of the tea plant *Camellia sinensis: assamica* and *sinensis*. A unique feature of tea leaves is that they contain significant quantities of polyphenolic phytochemicals termed flavonoids. More than 30% of the dry weight of the tea leaf comprises flavonoid compounds with the predominant form being the flavan-3-ols or catechins. The fresh tea leaf contains six catechins in significant concentration: (+)-catechin, (−)-epicatechin (EC), (−)-epicatechin gallate (ECG), (+)-gallocatechin, (−)-epigallocatechin (EGC), and (−)-epigallocatechin gallate (EGCG). Young tea leaves contain the highest catechin concentrations, with the leaf bud and the first leaf being especially rich in epigallocatechin gallate (Bhatia, 1968). Catechins are colorless, water-soluble compounds that function as secondary metabolites in the tea leaf, possibly as inhibitors of bacterial and viral infestation due to their astringent properties.

The majority (80%) of the world's tea harvest is processed to black tea, and processing of the fresh tea leaf is associated with significant changes in flavonoid composition (Fig. 1). Black tea typically contains one-third the level of catechins present in green tea and is manufactured via a polyphenol oxidase-catalyzed condensation of the catechins found in the fresh leaf. Oxidation of the colorless catechin molecules results initially in the formation of catechin quinones, which can undergo complex condensation and coupling reactions to form a range of orange-yellow to reddish-brown colored flavonoids and several volatile constituents (Graham, 1992). The best characterized of these fermentation products are the theaflavins, epitheaflavic

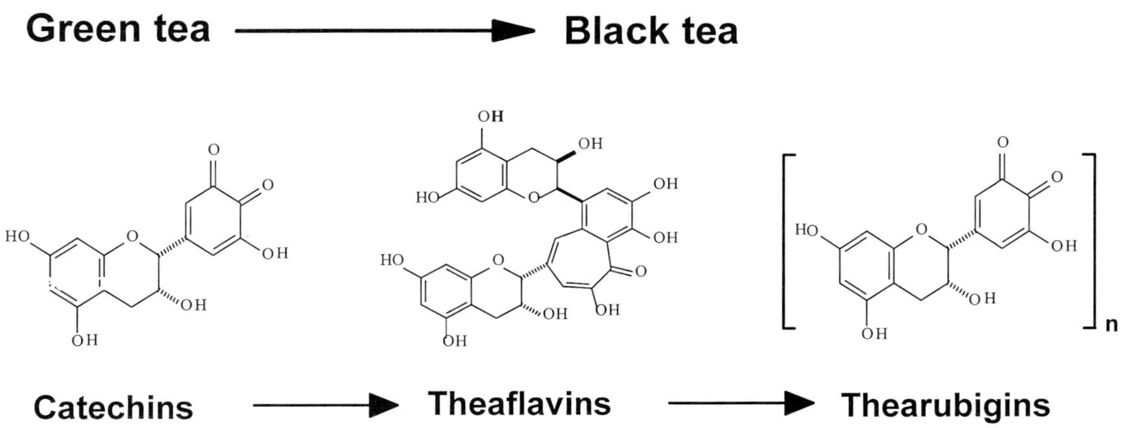

Figure 1 Chemical conversion of catechins during tea processing. Reprinted from Wiseman, S. A., *et al.* (1997). With permission.

acids, and bisflavanols. The majority of the products formed from the condensation reactions, however, comprise a group of uncharacterized complex polyphenols, known as thearubigins. Structures of the thearubigins and their synthetic pathways have not been identified, although they are known to display a wide molecular weight distribution of 1000–10,000. Thearubigins impart to black tea its unique color and taste characteristics. During green tea processing, the form most commonly consumed in southeast Asia and comprising about 20% of the total tea harvest, the enzymatic condensation of catechins is not allowed to occur. Polyphenol oxidase activity is inhibited by heat treatment of the fresh leaf immediately after picking, following which it is cut and ready for use.

Both green and black tea contain approximately 200 mg total flavonoids per cup, although this value is highly dependent on the brewing time and the amount of tea and water used. In addition, green and black tea also contain about 50 mg of caffeine per cup (about half that found in coffee). Vitamin C is present in quite significant amounts in the fresh tea leaf and in green tea but is almost completely lost as a consequence of processing to black tea.

EPIDEMIOLOGY OF TEA AND CARDIOVASCULAR DISEASE

The first indications of a protective effect of high black tea consumption on coronary heart disease (CHD) came from a case-control study of the Boston Collaborative Surveillance Program (1972). This study showed a nonsignificant risk reduction for myocardial infarction (MI) of 34% for subjects who consumed more than six cups of tea per day versus those who did not drink any tea. No adjustment was made for potentially confounding life-style and dietary factors. More recent prospective studies investigating the association between tea intake and subsequent CVD are not conclusive (Table I). In separate analyses from the same Californian cohort, tea intake was not related to either MI or to mortality from CVD (Klatsky *et al.*, 1990, 1993). In contrast, results from a Norwegian cohort demonstrated that the risk of dying from CHD was 36% lower for men consuming more than one cup of tea/day than for men who drank no tea, although this difference was not statistically significant (Stensvold *et al.*, 1992).

In the Zutphen Elderly Cohort, the risk of dying from CHD was significantly lower in men with a high tea intake; this effect was independent of the major established heart disease risk factors (Hertog *et al.*, 1993). In the same cohort, tea consumption was also associated with a lower incidence of stroke (Keli *et al.*, 1996). This Dutch population provided a wider range of tea consumption than the Californian or Norwegian populations,

Table I Prospective Studies on Black Tea and Cardiovascular Disease (CVD)

Cohort, country	No. of end points	End points	RR[a]	95% CI	Reference
Oppland County Study, Norway	141	CHD mortality	0.64	0.38–1.07	Stensvold *et al.* (1992)
Kaiser Permanente Hospital Cohort, United States[b]	1762	CVD mortality	0.98	NS[c]	Klatsky *et al.* (1993)
Zutphen Elderly Study, The Netherlands	43	CHD mortality	0.45	0.22–0.93	Hertog *et al.* (1993)
Zutphen Elderly Study, The Netherlands	42	Stroke incidence	0.31	0.12–0.84	Keli *et al.* (1996)
Caerphilly Study, Wales	131	CHD	2.3	1.0–5.1	Hertog *et al.* (1997)

[a] Relative risk for highest versus lowest category of intake in the fully adjusted model.
[b] Tea intake also not related to MI incidence in the same cohort (Klatsky *et al.*, 1990).
[c] Not significant.

which made it more likely that associations would be identified. Interestingly, however, in a Welsh population of heavy tea drinkers, the risk of death from CHD increased with tea intake (Hertog *et al.*, 1997). In continental Europe, tea drinking tends to be associated with a healthy life-style (Schwarz *et al.*, 1994); however, in the Welsh study, men with the highest intake of tea smoked more, consumed less alcohol, and ate more fat (Hertog *et al.*, 1997). Adjustment was made for these and other confounders, but factors such as smoking and social class were adjusted for only in broad categories and it is possible that all confounders were not completely eliminated. More prospective studies in cohorts with a wide range of tea consumption are required to confirm an association between tea consumption and reduced CVD risk.

LOW-DENSITY LIPOPROTEIN OXIDATION AND PROTECTION BY TEA ANTIOXIDANTS

Although oxidized LDL can be detected in atherosclerotic lesions (Holvoet *et al.*, 1995), the exact nature and source of the oxidants that initiate oxidation of LDL *in vivo* are not known. Cultured vascular cells—endothelial cells, smooth muscle cells, or macrophages—oxidatively modify LDL to an atherogenic form in the presence of low concentrations of transition metal ions (Heinecke *et al.*, 1984). Myeloperoxidase-derived intermediates, e.g., hypochlorous acid, and lipopoxygenase-derived products have also been suggested as potential modifiers of LDL *in vivo* (Heinecke, 1997). Additionally, nitric oxide—a major regulator of vascular tone secreted by endothelial cells—may stimulate LDL oxidation via peroxynitrite that is formed by the reaction of nitric oxide with superoxide radicals (Frei, 1995; Berliner and Heinecke, 1996). Antioxidants, including tea flavonoids, may inhibit the sequence of atherogenic events that arise as a consequence of LDL oxidation in the arterial wall by mechanisms that include increasing the cellular antioxidant capacity, thereby reducing the ability of the cell to oxidize LDL or by increasing the intrinsic antioxidant capacity of the LDL particle itself (Diaz *et al.*, 1997).

Radical Scavenging by Tea Antioxidants

Green and black tea beverages and tea flavonoids consistently demonstrate good antioxidant activity in *in vitro* systems based on scavenging of either stable-free radicals or of more short-lived reactive oxygen or nitrogen species (Wiseman *et al.*, 1997). Tea flavonoids are effective scavengers in the aqueous phase of the stable-free radical cation, $ABTS^{.+}$ [2,2'-azinobis(3-ethylbenzothiazoline-6-sulfonic acid)]. Both green and black tea beverages have potent antioxidant capacity relative to Trolox, the water-soluble vitamin E analog (Salah *et al.*, 1995), and antioxidants from tea are significantly more potent radical scavengers in this system than the well-recognized antioxidants, vitamin E and vitamin C. Catechins displayed up to 5 times the ABTS radical scavenging activity of Trolox, and the theaflavins were between 2.9 and 6.2 times as active, depending on the degree of gallate substitution in the theaflavin molecule (Miller *et al.*, 1996). Green and black tea demonstrated effective scavenging of the stable Fremy's and galvinoxyl radicals (aqueous and lipophilic) when electron spin resonance (ESR) detection was used (Gardner *et al.*, 1997). Green tea was more effective than black tea. In the lipophilic phase, tea flavonoids were better scavengers of the diphenyl -2-picrylhydrazyl (DPPH) radical than vitamin E (Hatano *et al.*, 1989; Hong *et al.*, 1994; Yoshida *et al.*, 1989; Nanjo *et al.*, 1996) and EGCG was the most effective catechin.

Extracts of green tea, pouchong tea, oolong tea, and black tea have been shown to be efficient scavengers of the superoxide-free radical and hydrogen peroxide (Yen and Chen, 1995). Catechins also actively scavenge superoxide radicals, with EGCG being the most reactive (Jovanovic *et al.*, 1995). Theaflavins were even more effective superoxide radical scavengers than catechins (Jovanovic *et al.*, 1997). Tea antioxidants also scavenge hydroxyl radicals (Wang *et al.*, 1993), peroxyl radicals (Terao *et al.*, 1994; Shiraki *et al.*, 1994; Yoshino *et al.*, 1994), singlet oxygen (Tournaire *et al.*,

1993), and peroxynitrite (Pannala *et al.*, 1997) in *in vitro* systems and therefore have the potential to provide effective antioxidant protection against a range of physiologically relevant reactive species.

Tea Antioxidants and LDL Oxidation Resistance

Green tea beverage is an effective inhibitor of copper-mediated LDL oxidation *in vitro* (Luo *et al.*, 1997). Catechin inhibited the formation of lipid peroxidation products in LDL exposed to copper (Mangiapane *et al.*, 1992), and the gallocatechin esters—EGCG and ECG—were more effective than their respective free forms (EGC and EC) in preventing LDL oxidation (Miura *et al.*, 1994; Vinson *et al.*, 1995). Gallocatechins inhibited LDL + VLDL oxidation at lower concentrations than vitamins E and C, β-carotene, or synthetic phenolic antioxidants (Vinson *et al.*, 1995). Theaflavins were also effective inhibitors of copper-mediated LDL oxidation *in vitro*, although the efficacy was less than monomeric gallocatechins (Miller *et al.*, 1996), possibly due to the greater polarity of theaflavins and a reduced ability to partition into lipid environments. No data are available on the inhibition of LDL oxidation by the undefined catechin condensation products of black tea (e.g., thearubigins).

There are several indications that tea flavonoids are able to inhibit the cell-mediated oxidation of LDL. A green tea polyphenol fraction was able to inhibit LDL oxidation by cultured macrophages (Zhenua *et al.*, 1991), but there is currently little information on the effectiveness of the individual tea flavonoids. Catechin and quercetin protected LDL from oxidation when incubated with various cell types, e.g., human monocyte-derived macrophages, human umbilical vein endothelial cells, or lymphoid cells (de Whalley *et al.*, 1990; Mangiapane *et al.*, 1992; Nègre-Salvayre *et al.*, 1991). In addition to inhibitory effects on cell-mediated LDL oxidation, catechin and quercetin also protected lymphoid cells against the cytotoxic effects of previously oxidized LDL. This effect may be due to an increased cellular antioxidant status in the presence of flavonoids (Nègre-Salvayre *et al.*, 1991; Nègre-Salvayre and Salvayre, 1992).

Inhibition of LDL oxidation *ex vivo* by tea antioxidants requires absorption and partitioning of the active compounds into the LDL particle. Consumption of six cups per day of green or black tea (900 ml/day) for 4 weeks had no significant effects on the resistance of LDL to copper-mediated *ex vivo* oxidation in nonsmokers (van het Hof *et al.*, 1997) or in smoking subjects (H. Princen, personal communication). In contrast, a small significant increase in *ex vivo* oxidation resistance compared to baseline was found after consumption of 750 ml/day of black tea for 4 weeks (Ishikawa *et al.*, 1997). Whether the differences between these studies can be attributed to differences in the intake of tea flavonoids or to other variables remains to be established.

The majority of significant effects on LDL oxidation *ex vivo* following dietary antioxidant interventions have been observed in studies using the lipophilic antioxidant, vitamin E (Princen *et al.*, 1995; Jialal *et al.*, 1995). Investigations involving the more hydrophilic flavonoid antioxidants have generally not demonstrated conclusive results, as illustrated earlier for studies on tea flavonoids and as indicated by conflicting results obtained using red wine (Fuhrman *et al.*, 1995; de Rijke *et al.*, 1996). The LDL particle is composed of a central hydrophobic core containing triglycerides and cholesterol esters that is surrounded by a monolayer of phospholipid and free cholesterol molecules (Esterbauer and Ramos, 1995). Apolipoprotein (apo)B, the protein of LDL, covers the bulk of the outer surface of the particle and is responsible for recognition and interaction with the LDL receptor. Association of antioxidants with LDL would theoretically increase the intrinsic antioxidant capacity of the particle and, via sparing of endogenous LDL antioxidants such as vitamin E or direct radical scavenging, oxidative modifications could be inhibited. In the case of flavonoid molecules, it is not likely that they will partition into the lipophilic core of the molecule—the octanol:water partition coefficient for catechin is 1.17 (Paganga *et al.*, 1996)—but interactions with apo(B) are feasible, given the affinity of flavonoids for proteins (Spencer *et al.*, 1988).

Surface interactions of tea flavonoids with LDL protein moieties are therefore possible, and *in vitro* studies using preincubation of tea beverage or purified catechins with plasma prior to isolation of LDL by ultracentrifugation have been used to investigate the degree of association between tea antioxidants and LDL constituents. The LDL isolation procedure would be expected to remove any molecules not firmly associated with the LDL particle, and any antioxidant effects remaining after this procedure may be attributed to bound tea-derived molecules. To investigate this, pooled human plasma was incubated for 30 min at room temperature with a range (0, 30, 50, 100, and 200 mg/liter) of green or black tea solids [U.S. Tea Association regular tea solids (freeze-dried), T. J. Lipton, Englewood Cliffs, NJ]. Six plasma samples were prepared at each tea concentration and lipoproteins were isolated from the tea plus plasma mixtures by density gradient ultracentrifugation (Redgrave *et al.*, 1978). Total catechin concentrations were determined in VLDL, LDL, and HDL using colorimetric methodology based on reaction with *p*-dimethylaminocinnamaldehyde (Kivits *et al.*, 1997).

Catechins could be detected in LDL isolated from plasma that had been incubated with only 10 mg/liter of green tea extract, although the concentration was very low (0.09 μmol/liter). For LDL derived from plasma incubated with black tea, catechins were only detectable in LDL after incubation with 30 mg/liter of black tea extract, reflecting the threefold higher catechin concentration in green tea. Following incubation of plasma with 100 mg/liter green tea extract, the highest catechin concentration was found in the lipoprotein fraction with the highest protein content: HDL. The catechin content in the HDL fraction was 1.75 μmol/liter plasma, with lower levels in LDL (0.95) and VLDL (0.27). The amount of catechins recovered in the lipoprotein fractions was only approximately 3% of the amount initially added to the plasma. It is probable that the bulk of the catechins associated with the plasma protein fraction.

The resistance of the LDL samples to oxidation was investigated by incubation with copper ions and monitoring the formation of conjugated dienes at 234 nm (Princen *et al.*, 1995). LDL samples derived from plasma incubated with 50 mg/liter green tea and 100 mg/liter black tea were significantly more resistant to oxidation than LDL isolated from plasma without tea (Fig. 2). Green tea is composed of approximately 30% catechins, and applying a mean molecular weight of 312 for catechins (Luo *et al.*, 1997) demonstrated that a plasma catechin concentration equivalent to 50 μM is required to significantly inhibit copper-mediated LDL oxidation.

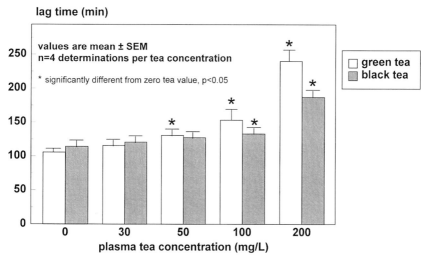

Figure 2   Oxidation resistance of LDL samples isolated from plasma preincubated (30 min at room temperature) with 0, 30, 50, 100, or 200 mg/liter green or black tea [U.S. Tea Association regular tea solids (freeze-dried), T. J. Lipton, Englewood Cliffs, NJ]. LDL samples were isolated by ultracentrifugation (Redgrave *et al.*, 1975) and subjected to copper-mediated oxidation with monitoring of conjugated diene formation 234 nm as described previously (Princen *et al.*, 1995).

A similar *in vitro* study was performed using isolated tea catechins (Ishikawa *et al.,* 1997). LDL isolated from plasma containing 200 μM EGCG or ECG was significantly more resistant to copper-mediated oxidation than LDL isolated from catechin-free plasma. Preincubation of plasma with EGCG prolonged the lag time to more than three times that of control LDL, whereas only a twofold increase was seen with an equivalent concentration of vitamin E. Higher concentrations (400 μM) of EC and EGC in plasma were required to achieve similar significant effects. Theaflavins were more effective than catechins in preventing LDL oxidation. These *in vitro* studies therefore confirm association of tea flavonoids with LDL and demonstrate the ability of tea catechins to inhibit LDL oxidation at concentrations in plasma between 50 and 400 μM.

TEA AND *IN VIVO* ANTIOXIDANT ACTIVITY

Catechin concentrations in plasma after single dose tea consumption were determined in a crossover design study in which six healthy subjects consumed 300 ml of black tea, black tea with milk, or mineral water. The black tea was prepared by boiling 300 ml mineral water with 6.0 g tea leaves (Twinings Earl Grey) for 1.5 min. The tea was filtered and presented to the subjects for consumption within 10 min. Venous blood samples were collected into precooled EDTA tubes before tea consumption and at 30, 50, and 80 min postingestion. Plasma was obtained by immediate centrifugation and assayed for plasma catechin concentration (Kivits *et al.,* 1997) and total plasma antioxidant activity by total radical-trapping antioxidant parameter (TRAP, Ghiselli *et al.,* 1995), oxygen radical absorbance capacity (ORAC, Cao *et al.,* 1996), and ferric iron-reducing ability of plasma, (FRAP, Benzie *et al.,* 1996) methodologies.

Catechin concentrations increased in plasma after black tea consumption and by 80 min posttea ingestion had reached a mean value of 1.41 $\pm$ 0.41 μM. No significant difference in the area under the plasma catechin response curve was observed after the addition of milk to the tea. Data from the *in vitro* studies described earlier suggest that plasma catechin concentrations of 50–400 μM are needed to achieve sufficiently high concentrations of catechins in LDL in order to increase the resistance of the particle to oxidative modification. Results from this study in six subjects indicated that even after a relatively high dose of tea, the total concentration of catechins that could be detected in plasma did not exceed 2 μM, a level that appears insufficient to significantly influence LDL oxidation resistance.

Plasma total antioxidant activity after tea ingestion was investigated using three methods (TRAP, ORAC, and FRAP). This type of assay is a convenient measure of the radical scavenging ability of complex antioxidant mixtures present in biological fluids. Plasma possesses a high background antioxidant activity due to constituents present in relatively high concentrations such as protein and uric acid. This high background may limit the sensitivity of antioxidant activity assays to detect low concentrations of antioxidants of dietary origin. The sensitivity of the TRAP, ORAC, and FRAP methods to detect enhanced plasma antioxidant activity following black tea consumption was investigated. The TRAP and the ORAC assays are based on the inhibition of β-phycoerythrin fluorescence by incubation with a peroxyl radical inducer, 2,2'-diazobis(2-amidinopropane) dihydochloride (ABAP). The addition of antioxidants to the assay system delays the loss of the fluorescent signal, which may be quantified as an increased lag time (TRAP) or increased area under the response curve (ORAC). The FRAP assay determines the ability of a biological sample to reduce a colorless ferric iron complex to a blue ferrous iron complex.

It was not possible to detect any increase in plasma antioxidant activity following black tea consumption (results not shown) using the TRAP and ORAC assays, although a previous report had suggested that the TRAP assay was sufficiently sensitive to detect such an effect (Serafini *et al.*, 1996). In contrast to the results obtained with the peroxyl radical scavenging assays, the ferric iron-reducing ability of plasma was increased significantly following black tea consumption (Fig. 3). The magnitude of the enhancement in plasma-reducing activity compared with the response following mineral water consumption was small (range 2–4% in six subjects), but this assay was highly reproducible and, due to the low reactivity with plasma proteins, allowed sensitive detection of the small increase in antioxidant activity due to the presence in plasma of tea antioxidants.

DISCUSSION

Evidence from recent prospective epidemiological studies suggests that tea consumption is associated with reduced cardiovascular disease risk (Hertog *et al.*, 1993; Keli *et al.*, 1996). Experimental data related to *in vivo* antioxidant effects of tea flavonoids in CVD are currently limited and relate mainly to processes affecting LDL oxidation resistance *in vitro* and *ex vivo*. Although *in vitro* data show green and black tea beverage and purified tea catechins to be effective inhibitors of LDL oxidation in transition metal ion-dependent (Vinson *et al.*, 1996) and -independent systems (Salah *et al.*, 1995), data related to effects on *ex vivo* LDL oxidation are currently inconclusive (van het Hof *et al.*, 1997; Ishikawa *et al.*, 1997). As suggested by data from *in vitro* studies, inhibition of LDL oxidation by increasing in-

Area under response curve (min.μmol/L)

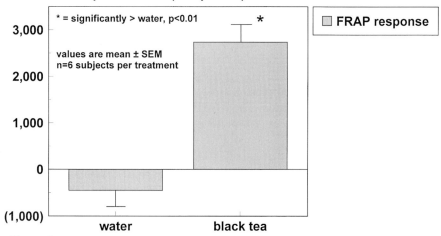

Figure 3 Ferric iron-reducing ability of plasma (FRAP) following ingestion of 300 ml of black tea or mineral water by six healthy volunteers. Blood samples were obtained at 0, 30, 50, and 80 min posttea consumption and plasma FRAP activity was determined as described (Benzie *et al.*, 1996). Results are expressed as area under the plasma FRAP response curve.

trinsic LDL antioxidant capacity may not in any case be the major protective mechanism of tea flavonoids in CVD. In order to achieve significant accumulation of catechins in LDL to enhance oxidation resistance, plasma catechin concentrations in the order of 50–400 μM appear to be required and it does not appear likely that such concentrations of tea catechins will be achievable in plasma following normal patterns of tea consumption.

Other protective mechanisms are, however, feasible. Enhancement of plasma antioxidant activity by approximately 3% following tea ingestion has been demonstrated in a study described in this chapter and further work will provide more conclusive data. For comparison, an increase of only 13% in LDL total antioxidant activity could be detected following 9 weeks of supplementation with 300 mg *dl-α*-tocopherol acetate daily (Miller *et al.*, 1995). The physiological relevance of small increases in plasma antioxidant activity after tea consumption needs to be defined in terms of reduction of oxidative damage to cells and cellular constituents using validated markers. The enhancement of vascular cell antioxidant activity by tea catechins may inhibit the ability of these cells to induce LDL oxidation in the subendothelium and may also determine the extent to which these cells may be damaged by the toxic effects of oxidized LDL present in the vessel wall (Diaz *et al.*, 1997). Although data now suggest that absorption of catechins in humans is rapid and in the order of 47–58% of a consumed dose (Hollman *et al.*, 1997), more data are required on the tissue localization of tea flavonoids and the extent of their cellular accumulation.

Epidemiological data on tea and CVD are strongest for coronary heart disease mortality and not morbidity. This suggests that tea may particularly influence the final thrombotic event that leads to occlusion of the atherosclerotic coronary artery. Flavonoid compounds inhibit platelet aggregation (Cook and Samman, 1996), and experimental evidence suggests that tea inhibits *in vivo* platelet activity and prevents experimental coronary thrombosis in dogs (Folts, 1996). Experimental data on the effects of tea on platelet activity in humans are currently limited and should be a focus for future work. Additionally, the role of tea antioxidants within the hierarchy of other plasma antioxidants needs to be established. Sparing of vitamin E in cell membranes and lipoproteins as a consequence of tea flavonoid antioxidant action early in the oxidation cascade could be an important protective mechanism related to the beneficial effects of tea consumption.

Tea beverage is currently consumed primarily on the basis of its taste properties. The body of evidence related to tea and cardiovascular diseases is still inconclusive but there are sufficient positive indications from epidemiological and experimental studies to suggest that tea may ultimately be consumed also on the basis of its functional health properties.

REFERENCES

Benzie, I. F. F., and Strain, J. J. (1996). The ferric reducing ability of plasma (FRAP) as a measure of "antioxidant power": The FRAP assay. *Anal. Biochem.* **239**, 70–76.

Berliner, J. A., and Heinecke, J. W. (1996). The role of oxidised lipoproteins in atherogenesis. *Free Radic. Biol. Med.* **20**, 707–727.

Bhatia, I., and Ullah, M. (1968). Qualitative and quantitative study of the polyphenols of different organs and some cultivated varieties of tea plant. *J. Sci. Food. Agric.* **19**, 535.

Boston Collaborative Surveillance Program. (1972). Coffee drinking and acute myocardial infarction. *Lancet* **ii**, 1278–1281.

Cao, G., Verdon, C. P., Wu, A. H. B., Wang, H., and Prior, R. L. (1996). Automated assay of oxygen radical absorbance capacity with the COBAS FARA II. *Clin. Chem.* **41**, 1738–1744.

Cook, N. C., and Samman, S. (1996). Flavonoids: Chemistry, metabolism, cardioprotective effects and dietary sources. *J. Nutr. Biochem.* **7**, 66–76.

de Rijke, Y. B., Demacker, P. N. M., Assen, N. A., Sloots, L. M., Katan, M. B., and Stalenhoef, A. F. H. (1996). Red wine consumption does not affect oxidizability of low-density lipoproteins in volunteers. *Am. J. Clin. Nutr.* **63**, 329–334.

de Whalley, C. V., Rankin, S. M., Hoult, J. R. S., Jessup, W., and Leake, D. S. (1990). Flavonoids inhibit the oxidative modification of low-density lipoproteins by macrophages. *Biochem. Pharmacol.* **39**, 1743–1750.

Diaz, M. N., Frei, B., Vita, J. A., and Keaney, J. F. (1997). Antioxidants and atherosclerotic heart disease. *N. Engl. J. Med.* **337**, 408–416.

Esterbauer, H., and Ramos, P. (1995). Chemistry and pathophysiology of oxidation of LDL. *Rev. Physiol. Biochem. Pharmacol.* **25**, 285–311.

Folts, J. D. (1996). Flavonoids in tea but not coffee given by gastric tube inhibit *in vivo* platelet activity and thrombus formation in stenosed dog coronary arteries. *FASEB J.* **10**, A793.

Frei, B. (1995). Cardiovascular disease and nutrient antioxidants: Role of low-density lipoprotein oxidation. *Crit. Rev. Food Sci. Nutr.* **35**, 83–98.

Fuhrman, B., Lavy, A., and Aviram, M. (1995). Consumption of red wine with meals reduces the susceptibility of human plasma and low-density lipoprotein to lipid peroxidation. *Am. J. Clin. Nutr.* **61**, 549–554.

Gardner, P. T., McPhail, D. B., and Duthie, G. C. (1998). Electron spin resonance spectroscopic assessment of the antioxidant potential of teas in aqueous and organic media. *J. Sci. Food Agric.,* **76**, 257–262.

Gey, K. F. (1995). Ten-year retrospective on the antioxidant hypothesis of atherosclerosis: threshold plasma levels of antioxidant micronutrients related to minimum cardiovascular risk. *J. Nutr. Biochem.* **6**, 206–236.

Ghiselli, A., Serafini, M., Maiani, G., Azzini, E., and Ferro-Luzzi, A. (1995). A fluorescence-based method for measuring total plasma antioxidant capability. *Free Radic. Biol. Med.* **18**, 29–36.

Graham, H. N. (1992). Green tea composition, consumption and polyphenol chemistry. *Prev. Med.* **21**, 334–350.

Hatano, T., Edamatsu, R., Hiramatsu, M., Mori, A., Fujita, Y., Yasahura, T., Yoshida, T., and Okuda, T. (1989). Effects of the interaction of tannins with co-existing substances. VI. Effects of tannins and related polyphenols on superoxide anion radical and on 1,1-diphenyl-2-picrylhydrazyl radical. *Chem. Pharm. Bull.* **37**, 2016–2021.

Heinecke, J. W. (1997). Mechanisms of oxidative damage of low density lipoprotein in human atherosclerosis. *Curr. Opin. Lipidol.* **8**, 266–274.

Heinecke, J. W., Rosen, H., and Chait, A. (1984). Iron and copper promote modification of low density lipoprotein by human arterial smooth muscle cells in culture. *J. Clin. Invest.* **74**, 1890–1894.

Hertog, M. G. L., Feskens, E. J. M., Hollman, P. C. H., Katan, M. B., and Kromhout, D. (1993). Dietary antioxidant flavonoids and the risk of coronary heart disease: The Zutphen elderly study. *Lancet* **342**, 1007–1011.

Hertog, M. G. L., Sweetnam, P. M., Fehily, A. M., Elwood, P. C., and Kromhout, D. (1997). Antioxidant flavonols and ischaemic heart disease in a Welsh population of men: The Caerphilly Study. *Am. J. Clin. Nutr.* **65**, 1489–1494.

Hollman, P. C. H., Tijburg, L. B. M., and Yang, C. S. (1997). Bioavailability of flavonoids from tea. *Crit. Rev. Food Sci. Nutr.* **37**, 719–738.

Holvoet, P., Perez, G., Zhao, Z., Brouwers, E., Bernar, H., and Collen, D. (1995). Malondialdehyde-modified low density lipoproteins in patients with atherosclerotic disease. *J. Clin. Invest.* **95**, 2611–2619.

Hong, C. Y., Wang, C. P., Lo, Y. C., and Hsu, F. L. (1994). Effect of flavan-3-ol tannins purified from *Camellia sinensis* on lipid peroxidation of rat heart mitochondria. *Am. J. Chin. Med.* **22**, 285–292.

Ishikawa, T., Suzukawa, M., Ito, T., Yoshida, H., Ayaori, M., Nishiwaki, M., Yonemura, A., Hara, Y., and Nakamura, H. (1997). Effect of tea flavonoid supplementation on the susceptibility of low-density lipoprotein to oxidative modification. *Am. J. Clin. Nutr.* **66**, 261–266.

Jialal, I., and Grundy, S. M. (1992). Effect of dietary supplementation with alpha-tocopherol on the oxidative modification of low density lipoprotein. *J. Lipid Res.* **33**, 899–906.

Jovanovic, S., Hara, Y., Steenken, S., and Simic, M. (1995). Antioxidant potential of gallo-catechins: A pulse radiolysis and laser photolysis study. *J. Am. Chem. Soc.* **117**, 9881–9888.

Jovanovic, S. V., Hara, Y., Steenken, S., and Simic, M. G. (1997). Antioxidant potential of theaflavins: A pulse radiolysis study. *J. Am. Chem. Soc.* **119**, 5337–5343.

Keli, S. O., Hertog, M. G. L., Feskens, E. J. M., and Kromhout, D. (1996). Dietary flavonoids, antioxidant vitamins and incidence of stroke. *Arch. Intern. Med.* **154**, 637–642.

Kivits, G. A. A., van der Sman, F. J. P., and Tijburg, L. B. M. (1997). Analysis of catechins from green and black tea in humans: A specific and sensitive colorimetric assay of total catechins in biological fluids. *Int. J. Food Sci. Nutr.* **48**, 387–392.

Klatsky, A. L., Armstrong, M. A., and Friedman, G. D. (1993). Coffee, tea and mortality. *Ann. Epidemiol.* **3**, 375–381.

Klatsky, A. L., Friedman, G. D., and Armstrong, M. A. (1990). Coffee use prior to myocardial infarction restudied: Heavier intake may increase the risk. *Am. J. Epidemiol.* **132**, 479–488.

Kushi, L. H., Folsom, A. R., Prineas, R. J., Mink, P. J., Wu, Y., and Bostick, R. M. (1996). Dietary antioxidant vitamins and death from coronary heart disease in postmenopausal women. *N. Engl. J. Med.* **334**, 1156–1162.

Luo, M., Kannar, L., Wahlqvist, M. L., and O'Brien, R. C. (1997). Inhibition of LDL oxidation by green tea extract. *Lancet* **349**, 360–361.

Mangiapane, H., Thomson, J., Salter, A., Brown, S., Bell, G. D., and White, D. (1992). The inhibition of the oxidation of low-density lipoprotein by (+)-catechin, a naturally occurring flavonoid. *Biochem. Pharmacol.* **43**, 445–450.

Miller, N., Castellucio, C., Tijburg, L., and Rice-Evans, C. (1996). The antioxidant properties of theaflavins and their gallate esters: Radical scavengers or metal chelators? *FEBS Lett.* **392**, 40–44.

Miller, N. J., Paganga, G., Wiseman, S., van Nielen, W., Tijburg, L., Chowienczyk, P., and Rice-Evans, C. A. (1995). Total antioxidant activity of low density lipoproteins and the relationship with α-tocopherol status. *FEBS Lett.* **365**, 164–166.

Miura, S., Watanabe, J., Tomita, T., Sano, M., and Tomita, I. (1994). The inhibitory effects of tea polyphenols (flavan-3-ol derivatives) on Cu^{2+} mediated oxidative modification of low density lipoprotein. *Biol. Pharm. Bull.* **17**, 1567–1572.

Nanjo, F., Goto, K., Seto, R., Suzuki, M., Sakai, M., and Hara, Y. (1996). Scavenging effects of tea catechins and their derivatives on 1,1-diphenyl-2-picrylhydrazyl radical. *Free Radic. Biol. Med.* **21**, 895–902.

Nègre-Salvayre, A., Alomar, Y., Troly, M., and Salvayre, R. (1991). Ultraviolet-treated lipoproteins as a model system for the study of the biological effects of lipid peroxides on cultured cells. III. The protective effects of antioxidants (probucol, catechin, vitamin E) against the cytotoxicity of oxidised LDL occurs in two different ways. *Biochim. Biophys. Acta* **1096**, 291–300.

Nègre-Salvayre, A., and Salvayre, R. (1992). Quercetin prevents the cytotoxicity of oxidised LDL on lymphoid cell lines. *Free Radic. Biol. Med.* **12**, 101–106.

Paganga, G., Al-Hashim, H., Khodr, H., Scott, B. C., Aruoma, O. I., Hider, R. C., Halliwell, B., and Rice-Evans, C. A. (1996). Mechanisms of antioxidant activities of quercetin and catechin. *Redox Rep.* **2**, 359–364.

Pannala, A., Rice-Evans, C. A., Halliwell, B., and Surinder, S. (1997). Inhibition of peroxynitrite-mediated tyrosine nitration by carechin polyphenols. *Biochem. Biophys. Res. Commun.* **232**, 164–168.

Princen, H. M. G., van Duyvenvoorde, W., Buytenhek, R., van der Laarse, A., van Poppel, G., Gevers Leuven, J. A., and van Hinsbergh, V. W. M. (1995). Supplementation with low doses of vitamin E protects LDL from lipid peroxidation in men and women. *Arterioscler. Thromb. Vasc. Biol.* **15**, 325–333.

Redgrave, T. G., Roberts, D. C. K., and West, C. E. (1975). Separation of plasma lipoproteins by density-gradient ultracentrifugation. *Anal. Biochem.* **65**, 42–49.

Salah, N., Miller, N., Paganga, G., Tijburg, L., Bolwell, G. P., and Rice-Evans, C. (1995). Polyphenolic flavanols as scavengers of aqueous phase radicals and as chain-breaking antioxidants. *Arch. Biochem. Biophys.* **322**, 339–346.

Schwarz, B., Bischof, H. P., and Kunze, M. (1994). Coffee, tea and lifestyle. *Prev. Med.* **23**, 377–384.

Serafini, M., Ghiselli, A., and Ferro-Luzzi, A. (1996). *In vivo* antioxidant effect of green and black tea in man. *Eur. J. Clin. Nutr.* **50**, 28–32.

Shiraki, M., Hara, Y., Osawa, T., Kumon, H., Nakayama, T., and Kawakishi, S. (1994). Antioxidative and antimutagenic effects of theaflavins from black tea. *Mutat. Res.* **323**, 29–34.

Spencer, C. M., Cai, Y., Martin, R., Gaffney, S. H., Goulding, P. N., Magnolato, D., Lilley, T. H., and Haslam, E. (1988). Polyphenol complexation: Some thoughts and observations. *Phytochemistry* **27**, 2397–2409.

Stensvold, I., Tverdal, A., Solvoll, K., and Foss, O. P. (1992). Tea consumption: Relationship to cholesterol, blood pressure and coronary and total mortality. *Prev. Med.* **21**, 546–553.

Stephens, N. G., Parsons, A., Schofield, P. M., Kelly, F., Cheeseman, K., Mitchinson, M. J., and Brown, M. J. (1996). Randomised controlled trial of vitamin E in patients with coronary heart disease: Cambridge Heart Antioxidant Study (CHAOS). *Lancet* **347**, 781–786.

Terao, J., Piskula, M., and Yao, Q. (1994). Protective effect of epicatechin, epicatechin gallate and quercetin on lipid peroxidation in phospholipid bilayers. *Arch. Biochem. Biophys.* **308**, 278–284.

Tournaire, C., Croux, S., Maurette, M., Beck, I., Hocquax, M., Braun, A. M., and Oliveros, E. (1993). Antioxidant activity of flavonoids: Efficiency of singlet oxygen ($^1\Delta_g$) quenching. *J. Photochem. Photobiol. B Biol.* **19**, 205–215.

van het Hof, K., de Boer, S. M., Wiseman, S. A., Lien, N., Weststrate, J. A., and Tijburg, L. B. M. (1997). Consumption of green or black tea does not increase the resistance of LDL to oxidation in humans. *Am. J. Clin. Nutr.* **66**, 1125–1132.

Vinson, J. A., Dabbagh, Y. A., Serry, M. M., and Jang, J. (1995). Plant flavonoids, especially tea flavanols, are powerful antioxidants using an in vitro oxidation model for heart disease. *J. Agric. Food Chem.* **43**, 2800–2802.

Wang, W. F., Luo, J., Yao, S. D., Lian, Z. R., Zhang, J. S., and Lin, N. Y. (1993). Interaction of phenolic antioxidants and hydroxyl radicals. *Radiat. Phys. Chem.* **42**, 985–987.

Wiseman, S. A., Balentine, D. A., and Frei, B. (1997). Antioxidants in tea. *Crit. Rev. Food Sci. Nutr.* **37**, 705–718.

Yen, G., and Chen, H. (1995). Antioxidant activity of various tea extracts in relation to their antimutagenicity. *J. Agric. Food Chem.* **43**, 27–32.

Yoshida, T., Mori, K., Hatano, T., Okomura, T., Uehara, I., Komagoe, K., Fujita, Y., and Okuda, T. (1989). Studies on inhibition mechanism of autoxidation by tannins and flavonoids. V. Radical-scavenging effects of tannins and related polyphenols on 1,1-diphenyl-2-picrylhydrazyl radical. *Chem. Pharm. Bull.* **37**, 1919–1921.

Zhenua, D., Yuan, C., Mei, Z., and Yunzhong, F. (1991). Inhibitory effect of China green tea polyphenol on the oxidative modification of low density lipoprotein by macrophages. *Med. Sci. Res.* **19**, 767–768.

30 Antioxidant Properties of *Crassostera gigas* Oyster Extract

Toshikazu Yoshikawa, Yuji Naito, Yasunari Masui, Takaaki Fujii, Yoshio Boku, Norimasa Yoshida, and Motoharu Kondo

First Department of Medicine
Kyoto Prefectural University of Medicine
Kyoto 602-8566, Japan

INTRODUCTION

Oxygen radicals have been implicated as a mediator of tissue injury associated with ischemia or inflammation. Studies using animal models have established that some antioxidants, including superoxide dismutase (SOD; superoxide radical scavenger), dimethyl sulfoxide (DMSO; hydroxyl radical scavenger), and several synthetic oxygen radical scavengers, are effective in the treatment and prevention of these disease (1–4). Therefore, a new therapeutic approach using agents with antioxidant properties has been proposed (5). Natural food with free radical scavenging activities is one of the candidates for preventing these tissue injuries (6). Japan Clinic oyster extract (JCOE) is an amino acid powder extracted from *Crassostera gigas,* which has a high content of glutamic acid and taurine. JCOE, prepared from fresh raw oyster heated at 80°C for 1 h, contains a mixture that is shown in Table I. According to the results of tests made by the Japan Food Analysis Center, JCOE contains a variety of substances having antioxidant activity, including thiol-containing amino acids, especially taurine. This chapter reviews previous data (7)

Antioxidant Food Supplements in Human Health

461

Table I   Ingredients of *Crassostera gigas* Extract Powder (JCOE)

Ingredient	%
Protein	23.50
Amino acid	
Glutamic acid	2.78
Proline	1.33
Alanine	1.23
Aspartic acid	1.20
Glycine	1.10
Lysine	0.54
Arginine	0.53
Threonine	0.51
Leucine	0.45
Serine	0.42
Valine	0.37
Phenylalanine	0.30
Isoleucine	0.30
Histidine	0.28
Taurine	5.11
Sugar	58.40
Lipid	0.20

on the free radical scavenging activity of JCOE and the protective effect of JCOE against gastric mucosal cell injury induced by reactive oxygen species and presents findings about the mechanism of cytoprotective effect by JCOE.

SUPEROXIDE RADICAL SCAVENGING ACTIVITY OF JCOE

Superoxide scavenging activity was measured by the electron paramagnetic resonance (EPR) spin-trapping method. Following a previous report (8), the superoxide generated by the hypoxanthine–xanthine oxidase enzyme system was trapped by 5,5-dimethyl-1-pyrroline-N-oxide (DMPO), and the scavenging activity of the sample was measured by comparing the inhibition rate of the resulting DMPO-OOH signal intensity caused by the sample to the inhibition rate with a standard superoxide dismutase. For actual measurements, 0.5 mM hypoxanthine, 0.1 mM diethylenetriaminepentaacetic acid (DETAPAC), 0.1 M DMPO, and the sample or superoxide dismutase was added to a 20 m*M* phosphate-buffered solution (pH 7.8). The reaction was started by the addition of xanthine oxidase. An aliquot of reaction mixture was then transferred into a quartz cell, and the EPR signal was measured 1 min after starting. The EPR instrument was a JEOL-JES-FR80 spectrometer (JEOL, Tokyo), and measurements were carried out under the following conditions: magnetic field, 335.8 ± 5 mT; microwave power, 8.0 mW; frequency, 9.420 GHz; modulation frequency, 100 kHz; modulation amplitude, 0.1 mT; sweep time, 5 mT/min; response time, 0.1 sec; receiver gain, × 500; and temperature, 18°C. When xanthine oxidase was added to the complete system containing hypoxanthine and DMPO in phosphate buffer, the EPR signal was detected by an EPR spectrometer. The 12 characteristic lines of the signal were observed 1 min after the addition of xanthine oxidase. The g value and hyperfine coupling constant were g = 2.006 G, a_N = 1.41 mT, $a_H(\beta)$ = 1.13 mT, and $a_H(\gamma)$ = 0.12 mT, which could be assigned to a DMPO-OOH adduct (spin adduct trapping superoxide). At final concentrations of JCOE of 0.1, 1.0, and 10 mg/ml, a marked dose-dependent scavenging activity was observed (Fig. 1). The addition of glutamate or taurine to the system did not affect the ERR signal intensity of DMPO-OOH. These findings indicate that the superoxide radical scavenging activity of JCOE is not attributable solely to the glutamate or taurine contained in JCOE.

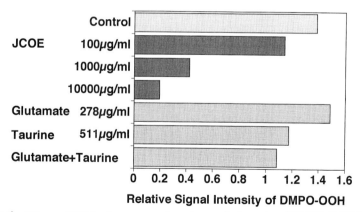

Figure 1   Effects of JCOE, glutamate, and taurine on the intensity of DMPO-OOH EPR signal generated from the hypoxanthine–xanthine oxidase system in the presence of DMPO. The intensity of the DMPO-OOH EPR signal was measured as a ratio to the intensity of Mn^{2+} used as an internal standard.

HYDROXYL RADICAL SCAVENGING ACTIVITY OF JCOE

The hydroxyl radical is known as a highly reactive oxygen species that reacts rapidly with biological materials causing oxidative damage. The EPR spin-trapping method using DMPO is a common method for the detection of hydroxyl radical. The Fenton system, containing DETAPAC, $FeSO_4$, and hydrogen peroxide (H_2O_2), was used as a hydroxyl radical generating system. As previously reported (9), 50 μM $FeSO_4$, 0.125 mM DETAPAC, the test sample, and 1.0 mM DMPO were added to a 50 mM phosphate-buffered solution (pH 7.8). The reaction was started by adding 1 mM H_2O_2. An aliquot of the reaction mixture was then transferred into a quartz cell, and the EPR signal was measured 1 min after the start. Measurements were carried out under the following conditions: magnetic field, 335.8 $\pm$ 5.0 mT; microwave power, 6.0 mW; frequency, 9.420 GHz; modulation frequency, 100 kHz; modulation amplitude, 0.1 mT; sweep time, 10 mT/min; response time, 0.1 sec; receiver gain, $\times$ 200; temperature 18°C. From the Fenton reaction mixture with DMPO, a four-line, 1 : 2 : 2 : 1, spectrum was detected with the EPR spectrometer. The hyperfine coupling constants were a_N = 1.49 mT and a_H = 1.49 mT, which could be assigned to a DMPO-OH adduct (spin adduct trapping hydroxyl radical). JCOE inhibited the DMPO-OH signal intensity in a dose-dependent manner (Fig. 2). The DMPO-OH EPR signal was decreased slightly by glutamic acid at a concentration of 278 μg/ml as well as by taurine at a concentration of 511 μg/ml. These findings indicate that the hydroxyl radical scavenging activity of JCOE is not attributable solely to the glutamate or taurine contained in JCOE. From previous data (10), the aromatic amino acid, sulfur-containing amino acid, and histidine have a high reactivity to the hydroxyl radical. Further studies will be necessary to establish which compounds in JCOE are responsible for the scavenging effect on these radicals.

REACTIVE OXYGEN SPECIES-INDUCED CELLULAR INJURY

The authors study demonstrated that H_2O_2 results in a decrease in cell viability of RGM-1 cells in a concentration-dependent manner, which is the same finding of previous studies using other cells (11–13). In their *in vitro* studies of oxygen radical-induced injury of vascular endothelial cells, colonic carcinoma cells, and gastric cells. Ager and Gordon (11), Watson *et al.* (12), and Hiraishi *et al.* (13) concluded that H_2O_2 must be the oxidant

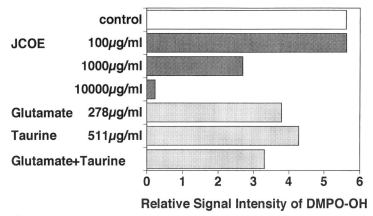

Figure 2 Effects of JCOE, glutamate, and taurine on the intensity of DMPO-OH EPR signal generated from the hydrogen peroxide–ferrous sulfate system in the presence of DMPO. The intensity of the DMPO-OH EPR signal was measured as a ratio to the intensity of Mn^{2+} used as an internal standard.

species primarily responsible for oxidative cellular injury. In these cellular injuries, H_2O_2 not only damages cells directly but can also act as a substrate for the generation of the highly toxic hydroxyl radical ($OH^·$) via the Fenton reaction (14). H_2O_2 formation in proximity to chloride ions, abundant in gastrointestinal secretions, and certain peroxidases, especially myeloperoxidase (MPO) secreted into the extracellular medium by activated leukocytes, results in the formation of hypochlorous acid (HOCl). Therefore, the effect of JCOE on H_2O_2- or HOCl-induced cellular injury was investigated further.

The rat gastric mucosal cell line RGM-1 (RCB-0876 at Riken Cell Bank, Tsukuba) (15, 16), established by Matui and Ohno, was used. RGM-1 cells were grown in a 1:1 mixture of Dulbecco's modified Eagle's medium (DMEM) and Ham's F12 medium supplemented with 20% fetal bovine serum, 2 mM glutamine, 100 units/ml penicillin, 100 units/ml streptomycin, and 0.25 μg/ml amphotericin. Cultures were maintained in a humidified atmosphere of 95% air and 5% CO_2 at 37°C. Cells were seeded onto 96-well plates at a density of 5×10^4/ml and cultured for 48 hr. At the end of a 1- or 24-h incubation with JCOE solution or vehicle, the incubation medium was removed from the wells. After exposure to 200 μM H_2O_2 or 100 μM HOCl for 4 hrs in Hank's balanced buffered solution, cell viability was measured by the modified MTT assay (WST-1 assay). Pretreatment with

JCOE (1000 μg/ml) solution alone for 1 or 24 hr did not affect cell viability. A H_2O_2- or HOCl-induced decrease in cell viability was significantly reversed by a 24-hr treatment with JCOE solution in a concentration-dependent manner, but not by a 1-hr treatment with the solution (Table II). Treatment with glutamic acid (27.8 μg/ml), taurine (51.1 μg/ml), or glutamic acid plus taurine for 24 hr did not reverse cellular injury in RGM-1 cells. The generation of hydroxyl radicals is supported by evidence that the Fe^{2+} chelator or hydroxyl radical scavenger provides a significant degree of protection against H_2O_2-induced injury (12, 17). The present study, however, has shown that H_2O_2-induced injury was not reversed by a 1-hr preincubation with 100–1000 μg/ml JCOE solution, which has high reactivity to hydroxyl radicals. This suggests the possibility that the active large molecules of JCOE, which have the scavenging action of the hydroxyl radical, did not penetrate cell membranes easily.

In contrast, the 24-hr preincubation with JCOE resulted in significant protection against H_2O_2-induced injury in a concentration-dependent manner. Although the precise mechanism for protection is not obvious, suggestive data have been reported by Tapiero and Tew (18). They demonstrated that statistically significant increases in glutathione occurred at 0.05 and 0.1% JCOE in HL60 cells (18). Reduced glutathione (GSH) has been shown to be an important cellular protectant against reactive oxygen species in cultured gastric mucous cells (19) as well as in endothelial cells

Table II   Effect of JCOE on Hydrogen Peroxide (H_2O_2)- and Hypochlorous Acid (HOCl)-Induced Injury in RGM-1 Cells

	Cell viability[a] (% to control)	
	200 μM H_2O_2	100 μM HOCl
Reactive oxygen species alone	13.5 ± 2.0	9.4 ± 1.9
+ JCOE (100 μg/ml, 1 hr)	6.4 ± 0.9	7.1 ± 2.3
+ JCOE (500 μg/ml, 1 hr)	6.5 ± 0.5	8.7 ± 2.1
+ JCOE (1000 μg/ml, 1 hr)	6.5 ± 0.8	5.2 ± 1.0
+ JCOE (100 μg/ml, 24 hr)	22.5 ± 2.6	8.9 ± 1.6
+ JCOE (500 μg/ml, 24 hr)	71.1 ± 5.4^b	36.2 ± 7.0^b
+ JCOE (1000 μg/ml, 24 hr)	88.4 ± 5.4^b	40.6 ± 2.9^b

[a] Cell viability was measured by the WST-1 test. Cells were incubated for 1 or 24 hr with 100–1000 μg/ml JCOE in culture medium and then incubated for 4 hr in HBSS plus 200 μM H_2O_2 or 100 μM HOCl. Each value indicates the mean 6 SE.
[b] $p < 0.01$ when compared with the value of H_2O_2 or HOCl alone.

(20) and hepatocytes (21). Glutathione peroxidase utilizes GSH as a reductant to reduce toxic peroxidase and GSH acts as a direct free radical scavenger in intracellular space. Depletion of GSH by electrophilic compounds such as diethyl maleate or by inhibition of γ-glutamylcysteine synthetase with buthionine sulfoximine (BSO) potentiates injury from H_2O_2 in cultured gastric mucosal cells (19). BSO has been known to inhibit GSH synthesis by irreversibly inhibiting γ-glutamylcysteine synthetase, a rate-limiting enzyme of GSH biosynthesis. The role of intracellular GSH in the cytoprotective effect of JCOE against H_2O_2-induced cellular injury using RGM-1 cells was also studied (Fig. 3). Treatment with JCOE at a dose of 1000 μg/ml for 24 hr increased the cellular GSH content. The presence of BSO (100 μM) during the preincubation period with JCOE prevented protection against H_2O_2, suggesting that protection afforded by preincubation with JCOE was mediated by γ-glutamylcysteine synthetase (Fig. 4). Further studies will be needed to examine whether such a mechanism of increasing the content of cellular GSH is present in cells treated with JCOE.

CONCLUSION

In summary, studies indicate that JCOE directly scavenges superoxide and hydroxyl radicals and that a 24-hr preincubation with JCOE significantly protects RGM-1 cells against H_2O_2- or HOCl-induced injury. Its

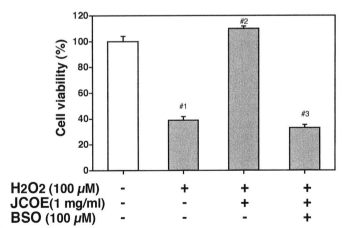

Figure 3   Role of cellular glutathione in the protective effect of JCOE against H_2O_2-induced injury in RGM-1 cells. Each value indicates the mean + SE. [#1]$p < 0.01$ vs control, [#2]$p < 0.01$ vs H_2O_2 alone, [#3]$p < 0.01$ vs H_2O_2 + JCOE.

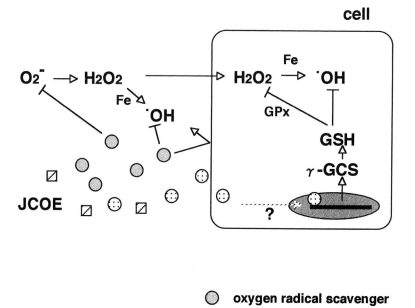

Figure 4   Antioxidative effects of *Crassostera gigas* extract powder (JCOE).

protective effect is not attributed to its direct free radical scavenging activity, but is mediated by accelerating intracellular GSH synthesis.

REFERENCES

1. Yoshikawa, T., Murakami, M., Seto, O., Kakimi, Y., Takemura, T., Tanigawa, T., Sugino, S., and Kondo, M. (1986). *J. Clin. Biochem. Nutr.* **1**, 165–170.
2. Yoshikawa, T., Ueda, S., Naito, Y., Takahashi, S., Oyamada, H., Morita, Y., Yoneta, T., and Kondo, M. (1989). *Free Radic. Res. Commun.* **7**, 285–291.
3. Yoshikawa, T., Naito, Y., Tanigawa, T., Yoneta, T., Oyamada, H., Ueda, S., Takemura, T., Sugino, S., and Kondo, M. (1989). *J. Clin. Biochem. Nutr.* **7**, 107–113.
4. Yoshikawa, T., Yasuda, M., Ueda, S., Naito, Y., Tanigawa, T., Oyamada, H., and Kondo, M. (1991). *Am. J. Clin. Nutr.* **53**, 210S–214S.
5. Yoshikawa, T., Naito, Y., and Kondo, M. (1993). *J. Nutr. Sci. Vitaminol.* **39**, S35–41.
6. Yoshikawa, T., Takahashi, S., Ichikawa, H., Takano, H., Tasaki, N., Yasuda, M., Naito, Y., Tanigawa, T., and Kondo, M. (1991). *J. Clin. Biochem. Nutr.* **10**, 189–196.
7. Yoshikawa, T., Naito, Y., Masui, Y., Fujii, T., Boku, Y., Nakagawa, S., Yoshida, N., and Kondo, M. (1997). *Biomed. Pharmacother.* **51**, 328–332.

8. Miyagawa, H., Yoshikawa, T., Tanigawa, T., Yoshida, N., Sugino, S., and Kondo, M. (1988). *J. Clin. Biochem. Nutr.* **5,** 1–7.
9. Tanigawa, T. (1990). *J. Kyoto Pref. Univ. Med.* **99,** 133–143.
10. Yoshikawa, T., Takahashi, S., Tanigawa, T., Naito, Y., Ichikawa, H., Takano, H., and Kondo, M. (1991). *J. Clin. Biochem. Nutr.* **11,** 161–169.
11. Ager, A., and Gordon, J. L. (1984). *J. Exp. Med.* **159,** 592–603.
12. Watson, A. M. J., Askew, J. N., and Sandle, G. I. (1994). *Gut* **35,** 1575–1581.
13. Hiraishi, H., Terano, A., Ota, S., Ivey, J. K., and Sugimoto, T. (1987). *Am. J. Physiol.* **253,** G40–G48.
14. Halliwell, B., and Gutteridge, M. C. (1990). *In* "Methods in Enzymology" (L. Packer and A. N. Glazer, eds.), Vol. 186, pp. 1–85. Academic Press, San Diego.
15. Kobayashi, I., Takei, Y., Kawano, S., Nagano, K., Tsuji, S., Fusamoto, H., Matsui, A., Ohno, T., Nakama, H., Fukutomi, H., Sawaoka, H., Kamada, T., and Masuda, E. (1996). *In Vitro Cell. Dev. Biol. Anim.* **32,** 259–261.
16. Hassan, S., Kinoshita, Y., Min, D., Nakata, H., Kishi, K., Matsushima, Y., Asahara, M., He-yao, W., Okada, A., Maekawa, T., Matsui, H., and Chiba, T. (1996). *Digestion* **57,** 196–200.
17. Hiraishi, H., Terano, A., Razandi, M., Sugimoto, T., Harada, T., and Ivey, K. J. (1993). *Gastroenterology* **104,** 780–788.
18. Tapiero, H., and Tew, K. D. (1996). *Biomed. Pharmacother.* **50,** 149–153.
19. Hiraishi, H., Terano, A., Ota, S., H., M., Sugimoto, T., Razandi, M., and Ivey, K. (1991). *Am. J. Physiol.* **261,** G921–G928.
20. Harlan, J. M., Levine, J. D., Callahan, K. S., and Schwarz, B. R. (1984). *J. Clin. Invest.* **73,** 706–713.
21. Starke, P. E., and Farber, J. L. (1985). *J. Biol. Chem.* **260,** 86–92.

31 Superoxide Anion Radical Scavenging Activity of Uyaku (*Lindera strychnifolia*), a Natural Extract Used in Traditional Medicine

Yasuko Noda,* Akitane Mori,* Kazunori Anzai,† and Lester Packer*

*Department of Molecular and Cell Biology
University of California
Berkeley, California 94720
†National Institute of Radiological Sciences
Chiba, Japan

Uyaku (*Lindera strychnifolia*: Sieb. et Zucc.) is well known as a traditional Asian medicine used in the treatment of stomach and renal diseases, neuralgia, rheumatism, and aging. However, the mechanism of its effectivity is not yet known. In this study, scavenging activities of superoxide anion and hydroxyl radicals were examined by electron spin resonance (ESR) using the spin-trapping reagent 5,5'-dimethyl-1-pyrroline-N-oxide (DMPO). Water-soluble components in both Uyaku root and leaf extracts showed potent superoxide anion scavenging activity. Uyaku seed extract showed weak superoxide anion scavenging activity. In all samples of Uyaku, scavenging activity was not diminished by ascorbate oxidase treatment. The filtrate, passed through a low molecular weight cutoff microcentrifuge filter (MW < 100,000 or <10,000), showed approximately half the activity of untreated extract samples of Uyaku roots and leaf extracts, indicating that approximately half of the scavenging activity is due to the low molecular weight components. To evaluate the hydroxyl radical scavenging activity, L-ascorbic acid 2-[3,4-dihydro-2,5,7,8-tetramethyl-2-(4,8,12-trimethyltridecyl)-2*H*-1-benzopyran-6-y1-hydrogen phosphate] potassium salt (EPC-K$_1$), which has vitamin E and vitamin C moieties and a high potency for hydroxyl radical scavenging, was used as a standard material. Results showed little hydroxyl radical scavenging activity in root and leaf extracts, although weak scavenging activity was demonstrated in the seed extract. These results demonstrated that Uyaku roots and leaves have specific and potent superoxide anion radical scavenging activity in comparison with several other natural extracts. Thus, Uyaku may have potential for beneficial effects for health, especially for diseases in which the production of superoxide anion radical is a factor.

INTRODUCTION

Uyaku (in Japanese, Wu Yao in Chinese, Oyak in Korean) is the dried roots of *L. strychnifolia* (Sieb. et Zucc.) F. Villar of the Lauraceae family (1–3) and a traditional Asian medicine used in China as an astringent, carminative, stomachic or tonic, for asthma, cholera, congestion, dyspepsia, dysmenorrhea, fluxes, gonorrhea, hernia, malaria, menorrhagia, stomach ache, stroke, and urinary difficulties (4).

According to a traditional story, the first Emperor of Qin Dynasty, about 2300 years ago, ordered Jo Fuku (Xu Fu in Chinese) to be brought to Horai, an island on the far east sea, looking for a drug effective for longevity. After a long sailing, he and his colleagues arrived at Shingu in Wakayama prefecture, Japan, in B.C. 219 and established "Tendai-uyaku" (another name for Uyaku, used especially in this district) (5–7). Uyaku might have been brought from China from 1716–1735, according to Makino (1). Uyaku has been used as a folk drug for good health and for the treatment of diseases such as stomach and renal diseases, neuralgia, and rheumatism in some districts, including Shingu, Wakayama prefecture, Japan (8, 9), although no scientific background for these treatments has been known until now. This study examines the effects of Uyaku (root, leaf, and seed extracts for their action) on reactive oxygen species using ESR spectrometry.

MATERIALS AND METHODS

Chemicals

Common chemicals were from Sigma Chemical Co. (St. Louis, MO). Xanthine oxidase (XOD) was from Boehringer Mannheim Corp. (Indianapolis, IN). The spin trap DMPO was from Labotec (Tokyo, Japan). All chemicals were the highest grade available.

Standard Solutions

SOD standard solutions were obtained as a kit (Dojindo Laboratories, Kumamoto, Japan). For measurements of hydroxyl radical scavenging activity, L-ascorbic acid 2-[3,4-dihydro-2,5,7,8-tetramethyl-2-(4,8,12-trimethyltridecyl)-2H-1-benzopyran-6-yl-hydrogen phosphate] potassium salt (EPC-K$_1$) (Senju Pharmaceutical Co. Ltd., Osaka, Japan) (8) was used as a standard. EPC-K$_1$, which is composed of vitamin E and vitamin C by phosphate diester linkage, is water soluble and the solution is stable at room temperature. In this study, a stock solution of 10 mM EPC-K$_1$ dissolved in 0.1 M phosphate buffer (pH 7.4) was prepared and diluted to obtain the appropriate concentrations (0–2 mM) before use. The stock solution was stored at 4°C.

Samples

Sources

Ground powder of Uyaku roots was kindly donated by Taikodo Pharmaceutical Co., Ltd. (Kobe, Japan). Dried leaves and seeds of Uyaku were collected in Wakayama, dried at room temperature for 1 month (leaves) or in a desiccator at 4°C (seeds), and ground with a mortar and pestle. To obtain the extract, the powder (5 g of Uyaku roots, 1.2 g of Uyaku leaves, or 10 g of Uyaku seeds) was added to 20 times the volume of water (1:20, by w/v), and the sample was extracted with boiling water for 10 min. After cooling to room temperature, the suspension was filtered (Whatman No. 2) and the recovered filtrate was freeze-dried. The sample was kept dry and at 4°C until use (Fig. 1).

Treatments

Freshly prepared sample solutions were used for assay. The extracted material was dissolved in 0.1 M phosphate buffer (pH 7.4) at concentrations from 0.1 to 0.5 mg/ml by vortex mixing (nontreatment). An aliquot of the sample was treated with ascorbate oxidase (20 units/ml of sample solution). The ascorbate oxidase stock solution dissolved in 0.1 M phosphate buffer, pH 7.4, was dialyzed at 4°C to eliminate possible contamination by sucrose. Another aliquot was passed through a centrifuge filter (Ultrafree-MC Filters: 10,000 NMWL regenerated cellulose membrane; 100,000 NMWL polysulfone membrane, Millipore, MA) by centrifugation (5000 g,

Procedure for Preparation of Uyaku Extracts

Dried Uyaku roots, leaves or seeds sample (1-10 g)
↓
The sample was ground with a mortar and pestle
↓
The powder was weighed
↓
Powder was extracted with boiling water (1:20, by $\overline{w}/\overline{v}$)
for 10 minutes at 100°C
↓
Extract was cooled to room temperature and filtrated using
Whatman No. 2 filter paper
↓
Preparation was freeze-dried
↓
The sample was stored at 4°C in a silica-gel desiccator until use

Figure 1   Procedure for preparation of Uyaku root, leaf, or seed extracts.

4°C) to obtain materials of smaller molecular weights: less than 10,000 or 100,000.

ESR Measurements

ESR Spectrometer

A new computerized, compact, highly sensitive ESR spectrometer (free radical monitor, JES FR-30, JEOL, Tokyo, Japan) was used. This ESR system has the ability of normalizing all spectra for accurate calculation using manganese dioxide as an internal standard. Other details have been described in a previous report (9). Conditions for measurements in this study were as follows: magnetic field, 335.5 ± 5 mT; microwave power, 4 mW; microwave frequency, 9.41 GHz; modulation width, 0.063 or 0.1 mT; sweep width, 5.0 mT; response time, 0.1 sec; amplitude, 1×200; and sweep time, 2 min. ESR spectra were measured at 23°C. Scavenging activities were estimated using the relative peak height of ESR spectra of DMPO-OOH and DMPO-OH spin adduct for superoxide anion and hydroxyl radical scavenging activities, respectively.

Measurement of Superoxide Anion Radical Scavenging Activity

Fifty microliters of hypoxanthine oxidase (2.7 mg/5 ml) was first added to a test tube, followed by DMSO (30 μl), sample solution (50 μl), 4.5 M DMPO (20 μl), and XOD (0.4 units/ml) (50 μl). After vortex mixing for 10 sec at room temperature, the sample solution was transferred immediately into a quartz flat cell. ESR recording was started exactly 30 sec after the addition of XOD.

Measurement of Hydroxyl Radical Scavenging Activity

Fifty microliters of sample solution was added to a test tube followed by 0.18 M DMPO (50 μl), 2 mM H_2O_2 (50 μl), and 0.2 mM $FeSO_4$ (50 μl). They were mixed by a vortex mixer for 10 sec. Exactly 30 sec after the addition of $FeSO_4$ solution, ESR recording was started.

Statistics

All data were expressed as means $\pm$ SEM. Statistical analysis was performed using the Student's t test.

RESULTS

Extraction of Materials

Samples were extracted from ground Uyaku roots, dried Uyaku leaves, or dried Uyaku seeds using boiling water for 10 min, and the filtrates were freeze-dried overnight. The yield of the dried extracted materials was approximately 5, 12, and 5% for roots, leaves, and seeds, respectively.

Superoxide Anion Radical Scavenging Activity (SOD-like activity)

Extracted materials were dissolved at concentrations of 0.1 to 0.5 mg/ml for untreated samples (control). Uyaku roots showed markedly potent scavenging activity (115.3 $\pm$ 5.4 SOD equivalent units/mg of freeze-dried extract, $n = 10$). Uyaku leaf extracts also showed moderately potent scavenging activity (88.5 $\pm$ 4.8 SOD equivalent units/mg of freeze-dried extract, $n = 5$), but the scavenging activity of Uyaku seed extracts was much less (16.2 $\pm$ 0.2 SOD equivalent units/mg of freeze-dried extract, $n = 4$). These superoxide scavenging activities were not changed by the treatment with ascorbate oxidase in all Uyaku extracts.

Filtered samples from Uyaku root and leaf extracts, which contain low molecular weight materials (less than 100,000 or 10,000) showed lower (roots: 56 or 48%, leaves: 57 or 51%) SOD-like activity when compared to untreated samples. The SOD-like activity of Uyaku seed extracts, was not changed by filtration (Fig. 2).

Extracted samples from Uyaku roots, leaves, and seeds were dissolved at a concentration of 0.25 mg/ml, and hydroxyl radical scavenging activity

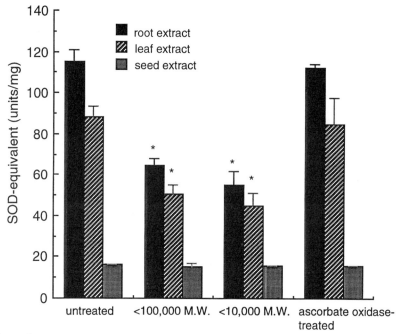

Figure 2   Superoxide anion radical scavenging activity of Uyaku root, leaf, and seed extracts. Values are untreated (roots: $n = 10$; leaves: $n = 5$; seeds: $n = 4$), filtrates [MW < 100,000 (roots: $n = 4$; leaves: $n = 4$; seeds: $n = 3$) and MW < 10,000 (roots: $n = 4$; leaves: $n = 4$; seeds: $n = 3$)], and ascorbate oxidase-treated (roots: $n = 3$; leaves: $n = 4$; seeds: $n = 4$) samples. *$p < 0.05$, compared to untreated samples. Data expressed as means ± SEM.

was evaluated in comparison with EPC-K$_1$. Weak scavenging activities for hydroxyl radical were observed in Uyaku roots, in leaves, and in seeds (Fig. 3).

DISCUSSION

Uyaku roots and leaves showed markedly higher SOD-like activities in comparison with several other natural extracts. A higher scavenging activity for free radicals in natural source antioxidants is sometimes known to depend on ascorbate, which may be a component of this material (9). Compared with other known natural source materials under the same conditions, Uyaku roots, especially ascorbate oxidase-treated samples, belong to the highest class of SOD-like activity, i.e., almost the same activity of green

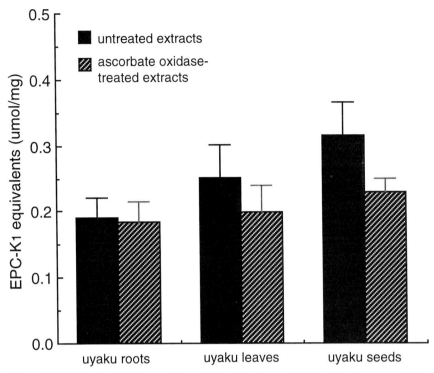

Figure 3 Hydroxyl radical scavenging activity of Uyaku root, leaf, and seed extracts. Values are untreated (roots: $n = 5$; leaves: $n = 5$; seeds: $n = 6$) and ascorbate oxidase-treated (roots: $n = 4$; leaves: $n = 4$; seeds: $n = 4$) samples. Data expressed as means $\pm$ SEM.

tea (Matsucha, Kyoto, Japan) (untreated: 189.7 ± 16.6, $n = 4$; ascorbate oxidase treated: 134.9 ± 26.4, $n = 2$) and pine bark extracts (Pycnogenol) (untreated: 115.1 ± 5.5, $n = 6$; ascorbate oxidase treated: 111.0 ± 8.5, $n = 5$) (9) (Fig. 4). However, the contents of effective components in herbs or natural source materials are known to vary depending on the production process, season of harvest, or storage conditions. Therefore, such comparative evaluation should be performed more carefully in the future. SOD-like activities in Uyaku samples are not dependent on the scavenging activity of ascorbate, as these activities were not changed by treatment with ascorbate oxidase.

Filtered samples from Uyaku roots and leaves, which contain low molecular weight materials (less than 10,000 or 100,000), showed approximately half of the SOD-like activity compared with untreated samples. These two filtrates had similar SOD-like activity, indicating that approxi-

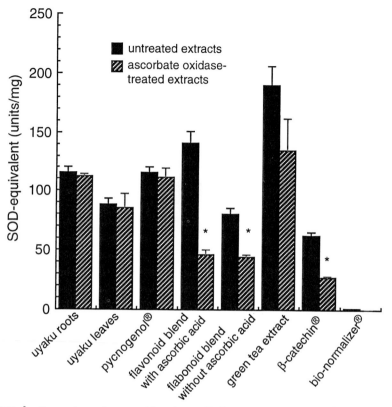

Figure 4 Comparison of superoxide anion radical scavenging activities of root, leaf, and other natural extracts. Values are untreated (roots: $n = 10$; leaves: $n = 5$; all other natural extracts: $n = 6$) and ascorbate oxidase-treated (roots: $n = 3$; leaves: $n = 4$; all other natural extracts: $n = 5$) samples. *$p < 0.05$, compared to untreated samples. Data expressed as means ± SEM.

mately half of the scavenging activity is probably due to the lower molecular weight materials. Another half of the SOD-like activity observed in untreated samples would be higher molecular weight proteins (MW >100,000), which may also contribute to SOD-like activity. No significant differences in SOD-like activities were observed between the fractions containing molecules of less than 10,000 and 100,000 in all samples of Uyaku. This finding suggests that SOD-like activities in Uyaku extracts were not dependent on the enzyme SOD itself, as the molecular weights of the SODs are known to be between 10,000 and 100,000 [Cu-SOD: 32,000; Mn-SOD: 42,000–45,000 or 85,000–90,000 depending on the monomer or dimer

(10–13)]. Because the SOD-like activity in Uyaku seed extracts was not changed by filtration, this suggests that all SOD-like activity may be dependent on lower molecular materials.

Many chemical constituents of Uyaku roots are known. Examples of substances reported to be present are borneol, linderane, linderalactone, isolinderalactone, neolinderalactone, linderstrenolide, linderene, lindenenone, lindestrene, linderene acetate, isolinderoxide, linderaic acid, linderazulene, chamazulene, laurolitsine, and isogermafurene (2, 3). The precise components responsible for SOD-like activity of Uyaku roots are presently unknown.

In conclusion, Uyaku roots have strong superoxide anion radical scavenging activity. SOD-like activity suggests that Uyaku extracts, both from roots and leaves, may have beneficial effects for health, especially for prevention or treatment of free radical-related diseases.

REFERENCES

1. Makino, T. (1989). *In* "Makino's New Illustrated Flora of Japan" (revised edition), p. 126. Hokuryukan, Tokyo. [in Japanese]
2. Hsu, H.-Y., Chen, Y.-P., Shen, S,-J., Hsu, C.-S., Chen, C.-C., and Chang, H.-C. (1986). *In* "Oriental Materia Medica: A Concise Guide," p. 420. Oriental Healing Arts Institute, Long Beach, CA.
3. Bensky, D., and Gamble, A. (1986). *In* "Chinese Herbal Medicine." p. 341. Eastland Press, Seattle.
4. Duke, J. A., and Ayensu, E. S. (1985). *In* "Medical Plants of China," Vol. 2, p. 390. Reference Publications, Algonac, Michigan.
5. Maekawa, M. (1926). "Jo Fuku." Shingu Kanko Kyokai, Shingu. [in Japanese]
6. Tsuji, S. (1926). *In* "Jo Fuku Densetsu," Kanko Kyokai, Shingu. [in Japanese]
7. Okuno, T. (1991). *In* "Roman no Hito: Jo Fuku." Gakken Okuno Tosho, Shingu. [in Japanese]
8. Mori, A., Edamatsu, R., Kohno, M., and Ohmori, S. (1989). A new hydroxyl radical scavenger: EPC-K$_1$. *Neurosciences* **15**, 371.
9. Noda, Y., Anzai, K., Mori, A., Kohno, M., Shimmei, M., and Packer, L. (1997). *Biochem. Mol. Biol. Int.* **42**, 35.
10. Steinman, H. M., Naik, V. R., Abernathy, J. L., and Hill, R. L. (1974). Bovine erythrocyte superoxide dismutase: Complete amino acid sequence. *J. Biol. Chem.* **249**, 7326.
11. Barra, D., Martini, F., Bannister, J. V., Schinia, M. E., Rotilio, G., Barrister, W. H., and Bossa, F. (1980). The complete amino acid sequence of human Cu/Zn superoxide dismutase. *FEBS Lett.* **120**, 53.
12. Keele, B. B., McCord, J. M., and Fridovich, I. (1970). Superoxide dismutase from *Escherichia coli B*: A new manganese-containing enzyme. *J. Biol. Chem.* **245**, 6176.
13. Salin, M. L., Day, E. D., and Crapo, J. D. (1978). Isolation and characterization of manganese-containing superoxide dismutase from rat liver. *Arc. Biochem. Biophys.* **187**, 223.

32 Biological Effects of the Fermentation Product of *Carica papaya* (Bio-Normalizer)

Hirotsugu Kobuchi, * *Nobuya Haramaki,*[†]
Lucia Marcocci,[†] *and Lester Packer*[†]

*Environmental Energy Technologies Division
University of California
Berkeley, California 94720
[†] Membrane Bioenergetics Group
Department of Molecular and Cell Biology
University of California
Berkeley, California 94720

INTRODUCTION

Bio-Normalizer (BN) is a functional health food supplement sold in Japan, the United States, and in other countries. This product is made from *Carica papaya* Linn. and other plants (*Pennisetum pupureum* Schum., *Sechium edule* Swartz) by yeast fermentation under strict quality control. Although ingredients of the product are still unknown in detail, carbohydrate (90%), protein, amino acids, and vitamins have been detected as main substances by chemical analysis (unpublished data). Bio-Normalizer has been proposed as a free radical modulating agent (Osato *et al.*, 1995; Santiago *et al.*, 1991). In biological studies, Bio-Normalizer has been found to scavenge hydroxyl radicals *in vitro* (Santiago *et al.*, 1991), and in animal studies it has been reported to protect the rat brain against oxidative damage caused by aging (Santiago *et al.*, 1993a), iron treatment (Santiago *et al.*, 1992), or ischemia–reperfusion (Santiago *et al.*, 1993b). From such reports, it has been proposed that beneficial effects of Bio-Normalizer might be due to its free radical scavenging properties. However, Bio-Normalizer has also been reported to upregulate phorbol ester-induced and zymosan-induced superoxide production in rat peritoneal macrophages (Osato *et al.*, 1995), natural killer cell activity (Okuda *et al.*, 1993), and the level of interferon (IFN-γ in human blood (Santiago *et al.*, 1994). Such evidence suggests that BN also possesses the ability to modulate immune effector cells in addition to its direct free radical scavenging activity. Despite accumulating data on the beneficial effects of Bio-Normalizer, the biological mechanisms responsible for the therapeutic activity of Bio-Normalizer are not well understood. The authors have evaluated its activity under various conditions in order to gain further insight into the biological mechanisms of Bio-Normalizer action and new possibilities for therapeutic applications.

ANTIOXIDANT PROPERTY OF BIO-NORMALIZER

Free radicals and other reactive oxygen species are formed constantly in the human body. They play a crucial role in a variety of human physiological functions. However, excess generation of reactive oxygen species can often give rise to oxidative stress that results from the imbalance in the human antioxidant/oxidant status. It has now been recognized that prolonging this imbalance is implicated in a number of human diseases (Halliwell, 1993). In fact, there is increasing evidence that reactive oxygen and nitrogen species are involved in the pathogenesis of a diverse range of chronic and degenerative disorders (aging, atherosclerosis, cancer, cataract), as well as in acute clinical conditions (ischemia–reperfusion injury) (Halliwell *et al.*, 1989). Therefore, substances with antioxidant properties have been used as possible treatments of these disorders (Aruoma, 1994; Maxwell, 1995). To better define the antioxidant properties of Bio-Normalizer, BN was administered to rats for up to 6 weeks and the consequences of oral supplementation on *in vitro* models of oxidative stress-induced damage were then investigated.

Myocardiac Ischemia–Reperfusion Injury

It is known that reactive oxygen species such as superoxide radical, hydroxyl radical, and hydrogen peroxide are implicated in cardiac ischemia–reperfusion injury (Chan, 1996; Ferrari, 1995). Agents with free radical scavenging ability have been shown to enhance functional recovery of the heart exposed to ischemia–reperfusion (Janero, 1995; Singh *et al.*, 1995). To delineate the antioxidant ability of BN against reactive oxygen radical-mediated injury, the influence of BN on cardiac ischemia–reperfusion injury was investigated (Haramaki *et al.*, 1995). Male Sprague–Dawley rats (250–300 g) were fed a normal diet, supplemented with or without 0.1% BN (w/v) in drinking water. Six weeks later, the rats were subjected to 40 min of myocardial ischemia followed by 20 min of reperfusion using the Langendolff perfusion technique. Lactate dehydrogenase (LDH) leakage in the coronary effluent was monitored as an index of cardiac ischemia–reperfusion damage. During the reperfusion period, following 40 min of global ischemia, a high LDH leakage was observed in the effluent of the control rat heart, showing ischemia–reperfusion damage (Fig. 1). In contrast, BN-supplemented rat hearts showed a significantly lower level of LDH leakage (*p* < 0.01), suggesting that BN or its bioactive metabolites reached the rat heart through oral supplementation and prevented the hearts from ischemia–reperfusion injury.

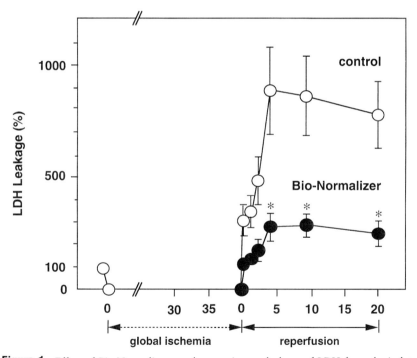

Figure 1 Effect of Bio-Normalizer supplementation on leakage of LDH from the isolated Langendolff rat during reperfusion. LDH leakage into the coronary effluent from control (○) or Bio-Normalizer-supplemented hearts (●) was measured. LDH leakage of the preischemic period was expressed as 100%. Values are expressed as mean ± SEM obtained in six different experiments. *p < 0.01 compared with values from control hearts.

Peroxyl Radical-Induced Oxidative Stress

To further define the protective action of BN in rat heart and other tissues, the effect of BN supplementation on the susceptibility of various rat tissues to oxidative stress induced by peroxyl radials was analyzed (Marcocci *et al.*, 1996). This radical, formed by the reaction of carbon-centered radicals with oxygen, has been known to be involved in the chain propagation step in lipid peroxidation (Halliwell and Gutteridge, 1989). The hydrophobic azo initiator [AMVN: 2,2′-azobis(2,4-dimethylvaleronitrile)] was used as a specific source of the peroxyl radicals generated at a constant rate on thermal decomposition of the azo initiator. The levels of thiobarbituric acid-reactive substance (TBARS) accumulation in homogenate samples were measured as an index of peroxyl radical-induced oxidative stress.

The level of TBARS in the absence of AMVN was very low in all homogenate samples (data not shown). In samples prepared from non-BN-supplemented animals, the exposure to AMVN resulted in higher levels of TBARS accumulation. In contrast, among the samples from BN-supplemented rats, significantly lower levels of TBARS were observed in kidney and heart (Fig. 2a). However, because the authors did not observe protection in either liver or brain homogenates, the uptake of BN to the tissue and its metabolism seems to be tissue specific. Moreover, oral supplementation of BN protected kidney homogenates from peroxyl radical-induced time-dependent accumulation of TBARS and depletion of α-tocopherol (Figs. 2b and 2c). These data further demonstrated BN to function as a free radical scavenger that protects the heart and cell membranes from free radical-induced oxidative damage. Thus one of the mechanisms in these protective effects of BN may be its antioxidant properties.

IMMUNOMODULATING ACTIVITY OF BIO-NORMALIZER

Nitric oxide (NO) is a versatile molecule that has many diverse biological functions. In addition to its function as a potent vasodilator and neu-

---

Figure 2   Effect of Bio-Normalizer supplementation on AMVN-induced TBARS accumulation in various rat organ homogenates. (A) TBARS accumulation in various rat organ homogenates. (B) Time-dependent accumulation of TBARS in rat kidney homogenates. (C) Time-dependent depletion of α-tocopherol in rat kidney homogenates. Rat tissues homogenates (10 mg protein/ml) prepared from control bars or from Bio-Normalizer-supplemented animals bars were incubated in PBS at 42°C for 1 h or at different periods in the absence or presence of 5 mM AMVN. Levels of TBARS and α-tocopherol were measured using 1,1,3,3-tetramethoxypropane (TMP) as standard or the HPLC electrochemical method, respectively, and given as mean $\pm$ SEM of six different samples.

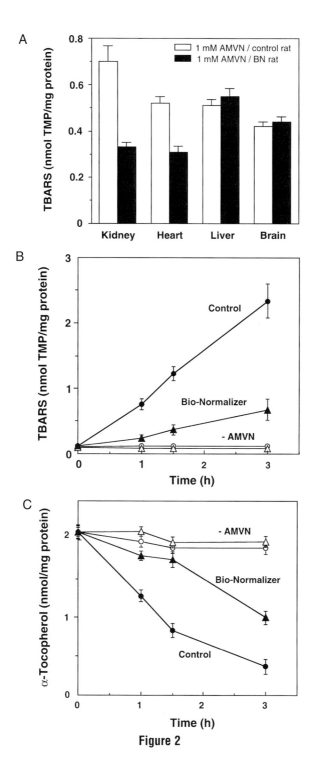

Figure 2

rotransmitter, it is well established that NO plays a crucial role in the immune system as a cytostatic and cytolytic agent against tumorigenic cells (Nathan, 1992) or pathogenic organisms, including parasites (Green *et al.*, 1990), viruses (Karupiah *et al.*, 1993, 1995), and fungi (Granger *et al.*, 1988). NO is synthesized by means of arginine oxidation by a family of nitric oxide synthases. Until now, three distinct isoforms of NO synthase (NOS, EC 1.14.13.39), neuronal NOS (nNOS), endothelial NOS (eNOS), and inducible NOS (iNOS), have been isolated and characterized extensively (Nathan *et al.*, 1994). Among these enzymes, eNOS contributes to the vasodilatation and adherence of platelets and leukocytes to endothelial cells; iNOS mainly participates in the host defense mechanism. Since the initial evidence of iNOS expression in murine macrophages was published, it has become apparent that iNOS is expressed by several cell types, including hepatocytes, endothelial cells, and smooth muscle cells. These cells have also been demonstrated to exert host defense activity by having iNOS activity in response to the stimulation of endotoxin and cytokines such as IFN-γ, interleukin-1β (IL-1β), and tumor necrosis factor-α (TNF-α) (Nathan, 1992).

Direct Interaction with Nitric Oxide

The authors demonstrated that BN scavenges oxygen radical species induced by either ischemia–reperfusion or a chemical system. In order to characterize the action of BN against NO radicals, the authors investigated whether BN interacted with NO directly. This evaluation was performed using the NO donor, sodium nitroprusside. This compound is known to decompose in aqueous solution at physiological pH to produce NO. Accumulated nitrite, which is the stable product of NO, was followed by reaction with the Griess reagent. The sodium nitroprusside solution was incubated with various amounts of BN (0–3 mg/ml) at 25°C for different time periods (0–120 min). BN had no effect on the level of nitrite accumulation in the reaction mixture (Fig. 3). This result suggests that BN itself does not react with NO in a chemical system.

Modulation of Nitric Oxide Production in Macrophages by Bio-Normalizer

BN has been used for a variety of pathological conditions, and now BN has been reported to have beneficial effects against different diseases, such as cancer, hepatitis, and bacterial infection. BN has been reported to induce natural killer cell activity (Okuda *et al.*, 1993) and to enhance the capacity of respiratory burst in neutrophils (Osato *et al.*, 1995). Such evidence encouraged us to evaluate BN as an immune modulator agent. The authors investigated the efficacy of BN to influence NO production in macrophages

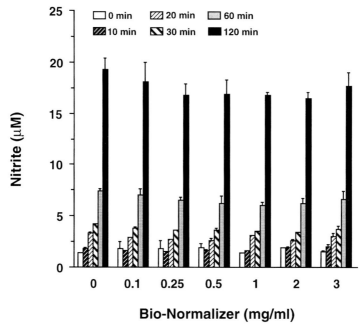

Figure 3 Effect of Bio-Normalizer on NO produced by sodium nitroprusside. Sodium ni-troprusside (5 mM) was incubated with various doses of BN as indicated in phosphate-buffered salin at 25°C. After incubation for different time periods, the aliquot of reactions was mixed with an equal volume of the Griess reagent and then the absorbance at 550 nm was measured. All values represent the mean ± SD of three independent experiments.

and characterized the mechanism by which BN affects cellular NO metabolism (Kobuchi and Packer, 1997). The mouse macrophage cell line RAW 264.7, which has been well characterized in NO studies and analyzed for the accumulation of nitrite/nitrate in culture medium using the Griess reagent as an index for NO synthesis, was used. Treatment with BN alone failed to induce an appreciable level of NO production. A major increase of NO production was observed when cells were treated with a combination of IFN-γ plus BN; the enhancement of NO production by BN was dose dependent (Fig. 4a). In addition, when macrophages were incubated with various concentrations of IFN-γ, a similar effect on NO production was observed (Fig. 4b). IFN-γ allows the production of appropriate levels of NO at concentrations above 5 U/ml. When 3 mg/ml BN was added into this test system, the IFN-γ-induced NO production in macrophages was further up-regulated by BN. These results suggest that BN itself does not provide a signal that triggers induction of the NO pathway; however, BN possesses the ability to enhance the production of NO from activated macrophages.

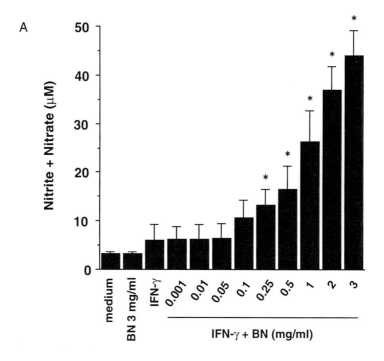

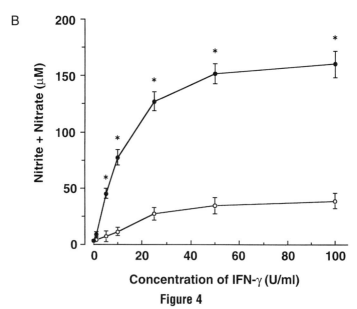

Figure 4

In light of the unique properties of BN to act both as a direct free radical scavenger and as an upregulator of NO production, it was of interest to determine how it regulates cellular NO metabolism. In general, iNOS is modulated at different regulatory sites, including transcriptional, posttranscriptional, translational, and posttranslational control (Nathan and Xie, 1994). To characterize the mechanism by which BN upregulates IFN-γ-induced NO production in macrophages, the authors tested the effect of BN on the different steps that regulate NO production.

iNOS Enzyme Activity The ability of BN to affect the enzyme activity of iNOS directly was investigated (Fig. 5). NOS activity was determined by monitoring the conversion of $[^{14}C]$arginine to citrulline using a cell-free preparation from activated macrophages as a source of iNOS. The direct modulatory action of BN on iNOS enzyme activity was not observed, even at high amounts of BN. These results suggest that the NO upregulation by BN is not because of its direct activation of iNOS enzyme activity.

iNOS mRNA Expression Macrophages did not express iNOS mRNA either constitutively or after treatment with BN alone. However, a low level of iNOS mRNA was observed following the stimulation of macrophages with IFN-γ alone. In contrast, treatment of macrophages with a combination of IFN-γ *a*nd BN resulted in a synergistic induction of iNOS mRNA, which was evident at a minimum concentration of 0.1 mg/ml BN (Fig. 6a). The augmentation was threefold greater than that induced by IFN-γ alone. These results were consistent with data on NO production, indicating that its synergistic interaction with IFN-γ to enhance NO production is due to the augmentation of iNOS mRNA expression. This iNOS gene upregulation effect of BN was also observed in a time-dependent manner (Fig. 6b). The IFN-γ-dependent induction in iNOS mRNA was not observed, even 2 h after stimulation. In contrast, the addition of 3 mg/ml BN caused a ma-

Figure 4 Bio-Normalizer synergizes with IFN-γ for NO production in RAW 264.7 macrophages. (A) Dose-dependent effect of Bio-Normalizer on NO production. Macrophages were incubated with medium alone, Bio-Normalizer (3 mg/ml) alone, IFN-γ (5 U/ml) alone, or IFN-γ in the presence of the indicated concentrations of Bio-Normalizer for 24 hr. All values represent the mean ± SD of three independent experiments. *$p < 0.05$ compared with that of IFN-γ alone treatment. (B) Efficacy of IFN-γ on NO production by Bio-Normalizer. Macrophages were incubated with the indicated concentrations of IFN-γ in either the absence (○) or the presence (●) of 3 mg/ml of Bio-Normalizer for 24 hr. All values represent the mean ± SD of three independent experiments. *$p < 0.05$ compared with that of IFN-γ alone treatment.

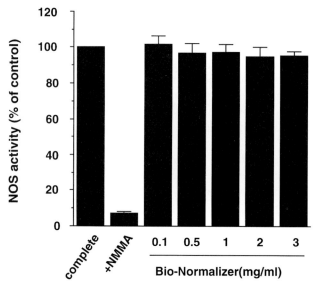

Figure 5   Bio-Normalizer does not affect iNOS enzyme activity. NOS activity was determined by the conversion of radiolabeled arginine to citrulline using a cytosolic preparation from macrophages. Complete indicates the value of iNOS activity in the reaction mixture that contains appropriate cofactors as a positive control. N^G-monomethylarginine (NMMA) was present in the reaction mixture at 100 μM. Bio-Normalizer was added to the complete assay mixture as indicated. All values are expressed as a percentage of the control (100%: 178 ± 3 pmol/mg protein/min) and represent the mean ± SD of three independent experiments.

jor increase of IFN-γ-induced iNOS mRNA expression that was already detectable after 2 h of treatment. After 6 h of incubation, BN caused about a threefold augmentation of iNOS mRNA expression over that induced by IFN-γ alone, similar to that observed in Fig. 6a. These data demonstrate that BN acted synergistically with IFN-γ in a dose- and time-dependent fashion for the induction of iNOS mRNA expression in macrophages.

Stability of iNOS mRNA   Upregulation of iNOS mRNA by BN is most likely caused by decreased mRNA destabilization and/or increased transcription rate. Weisz *et al.* (1994) reported that the half-life of iNOS mRNA in macrophages stimulated with a combination of LPS and IFN-γ was prolonged as compared to IFN-γ stimulation alone. In an attempt to determine the mechanism that may be involved in the augmentation of iNOS mRNA expression by BN plus IFN-γ, the stability of iNOS mRNA was measured using the *de novo* RNA synthesis inhibitor, actinomycin D. These experiments showed that there was no significant difference on iNOS

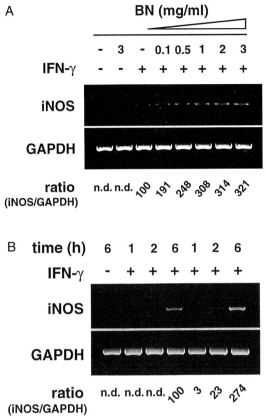

Figure 6 Synergistic induction of iNOS mRNA expression by IFN-γ plus Bio-Normalizer. (A) Dose-dependent induction. Macrophages were incubated with medium alone, Bio-Normalizer (3 mg/ml) alone, IFN-γ (5 U/ml) alone, or IFN-γ in the presence of the indicated concentrations of Bio-Normalizer for 6 hr, and total cellular RNA was applied to RT-PCR using the iNOS-specific primer. All values are expressed as a percentage of the control (IFN-γ alone) for the iNOS/GAPDH ratio from results obtained by RT-PCR. (B) Time course induction. Macrophages were incubated with IFN-γ (5 U/ml) alone or with IFN-γ plus Bio-Normalizer (3 mg/ml) for various periods as indicated, and total cellular RNA was applied to RT-PCR. All values are expressed as a percentage of the control (IFN-γ alone, 6 hr) for the iNOS/GAPDH ratio from results obtained by RT-PCR.

mRNA stability between IFN-γ plus BN-treated cells and IFN-γ alone treated cells (Fig. 7). These results indicate that the mechanism for augmentation of iNOS mRNA expression is not due to an increase in the stability of iNOS mRNA, but rather to an increase in the transcription rate or other steps.

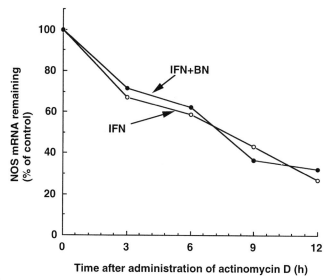

Figure 7   Effect of Bio-Normalizer on iNOS mRNA stability in IFN-γ-treated RAW 264.7 macrophages. Macrophages were incubated with IFN-γ (5 U/ml) alone (○) or with IFN-γ plus Bio-Normalizer (3 mg/ml) (●) for 6 hr. Actinomycin D (5 μg/ml) was then added to the culture, and total cellular RNA was extracted at the indicated time and applied to RT-PCR. Data are presented as the relative amount of iNOS mRNA remaining after the addition of actinomycin D and normalization to the respective amount of GAPDH mRNA.

Upregulation of IL-1β and TNF-α mRNA Expression by Bio-Normalizer
It is well documented that the induction of iNOS gene expression is regulated tightly by cytokines (Oswald *et al.,* 1996). For example, IFN-γ, TNF-α, and IL-1β upregulate iNOS expression. Other cytokines, such as transforming growth factor-β, IL-4, and IL-10, have been shown to block the signal in the NO pathway. The ability of BN to enhance the expression of IL-1β or TNF-α genes in macrophages was investigated because these cytokines are implicated in the induction of the iNOS gene as well as in tumoricidal activity. This macrophage cell line constitutively expresses IL-1β and TNF-α mRNA, and these mRNAs were slightly enhanced by the treatment with IFN-γ (Fig. 8). In the case of IL-1β, BN caused an approximately twofold augmentation over mRNA levels induced by IFN-γ alone. TNF-α mRNA levels were further enhanced by treatment with BN. These cytokines may be partially involved in the augmentation of iNOS mRNA expression in macrophages and in other cellular functions.

Taking these results together, BN was not involved directly in the pathway for iNOS induction, but showed synergistic interaction with IFN-γ to

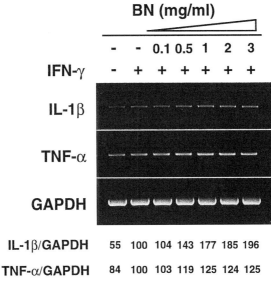

Figure 8 Dose-dependent induction of IL-1β, TNF-α mRNA by IFN-γ plus Bio-Normalizer. Macrophages were incubated with medium alone, IFN-γ (5 U/ml) alone, or IFN-γ in the presence of the indicated concentrations of Bio-Normalizer for 6 hr. Total cellular RNA was then extracted and applied to RT-PCR using IL-1β- or TNF-α-specific primers. All values are expressed as a percentage of the control (IFN-γ alone) from results obtained by RT-PCR.

induce NO synthesis in macrophages, suggesting that BN may have indirect mechanisms of microbicidal, tumoricidal activity because of its ability to modulate immune effector cells. In fact, macrophage-derived NO has been shown to induce hypoxia and apoptosis in tumor cells and to inhibit metastasis through a decrease in tumor cell-induced platelet aggregation. On the other hand, the release of high amounts of NO may damage healthy tissues and contribute to the pathogenesis of a wide range of diseases by the formation of peroxynitrite. However, the physiological function of NO appears to be determined primarily by the site and the quantity of its production. BN itself does not exert the upregulatory action on NO production unless iNOS inducers such as cytokines are present. Therefore, it would be interesting to investigate the antitumor effect in combined treatment of BN and IFN-γ in an animal model in order to understand further the mechanisms involved in the modulation of the immune system by BN. Further studies are needed to elucidate the bioavailability of BN to the human body, as well as the substances that are responsible for the free radical scavenging property and/or immunomodulating properties of BN.

SUMMARY

To better define the biological properties of Bio-Normalizer, the efficacy of Bio-Normalizer on the antioxidant and immunomodulating activities using different model systems was investigated. In an animal supplementation study, Bio-Normalizer protected the isolated rat heart from ischemia–reperfusion injury and suppressed AMVN-induced TBARS accumulation and depletion of α-tocopherol content in various tissue homogenates by its free radical scavenging property. Although Bio-Normalizer failed to scavenge NO in a chemical system, in experiments using a macrophage cell line, Bio-Normalizer upregulated IFN-γ-induced NO production in these macrophages in a dose-dependent manner. Such an effect of Bio-Normalizer on NO production was not due to changes in the activity of iNOS, but rather to the augmentation of iNOS mRNA expression without any alteration in the stability of mRNA. These results show that Bio-Normalizer appears to be a free radical modulator that may regulate the antioxidant/oxidant status in the body through biological events such as radical scavenging effects and activation of immune effector cells. These properties of Bio-Normalizer may be, in part, responsible for its reported therapeutic activity toward various diseases.

REFERENCES

Aruoma, O. I. (1994). Nutrition and health aspects of free radicals and antioxidants. *Food Chem. Toxicol.* **32**, 671–683.

Chan, P. H. (1996). Role of oxidants in ischemic brain damage. *Stroke* **27**, 1124–1129.

Ferrari, R. (1995). Metabolic disturbances during myocardial ischemia and reperfusion. *Am. J. Cardiol.* **76**, 17B–24B.

Granger, D. L., Hibbs, J. B., Jr., Perfect, J. R., and Durack, D. T. (1988). Specific amino acid (L-arginine) requirement for the microbiostatic activity of murine macrophages. *J. Clin. Invest.* **81**, 1129–1136.

Green, S. J., Meltzer, M. S., Hibbs, J. B., Jr., and Nacy, C. A. (1990). Activated macrophages destroy intracellular *Leishmania major* amastigotes by an L-arginine-dependent killing mechanism. *J. Immunol.* **144**, 278–283.

Halliwell, B. (1993). The role of oxygen radicals in human disease, with particular reference to the vascular system. *Haemostasis* **23**(Suppl. 1), 118–126.

Halliwell, B., and Gutteridge, J. M. (1989). "Free Radicals in Biology and Medicine." Oxford University Press, Oxford.

Haramaki, N., Marcocci, L., D'Anna, R., Yan, L. J., Kobuchi, H., and Packer, L. (1995). Bio-Catalyzer $\alpha \cdot \rho$ No. 11 (Bio-Normalizer) supplementation: Effect on oxidative stress to isolated rat hearts. *Biochem. Mol. Biol. Int.* **36**, 1263–1268.

Hogg, N., Kalyanaraman, B., Joseph, J., Struck, A., and Parthasarathy, S. (1993). Inhibition of low-density lipoprotein oxidation by nitric oxide: Potential role in atherogenesis. *FEBS Lett.* **334**, 170–174.

Janero, D. R. (1995). Ischemic heart disease and antioxidants: Mechanistic aspects of oxidative injury and its prevention. *Crit. Rev. Food Sci. Nutr.* **35**, 65–81.

Karupiah, G., and Harris, N. (1995). Inhibition of viral replication by nitric oxide and its reversal by ferrous sulfate and tricarboxylic acid cycle metabolites. *J. Exp. Med.* **181**, 2171–2179.

Karupiah, G., Xie, Q. W., Buller, R. M., Nathan, C., Duarte, C., and MacMicking, J. D. (1993). Inhibition of viral replication by interferon-gamma-induced nitric oxide synthase. *Science* **261**, 1445–1448.

Kobuchi, H., and Packer, L. (1997). Bio-Normalizer modulates interferon-γ-induced nitric oxide production in the mouse macrophage cell line RAW 264.7. *Biochem. Mol. Biol. Int.* **43**, 141–152.

Marcocci, L., D'Anna, R., Yan, L. J., Haramaki, N., and Packer, L. (1996). Efficacy of Bio-Catalyzer α•ρ No. 11 (Bio-Normalizer) supplementation against peroxyl radical-induced oxidative damage in rat organ homogenates. *Biochem. Mol. Biol. Int.* **38**, 535–541.

Maxwell, S. R. (1995). Prospects for the use of antioxidant therapies. *Drugs* **49**, 345–361.

Nathan, C. (1992). Nitric oxide as a secretory product of mammalian cells. *FASEB J.* **6**, 3051–3064.

Nathan, C., and Xie, Q. W. (1994). Regulation of biosynthesis of nitric oxide. *J. Biol. Chem.* **269**, 13725–13728.

Okuda, H., Ominami, H., Zhou, A., Osato, A., and Santiago, L. A. (1993). Studies on biological activities of Bio-normalizer. *Clin. Rep.* **27**, 4249–4258.

Osato, J. A., Korkina, L. G., Santiago, L. A., and Afanas'ev, I. B. (1995). Effects of bio-normalizer (a food supplementation) on free radical production by human blood neutrophils, erythrocytes, and rat peritoneal macrophages. *Nutrition* **11**, 568–572.

Oswald, I. P., and Stephanie, J. L. (1996). Nitrogen oxide in host defense against parasites. *Methods* **10**, 8–14.

Santiago, L. A., Osato, J. A., Hiramatsu, M., Edamatsu, R., and Mori, A. (1991). Free radical scavenging action of Bio-catalyzer alpha.rho No. 11 (Bio-normalyzer) and its by-product. *Free Radic. Biol. Med.* **11**, 379–383.

Santiago, L. A., Osato, J. A., Kabuto, H., and Mori, A. (1992). Decreased release of monoamine metabolites in iron-induced epileptogenic focus in the rat following administration of Bio-normalizer. *Med. Sci. Res.* **21**, 139–141.

Santiago, L. A., Osato, J. A., Liu, J., and Mori, A. (1993a). Age-related increases in superoxide dismutase activity and thiobarbituric acid-reactive substances: Effect of bio-catalyzer in aged rat brain. *Neurochem. Res.* **18**, 711–717.

Santiago, L. A., Osato, J. A., Ogawa, N., and Mori, A. (1993b). Antioxidant protection of bio-normalizer in cerebral ischaemia-reperfusion injury in the gerbil. *Neuroreport* **4**, 1031–1034.

Santiago, L. A., Uno, K., Kishida, T., Miyagawa, F., Osata, J. A., and Mori, A. (1994). Effect of Bio-normalizer on serum components and immunological function in humans. *Neurosciences* **20**, 149–152.

Singh, N., Dhalla, A. K., Seneviratne, C., and Singal, P. K. (1995). Oxidative stress and heart failure. *Mol. Cell Biochem.* **147**, 77–81.

Weisz, A., Oguchi, S., Cicatiello, L., and Esumi, H. (1994). Dual mechanism for the control of inducible-type NO synthase gene expression in macrophages during activation by interferon-gamma and bacterial lipopolysaccharide: Transcriptional and post-transcriptional regulation. *J. Biol. Chem.* **269**, 8324–8333.

Wink, D. A., Hanbauer, I., Krishna, M. C., DeGraff, W., Gamson, J., and Mitchell, J. B. (1993). Nitric oxide protects against cellular damage and cytotoxicity from reactive oxygen species. *Proc. Natl. Acad. Sci USA* **90**, 9813–9817.

Index

Absorption
 aglycone, 262, 290
 carotenoids, 194
 catechins, 262, 290–295, 300–302, 304
 flavonoids, 260–262, 290
 flavonols, 290–295
 quercetin, 290–293
 rutin, 261
 tocopherols, 59–62
 vitamin E, 59–60
Aconitine, chemical structure, 423
Aconiti tuber, 423
Adhesion molecules, in ROS-stimulated
  neutrophil adherence, 371–380
Age-related macular degeneration,
  carotenoids and, 188, 226
Aging, mitochondria and, 346–347, 352
Aglycone, 255
 absorption and metabolism, 262, 290
 chemical structure, 288
 in foods, 272–275
Alismatis rhizoma, 416, 418, 422
Alisol, chemical structure, 418, 422
Allopurinol, chemical structure, 399, 400

Allyl sulfide, tea polyphenols and,
  439, 440
Alpha-Tocopherol, Beta-Carotene Cancer
  Prevention study, 4, 8, 11, 13–14,
  15, 188
Alzheimer's disease
 clinical trials for, 424, 425
 cognitive enhancement therapy for, 421
 Japanese herbs and, 411–425
Angelicae radix, 416, 419
Animal tissue, lipollysine assay in, 127–132,
  137–141
Anomalin, chemical structure, 414
Anthocyanidins
 chemical structure, 287
 in foods, 271, 272–273, 286
Anthocyanins, 286
 chemical structure, 287
 in grapes, 388
Anti-inflammatory properties
 carotenoids, 220–221
 of pycnogenol, 319
 willow bark, 312, 317
Anti-L-selectin, 375